AF842845

# Mathematics for Sustainable Developments


**Managing Editors**

Agida Manizade, College of Graduate Studies, Radford University, Radford, USA

Anjum Halai, Institute for Educational Development, Aga Khan University, Karachi City, Pakistan

Chris Budd, Department of Mathematical Sciences, University of Bath, Bath, UK

Evans Osabuohien, Department of Economics & Development Studies, Covenant University, Ota, Nigeria

Huaxiong Huang, Department of Mathematics and Statistics, York University, Toronto, ON, Canada

Merrilyn Goos, School of Education, The University of Queensland, Brisbane, QLD, Australia

Nitu Kumari, Department of Mathematics, Indian Institute of Technology Mandi, Mandi, Himachal Pradesh, India

Praveen Agarwal, Anand International College of Engineering, Jaipur, India

Zhenjum Ma, Sustainable Buildings Research Centre, University of Wollongong, North Wollongong, Australia

Stamatina Th. Rassia, University of Macedonia, Thessaloniki, Greece

José Alberto Cuminato, ICMC, University of São Paulo, São Carlos, São Paulo, Brazil

Aligned with the Springer Nature's SDGs program, this series aims to publish books on diverse fields exploring the intersection of mathematics, statistics, and sustainable societal developments. It will not only cover emerging mathematical trends but also address essential topics for human and societal well-being. The goal is to advance sustainability science globally by transcending traditional research boundaries and addressing complex challenges. Scholars from across disciplines will be harnessed to tackle obstacles across all human endeavors, fostering a holistic understanding of sustainability and promoting innovative solutions for a resilient future.

Millie Pant · Kusum Deep · Atulya K. Nagar
Editors

# Real-World Applications of Machine Learning

Springer

*Editors*
Millie Pant
Department of Applied Mathematics
and Scientific Computing
Indian Institute of Technology Roorkee
Roorkee, Uttar Pradesh, India

Kusum Deep
Department of Mathematics
Indian Institute of Technology Roorkee
Roorkee, Uttarakhand, India

Atulya K. Nagar
School of Mathematics,
Computer Science and Engineering
Liverpool Hope University
Liverpool, UK

ISSN 3004-9016 ISSN 3004-9024 (electronic)
Mathematics for Sustainable Developments
ISBN 978-981-95-7133-8 ISBN 978-981-95-7134-5 (eBook)
https://doi.org/10.1007/978-981-95-7134-5

Mathematics Subject Classification: 68Q25, 68U20, 92D20, 93B40, 93E35, 92C55, 68M10

© The Editor(s) (if applicable) and The Author(s), under exclusive license to Springer Nature Singapore Pte Ltd. 2026

This work is subject to copyright. All rights are solely and exclusively licensed by the Publisher, whether the whole or part of the material is concerned, specifically the rights of translation, reprinting, reuse of illustrations, recitation, broadcasting, reproduction on microfilms or in any other physical way, and transmission or information storage and retrieval, electronic adaptation, computer software, or by similar or dissimilar methodology now known or hereafter developed.
The use of general descriptive names, registered names, trademarks, service marks, etc. in this publication does not imply, even in the absence of a specific statement, that such names are exempt from the relevant protective laws and regulations and therefore free for general use.
The publisher, the authors and the editors are safe to assume that the advice and information in this book are believed to be true and accurate at the date of publication. Neither the publisher nor the authors or the editors give a warranty, expressed or implied, with respect to the material contained herein or for any errors or omissions that may have been made. The publisher remains neutral with regard to jurisdictional claims in published maps and institutional affiliations.

This Springer imprint is published by the registered company Springer Nature Singapore Pte Ltd.
The registered company address is: 152 Beach Road, #21-01/04 Gateway East, Singapore 189721, Singapore

If disposing of this product, please recycle the paper.

# Preface

Machine learning (ML) has emerged as a cornerstone of modern computing, transforming how data-driven decisions are made across diverse sectors—from finance and healthcare to e-commerce, cybersecurity, and environmental monitoring. As this field continues to evolve, it enables not only intelligent automation but also deeper insights through predictive analytics, classification, pattern recognition, and natural language processing.

This book brings together a curated selection of contemporary research and innovative implementations that reflect the breadth and depth of ML in real-world contexts. The chapters explore a variety of themes including facial expression recognition, explainable AI, financial forecasting, sentiment analysis, biomedical diagnostics, recommendation systems, signal processing, and sustainable inventory models. Techniques such as convolutional neural networks, differential evolution, fuzzy logic, and hybrid deep learning architectures are examined with applications that are both socially relevant and technically sophisticated.

This volume is intended for students, researchers, professionals, and practitioners in the fields of computer science, data science, and AI. It aims to provide a balanced perspective—bridging theoretical foundations with practical use cases—and serves as a useful reference for exploring how machine learning can be harnessed to solve complex challenges.

We are deeply grateful to the authors for their valuable contributions and to the reviewers for their constructive insights. Our sincere thanks also go to our academic institutions and publishing partners for their support throughout this endeavor. We hope that this book will inspire further research and innovation in the ever-expanding field of machine learning.

Saharanpur, India — Millie Pant
Roorkee, India — Kusum Deep
Liverpool, UK — Atulya K. Nagar

# Contents

# A Comprehensive Review of Vision Language Models in the Realm of Full Self-Driving Systems

**Jayant Toleti, V. Amirthavarshini, and Sreenivasa Chakravarthi Sangapu**

**Abstract** This paper presents a comprehensive review of the advancements in the context of autonomous driving (AD) systems, exploring the potential and current applications of Large Language Models and Vision Language Models in enhancing Full Self-Driving (FSD) capabilities. As AD systems evolve, the integration of multi-modal inputs, particularly visual and textual data, through advanced VLMs has become a critical factor in improving vehicle perception, decision-making, and interaction with complex environments. This survey addresses six key research questions that assess the progress of VLMs in various aspects, including object detection, scene understanding, natural language instructions, and safety-critical decision processes in FSD systems. By synthesizing recent developments and ongoing research, the objective of this paper is to provide a clear overview of the current state of VLM integration in autonomous driving and to outline future directions for advancing these technologies in real-world applications.

**Keywords** Autonomous driving systems · Large language model · Vision language models

## 1 Introduction

A critical component in autonomous systems are VLMs, which combine visual data with textual or linguistic information and are key enablers in improving how AVs comprehend and respond to dynamic and multifaceted environments.

VLMs are designed to allow AVs to understand and reason about their surroundings in real-time. By linking what the vehicle perceives visually with linguistic interpretations (such as road signs, traffic rules, or verbal instructions), these systems aim

J. Toleti (✉) · V. Amirthavarshini · S. C. Sangapu
Department of Computer Science and Engineering, Amrita School of Computing, Amrita Vishwa Vidyapeetham, Chennai, India
e-mail: jtoleti@gmail.com

S. C. Sangapu
e-mail: ss_chakravarthi@ch.amrita.edu

© The Author(s), under exclusive license to Springer Nature Singapore Pte Ltd. 2026 
M. Pant et al. (eds.), *Real-World Applications of Machine Learning*, Mathematics for Sustainable Developments, https://doi.org/10.1007/978-981-95-7134-5_1

to improve decision-making, safety, and overall performance. VLMs are particularly promising for handling edge cases, such as unusual road conditions, pedestrian behavior, or novel driving scenarios that require an understanding beyond raw visual perception.

## 2 Research Methodology

The research methodology adopted in this study investigates the advancements in the rapidly evolving field of autonomous driving, with a focus on VLMs and FSD systems.

Our research systematically reviews academic publications from leading sources and presents only the most relevant and impactful works, ensuring that our study rests on a strong and comprehensive foundation. This research has been structured around six key research questions, which serve as a framework to guide and focus the survey.

- *Research Question 1: What are the limitations and challenges faced by the current generation of FSD systems in achieving robust AV operation?*
- *Research Question 2: What are the methodologies being employed to address the limitations of FSD systems, and how viable are these techniques in achieving fully autonomous and reliable vehicle operation?*
- *Research Question 3: How can LLMs be efficiently integrated into the perception systems of AVs to improve decision-making in real-time?*
- *Research Question 4: What are the prominent datasets being utilized for the development and training of VLMs in autonomous driving systems, and how do these datasets contribute to enhancing their performance in real-world scenarios?*
- *Research Question 5: What are the challenges and potential solutions for ensuring the safety and robustness of VLMs in autonomous driving, particularly in adversarial conditions?*

## 3 Literature Survey

### 3.1 What Are the Limitations and Challenges Faced by the Current Generation of FSD Systems in Achieving Robust and Reliable Autonomous Vehicle Operation?

This research highlights significant challenges in FSD systems, particularly in addressing driver behavior, environmental conditions, technological constraints, and regulatory uncertainties. Drivers using systems like Tesla's FSD Beta and Autopilot are prone to risky behaviors such as hands-free driving, while constant supervision increases stress and mental workload, raising safety concerns [1]. Similarly, adverse weather, technological limitations, and the need for robust data security remain critical issues [2]. Cyber-sickness, motion sickness, and high-speed network reliability further complicate immersive media consumption in autonomous vehicles [3].

Simulations, like those conducted by Waymo, face criticism for limited real-world applicability and outdated vehicle comparisons, which obscure the relative safety of FSD systems [4]. Object detection algorithms such as YOLOv8 struggle with environmental generalization and detecting small-scale objects [5]. Additionally, challenges persist in sensor reliability, legal regulations, and the integration of multi-modal sensor inputs, which hinder consistent performance across scenarios [6, 7]. AI black-box models exacerbate transparency and accountability concerns, limiting public acceptance [8]. Furthermore, real-time decision-making remains constrained by computational demands, sensor limitations, and inadequate datasets, especially for driver behavior modeling [9, 10]. Ethical and legal concerns, such as cyber-attacks and inadequate machine-readable traffic laws, further complicate the adoption of FSD systems [11, 12]. Issues related to human-machine communication, control transitions, and situational awareness amplify risks in hybrid operation scenarios [13, 14].

### *3.2 What Are the Current Approaches and Methodologies Being Employed to Address the Limitations of FSD Systems, and How Viable Are These Techniques in Achieving Fully Autonomous and Reliable Vehicle Operation?*

A key challenge in FSD systems is accurate perception, especially in complex scenarios, which has led to advancements in object detection and segmentation through multi-sensor fusion. For example, LWSIS uses 3D point cloud data to improve weakly supervised instance segmentation, reducing the need for costly 2D annotations, though sensor calibration inaccuracies and noisy LiDAR data remain issues [15]. Additionally, multi-sensor fusion SLAM methods have improved mapping and localization accuracy to 5–10 cm and 20–30 cm, respectively, but still face challenges in large-scale environments due to sensor dependence and high computational demands [16].

Adverse weather conditions are a significant challenge for FSD systems, particularly in LiDAR-based object detection scenarios. A study evaluating LiDAR sensors under various rainfall conditions highlights the need for improved sensor performance in adverse conditions to enhance the reliability of autonomous systems [17]. Similarly, the MaskPLS approach proposes an end-to-end method for panoptic segmentation of LiDAR scans, which outperforms heuristic approaches but still struggles with noisy supervision signals [18]. Object detection, particularly for small objects in real-time, is crucial in autonomous driving. An improved YOLOv5 model enhances

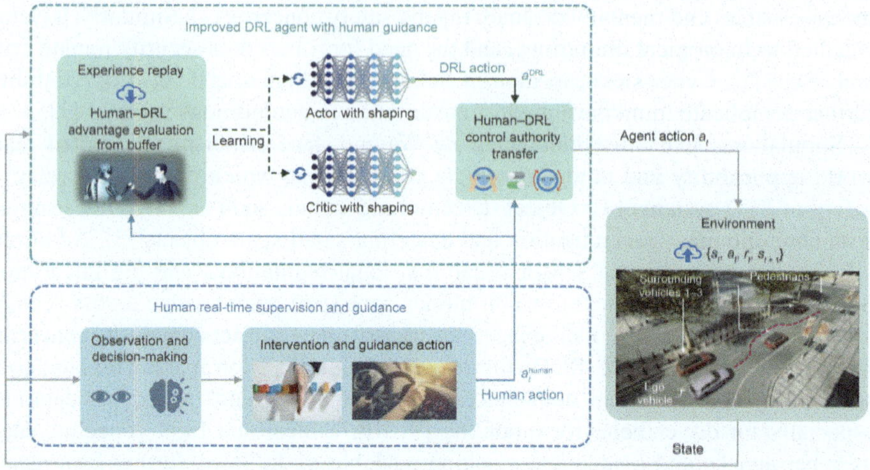


**Fig. 1** High-level architecture of the proposed Hug-DRL method with real-time human guidance

detection accuracy and speed but faces high computational demands [19]. Multi-sensor fusion, like the KAN-RCBEVDepth algorithm, combines LiDAR, cameras, and radar for better 3D detection but still struggles with context loss and could benefit from more annotations [20].

Recent research has also highlighted innovative methodologies that leverage reinforcement learning (RL) and human-like decision-making frameworks. One significant study presents a fear-neuro-inspired reinforcement learning (FNI-RL) framework. It mimics the functionality of the amygdala to instill defensive behaviors in autonomous agents. This approach aims to enhance safety in autonomous driving by reducing safety violations. The authors emphasize that further refinement of these models is necessary [21]. Another research effort focuses on conditional predictive behavior planning through inverse reinforcement learning (IRL). This framework enables autonomous vehicles to replicate human-like decision-making by learning from extensive datasets of human driving behavior. This method demonstrates superior performance in generating trajectories that closely resemble human driving patterns when validated on urban datasets. It faces challenges in closed-loop scenarios and struggles with reactive predictions when confronted with atypical behaviors [22]. Additionally, a study proposes Hug-DRL, it integrates real-time human guidance into deep reinforcement learning processes. Hug-DRL uses human input to improve efficiency, decision-making, and scene understanding, enabling faster convergence and stronger autonomous performance, as shown in Fig. 1 [23].

These studies highlight the potential of integrating biological and human-guided approaches within reinforcement learning frameworks to advance the safety and effectiveness of autonomous driving technologies. Graph-based models like ISG enhance event detection, perception, prediction, and planning by modeling interactions among vehicles, agents, and maps [24]. However, the effectiveness of these graph-based models heavily relies on the quality of input data, and interpreting the decision-making process remains challenging.

Current approaches to improving FSD systems, including sensor fusion, reinforcement learning, and end-to-end learning, show promise in tackling challenges like perception accuracy, adverse weather, and human-like decision-making. However, achieving full autonomy depends on addressing issues with sensor calibration, computational demands, data quality, and model integration. Continued research is crucial to bridging these gaps.

### 3.3 How Can LLMs Be Efficiently Integrated Into the Perception Systems of Autonomous Vehicles to Improve Decision-Making in Real-Time?

One study introduced a framework that employs RLHF to optimize driving models based on human preferences and safety standards. This framework uses multi-agent simulations, where LLMs act as agents processing multi-modal data, it aligns autonomous systems to human comfort and is safety-validated using real-world testbeds [25]. LLMs have been integrated into autonomous driving systems to address the issue of perception attacks. A driving reasoning agent called the HUDSON system extends LLM-based systems by using domain-specific language (DSL) to detect and respond to adversarial attacks in real-time. HUDSON was tested in various scenarios using LLMs like GPT-4 and Llama, achieving high-attack detection accuracy and making safe control decisions, thus enhancing the safety of autonomous systems during perception attacks [26]. OmniDrive demonstrates how LLMs can handle 3D perception, reasoning, and planning in autonomous driving. It uses a 3D vision-language model to compress visual data into a world model that integrates perception with action in 3D space. It also uses sparse queries to process high-resolution, multi-view video inputs, enabling long-range perception and decision-making. OmniDrive's ability has been bench-marked using the nuScenes dataset [27]. The DriVLMe framework, similar to OmniDrive, utilizes video tokenizers and symbolic planning to improve route planning for long-horizon goals and enhances LLM-based agents with embodied and social experiences. DriVLMe enhances adaptability in dynamic scenarios, such as weather changes and sudden goal shifts [28]. The LimSim++ platform, which provides a closed-loop environment, enables continuous learning for LLM-based systems by incorporating memory and reflection mechanisms. This allows for better decisions by learning from past experiences, thereby improving generalization and safety in real-time driving conditions [29].

LLMs also play a critical role in multi-modal traffic accident forecasting, as seen in a driving LLM Agent in Fig. 2. In real-time scenarios, LLMs like LLaMA-2 and Zephyr-7b adjust driving behavior based on the likelihood of accidents. The integration of visual and linguistic cues through VLMs allows these systems to dynamically adapt to potential hazards by analyzing historical traffic data. de Zarzà et

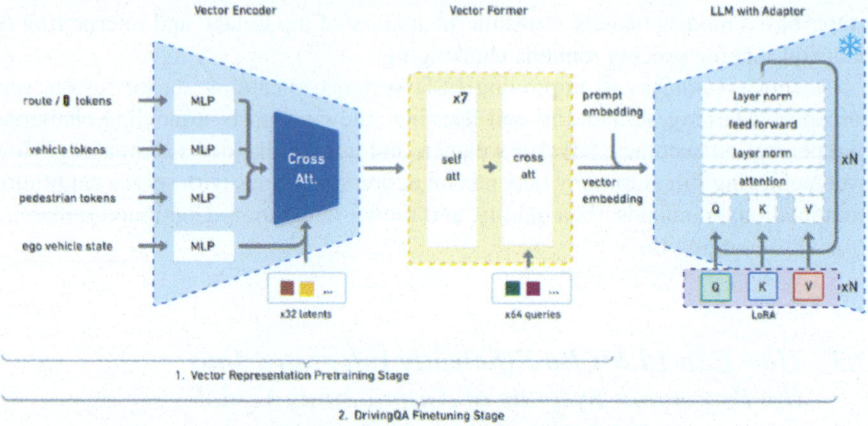


**Fig. 2** Architecture of the driving LLM agent. Adapted from Chen et al. [30]

al. found that Bayesian feature extraction methods enhance decision-making by providing uncertainty estimates, thus helping LLMs make more reliable real-time decisions [31].

Human-centric designs are focused particularly in improving interaction between autonomous vehicles and drivers. LLMs have been shown to interpret verbal commands and adapt driving behavior based on environmental data, making interactions more intuitive. This is demonstrated in an interactive driver education system that utilizes ChatGPT to provide real-time, conversation-based training, enhancing drivers' understanding of Advanced Driver Assistance Systems (ADAS) [32].

Meanwhile, the DriveGenVLM framework focuses on video generation for training autonomous systems. By using De-noising Diffusion Probabilistic Models (DDPM) to create realistic driving videos, it improves LLMs' ability to understand and predict real-world driving scenarios [33]. LLMs' ability to interpret and respond to complex driving instructions is further emphasized in the LaMPilot system. LaMPilot uses multi-modal learning to analyze real-time driving scenarios and respond to natural language instructions. It emphasizes continuous learning from new data and error analysis [34].

Several studies have explored how LLMs can improve hierarchical data structures for better scene analysis and reasoning in corner cases. Automated benchmarks such as CODA-LM offer a framework for assessing the ability of LLMs to process complex driving scenes. Knowledge transfer and fine-tuning further enhance the models' recognition and planning capabilities [35]. HiLM-D, introduces a two-branch system for processing high- and low-resolution data. This enhances the detection of risk objects in driving scenes by comparing objects of different resolutions through cross-attention mechanisms [36].

The work on GPT-4V(ision) explores its application as an autonomous driving agent. It solely focuses on its ability to recognize environmental factors such as weather, illumination, traffic lights, and signs. It faces challenges in spatial reasoning, traffic light recognition, and handling nuanced details, particularly in non-English contexts. The model processes frame-by-frame driving videos, vehicle speed, and other data to make decisions and generate explanations [37].

In conclusion, the literature on LLM-enhanced autonomous driving systems presents a wide array of research contributions that span from improving human-centric design and interaction to optimizing real-time decision-making through multi-modal learning. These approaches collectively demonstrate the potential of LLMs to revolutionize autonomous driving by enhancing safety, perception, reasoning, and adaptability in real-world driving scenarios.

## *3.4 What Are the Prominent Datasets Currently Being Utilized for the Development and Training of VLMs in Autonomous Driving Systems, and How Do These Datasets Contribute to Enhancing Their Performance in Real-World Scenarios?*

The Rank2Tell dataset, developed by Honda Research Institute and Stanford, ranks object importance in driving scenarios with natural language explanations, bridging visual perception and reasoning. DriveLM, from Stanford and UC Berkeley, expands on the nuScenes dataset to test autonomous models in perception, prediction, and planning. WEDGE, by Carnegie Mellon and Symbiosis International, evaluates system robustness in diverse weather using vision-language models, advancing generalization for real-world conditions. Together, these datasets enhance benchmarks for autonomous driving [38–40]. NuScenes-QA, by Fudan University, offers a large-scale VQA dataset with 459,941 question-answer pairs across 34,149 scenes, using BEV features for scene understanding and reasoning [41]. DRAMA focuses on real-time risk assessment, localizing hazards, and providing linguistic descriptions to help vehicles identify and communicate threats [42]. Nuprompt explores natural language prompts for generating driving instructions and actionable guidance, bridging language models and autonomous systems [43]. The Driving with LLMs dataset integrates LLMs with object-level modalities for explainable decision-making using reinforcement learning experts [44]. BDD-X, developed by UC Berkeley and collaborators, links vehicle behavior with textual explanations to enhance interpretability [45]. HAD, from UC Berkeley and Honda, translates human advice into actionable driving commands, improving human-vehicle communication for safer, more intuitive interactions [46]. A summary of all these details can be found in Table 1.

**Table 1** Prominent datasets

| Author and year | Dataset | Objective |
|---|---|---|
| Sachdeva, Enna, et al. (2024) | Rank2Tell | Object importance ranking and natural language explanations for driving scenarios |
| Sima, Chonghao, et al. (2023) | DriveLM | Benchmarking perception, prediction, and planning for autonomous driving |
| Marathe, Aboli, et al. (2023) | WEDGE | Testing autonomous driving in various weather conditions using vision-language models |
| Qian, Tianwen, et al. (2024) | NuScenes-QA | Visual question-answering (VQA) focused on scene understanding for autonomous driving |
| Malla, Srikanth, et al. (2023) | DRAMA | Identifying and explaining driving risks through real-time captions |
| Wu, Dongming, et al. (2023) | Nuprompt | Generating driving instructions and reasoning with language prompts |
| Chen, Long, et al. (2024) | Driving with LLMs | Integrating large language models for explainability and reasoning in driving scenarios |
| Kim, Jinkyu, et al. (2018) | BDD-X | Linking vehicle decision-making with verbal reasoning for interpretability |
| Kim, Jinkyu, et al. (2019) | HAD | Improving human-to-vehicle communication and actionable driving commands |

## 3.5 What Are the Challenges and Potential Solutions for Ensuring the Safety and Robustness of VLMs in Autonomous Driving, Particularly in Adversarial Conditions?

VLMs have shown remarkable advances in multi-modal understanding, merging visual and textual inputs to tackle a wide range of tasks. However, despite their promising capabilities, VLMs still exhibit several key limitations that constrain their performance and generalizability across different tasks and settings.

One of the most prominent vulnerabilities of VLMs is their susceptibility to adversarial attacks, which can deceive these models despite the incorporation of localization and attention mechanisms. These attacks have been shown to fool VLMs more than 90% of the time, which raises concerns about their robustness and security in practical applications [47]. This issue is exacerbated by task biases, where models favor certain visual representations or nouns over more complex reasoning patterns or contextual cues, as observed in visual-language navigation tasks [48, 49].

When dealing with out-of-distribution data, VLMs struggle with the problem of generalization. Studies have shown a generalization gap between training data and a variety of real-world images, which results in performance degradation when these models are exposed to unique scenarios [50]. This gap depends on their heavy reliance on large-scale high-quality labeled data, without which their generalization abilities falter [51, 52].

Although VLMs can retain visual features, these models often lack commonsense knowledge to make correct inferences about time, location, or logical sequences [53]. And, in tasks that require complex reasoning, VLMs perform poorly without bimodal supervision, emphasizing their dependence on aligned image-text pairs [54]. This limitation highlights the need for improved knowledge injection mechanisms to improve the reasoning capacity of VLMs [55].

In addition, VLMs frequently show inconsistencies across heterogeneous tasks, making them unreliable for applications that require high consistency in outputs across different domains. Such inconsistencies make it complex to integrate them into larger systems that rely on accurate, multi-task processing systems [56]. While VLMs may learn useful primitive concepts for visual recognition tasks, they often fail to provide interpretable models for more complex compositions, which limits their broader applicability [57].

VLMs also struggle from performance biases, especially when dealing with imbalanced datasets or tasks with varying semantic complexity. Studies show that VLMs tend to perform poorly on long-tailed data and struggle with tasks requiring nuanced semantic granularity [58, 59]. Their limitations are not restricted to recognition tasks, but extend to language generation, where they frequently produce overly generic text that lacks specificity and context, especially with named entities [60].

Another major challenge for VLMs is their vulnerability to hallucinations, where models generate plausible-sounding but factually incorrect outputs. This is a problem in applications that require higher levels of faithfulness and informativeness [61]. The generation of biased predictions further worsens the situation, as VLMs often pick up unwanted biases from image features, which skew their understanding and outputs [62].

In conclusion, even though VLMs represent a significant advancement in the field of multi-modal learning, they still face notable limitations in robustness, generalization, reasoning, consistency, interpretability, bias, and hallucination. Addressing these challenges will require further research into knowledge integration, improved pre-training objectives, and more sophisticated evaluation metrics.

## 4 Conclusion

The advancements in AI through the use of VLMs and LLMs have influenced the progress of autonomous vehicles. VLMs help the FSD system by combining what they see from sensors with understanding language in context. This helps vehicle systems to spot things and make smart choices, such as stopping when people are using crosswalks. LLMs contribute by handling real-time information and guessing what might happen next based on what they've learned. This leads to better decisions in tricky traffic situations. As self-driving cars need to handle public roads with little room for mistakes safety is paramount. The current versions of VLMs and LLMs can struggle in unusual situations, such as bad weather or strange road conditions. The efficiency of these models work depends on how good and varied their training data is, which might not cover all the complex situations in the real world. Research in this field is essential to refine both the datasets and the algorithms to enhance the robustness of FSD systems. New datasets are required and they should include various driving scenarios and improved annotation practices. Domains like RL and continuous learning approaches could allow FSD systems to dynamically adapt to real-world experiences, helping them improve safety and reliability. Addressing these technical challenges alongside social and ethical considerations will be crucial for the successful integration of FSD technology into everyday transportation.

## References

1. Nordhoff, S., et al.: (Mis-) use of standard autopilot and full self-driving (FSD) beta: results from interviews with users of Tesla's FSD Beta. Front. Psychol. **14**, 1101520 (2023)
2. Chougule, A., et al.: A comprehensive review on limitations of autonomous driving and its impact on accidents and collisions. IEEE Open J. Veh, Technol (2023)
3. Anwar, M.S., et al.: A moving Metaverse: QoE challenges and standards requirements for immersive media consumption in autonomous vehicles. Appl. Soft Comput. **159**, 111577 (2024)
4. Lundgren, B.: A new standard for accident simulations for self-driving vehicles: can we use Waymo's results from accident simulations? AI & SOCIETY **39**(2), 669–673 (2024)
5. Renuka, O.: A YOLOv8-based approach for multi-class traffic sign detection. Int. J. Sci. Res. Arch. **11**(2), 824–829 (2024)
6. Högye-Nagy, Á., Kovács, G., Kurucz, G.: Acceptance of self-driving cars among the university community: effects of gender, previous experience, technology adoption propensity, and attitudes toward autonomous vehicles. Transport. Res. F: Traffic Psychol. Behav. **94**, 353–361 (2023)
7. Islam, M.M.: Autonomous systems revolution: exploring the future of self-driving technology. J. Artif. Intell. Gen. Sci. (JAIGS) **3**(1), 16–23 (2024). ISSN: 3006–4023
8. Kotseruba, I., Tsotsos, J.K.: Data limitations for modeling top-down effects on drivers' attention. arXiv preprint arXiv:2404.08749 (2024)
9. Kosuru, V.S.R., Venkitaraman, A.K.: Advancements and challenges in achieving fully autonomous self-driving vehicles. World J. Adv. Res. Rev. **18**(1), 161–167 (2023)
10. Chen, L., et al.: End-to-end autonomous driving: challenges and frontiers. IEEE Trans. Pattern Anal. Mach, Intell (2024)
11. Atakishiyev, S., et al.: Explainable artificial intelligence for autonomous driving: a comprehensive overview and field guide for future research directions. IEEE Access (2024)

12. Constantinides, P., Monteiro, E., Mathiassen, L.: Human-AI joint task performance: learning from uncertainty in autonomous driving systems. Inf. Organ. **34**(2), 100502 (2024)
13. Tan, K., et al.: Integrating advanced computer vision and AI algorithms for autonomous driving systems. J. Theory Pract. Eng. Sci. **4**(01), 41–48 (2024)
14. El-Kady, R.: Navigating the legal maze: electric self-driving cars and the complexities of criminal law. In: Solving Fundamental Challenges of Electric Vehicles, pp. 215–247. IGI Global (2024)
15. Li, X., et al.: Lwsis: LiDAR-guided weakly supervised instance segmentation for autonomous driving. In: Proceedings of the AAAI Conference on Artificial Intelligence, vol. 37, no. 2 (2023)
16. Dai, K., et al.: LiDAR-based sensor fusion slam and localization for autonomous driving vehicles in complex scenarios. J. Imag. **9**(2), 52 (2023)
17. Choe, J., Cho, H., Chung, Y.: Performance verification of autonomous driving LiDAR sensors under rainfall conditions in darkroom. Sensors **24**(1), 14 (2023)
18. Marcuzzi, R., et al.: Mask-based panoptic LiDAR segmentation for autonomous driving. IEEE Robot. Autom. Lett. **8**(2), 1141–1148 (2023)
19. Mahaur, B., Mishra, K.K., Kumar, A.: An improved lightweight small object detection framework applied to real-time autonomous driving. Expert Syst. Appl. **234**, 121036 (2023)
20. Lai, Z., et al.: KAN-RCBEVDepth: a multi-modal fusion algorithm in object detection for autonomous driving. arXiv preprint arXiv:2408.02088 (2024)
21. He, X., et al.: Fear-neuro-inspired reinforcement learning for safe autonomous driving. IEEE Trans. Pattern Anal. Mach. Intell. (2023)
22. Huang, Z., et al.: Conditional predictive behavior planning with inverse reinforcement learning for human-like autonomous driving. IEEE Trans. Intell. Transp. Syst. **24**(7), 7244–7258 (2023)
23. Wu, J., et al.: Toward human-in-the-loop AI: enhancing deep reinforcement learning via real-time human guidance for autonomous driving. Engineering **21**, 1–75 (2023)
24. Zhang, Y., et al.: Graphad: interaction scene graph for end-to-end autonomous driving. arXiv preprint arXiv:2403.19098 (2024)
25. Sun, Y., et al.: Optimizing autonomous driving for safety: a human-centric approach with LLM-enhanced RLHF. arXiv preprint arXiv:2406.04481 (2024)
26. Song, R., et al.: Enhancing LLM-based autonomous driving agents to mitigate perception attacks. arXiv preprint arXiv:2409.14488 (2024)
27. Wang, S., et al.: OmniDrive: a holistic LLM-agent framework for autonomous driving with 3D perception, reasoning and planning. arXiv preprint arXiv:2405.01533 (2024)
28. Huang, Y., et al.: DriVLMe: enhancing LLM-based autonomous driving agents with embodied and social experiences. arXiv preprint arXiv:2406.03008 (2024)
29. Fu, D., et al.: LimSim++: a closed-loop platform for deploying multimodal LLMs in autonomous driving. arXiv preprint arXiv:2402.01246 (2024)
30. Chen, L., et al.: Driving with LLMs: fusing object-level vector modality for explainable autonomous driving. In: 2024 IEEE International Conference on Robotics and Automation (ICRA). IEEE (2024)
31. de Zarzà, I., et al.: LLM multimodal traffic accident forecasting. Sensors **23**(22), 9225 (2023)
32. Murtaza, M., et al.: Transforming driver education: a comparative analysis of LLM-augmented training and conventional instruction for autonomous vehicle technologies. Int. J. Artif. Intell. Educ. 1–38 (2024)
33. Fu, Y., et al.: DriveGenVLM: real-world video generation for vision language model based autonomous driving. arXiv preprint arXiv:2408.16647 (2024)
34. Ma, Y., et al.: Lampilot: an open benchmark dataset for autonomous driving with language model programs. In: Proceedings of the IEEE/CVF Conference on Computer Vision and Pattern Recognition (2024)
35. Li, Y., et al.: Automated evaluation of large vision-language models on self-driving corner cases. arXiv preprint arXiv:2404.10595 (2024)
36. Ding, X., et al.: Hilm-d: towards high-resolution understanding in multimodal large language models for autonomous driving. arXiv preprint arXiv:2309.05186 (2023)

37. Wen, L., et al.: On the road with GPT-4V (ision): explorations of utilizing visual-language model as autonomous driving agent. In: ICLR 2024 Workshop on Large Language Model (LLM) Agents (2024)
38. Sachdeva, E., et al.: Rank2tell: a multimodal driving dataset for joint importance ranking and reasoning. In: Proceedings of the IEEE/CVF winter conference on applications of computer vision (2024)
39. Sima, C., et al.: Drivelm: driving with graph visual question answering. arXiv preprint arXiv:2312.14150 (2023)
40. Marathe, A., et al.: Wedge: a multi-weather autonomous driving dataset built from generative vision-language models. In: Proceedings of the IEEE/CVF Conference on Computer Vision and Pattern Recognition (2023)
41. Qian, T., et al.: Nuscenes-qa: a multi-modal visual question answering benchmark for autonomous driving scenario. In: Proceedings of the AAAI Conference on Artificial Intelligence, vol. 38, no. 5 (2024)
42. Malla, S., et al.: Drama: joint risk localization and captioning in driving. In: Proceedings of the IEEE/CVF Winter Conference on Applications of Computer Vision (2023)
43. Wu, D., et al.: Language prompt for autonomous driving. arXiv preprint arXiv:2309.04379 (2023)
44. Chen, L., et al.: Driving with LLMs: fusing object-level vector modality for explainable autonomous driving. In: 2024 IEEE International Conference on Robotics and Automation (ICRA). IEEE (2024)
45. Kim, J., et al.: Textual explanations for self-driving vehicles. In: Proceedings of the European Conference on Computer Vision (ECCV) (2018)
46. Kim, J., et al.: Grounding human-to-vehicle advice for self-driving vehicles. In: Proceedings of the IEEE/CVF Conference on Computer Vision and Pattern Recognition (2019)
47. Xu, X., et al.: Fooling vision and language models despite localization and attention mechanism. In: Proceedings of the IEEE Conference on Computer Vision and Pattern Recognition (2018)
48. Hahn, M., Raj, A., Rehg, J.M.: Which way isright'? Uncovering limitations of vision-and-language navigation model. arXiv preprint arXiv:2312.00151 (2023)
49. Menon, S., Chandratreya, I.P., Vondrick, C.: Task bias in vision-language models. arXiv preprint arXiv:2212.04412 (2022)
50. Zhou, W., et al.: Vlue: a multi-task benchmark for evaluating vision-language models. arXiv preprint arXiv:2205.15237 (2022)
51. Du, Y., et al.: A survey of vision-language pre-trained models. arXiv preprint arXiv:2202.10936 (2022)
52. Li, F., et al.: Vision-language intelligence: tasks, representation learning, and large models. arXiv preprint arXiv:2203.01922 (2022)
53. Zhang, G., et al.: Can vision-language models be a good guesser? Exploring VLMs for times and location reasoning. In: Proceedings of the IEEE/CVF Winter Conference on Applications of Computer Vision (2024)
54. Segal, E., Bogin, B., Berant, J.: Training vision-language models with less bimodal supervision. arXiv preprint arXiv:2211.00262 (2022)
55. Shevchenko, V., et al.: Reasoning over vision and language: exploring the benefits of supplemental knowledge. arXiv preprint arXiv:2101.06013 (2021)
56. Maharana, A., et al.: Exposing and addressing cross-task inconsistency in unified vision-language models. arXiv preprint arXiv:2303.16133 (2023)
57. Yun, T., et al.: Do vision-language pretrained models learn composable primitive concepts? arXiv preprint arXiv:2203.17271 (2022)
58. Wang, Y., et al.: Exploring vision-language models for imbalanced learning. Int. J. Comput. Vis. **132**(1), 224–237 (2024)
59. Xu, Z., et al.: Challenges of zero-shot recognition with vision-language models: granularity and correctness. arXiv preprint arXiv:2306.16048 (2023)
60. Robbins, W.: Towards multimodal vision-language models generating non-generic text. In: Proceedings of the AAAI Conference on Artificial Intelligence, vol. 36, no. 11 (2022)

61. Qiu, H., et al.: VALOR-EVAL: holistic coverage and faithfulness evaluation of large vision-language models. arXiv preprint arXiv:2404.13874 (2024)
62. Weng, Z., et al.: Images speak louder than words: understanding and mitigating bias in vision-language model from a causal mediation perspective. arXiv preprint arXiv:2407.02814 (2024)

# Optimal Defense Strategies for Security Agencies: A Dual Hesitant Fuzzy Matrix Game Approach

**Sandeep Kumar and Aashish**

**Abstract** Up till now, Game theory has been modified with the concepts of various kinds of uncertainties. Dual Hesitant Fuzzy Set (DHFS) provides a noteworthy tool for expressing the uncertain quantitative aspects effectively in two-player zero-sum matrix game problems (TMGPs) with uncertain pay-offs. In the present work, we deal with TMGP with pay-offs given by DHFSs. For solving such problems, first, we transform a dual hesitant fuzzy element into a Dual Hesitant Fuzzy Element with Possibility (DHFEP) for equating the lengths of two dual hesitant fuzzy element. Then, we define value and ambiguity functions for these elements with comparative laws. Further, four-steps algorithm is developed to solve dual hesitant fuzzy matrix game problem based on the suggested value and ambiguity functions. To demonstrate the relevance of the proposed work, a real-world event-based problem-how to counter a terrorist attack is considered in order to calculate the optimal defense strategies for the security agencies.

**Keywords** Dual hesitant fuzzy set · Value function · Ambiguity function · Two-person zero-sum matrix game · Terrorism

## 1 Introduction

Game theory is a mathematical framework for analyzing situations in which multiple decision-makers (referred to as "players") interact strategically. It provides tools to predict the outcomes of these interactions, assuming that players are rational and seek to maximize their own pay-offs. In non-cooperative game problems, if gain of one player is exactly opposite to other player's loss then such problems are known as TMGPs. The class of TMGP under crisp and fuzzy environments is widely used to solve real-world research problems [1–3].

S. Kumar (✉) · Aashish
Department of Mathematics, Ch. Charan Singh University, Meerut 250004, Uttar Pradesh, India
e-mail: drsandeepmath@gmail.com

Aashish
e-mail: aashish580004@gmail.com

© The Author(s), under exclusive license to Springer Nature Singapore Pte Ltd. 2026

M. Pant et al. (eds.), *Real-World Applications of Machine Learning*, Mathematics for Sustainable Developments, https://doi.org/10.1007/978-981-95-7134-5_2

Practically, it is difficult for players to express their opinions in a precise manner. To overcome this difficulty, fuzzy set theory and its extensions [4–6] play a significant role to various research problems, such as decision making problems [7–12], duality problems [13, 14], transportation problem [15, 16]. This theory is also widely applied to solve fuzzy TMGPs. DHFS [6] is one of the extensions of fuzzy sets which not only considers a set of membership grades of the elements in a given set but also a set of non-membership grades which makes it more efficient. In this paper, we solve TMGP with data represented by DHFSs.

## *1.1 Literature Review*

Campos [17] studied a zero-sum two-person game with imprecise matrix pay-offs and proposed a generalization of traditional method to solve a classical matrix game problem. Li [18] considered the value of a fuzzy matrix game as a fuzzy number and established fuzzy multi-objective linear programming models for each player, along with effective solution methods. Chen and Larbani [19] formulated a multi-attribute decision making problem with a fuzzy decision matrix as a TMG with an uncertain pay-off matrix to obtain the weights. Pandey and kumar [20] provided some modifications for solving multi-objective matrix games with vague data by using a new order function. Seikh et al. [21] developed a solution methodology for solving TMGPs with pay-offs triangular intuitionistic fuzzy numbers. In [22], the author put forward an approach to solve matrix game in which goals of both players are given in terms of fuzzy sets. Kumar [23] established the similarities between intuitionistic fuzzy programming and goal programming and also, the author transformed the intuitionistic fuzzy programming to a deterministic program by using the max and min operator. Kumar [24] outlined solution approaches for a two-player zero-sum matrix game with fuzzy goals, utilizing piecewise linear membership functions. Jana and Roy [25] proposed a similarity measure and applied this in a bidirectional approximate reasoning system into matrix game with pay-offs DHFSs. Kumar [26] studied a multi-objective TMG with intuitionistic fuzzy goals and presented max-min and min-max solution of the considered problem. Xue et al. [3] constructed a novel algorithm for solving a matrix game in which the pay-offs are given by hesitant fuzzy sets. Brikaa et al. [27] constructed an effective methodology to solve a matrix game with pay-offs of triangular dual hesitant fuzzy numbers. Yang and Xu [28] proposed a novel comprehensive rank method to measure probabilistic triangular intuitionistic hesitant fuzzy sets, the authors used the concept of the centroid and the ambiguity value to achieve their aim. Jana and Roy [29] proposed a game-theoretic linguistic pythagorean hesitant fuzzy distance measure to resolve the cross-influence problem. Karmakar and Seikh [2] developed a solution approach to matrix game problem in which the pay-offs are represented by interval-valued intuitionistic hesitant fuzzy elements.

### 1.2 Motivation and the Present Research Work

After reviewing the above literature, it is evident that literature related to dual hesitant fuzzy TMGPs are very few. Due to this reason, In the current study, we introduce an algorithm to solve TMGPs in which data are represented by DHFSs. To equalize the lengths of two DHFSs, DHFEPs are introduced. Value and Ambiguity functions, along with their comparative laws, are then defined to facilitate the comparison of DHFEPs. The applicability of these comparative laws is demonstrated through two examples. A matrix game with DHFSs pay-offs is formulated, followed by a four-steps algorithm to solve the game and determine optimal strategies for both players. The proposed algorithm is applied to a real-world scenario, where security agencies obtain their optimal defense strategy.

The following summarizes the primary contributions of this paper:

(i) First, we define DHFEP to equalize the lengths of two DHFSs with unequal lengths. Then, to compare DHFEPs, we construct a value function and an ambiguity function.
(ii) To construct an algorithm for solving TMGP with pay-offs as DHFSs.
(iii) To demonstrate the applicability of the proposed algorithm through a real-world numerical example.

This is how the remaining paper is arranged:

In upcoming section, definition of DHFS and some basic operations and results are mentioned. In Sect. 3, we propose a DHFEP to tackle the situation of DHFSs with different lengths. After that value and ambiguity functions for DHFEPs with their comparative laws are proposed. In Sect. 4, an algorithm is constructed to solve a matrix game in which pay-offs are given by DHFSs. Section 5 provides the applicability of the proposed work through a numerical example. In Sect. 6, conclusions are described about the present work with the future scope.

## 2 Preliminaries

In this section, we briefly review some basic definitions and operations related to DHFS.

**Definition 1** (DHFS)[6] Suppose $h, g$ are two functions defined on a fixed set $X$, which applied on $X$ returns to some subsets of $[0, 1]$. Therefore, a DHFS is defined as:

$$D = \{\langle x, h(x), g(x)\rangle | x \in X\}$$

where $h = \{\zeta_1, \zeta_2, \ldots, \zeta_\ell\}$ denotes possible membership degrees and $g = \{\xi_1, \xi_2, \ldots, \xi_\ell\}$ denotes possible non-membership degrees of $x \in X$ and $\ell$ is the length of DHFS, with the conditions:

$$0 \leq \zeta_i, \xi_i \leq 1, 0 \leq \zeta^+ + \xi^+ \leq 1; (i = 1, 2, \ldots, \ell)$$

where $\zeta_i \in h(x), \xi_i \in g(x), \zeta^+ \in h^+(x) = \cup_{\zeta_i \in h(x)} \max\{\zeta_i\}$, and $\xi^+ \in g^+(x) = \cup_{\xi_i \in g(x)} \max\{\xi_i\}$ for all $x \in X$. The pair $d = (h(x), g(x))$ is referred to as dual hesitant fuzzy element for convenience, with the conditions: $\zeta_i \in h(x), \xi_i \in g(x), \zeta^+ \in h^+(x) = \cup_{\zeta_i \in h(x)} \max\{\zeta_i\}, \xi^+ \in g^+(x) = \cup_{\xi_i \in g(x)} \max\{\xi_i\}, 0 \leq \zeta_i, \xi_i \leq 1$, and $0 \leq \zeta^+ + \xi^+ \leq 1$.

**Definition 2** (Basic operations for DHFSs)[6] Let $d_1$ and $d_2$ two dual hesitant fuzzy elements defined on a fixed set $X$, then

(i) $\oplus$-union: $d_1 \oplus d_2 = \{h_{d_1} \oplus h_{d_2}, g_{d_1} \otimes g_{d_2}\} = \cup_{\zeta_{d_1} \in h_{d_1}, \xi_{d_1} \in g_{d_1}, \zeta_{d_2} \in h_{d_2}, \xi_{d_2} \in g_{d_2}} \{\{\zeta_{d_1} + \zeta_{d_2} - \zeta_{d_1}\zeta_{d_2}\}, \{\xi_{d_1}\xi_{d_2}\}\}$

(ii) $\otimes$-intersection: $d_1 \otimes d_2 = \{h_{d_1} \otimes h_{d_2}, g_{d_1} \oplus g_{d_2}\} = \cup_{\zeta_{d_1} \in h_{d_1}, \xi_{d_1} \in g_{d_1}, \zeta_{d_2} \in h_{d_2}, \xi_{d_2} \in g_{d_2}} \{\{\zeta_{d_1}\zeta_{d_2}\}, \{\xi_{d_1} + \xi_{d_2} - \xi_{d_1}\xi_{d_2}\}\}$

(iii) $\lambda d_1 = \cup_{\zeta_d \in h_d, \xi_d \in g_d} \{1 - (1 - \zeta_d)^\lambda, (\xi_d)^\lambda\}$;

(iv) $d^\lambda = \cup_{\zeta_d \in h_d, \xi_d \in g_d} \{(\zeta_d)^\lambda, 1 - (1 - \xi_d)^\lambda\}$; where $\lambda \in \mathbb{Z}^+$ and all the results are also dual hesitant fuzzy elements.

## 3 Proposed Value and Ambiguity Functions for DHFSs

In this section, the first aim of this section is to introduce a process to equalize the length of two different dual hesitant fuzzy elements. Then, we develop value and ambiguity functions for these extended elements with the comparative laws. Inspired by the work in [3], we extend the concept of extension with possibility to DHFSs.

**Definition 3 (DHFEP)** Consider two dual hesitant fuzzy elements $d_1 = (h_1, g_1)$ and $d_2 = (h_2, g_2)$ whose lengths are $\ell_1, \ell_2$, respectively, and also assume that $\ell_1 < \ell_2$. Now to accomplish our aim that is to make lengths equal, we extend $d_1$ (as it is shorter one) and transform it into $\tilde{d}_1 = (\tilde{h}_1, \tilde{g}_1) = (\{{}^1\tilde{\zeta}_{11}, {}^1\tilde{\zeta}_{12}, \ldots, {}^{\frac{1}{\ell_2-\ell_1+1}}\tilde{\zeta}_{1\ell_1}, \ldots, {}^{\frac{1}{\ell_2-\ell_1+1}}\tilde{\zeta}_{1\ell_2}\}, \{{}^1\tilde{\xi}_{11}, {}^1\tilde{\xi}_{12}, \ldots, {}^{\frac{1}{\ell_2-\ell_1+1}}\tilde{\xi}_{1\ell_1}, \ldots, {}^{\frac{1}{\ell_2-\ell_1+1}}\tilde{\xi}_{1\ell_2}\})$ and just to keep the same notation, we express $d_2$ as $\tilde{d}_2 = (\tilde{h}_2, \tilde{g}_2) = (\{{}^1\tilde{\zeta}_{21}, {}^1\tilde{\zeta}_{22}, \ldots, {}^1\tilde{\zeta}_{2\ell_2}\}, \{{}^1\tilde{\xi}_{21}, {}^1\tilde{\xi}_{22}, \ldots, {}^1\tilde{\xi}_{2\ell_2}\})$. $\tilde{d}_1$ and $\tilde{d}_2$ are defined as DHFEPs whose expression form can be denoted as $\tilde{d}_1 = (\tilde{h}_1, \tilde{g}_1) = (\cup_{\tilde{\zeta}_1 \in \tilde{h}_1}\{{}^{p_1}\tilde{\zeta}_{11}\}, \cup_{\tilde{\xi}_1 \in \tilde{g}_1}\{{}^{p_1}\tilde{\xi}_{11}\}), \tilde{d}_2 = (\tilde{h}_2, \tilde{g}_2) = (\cup_{\tilde{\zeta}_2 \in \tilde{h}_2}\{{}^{p_2}\tilde{\zeta}_{21}\}, \cup_{\tilde{\xi}_2 \in \tilde{g}_1}\{{}^{p_2}\tilde{\xi}_{21}\})$, where ${}^{p_1}\tilde{\zeta}_{11}, {}^{p_1}\tilde{\xi}_{11}$ and ${}^{p_2}\tilde{\zeta}_{21}, {}^{p_2}\tilde{\xi}_{21}$ are the elements of DHFEPs and $p_1, p_2 (0 \leq p_1, p_2 \leq 1)$ are denoting the corresponding possibilities of $\tilde{\zeta}_{11}, \tilde{\xi}_{11}$ & $\tilde{\zeta}_{21}, \tilde{\xi}_{21}$, respectively.

Three types of position functions and a weight function are mentioned in the next definition for assigning weights to DHFEP.

**Definition 4** (Weight function) Consider a DHFS $d = (h, g) = (\{\zeta_1, \zeta_2, \ldots, \zeta_\ell\}, \{\xi_1, \xi_2, \ldots, \xi_\ell\})$ of length $\ell$. Three position function are given as:

(i) the progressive position function $f_k^p(d) = \sin(\frac{\pi k}{2(\ell+1)})$; $k = 1, 2, \ldots, \ell$.
(ii) the position function $f_k(d) = \sin(\frac{\pi k}{\ell+1})$; $k = 1, 2, \ldots, \ell$.
(iii) the regressive position function $f_k^r(d) = \sin(\frac{\pi(k+\ell+1)}{2(\ell+1)})$; $k = 1, 2, \ldots, \ell$.

If we transform a DHFS $d$ of length $\ell$ to DHFEPs $\tilde{d}$ of length $\tilde{\ell}(\tilde{\ell} > \ell)$for increasing the length, then $f_k(\tilde{d}) = f_k(d), k = 1, 2, \ldots, \ell$; and $f_{\ell+1}(\tilde{d}) = \ldots = f_{\tilde{\ell}}(\tilde{d}) = f_\ell(d), k = \ell + 1, \ldots, \tilde{\ell}$. Following the completion of all such preparation, the position weight function is defined as $w_k(\tilde{d}) = \frac{\tilde{\ell}.f_k(\tilde{d})}{\sum_{k=1}^{\tilde{\ell}} f_k(\tilde{d})}$; $k = 1, 2, \ldots, \tilde{\ell}$.

To compare DHFSs of different lengths, elements are added to the shorter DHFS by transforming it into DHFEP. After transforming DHFSs into DHFEPs, value and ambiguity functions are introduced as follows:

**Definition 5** (Value and ambiguity functions) Combining the position weight function, the value function for DHFEP $\tilde{d} = (\tilde{h}, \tilde{g}) = (\{\tilde{\zeta}_1, \tilde{\zeta}_2, \ldots, \tilde{\zeta}_{\tilde{\ell}}\}, \{\tilde{\xi}_1, \tilde{\xi}_2, \ldots, \tilde{\xi}_{\tilde{\ell}}\})$ is defined as:

$$V(\tilde{d}) = \sum_{k=1}^{\tilde{\ell}} \frac{w_k(\tilde{d})\frac{({}^{p_k}\tilde{\zeta}_k + (1 - {}^{p_k}\tilde{\xi}_k))}{2} p_k}{\sum_{k=1}^{\tilde{\ell}} p_k} \tag{1}$$

and the ambiguity function is defined as:

$$A(\tilde{d}) = \sum_{k=1}^{\tilde{\ell}} \frac{w_k|{}^{p_k}\tilde{\zeta}_k - V(\tilde{d})|p_k}{\sum_{k=1}^{\tilde{\ell}} p_k} + \sum_{k=1}^{\tilde{\ell}} \frac{w_k|{}^{p_k}\tilde{\xi}_k - V(\tilde{d})|p_k}{\sum_{k=1}^{\tilde{\ell}} p_k}. \tag{2}$$

**Definition 6** (Comparative laws) Let $\tilde{d}_1, \tilde{d}_2$ be two DHFEPs defined on a fixed set $X$. then, to compare DHFEPs, comparative laws are given as:

(i) if $V(\tilde{d}_1) < V(\tilde{d}_2) \Longrightarrow \tilde{d}_1 < \tilde{d}_2$.
(ii) if $V(\tilde{d}_1) = V(\tilde{d}_2)$ then
  (a) $A(\tilde{d}_1) < A(\tilde{d}_2) \Longrightarrow \tilde{d}_1 > \tilde{d}_2$
  (b) $A(\tilde{d}_1) = A(\tilde{d}_2) \Longrightarrow \tilde{d}_1 \approxeq \tilde{d}_2$

***Example 1*** Let $d_1 = (\{0.2, 0.3\}, \{0.3, 0.5\}), d_2 = (\{0.4, 0.5, 0.6\}, \{0.2, 0.25, 0.3\})$ be two DHFSs. Now, we transform these into DHFEPs for equalizing the lengths. Thus, we have $\tilde{d}_1 = (\{{}^1 0.2, {}^{0.5} 0.3, {}^{0.5} 0.3\}, \{{}^1 0.3, {}^{0.5} 0.5, {}^{0.5} 0.5\}), \tilde{d}_2 = (\{{}^1 0.4, {}^1 0.5, {}^1 0.6\}, \{{}^1 0.2, {}^1 0.25, {}^1 0.3\})$. Now, for three different types of positions functions, we have

(i) **For progressive position function** $w_1(\tilde{d}_1) = \frac{2}{3}$, $w_2(\tilde{d}_1) = \frac{4}{3}$, $w_3(\tilde{d}_1) = \frac{4}{3}$, $w_1(\tilde{d}_2) = \frac{1}{2}$, $w_2(\tilde{d}_2) = 1$, $w_3(\tilde{d}_2) = \frac{3}{2}$, $V(\tilde{d}_1) = 0.4166$, $V(\tilde{d}_2) = 0.6333$ $\Longrightarrow \tilde{d}_1 < \tilde{d}_2$.

(ii) **For position function** $w_1(\tilde{d}_1) = 1, w_2(\tilde{d}_1) = 1, w_3(\tilde{d}_1) = 1, w_1(\tilde{d}_2) = \frac{3}{4}, w_2(\tilde{d}_2) = \frac{3}{2}, w_3(\tilde{d}_2) = \frac{3}{4}$ $V(\tilde{d}_1) = 0.425, V(\tilde{d}_2) = 0.625 \Longrightarrow \tilde{d}_1 < \tilde{d}_2$.

(iii) **for regressive position function** $w_1(\tilde{d}_1) = \frac{4}{3}, w_2(\tilde{d}_1) = \frac{2}{3}, w_3(\tilde{d}_1) = \frac{2}{3}, w_1(\tilde{d}_2) = \frac{3}{2}, w_2(\tilde{d}_2) = 1, w_3(\tilde{d}_2) = \frac{1}{2}$ $V(\tilde{d}_1) = 0.4333, V(\tilde{d}_2) = 0.3111 \Longrightarrow \tilde{d}_1 < \tilde{d}_2$.

Now, to show applicability of the ambiguity function, there is given one more example.

***Example 2*** Let $d_1 = (\{0.3, 0.4, 0.45\}, \{0.2, 0.3, 0.4\}), d_2 = (\{0.4, 0.45, 0.55\}, \{0.3, 0.35, 0.5\})$ be two DHFSs. Here, lengths of the given sets are equal. So, for three different types of position function, we have

(i) **For progressive position function** $w_1(d_1) = \frac{1}{2} = w_1(d_2), w_2(d_1) = 1 = w_2(d_1), w_3(d_1) = \frac{3}{2} = w_3(d_2), V(d_1) = 0.5375 = V(d_2)$. So, we use ambiguity function (2). We note $A(d_1) = 0.3333, A(d_2) = 0.1791 \Longrightarrow d_1 < d_2$.
(ii) **For position function** $w_1(d_1) = \frac{3}{4} = w_1(d_2), w_2(d_1) = \frac{3}{2} = w_2(d_2), w_3(d_1) = \frac{3}{4} = w_3(d_1), V(d_1) = 0.5437 = V(d_2)$. So, we use ambiguity function(2). We note $A(d_1) = 0.4, A(d_2) = 0.2531 \Longrightarrow d_1 < d_2$.
(iii) **For regressive position function** $w_1(d_1) = \frac{3}{2} = w_1(d_2), w_2(d_1) = 1 = w_2(d_2), w_3(d_1) = \frac{1}{2} = w_3(d_2), V(d_1) = 0.5458 = V(d_2)$. So, we use ambiguity function(2). We note $A(d_1) = 0.4666, A(d_2) = 0.3013 \Longrightarrow d_1 < d_2$.

# 4 Formulation and Algorithm for the Proposed Dual Hesitant Fuzzy Matrix Game Model

In this section, first, we consider a TMGP under dual hesitant fuzzy environment. In this problem, the pay-offs of both players are given by DHFSs. Then, we introduce an algorithm for solving such game problems.

## *4.1 Formulation of Dual Hesitant Fuzzy Matrix Game*

Consider a TMGP, in which utility function values are given by DHFSs. In it, players I and II are denoted by $T1$ and $T2$, respectively. Here, this problem can be defined as

$$DHG = (T, \Delta(T1), \Delta(T2)).$$

Here, $R_1 = \{\kappa_1, \kappa_2, \ldots, \kappa_m\}$ & $R_2 = \{\varrho_1, \varrho_2, \ldots, \varrho_n\}$ represents the pure strategy spaces for $T1$ and $T2$, respectively. While, $\Delta(T1) = \{\acute{r} = (\acute{r}_i) : \sum_{i=1}^{m} \acute{r}_i = 1, \acute{r}_i \geq 0, i = 1, 2, \ldots, m\}$ and $\Delta(T2) = \{\acute{s} = (\acute{s}_j) : \sum_{j=1}^{n} \acute{s}_j = 1, \acute{s}_j \geq 0, j = 1, 2, \ldots, n\}$ are the mixed strategy spaces for $T1$ and $T2$, respectively. Assuming that $T1$ is maximizer while $T2$ is treated as minimizer. Then, the pay-off matrix $T$ for DHG is given by the following matrix

$$T = [d_{ij}]_{m \times n},$$

where $d_{ij} = (h_{d_{ij}}, g_{d_{ij}})$ is a $ij^{th}$ dual hesitant fuzzy element with $h_{d_{ij}} = \{\zeta_{d_{ij1}}, \zeta_{d_{ij2}}, \ldots, \zeta_{d_{ij\ell}}\}$ &; $g_{d_{ij}} = \{\xi_{d_{ij1}}, \xi_{d_{ij2}}, \ldots, \xi_{d_{ij\ell}}\}$. That is, for $i^{th}$ strategy of $T1$ and $j^{th}$ strategy for $T2$, $h_{d_{ij}}$ shows possible membership grades of $T1$ and $g_{d_{ij}}$ represents the possible non-membership grades of $T1$. Since $DHG$ is TMG, so $h_{d_{ij}}$ represents possible non-membership grades of $T2$ and $g_{d_{ij}}$ represents the possible membership grades of $T2$.

The algorithm steps for the proposed $DHG$ are outlined as follows:

## 4.2 Proposed Algorithm

**Step-I:**

(i) If the lengths of the elements in the pay-off matrix $T$ are equal, then proceed to Step-II.

(ii) If the elements of matrix $T$ are not of equal lengths, then transform the given elements into DHFEPs by applying the concept given in Sect. 3. Suppose that the largest length of an element in $T$ is $\ell^*$. Then we transform the $ij^{th}$ element $d_{ij}$ into $\tilde{d}_{ij}$ by extending the length from $\ell$ to $\ell^*$. Thus, we obtain $\tilde{T} = [\tilde{d}_{ij}]_{m \times n}$, where $\tilde{d}_{ij} = (\tilde{h}_{d_{ij}}, \tilde{g}_{d_{ij}}) = (\{\zeta_{d_{ij1}}, \zeta_{d_{ij2}}, \ldots, \zeta_{d_{ij\ell}}, \zeta_{d_{ij(\ell+1)}} = \zeta_{d_{ij\ell}}, \ldots, \zeta_{d_{ij\ell^*}} = \zeta_{d_{ij\ell}}\}, \{\xi_{d_{ij1}}, \xi_{d_{ij2}}, \ldots, \xi_{d_{ij\ell}}, \xi_{d_{ij(\ell+1)}} = \xi_{d_{ij\ell}}, \ldots, \xi_{d_{ij\ell^*}} = \xi_{d_{ij\ell}}\})$.

**Step-II**: In this step, we calculate two crisp matrices from the calculated matrix $\tilde{T}$ by using the proposed functions (1) and (2), named as value and ambiguity matrices. These are obtained as follows:

**Value matrix**
$V(\tilde{T}) = [V(\tilde{d}_{ij})]_{m \times n}$, where

$$V(\tilde{d}_{ij}) = \sum_{k=1}^{\ell^*} \frac{w_k(\tilde{d}_{ij})^{\frac{(p_k \tilde{\zeta}_{d_{ijk}} + (1 - p_k \tilde{\xi}_{d_{ijk}}))}{2}} p_k}{\sum_{k=1}^{\ell^*} p_k}$$

and, **Ambiguity matrix**
$A(\tilde{T}) = [A(\tilde{d}_{ij})]_{m \times n}$, where

$$A(\tilde{d}_{ij}) = \sum_{k=1}^{\ell^*} \frac{w_k |^{p_k} \tilde{\zeta}_{d_{ijk}} - V(\tilde{d})|_{p_k}}{\sum_{k=1}^{\ell^*} p_k} + \sum_{k=1}^{\ell^*} \frac{w_k |^{p_k} \tilde{\xi}_{d_{ijk}} - V(\tilde{d})|_{p_k}}{\sum_{k=1}^{\ell^*} p_k}$$

**Step-III**: With respect to calculated matrices $V(\tilde{T})$ and $A(\tilde{T})$, we derive the optimization models for $T1$ as follows:

$PM1$:
$\max E_1(\acute{r}, \acute{s}) = \max \sum_{i=1}^{m} \sum_{j=1}^{n} \acute{r}_i V(\tilde{d}_{ij}) \acute{s}_j$
s.t.
$\sum_{i=1}^{m} \acute{r}_i = 1$
$\sum_{j=1}^{n} \acute{s}_j = 1$
$\acute{r}_i \geq 0; i = 1, 2, \ldots, m,$
$\acute{s}_j \geq 0; j = 1, 2, \ldots, n,$

$PM2$:
$\min E_2(\acute{r}, \acute{s}) = \min \sum_{i=1}^{m} \sum_{j=1}^{n} \acute{r}_i A(\tilde{d}_{ij}) \acute{s}_j$
s.t.
$\sum_{i=1}^{m} \acute{r}_i = 1$
$\sum_{j=1}^{n} \acute{s}_j = 1$
$\acute{r}_i \geq 0; i = 1, 2, \ldots, m,$
$\acute{s}_j \geq 0; j = 1, 2, \ldots, n$

If we consider $\acute{r}_i's$ as decision variables then above models take the following form:

$PM3$:
$\max\{V_1\}$
s.t.
$\sum_{i=1}^{m} \acute{r}_i V(\tilde{d}_{ij}) \geq V_1; j = 1, 2, \ldots, n$
$\sum_{i=1}^{m} \acute{r}_i = 1, \acute{r}_i \geq 0; i = 1, 2, \ldots, m.$

$PM4$:
$\min\{V_2\}$
s.t.
$\sum_{i=1}^{m} A(\tilde{d}_{ij}) \acute{r}_i \leq V_2; j = 1, 2, \ldots, n$
$\sum_{i=1}^{m} \acute{r}_i = 1, \acute{r}_i \geq 0; i = 1, 2, \ldots, m.$

Similarly, we can calculate the models for $T2$ as follows:

$PM5$:
$\min\{W_1\}$
s.t.
$\sum_{j=1}^{n} V(\tilde{d}_{ij}) \acute{s}_j \leq W_1; i = 1, 2, \ldots, m$
$\sum_{j=1}^{n} \acute{s}_j = 1, \acute{s}_j \geq 0; j = 1, 2, \ldots, n.$

$PM6$:
$\max\{W_2\}$
s.t.
$\sum_{j=1}^{n} A(\tilde{d}_{ij}) \acute{s}_j \geq W_2; i = 1, 2, \ldots, m$
$\sum_{j=1}^{n} \acute{s}_j = 1, \acute{s}_j \geq 0; j = 1, 2, \ldots, n.$

**Step-IV**: For $T1$, the optimal solutions for models $PM3$ and $PM4$ are, respectively, $(\acute{r}^*_{V_1}, V_1^*)$ and $(\acute{r}^*_{V_2}, V_2^*)$. Since $T1$ is maximizer, aiming to find minimum expected gain so $\min\{V_1^*, V_2^*\} = V^*$ will be the game value and corresponding strategy are the optimal strategy for $T1$. Similarly, we can find the optimal strategy for $T2$ by solving the models $PM5$ and $PM6$.

**Note**:- It is notable that models $PM3$ and $PM5$ (Models $PM4$ and $PM6$) are dual to each other in crisp sense.

## 5 Application: Optimal Defense Strategies Against Terrorism

Terrorism is the unlawful use of violence and intimidation, especially against civilians, to achieve political, ideological, or religious goals. It aims to create fear and chaos to influence governments, societies, or specific groups, often by targeting public places. Recently, Israel faced an attack by the terrorist on Oct 7, 2023, targeting civilians and military personnel. The situation has escalated tensions in the region, leading to military responses and heightened security measures. The global threat of terrorism endangers lives and property, requiring constant vigilance from security agencies despite limited resources. Studies indicate that the zero-sum game framework is a useful tool for analyzing the risks associated with terrorist attacks. In the context of security, game theory shows that both players can make strategic decisions and anticipate each other's actions. The interaction between attackers and defenders can be modeled as a two-person zero-sum game, where attackers represent the terrorists and defenders are the opposing side. By applying game theory, Terrorists aim to maximize their gain by inciting fear and causing property damage. Contrary to this, defenders seek to minimize the impact of terrorist's attack. In a TMG, attackers and defenders can be viewed as $T1$ and $T2$, respectively, each possessing specific attack or defense resources. The game model between the two players is generalized to help them develop strategies that maximize their benefits while managing multiple potential targets. Due to factors such as changing pedestrian traffic patterns and the unpredictability of attacks, it is challenging to precisely assess the value of each potential target. As a result, it is impractical for decision-makers to acquire precise and complete information for the pay-off matrix. To address this uncertainty, DHFSs are utilized to represent the pay-offs in the analysis. Imagine a scenario where people are eagerly preparing for an upcoming festival, creating a lively and festive atmosphere. However, attackers see this as an opportunity to exploit the situation. Fortunately, security agencies receive critical intelligence indicating a potential terror threat. In response, to ensure public safety and thwart any malicious plans, decision-makers swiftly evaluate four high-priority potential targets: $\kappa_1$ : Metro, $\kappa_2$ : Worship place, $\kappa_3$ : Theater, and $\kappa_4$ : Shopping Mall. The given pay-off matrix for $T1$ (the attacker) helps in formulating an optimal defense strategy.

$$T = \begin{array}{c} \\ \kappa_1 \\ \\ \kappa_2 \\ \\ \kappa_3 \\ \\ \kappa_4 \end{array} \begin{array}{c} \begin{array}{cccc} \varrho_1 & \varrho_2 & \varrho_3 & \varrho_4 \end{array} \\ \begin{pmatrix} (\{0.20,0.30\}, & (\{0.60,0.70,0.80\}, & (\{0.50,0.60\}, & (\{0.40,0.60\}, \\ \{0.60,0.65\}) & \{0.12,0.15,0.18\}) & \{0.35,0.40\}) & \{0.30,0.35\}) \\ (\{0.50,0.70\}, & (\{0.20,0.30\}, & (\{0.40,0.60\}, & (\{0.40,0.50\}, \\ \{0.25,0.30\}) & \{0.60,0.70\}) & \{0.30,0.35\}) & \{0.40,0.50\}) \\ (\{0.40,0.50\}, & (\{0.70,0.80\}, & (\{0.10,0.20,0.30\}, & (\{0.40,0.60\}, \\ \{0.30,0.40\}) & \{0.15,0.18\}) & \{0.50,0.60,0.65\}) & \{0.30,0.35\}) \\ (\{0.50,0.60,0.80\}, & (\{0.70,0.90\}, & (\{0.80,0.90\}, & (\{0.30,0.40\}, \\ \{0.05,0.10,0.15\}) & \{0.05,0.08\}) & \{0.05,0.08\}) & \{0.50,0.55\}) \end{pmatrix} \end{array}$$

To solve this problem, the steps of the proposed algorithm are given as follows:
**Step-I:** By utilizing the concepts of extension with possibility to equalize the length, we calculate the transformed matrix $\tilde{T}$ as follows:

$$\tilde{T} = \begin{array}{cc} & \begin{array}{cccc} \varrho_1 & \varrho_2 & \varrho_3 & \varrho_4 \end{array} \\ \begin{array}{c} \kappa_1 \\ \\ \kappa_2 \\ \\ \kappa_3 \\ \\ \kappa_4 \end{array} & \left( \begin{array}{cccc} \begin{array}{l} (\{{}^{1}0.20, {}^{0.5}0.30, {}^{0.5}0.30\}, \\ \{{}^{1}0.60, {}^{0.5}0.65, {}^{0.5}0.65\}) \end{array} & \begin{array}{l} (\{{}^{1}0.60, {}^{1}0.70, {}^{1}0.80\}, \\ \{{}^{1}0.12, {}^{1}0.15, {}^{1}0.18\}) \end{array} & \begin{array}{l} (\{{}^{1}0.50, {}^{0.5}0.60, {}^{0.5}0.60\}, \\ \{{}^{1}0.35, {}^{0.5}0.40, {}^{0.5}0.40\}) \end{array} & \begin{array}{l} (\{{}^{1}0.40, {}^{0.5}0.60, {}^{0.5}0.60\}, \\ \{{}^{1}0.30, {}^{0.5}0.35, {}^{0.5}0.35\}) \end{array} \\ \begin{array}{l} (\{{}^{1}0.50, {}^{0.5}0.70, {}^{0.5}0.70\}, \\ \{{}^{1}0.25, {}^{0.5}0.30, {}^{0.5}0.30\}) \end{array} & \begin{array}{l} (\{{}^{1}0.20, {}^{0.5}0.30, {}^{0.5}0.30\}, \\ \{{}^{1}0.60, {}^{0.5}0.70, {}^{0.5}0.70\}) \end{array} & \begin{array}{l} (\{{}^{1}0.40, {}^{0.5}0.60, {}^{0.5}0.60\}, \\ \{{}^{1}0.30, {}^{0.5}0.35, {}^{0.5}0.35\}) \end{array} & \begin{array}{l} (\{{}^{1}0.40, {}^{0.5}0.50, {}^{0.5}0.50\}, \\ \{{}^{1}0.40, {}^{0.5}0.50, {}^{0.5}0.50\}) \end{array} \\ \begin{array}{l} (\{{}^{1}0.40, {}^{0.5}0.50, {}^{0.5}0.50\}, \\ \{{}^{1}0.30, {}^{0.5}0.40, {}^{0.5}0.40\}) \end{array} & \begin{array}{l} (\{{}^{1}0.70, {}^{0.5}0.80, {}^{0.5}0.80\}, \\ \{{}^{1}0.15, {}^{0.5}0.18, {}^{0.5}0.18\}) \end{array} & \begin{array}{l} (\{{}^{1}0.10, {}^{1}0.20, {}^{1}0.30\}, \\ \{{}^{1}0.50, {}^{1}0.60, {}^{1}0.65\}) \end{array} & \begin{array}{l} (\{{}^{1}0.40, {}^{0.5}0.60, {}^{0.5}0.60\}, \\ \{{}^{1}0.30, {}^{0.5}0.35, {}^{0.5}0.35\}) \end{array} \\ \begin{array}{l} (\{{}^{1}0.50, {}^{1}0.60, {}^{1}0.80\}, \\ \{{}^{1}0.05, {}^{1}0.10, {}^{1}0.15\}) \end{array} & \begin{array}{l} (\{{}^{1}0.70, {}^{0.5}0.90, {}^{0.5}0.90\}, \\ \{{}^{1}0.05, {}^{0.5}0.08, {}^{0.5}0.08\}) \end{array} & \begin{array}{l} (\{{}^{1}0.80, {}^{0.5}0.90, {}^{0.5}0.90\}, \\ \{{}^{1}0.05, {}^{0.5}0.08, {}^{0.5}0.08\}) \end{array} & \begin{array}{l} (\{{}^{1}0.30, {}^{0.5}0.40, {}^{0.5}0.40\}, \\ \{{}^{1}0.50, {}^{0.5}0.55, {}^{0.5}0.55\}) \end{array} \end{array} \right) \end{array}$$

**Step-II** By using value (1) and ambiguity functions (2), we get

$$V(\tilde{T}) = \begin{array}{cc} & \begin{array}{cccc} \varrho_1 & \varrho_2 & \varrho_3 & \varrho_4 \end{array} \\ \begin{array}{c} \kappa_1 \\ \kappa_2 \\ \kappa_3 \\ \kappa_4 \end{array} & \left( \begin{array}{cccc} 0.3166 & 0.7866 & 0.5916 & 0.6 \\ 0.675 & 0.3 & 0.6 & 0.5 \\ 0.55 & 0.7983 & 0.3125 & 0.6 \\ 0.7833 & 0.8816 & 0.8983 & 0.4166 \end{array} \right) \end{array}, \quad A(\tilde{T}) = \begin{array}{cc} & \begin{array}{cccc} \varrho_1 & \varrho_2 & \varrho_3 & \varrho_4 \end{array} \\ \begin{array}{c} \kappa_1 \\ \kappa_2 \\ \kappa_3 \\ \kappa_4 \end{array} & \left( \begin{array}{cccc} 0.3666 & 0.6933 & 0.2444 & 0.3333 \\ 0.4666 & 0.4 & 0.3333 & 0.0666 \\ 0.2666 & 0.6622 & 0.375 & 0.3333 \\ 0.7833 & 0.8844 & 0.8622 & 0.1666 \end{array} \right) \end{array}$$

**Step-III** Optimization models for $T1$ is given as: $PM3$ and $PM4$

$$\begin{array}{ll} \max\{V_1\} & \min\{V_2\} \\ \text{s.t.} & \text{s.t.} \\ 0.3166\acute{r}_1 + 0.675\acute{r}_2 + 0.55\acute{r}_3 + 0.7833\acute{r}_4 \geq V_1 & 0.3666\acute{r}_1 + 0.4666\acute{r}_2 + 0.2666\acute{r}_3 + 0.7833\acute{r}_4 \leq V_2 \\ 0.675\acute{r}_1 + 0.3\acute{r}_2 + 0.7983\acute{r}_3 + 0.8816\acute{r}_4 \geq V_1 & 0.4666\acute{r}_1 + 0.4\acute{r}_2 + 0.3333\acute{r}_3 + 0.0666\acute{r}_4 \leq V_2 \\ 0.55\acute{r}_1 + 0.7983\acute{r}_2 + 0.3125\acute{r}_3 + 0.6\acute{r}_4 \geq V_1 & 0.2666\acute{r}_1 + 0.6622\acute{r}_2 + 0.375\acute{r}_3 + 0.3333\acute{r}_4 \leq V_2 \\ 0.7833\acute{r}_1 + 0.8816\acute{r}_2 + 0.7893\acute{r}_3 + 0.4166\acute{r}_4 \geq V_1 & 0.7833\acute{r}_1 + 0.8844\acute{r}_2 + 0.8622\acute{r}_3 + 0.1666\acute{r}_4 \leq V_2 \\ \acute{r}_1 + \acute{r}_2 + \acute{r}_3 + \acute{r}_4 = 1 & \acute{r}_1 + \acute{r}_2 + \acute{r}_3 + \acute{r}_4 = 1 \\ \acute{r}_1, \acute{r}_2, \acute{r}_3, \acute{r}_4 \geq 0 & \acute{r}_1, \acute{r}_2, \acute{r}_3, \acute{r}_4 \geq 0 \end{array}$$

**Step-IV:** After solving $PM3$, we get optimal solution for the $T1 = (\acute{r}^*_{V_1}, V_1^*) = (0.2749, 0, 0.4510, 0.2739, 0.5497)$. And after solving $PM4$, optimal solution for $T1 = (\acute{r}^*_{V_2}, V_2^*) = (0, 0.8559, 0.1440, 0, 0.4377)$. Since $\min\{V_1^* = 0.5497, V_2^* = 0.4377\} = 0.4377 = V^*$. Therefore, optimal strategy and value of the game for $T1$ is given as $(\acute{r}^*_V, V^*) = (0, 0.8559, 0.1440, 0, 0.4377)$. Similarly, we get the optimal defense strategy for $T2$ given as $(\acute{s}^*_W, W^*) = (0.1730, 0, 0.1446, 0.6822, 0.5497)$.

# 6 Conclusion

This paper presents an algorithm for solving TMGP in which pay-offs are represented by dual hesitant fuzzy elements. For it, the concept of DHFEP is proposed to handle the considered elements of unequal lengths. Then, the value and ambiguity functions are introduced to compare DHFSs of varying lengths. Using these functions, two dual linear programming models are developed to find optimal strategies for players in the proposed algorithm. To check the performance of the proposed algorithm, it is applied to obtain the optimal defense strategies for security agencies to combat terrorist threats. Wolfram Mathematica 9 software is used for calculation throughout the paper.

In the future work, the authors intend to extend this work in other fuzzy environments such as hesitant picture fuzzy set.

# References


1. Bauso, D.: Game Theory with Engineering Applications. SIAM (2016)
2. Karmakar, S., Seikh, M.R.: A nonlinear programming approach to solving interval-valued intuitionistic hesitant noncooperative fuzzy matrix games. Symmetry **16**(5), 573 (2024)

3. Xue, W., Xu, Z., Zeng, X.J.: Solving matrix games based on Ambika method with hesitant fuzzy information and its application in the counter-terrorism issue. Appl. Intell. **51**, 1227–1243 (2021)
4. Torra, V.: Hesitant fuzzy sets. Int. J. Intell. Syst. **25**(6), 529–539 (2010)
5. Zadeh, L.A.: Fuzzy sets. Inf. Control **8**(3), 338–353 (1965)
6. Zhu, B., Xu, Z., Xia, M.: Dual hesitant fuzzy sets. J. Appl. Math. **2012**(1), 879629 (2012)
7. Kumar, S., Garg, H.: Some novel point operators and multiple rounds voting process based decision-making algorithm under picture fuzzy set environment. Adv. Eng. Softw. **174**, 103274 (2022)
8. Kumar, S., Kumar, M.: A game theoretic approach to solve multiple group decision making problems with interval-valued intuitionistic fuzzy decision matrices. Int. J. Manag. Sci. Eng. Manag. **16**(1), 34–42 (2021)
9. Kumar, S., Kumar, M.: A new approach to solve group decision making problems with attribute values and attribute weights represented by interval-valued intuitionistic fuzzy numbers. Int. J. Appl. Comput. Math. **7**, 1–16 (2021)
10. Kumar, S., Rani, S., Kumar, M.: A new order function for interval-valued intuitionistic fuzzy numbers and its application in group decision making. Fuzzy Inf. Eng. **13**(1), 111–126 (2021)
11. Kumar, S., Tyagi, R.: A cutting edge algorithm for interval-valued intuitionistic fuzzy decision making based on mean, variance of alternative score matrices and a new score function. In: International Conference on Soft Computing: Theories and Applications, pp. 367–379. Springer (2023)
12. Kumar, S., Tyagi, R.: A novel score function for picture fuzzy numbers and its based entropy method to multiple attribute decision-making. In: Soft Computing for Problem Solving: Proceedings of the SocProS 2022, pp. 719–729. Springer (2023)
13. Kumar, S.: Duality results in fuzzy linear programming problems based on the concept of goal programming. Int. J. Syst. Sci. Oper. Logist. **7**(2), 206–216 (2020)
14. Pandey, D., Kumar, S.: Fuzzy optimization of primal-dual pair using piecewise linear membership functions. Yugosl. J. Oper. Res. **22**(1) (2011)
15. Kumar, S., Pandey, D.: Fuzzy programming approach to solve multi-objective transportation problem. In: Proceedings of the International Conference on Soft Computing for Problem Solving (SocProS 2011) December 20–22, 2011, vol. 1, pp. 525–533. Springer (2012)
16. Pandey, D., Kumar, S.: Fuzzy multi-objective fractional goal programming using tolerance. Int. J. Manag. Sci. Eng. Appl. **5**(1), 175–187 (2011)
17. Campos, L.: Fuzzy linear programming models to solve fuzzy matrix games. Fuzzy Sets Syst. **32**(3), 275–289 (1989)
18. Dengfeng, L.: A fuzzy multiobjective programming approach to solve fuzzy matrix games. J. Fuzzy Math. **7**(4), 907–912 (1999)
19. Chen, Y.W., Larbani, M.: Two-person zero-sum game approach for fuzzy multiple attribute decision making problems. Fuzzy Sets Syst. **157**(1), 34–51 (2006)
20. Pandey, D., Kumar, S.: Modified approach to multiobjective matrix game with vague payoffs. J. Int. Acad. Phys. Sci. **14**(2), 149–157 (2010)
21. Seikh, M.R., Nayak, P.K., Pal, M.: Matrix games with intuitionistic fuzzy pay-offs. J. Inf. Optim. Sci. **36**(1–2), 159–181 (2015)
22. Kumar, S.: Max-min solution approach for multi-objective matrix game with fuzzy goals. Yugosl. J. Oper. Res. **26**(1), 51–60 (2016)
23. Kumar, S.: The relationship between intuitionistic fuzzy programming and goal programming. In: Proceedings of Sixth International Conference on Soft Computing for Problem Solving: SocProS 2016, vol. 1, pp. 220–229. Springer (2017)
24. Kumar, S.: To solve matrix games with fuzzy goals using piecewise linear membership functions. In: Proceedings of the Jangjeon Mathematical Society, vol. 21, pp. 627–636 (2018)
25. Jana, J., Roy, S.K.: Dual hesitant fuzzy matrix games: based on new similarity measure. Soft. Comput. **23**, 8873–8886 (2019)
26. Kumar, S.: Piecewise linear programming approach to solve multi-objective matrix games with i-fuzzy goals. J. Control. Decis. **8**(1), 1–13 (2021)

27. Brikaa, M., Zheng, Z., Ammar, E.S.: Mehar approach for solving matrix games with triangular dual hesitant fuzzy payoffs. Granul. Comput. 1–20 (2022)
28. Yang, J., Xu, Z.: Matrix game-based approach for MADM with probabilistic triangular intuitionistic hesitant fuzzy information and its application. Comput. Ind. Eng. **163**, 107787 (2022)
29. Jana, J., Roy, S.K.: Linguistic pythagorean hesitant fuzzy matrix game and its application in multi-criteria decision making. Appl. Intell. **53**(1), 1–22 (2023)

# Crop Type Classification Using Satellite Images

**Sumanth Kompelly, B. V. Mralath Kumar, V. Nivethitha, and Suthir Sriram**

**Abstract** Novel progress in hyperspectral imaging and machine learning are revolutionizing the assessment of crop species, for which identification is central to feeding the world population. This research uses images obtained from the Deutsches Zentrum Earth Sensing Imaging Spectrometer (DESIS) to identify key arable crops in southeastern Hungary including hybrid corn, soybean, sunflower, and winter wheat. In this work, WA-CNN, Random Forest, and SVM machine learning algorithms are used to show how these models can improve the remote sensing of agricultural areas with precision thereby optimizing agricultural output and land usage. A notable aspect of this methodology is the application of cross-validation (CV), which plays a dual role: it not only compares the performance of different models but also optimizes weights of the ensemble, which are given to classification to enhance the reliability of the same. The results show that employing DESIS data allows for accurate determinations of crop development and yield prediction valuable information for farmers and authorities. Furthermore, since landscapes may be fragmented and fields differ in size, this study combines high spatial resolution imagery with high temporal frequency data. The study raises the argument that effective crop surveillance systems should be developed due to the beneficial effects they could have on agriculture as long as they increase the quality of decision-making in farming processes which will be crucial in meeting the ever-growing needs of the global population.

**Keywords** Crop classification · Hyperspectral imaging · Machine learning

---

S. Kompelly (✉) · B. V. Mralath Kumar · V. Nivethitha · S. Sriram
Department of Computer Science and Engineering, Amrita School of Computing, Amrita Vishwa Vidyapeetham, Chennai, India
e-mail: kompellysumanth29@gmail.com

B. V. Mralath Kumar
e-mail: bachupallymralath@gmail.com

V. Nivethitha
e-mail: v_nivethitha@ch.amrita.edu

S. Sriram
e-mail: s_suthir@ch.amrita.edu

© The Author(s), under exclusive license to Springer Nature Singapore Ptè Ltd. 2026

M. Pant et al. (eds.), *Real-World Applications of Machine Learning*, Mathematics for Sustainable Developments, https://doi.org/10.1007/978-981-95-7134-5_3

## 1 Introduction

Accurate and timely classification of crop types is a key practice required to enable efficient agricultural management, optimal allocation of resources, and addressing global food security challenges [1, 2]. Understanding the spatial distribution of various crops over large areas aids in land management, cropping area estimation, and yield prediction, while also enhancing farming practices using information technology. Traditional crop monitoring and yield forecasting relied on physical interventions, which were often arduous and costly, especially for extensive agricultural fields [3, 4].

The use of satellites as remote sensors has gained popularity as an effective solution, providing high-resolution images of crops with different spectral characteristics and enabling the analysis of changes across large areas in real time [5, 6]. This research leverages remote sensing to identify selected strategic crops, including hybrid maize, soybeans, sunflowers, and winter wheat, grown in specific regions using hyperspectral data obtained from the DLR Earth Sensing Imaging Spectrometer (DESIS) [7, 8]. Due to its potential application in precision farming, hyperspectral imaging is often considered a key technology [9]. Its high spectral resolution capability allows distinguishing different crops based on their spectral patterns, thereby achieving high classification accuracy [10]. When integrated with machine learning (ML) techniques, hyperspectral remote sensing enables high-quality, dense data acquisition for crop monitoring and yield estimation [11, 12].

Several experiments were conducted to assess the effectiveness of the proposed system, with successful classification of crop types achieved using multiple machine learning algorithms. These include the advanced Wavelet-Attention Convolutional Neural Network (WA-CNN) as well as traditional models such as Support Vector Machine (SVM) and Random Forest. The WA-CNN method is more complex than traditional approaches and integrates precise spectral discrepancies through wavelet transformations and attention mechanisms, which enhance feature extraction and outperform conventional models [13].

Cross-validation (CV) is employed in this study to minimize overfitting and evaluate model performance on different data segments. Furthermore, CV is used to determine the optimal weightage for combining models in an ensemble. Weighted averaging of model outputs in the ensemble leverages the strengths of individual models, leading to more accurate crop type classification [14]. The combination of hyperspectral remote sensing and machine learning enhances the accuracy of crop type prediction while addressing challenges such as intra-annual variability, crop growth stages, and field-scale differences. As global populations continue to grow, the need to increase crop yields becomes increasingly urgent. Coupled with challenges associated with climate change, this underscores the necessity for improved crop assessment.

This paper focuses on the identification of the four crops using DESIS hyperspectral imagery, comparing the performance of the WA-CNN model with traditional models such as SVM and Random Forest. By employing up-to-date remote sensing methods, the study provides a scalable and cost-effective solution for monitoring agricultural fields and forecasting yields, thereby supporting sustainable and productive farming practices [15].

## 2 Literature Review

Recent advances in satellite technology and machine learning have significantly improved the accuracy and scalability of crop type classification. High-resolution satellite imagery from Sentinel-1, Sentinel-2, Landsat, and hyperspectral data from the DESIS mission enables researchers to capture detailed spatial characteristics of crops. Convolutional Neural Networks (CNNs) have been widely applied to enhance classification rates by extracting spatial features from these images. The inclusion of additional features and time series information further improves classification performance, as crops with similar spectra at certain periods can be distinguished based on differences in their phenological cycles. For example, Long Short-Term Memory (LSTM) networks are particularly effective for analyzing time series data, enabling the identification of crop types based on temporal variations in spectral signatures [1–6].

Google Earth Engine is a prominent large-scale platform for analyzing agricultural crops, allowing researchers to apply machine learning models to extensive datasets. Commonly used indices and data include the Normalized Difference Vegetation Index (NDVI) and Synthetic Aperture Radar (SAR) data, often combined with classifiers such as Random Forest (RF) and Support Vector Machines (SVM). SAR data, in particular, is valuable for cloud-free observations in regions where persistent cloud cover hinders optical imaging. Preprocessing steps, including image filtering, georeferencing, and topographic phase removal, are essential to condition satellite data for machine learning, improving model precision [7–10].

Machine learning-based crop classification models consistently outperform traditional survey methods. Evaluation metrics such as overall accuracy, Kappa coefficient, and $F1$-score are commonly used to compare model performance. Integrating machine learning with satellite data has demonstrated improvements in classification accuracy, with some studies reporting up to a 20% increase when high spatial resolution data are combined with temporal datasets. Classification performance is further enhanced through data fusion techniques that integrate multiple sensor sources. Comparisons of the Spectral Angle Mapper (SAM) technique with spectral reflectance values and data interpolation from other sensors have shown improved crop type identification in both synthetic and real-world analyses [11–13].

Overall, the combination of spatial and temporal satellite data with machine learning has led to substantial improvements in crop type classification. Recent advancements in satellite sensors, cloud-based GIS, and data fusion technologies have made crop mapping more accurate and scalable. These methods are expected to play an increasingly important role in addressing challenges related to food security and sustainable agriculture as satellite and computational capabilities continue to advance [14, 15].

# 3 Data Generation and Preprocessing

## 3.1 Problem Definition

The primary objective of this research is to identify multiple crop types to develop a comprehensive agricultural baseline using hyperspectral satellite imagery. The proposed approach involves extracting spectral features from the biosphere via satellite data and applying statistical methods, particularly artificial neural networks, to classify crops with high accuracy. This methodology addresses the need for improved agricultural monitoring and provides a foundation for enhancing the robustness and applicability of crop differentiation systems in future studies.

## 3.2 Satellite Data

Hyperspectral data for this study were obtained from a high spatial resolution satellite imaging spectrometer covering the Visible and Near-Infrared (VNIR) region. The data provide specific reflectance indices for crops, facilitating the identification of different crop types. The survey was conducted in Mezohegyes, Hungary, focusing on crops such as corn, soybean, sunflower, and wheat. The spectral characteristics of hyperspectral imagery enabled precise discrimination between these crops.

## 3.3 Data Cleansing and Preprocessing

Several preprocessing steps were applied to segment and prepare the hyperspectral data for machine learning. Initially, noise reduction was performed using first-order atmospheric correction to minimize the influence of atmospheric effects and produce standardized reflectance values across all bands. Due to the high dimensionality of hyperspectral images, dimensionality reduction techniques, including Principal Component Analysis (PCA) and Factor Analysis, were employed to retain essential spectral information while reducing data size. For feature extraction, covariance-based methods were applied to accelerate processing and capture fine variations in spectral signatures among crops. These feature engineering steps produced a dataset better suited for machine learning models. Finally, feature scaling was applied during normalization to ensure all features contributed equally to model performance. After preprocessing, the dataset was partitioned, with 80% used for training the classification models.

# 4 Methodology

## 4.1 Data Preparation

The data used in this study includes both synthetic and real satellite data, which undergo the same model pipeline during training and testing. Synthetic data mimics real-world agricultural conditions, with varying field sizes, shapes, and spectral characteristics, creating a controlled development environment. Real satellite data from Landsat and MODIS satellites provide multispectral images at multiple time instances, allowing evaluation of the proposed model on actual agricultural fields. Preprocessing operations are applied to achieve spatial and temporal standardization, ensuring that the data is suitable for model input.

The frequency distribution of measurements in the example of Crop ID 1 is shown in Fig. 1 and it depicts the spectral variability available between diverse agricultural samples. This distribution can be used as a gauge of whether the preprocessing and normalization processes have been effective since we can establish that the dataset represents realistic variations in reflectance like those which are observed in the real satellite images. Such spectral consistency between synthetic and real data is important in the maintenance of model consistency and also the delivery of robust classification in a variety of agricultural settings.

## 4.2 Synthetic Data Generation

Synthetic data generation replicates agricultural conditions at different levels of complexity. Field sizes and patterns are randomly set to simulate variability, while realistic reflectance values are assigned based on MODIS data to represent crop growth seasonally. Environmental fluctuations and sensor noise are modeled using Gaussian noise. Low-resolution images are generated by averaging high-resolution pixel values ($64 \times 64$) to simulate MODIS spatial resolution. Additional Gaussian noise (standard deviation $= 0.15$) is added to enhance realism, ensuring that synthetic data closely resembles actual satellite images.

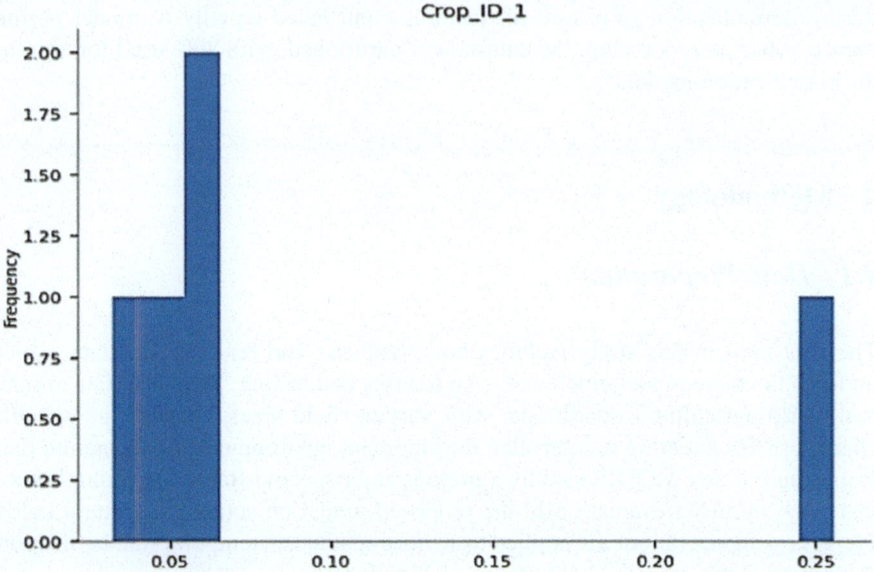


**Fig. 1** Frequency distribution of measurements for Crop ID 1.

## 4.3 Model Development

Crop classification employs two CatBoost classifiers:

1. CatBoost Classifier (cb) trained on balanced datasets.
2. CatBoost Classifier with Class Weights (cb2) to address under-represented crops.

The predictions of these classifiers are aggregated using weighted averaging based on validation performance, enhancing classification accuracy and mitigating bias or overfitting. This approach ensures minority crops receive appropriate attention without reducing the classification of dominant crops.

## 4.4 Implementation

The methodology integrates real satellite scenes with sensors of varying spatial and temporal resolutions. Landsat images (30 m, every 16 days) detect small fields, while MODIS images (500 m, daily) observe crop phenology. USDA NASS data provides reference crop maps for corn, cotton, rice, and soybean. A supervised decision tree classifier categorizes real datasets, establishing a benchmark for evaluating synthetic data.

As shown in Fig. 2, the values of Crop ID 1 in various Field IDs and time points show different changes, which portrays the fluctuation of crop reflectance pattern over

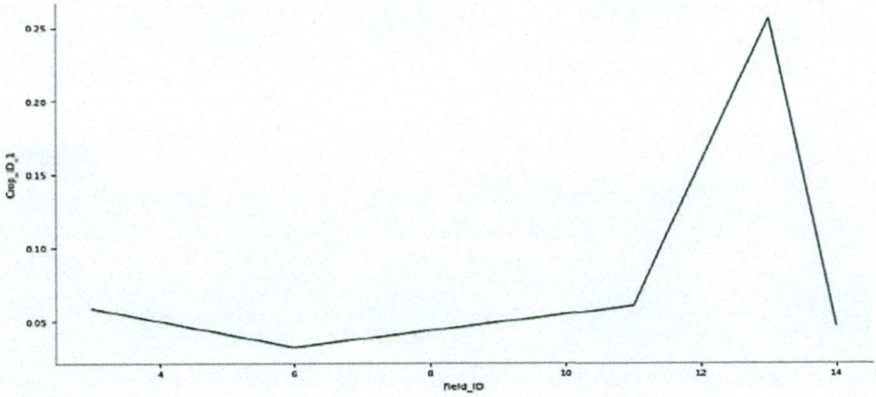


**Fig. 2** Measures of Crop ID 1 across different Field IDs at different times.

time. These changes can be explained by the fact that vegetation changes over time with growth stages and field conditions leading to these variations. These time trends can be regarded as a valuable source of input to prove the compatibility between real and synthetic data, as well as to increase the accuracy of the classification structure. The framework includes data creation, preprocessing, spatial statistics, feature and label generation, sample weighting, model creation, cross-validation, training, and prediction.

## *4.5 Algorithm Parameter Tuning*

Parameter tuning optimizes classification performance:

- Neighbor Lambda Coefficient: Adjusted via gradient descent to account for relationships between neighboring pixels.
- Synthetic Data Tuning: Introduces noise to test model stability.
- Real Data Tuning: Adjusts parameters based on varying soil, weather, crop growth stages, and climatic conditions.
- Low-Resolution Tuning: Balances differences between MODIS and Landsat spatial resolutions.

## *4.6 Classification Testing*

The model's performance is evaluated through several experiments:

- Baseline Tests: Compare CatBoost-based models with Maximum Likelihood Classification (MLC).

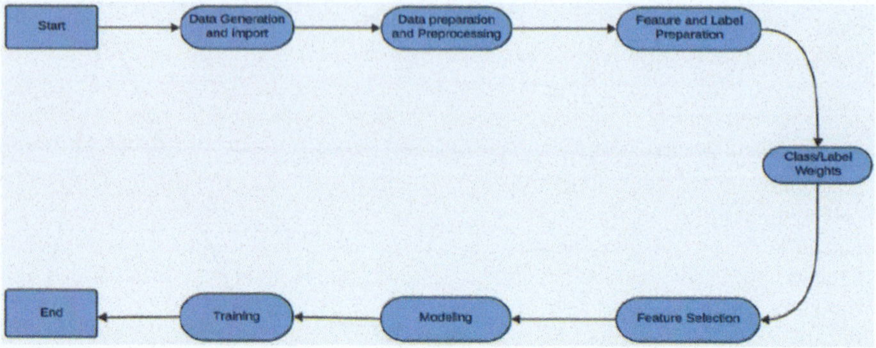


**Fig. 3** Methodology flowchart detailing the end-to-end process from data acquisition to model evaluation in satellite-based crop type classification.

- Enhanced Tests: Incorporate temporal data for serial crop identification.

Post-classification processing includes segmentation, smoothing, and majority-rule polygonal partitioning to remove noisy classifications and improve continuity in crop maps.

## 4.7 *Performance Evaluation*

Classification accuracy is assessed using overall accuracy, Kappa coefficient, and $F1$-score. The evaluation compares performance on synthetic and real datasets and demonstrates model generalization. Accuracy plots illustrate improvements across different environments, classifiers, and data resolutions, providing strong validation of model reliability.

The methodological flowchart in Fig. 3 illustrates the sequential steps of the proposed framework including the generation and preprocessing of the data, feature selection, training, and evaluation of the model as well. This parameter adjustment process is organized to support the systematic adjustment of parameters to make sure that the spatial and temporal variations are properly considered within a training pipeline.

# 5 Results and Discussion

The experiment outcomes related to crop type classification substantiate the general assumption that superior accuracy is achievable through the incorporation of advanced machine learning algorithms and multiscale satellite imagery data. The classification pipeline employed two versions of the CatBoost classifier: one without class weights (cb) and one with class weights (cb2). To combine their performances for each dataset, a weighted average approach was used. The best-performing combinations were 72% cb and 28% cb2 for the first dataset, and 76% cb and 24% cb2 for the second dataset. This weighted combination formalized the handling of class imbalance and provided more nuanced estimates.

Using low-resolution satellite images in the classification pipeline also contributed to substantial improvements in accuracy. When high-resolution images were scarce, incorporating low-resolution images increased class accuracy by 10–40%, depending on the temporal and spatial characteristics of the images. This improvement was most pronounced when only a single high-resolution image was available, demonstrating the efficacy of combining low- and high-resolution inputs. The largest accuracy gains were observed when high-resolution images were acquired during early or late stages of vegetation growth, where spectral responses were similar. In these cases, low-resolution images increased classification accuracy by up to 20% for crops that were otherwise difficult to distinguish. For instance, integrating ten low-resolution images with a single high-resolution image obtained early in the growing season raised accuracy by 22%, highlighting the benefits of multi-temporal imagery.

During mid-season, when crop spectra diverged significantly, improvements were marginal (around 10%) but still valuable. When multiple high-resolution images were available, further gains from adding low-resolution data were minimal (approximately 3%), indicating that high-resolution data alone can often capture essential crop differences. The integration of machine learning techniques with multiscale satellite imagery also improved crop-specific classification. For example, cotton was initially misclassified as corn due to spectral similarities, but the inclusion of low-resolution images increased cotton's producer accuracy from 0.01 to 72.77%. Similarly, corn's user accuracy improved from 24.53 to 61.58%. Rice, benefiting from its temporal and spectral characteristics, also demonstrated significant gains: producer accuracy increased from 52.95 to 70.11%, while user accuracy increased from 68.83 to 76.97%.

Overall, the results validate that combining high- and low-resolution satellite imagery with advanced machine learning models enhances crop type classification accuracy. This hybrid approach is particularly useful when only limited high-resolution data are available, allowing for robust and reliable crop classification across different datasets and crop types. The application of modeling architecture based on the described CatBoost Classifier sheds light on the prospect of crop type classification. Two methods for the CatBoost class weights were used: one with class weights disabled (cb) and the other with class weights enabled (cb2). The best results of both approaches were achieved across the two datasets. One fundamental observation is that using low-resolution images to augment a few high-resolution images leads to higher class accuracy. Analysis of results showed that the update mechanism, based on a combination of two CatBoost models with weights 72/28 for cb and cb2 for dataset 1, and 76/24 for dataset 2, allows the model to leverage extra information encoded in the lower-resolution data. This two-model approach skillfully utilizes low-resolution data to address the scarcity of high-resolution inputs, achieving significant accuracy enhancements.

The comparative analysis between synthetic and real data further underscores the difficulty in crop classification. Variation in accuracy improvement between synthetic and authentic data emphasizes the importance of carefully selected training datasets. Variability in performance across datasets necessitates hyperparameter tuning and optimal class weight assignment. Using class weights helps reduce potential bias arising from differences in input data quality and size. Furthermore, the timing of image acquisition affects the classification of normal versus tumor tissue. Preventing contrast invariance by obtaining high-quality images under varying light conditions improves model differentiation. Experiments showed that using low-resolution images effectively in certain areas increased accuracy, highlighting the importance of time synchronization in crop monitoring.

The choice of CatBoost Classifier allows feature selection, combining high dimensionality with low signal-to-noise ratios typical of low-quality images. By employing two classifiers, we benefit from the strengths of both models without being significantly affected by noisy data. Weighted averaging is proposed as a valid solution, as agricultural data is unpredictable and subject to atmospheric or sensor fluctuations. This approach minimizes overall noise while emphasizing high-resolution signals of interest. Although neighbor heuristics were not used in the current implementation, exploring them could enhance the model further. Neighbor heuristics exploit the likelihood that nearby pixels belong to the same class, improving spatial consistency and crop outline detection.

The findings indicate substantial potential for applying the proposed modeling architecture across different agricultural landscapes. Requiring only one high-resolution image alongside multiple low-resolution images makes our method particularly useful in situations where high-resolution inputs are limited or expensive. With low-resolution satellites like MODIS available since 2000, our solution can support crop monitoring and management.

However, there are certain limitations. The current Python implementation may face efficiency issues in processing time and memory compared to languages like FORTRAN. Future work could focus on optimizing the search process or exploring alternative algorithms such as stochastic gradient descent to improve speed. Additionally, all insights are dependent on the datasets used; replication across regions and crops is necessary to validate the results. In conclusion, our classifier architecture, combining multiple classifiers with different strategies, lays the foundation for developing crop type classification systems in diverse agricultural contexts. Integrating high-resolution with low-resolution imagery and model weights shows great potential in enhancing accuracy and effectiveness, supporting better evaluations of crop production status and location.

## 6 Conclusion

This study aimed to establish a novel approach for predicting crop types using various spatial and temporal images. Results suggest that integrating high-resolution images with low-resolution data improves crop type mapping accuracy more than using high-resolution images alone. It is recommended to obtain multiple color images during periods when crops are most distinguishable, ideally including at least one representing maximum crop growth. Incorporating approximately ten low-resolution images improves classification potential similarly to adding an extra high-resolution image. This integration enhances class mean performance while maintaining favorable error distribution provided by low-resolution inputs. Although the method has not been rigorously cross verified across locations or datasets, the demonstrated ability to improve crop type mapping is significant, especially for areas with small crop fields. Challenges for extending the approach include ensuring similar spectral band responses and sufficient coverage during vegetation development.

The developed architecture allows incorporation of diverse image types through CatBoost Classifier. Using weighted averages from two classifiers one with class weights and the other without—enhances predictive capacity. This approach improves classification accuracy and broadens the applicability of remote sensing data in agriculture. Growth of such methodologies is essential for efficiently utilizing the wealth of remotely sensed data available. The approach detailed here can serve as a foundation for further crop type classification development, advancing agricultural landscape understanding and promoting accurate high-resolution crop map production.

**Acknowledgements** The authors express their gratitude to their mentors for kind help as well as support received from Amrita Vishwa Vidyapeetham management up to the time the research is conducted.

## References

1. Momm, H.G., ElKadiri, R., Porter, W.: Crop-type classification for long-term modeling: an integrated remote sensing and machine learning approach. Remote Sens. **12**(3), 449 (2020). https://doi.org/10.3390/rs12030449
2. Sun, Z., Wang, D., Zhong, G.: A review of crop classification using satellite-based polarimetric SAR imagery. In: 2018 7th International Conference on Agro-Geoinformatics (Agro-Geoinformatics), Hangzhou, China (2018)
3. Tsuchiya, Y., Sonobe, R.: Crop classification using time-series Sentinel-1 SAR data: a comparison of LSTM, GRU, and TCN with attention. Remote Sens. **15**(2), 385 (2023). https://doi.org/10.3390/rs15020385
4. Joshi, A., Pradhan, B., Gite, S., Chakraborty, S.: Remote-sensing data and deep-learning techniques in crop mapping and yield prediction: a systematic review. Remote Sens. **16**(4), 697 (2024). https://doi.org/10.3390/rs16040697

5. Wang, H., Ye, Z., Wang, Y., Liu, X., Zhang, X., Zhao, Y., Li, S., Liu, Z.: Improving the crop classification performance by unlabeled remote sensing data. Expert Syst. Appl. **236**, 121283 (2024). https://doi.org/10.1016/j.eswa.2023.121283
6. Qu, T., Wang, H., Li, X., Luo, D., Yang, Y., Liu, J., Zhang, Y.: A fine crop classification model based on multitemporal Sentinel-2 images. Int. J. Appl. Earth Obs. Geoinform. **114**, 104172 (2024). https://doi.org/10.1016/j.jag.2024.104172
7. Li, W., Liang, S., Zhang, Y., Liu, L., Chen, K., Chen, Y., Ma, H., Xu, J., Ma, Y., Guan, S., Shi, Z.: Fine-grained hierarchical crop type classification from integrated hyperspectral EnMAP data and multispectral Sentinel-2 time series: a large-scale dataset and dual-stream transformer method. arXiv preprint arXiv:2506.06155 (2025)
8. Bourriz, M., Hajji, H., Laamrani, A., Elbouanani, N., Ait Abdelali, H., Bourzeix, F., El-Battay, A., Amazirh, A., Chehbouni, A.: Integration of hyperspectral imaging and AI techniques for crop type mapping: present status, trends, and challenges. Remote Sens. **17**(9), 1574 (2025). https://doi.org/10.3390/rs17091574
9. Zhang, Z.C., Kerner, H., Wang, S., Hao, P., Li, Z., Hunt, K.A., Abernethy, J., Zhao, H., Gao, F., Di, L., Guo, C., Liu, Z., Yang, Z., Mueller, R., Boryan, C., Chen, Q., Beeson, P.C., Zhang, H.K., Shen, Y.: Remote sensing for crop mapping: a perspective on current and future crop-specific land cover data products. Remote Sens. Environ. **330**, 114995 (2025). https://doi.org/10.1016/j.rse.2025.114995
10. Cheng, G., Ding, H., Yang, J., Cheng, Y.: Crop type classification with combined spectral, texture, and radar features of time-series Sentinel-1 and Sentinel-2 data. Int. J. Remote Sens. **44**(4), 1215–1237 (2023). https://doi.org/10.1080/01431161.2023.2176723
11. Mazzia, V., Khaliq, A., Chiaberge, M.: Improvement in land cover and crop classification based on temporal features learning from Sentinel-2 data using recurrent-convolutional neural network (R-CNN). Appl. Sci. **10**(1), 238 (2020). https://doi.org/10.3390/app10010238
12. Fan, J., Zhang, X., Zhao, C., Qin, Z., De Vroey, M., Defourny, P.: Evaluation of crop type classification with different high-resolution satellite data sources. Remote Sens. **13**(5), 911 (2021). https://doi.org/10.3390/rs13050911
13. Siesto, G., Fernández-Sellers, M., Lozano-Tello, A.: Crop classification of satellite imagery using synthetic multitemporal and multispectral images in convolutional neural networks. Remote Sens. **13**(17), 3378 (2021). https://doi.org/10.3390/rs13173378
14. Barriere, V., Claverie, M., Schneider, M., Lemoine, G., d'Andrimont, R.: Boosting crop classification by hierarchically fusing satellite, rotational, and contextual data. Comput. Electron. Agric. **221**, 109792 (2024). https://doi.org/10.1016/j.compag.2024.109792
15. Siesto, G., Sellers, M.F., Lozano-Tello, A.: Crop classification of satellite imagery using synthetic multitemporal and multispectral images in convolutional neural networks. Remote Sens. **13**(17), 3378 (2021). https://doi.org/10.3390/rs13173378

# Fraud Detection Using ML Algorithms

**Prerna Upadhyay, Jyotsna Koul, Shahin Saifi, Shailesh Kamble, and Nisha**

**Abstract** The increase in digital transactions has created an urgent need for reliable fraud detection. This study evaluates the performance of various machine learning models—Logistic Regression, K-Nearest Neighbors (KNN), Support Vector Classifier (SVC), Decision Trees, Random Forests, and Neural Networks—alongside a Graph Neural Network (GNN) to detect fraudulent transactions. Recognizing the inherent class imbalance in fraud detection datasets, we applied Random Undersampling and the Synthetic Minority Over-sampling Technique (SMOTE) to create balanced training sets and optimize model performance. Our results revealed that Neural Networks achieved the highest accuracy, reaching 98.22% with undersampling and 100.00% with SMOTE. Logistic Regression also demonstrated strong performance with a 98.67% accuracy on SMOTE-balanced data, followed by KNN (97.46%) and SVC (98.51%). Although the GNN model leverages advanced graph-based representation learning, it achieved an accuracy of 70.38%, suggesting limitations in Graph Neural Networks for this specific task. We evaluated model performance using accuracy, precision, recall, and $F1$-score metrics, which provided comprehensive insights through confusion matrices. The findings emphasize the critical role of data handling techniques, as balancing the dataset significantly enhanced model sensitivity and accuracy. This study underscores the efficacy of machine learning in fraud detection and highlights areas for improvement, including exploring optimized architectures and real-time data analysis for GNNs. In conclusion, selecting appropriate models and data balancing techniques is essential in developing effective fraud detection solutions, with Neural Networks and SMOTE proving particularly successful in identifying fraudulent transaction.

**Keywords** SMOTE · Oversampling · Classifiers · Decision Tree · Neural Network · PCA · GNN · SVM · Fraud

P. Upadhyay · J. Koul · S. Saifi (✉) · S. Kamble · Nisha
Department of Computer Science and Artificial Intelligence, Indira Gandhi Delhi Technical University of Women, New Delhi, Delhi, India
e-mail: shahin146btcsai21@igdtuw.ac.in

© The Author(s), under exclusive license to Springer Nature Singapore Pte Ltd. 2026

M. Pant et al. (eds.), *Real-World Applications of Machine Learning*, Mathematics for Sustainable Developments, https://doi.org/10.1007/978-981-95-7134-5_4

# 1 Introduction

Fraud detection in financial transactions is a major challenge in the digital age, as electronic payments and online financial activities continue to grow at a rapid pace. Traditional methods usually fail to detect fraudulent activities because fraud datasets are highly skewed, with fraudulent transactions occurring seldom in compared to genuine ones. This study looks into the use of advanced machine learning techniques to improve fraud detection systems. We evaluate the performance of various machine learning models, such as Logistic Regression, K-Nearest Neighbors, Support Vector Classifier, and Decision Trees, in detecting fraudulent transactions. To address class imbalance, we use Random Undersampling and Synthetic Minority Over-sampling Technique (SMOTE) to improve the performance of our models. Our data show that Logistic Regression, when optimized with SMOTE, has the maximum accuracy of 98.66%, proving its superiority over other models. This study emphasizes the importance of using advanced sampling approaches to address class imbalance and increase fraud detection skills, providing useful insights for designing more successful financial security systems. Through evaluating the models using confusion matrices and performance metrics of accuracy, precision, recall, and $F1$-score, this study gives a thorough comparison to emphasize the effects of data resampling techniques on model performance.

## *1.1 Significance of Fraud Detection*

It can be understood that in the present, with ever-increasing occurrences of various kinds of fraud, mostly done through online transactions, fraud detection becomes very important today. It has also been determining the losses that future electronically-ridden societies might incur, losing sensitive personal data as well as to preserve pleasing market conditions for future transfers of value. Advanced machine learning algorithms must be used to maximize accurate detection given the sophistication of new fraud schemes. Different strategies and techniques are employed in this work so as to enhance fraud detection capabilities, thereby significantly improving the overall accuracy and ensuring safer digital financial transactions.

## *1.2 Applications*

This model for fraud detection is flexible and has applications in e-banking, e-commerce, and digital payment systems. It helps improve fraud detection accuracy, therefore, preventing unauthorized transactions and securing user accounts. This will be applicable to detecting credit card fraud, preventing identity theft, and monitoring suspicious activities. This will greatly inform security policies and risk management strategies, thus improving overall financial security and trust in digital transactions. Such models and projects have championed in making a huge difference to the safety features of online transaction systems.

### 1.3 Motivation Behind Work

Our endeavor is driven by the growing danger of financial transaction fraud, leading to substantial global monetary losses. Unbalanced data sets and evolving fraud tactics often present obstacles for conventional detection methods. In this research, cutting-edge techniques such as SMOTE for address these challenges. This study seeks to enhance fraud protection in financial systems by increasing detection accuracy and decreasing false positives.

### 1.4 Problem Statement

Fraudulent transactions in financial systems pose terrifying threats. This research addresses the fraud detection and prevention problem by exploiting the following modern machine learning approaches: Logistic Regression, K-Nearest Neighbors, Support Vector Classifier, and Neural Networks. They significantly improve the detection rate and reduce false alarms for a set of dense transaction data.

### 1.5 Organization of Paper

The organization includes the following sections: 2. Literature Review, 3. Methodologies Involved, 4. Experimental, 5. Results, and 6. Conclusion.

## 2 Literature Review

We have a thorough and exhaustive review of other works done under this topic and a comprehensive detail of works studied has been state in Table 1. We have mainly included paper between 2010 and 2024 on fraud detection and Machine Learning.

Recently, credit card fraud detection has transitioned to more advanced machine learning models and the use of deep learning. Traditionally, constraints in machine learning led to the popularity of traditional models such as Logistic Regression, basic Decision Trees, and Support Vector Machines that used transactional behavioral patterns to detect fraudulent transactions. However, advances in big data and increased processing needs in real time has led to increased application of ensemble

**Table 1** Literature table

| Title of paper | Methodologies | Dataset | Result | Remarks |
|---|---|---|---|---|
| Alarfaj et al. [1] | The methodologies used include machine learning (Random Forest, SVM, XGBoost) and deep learning (CNN architectures) with data balancing techniques | The dataset contains 284,807 transactions over two days, with 492 fraudulent ones, using PCA for dimensionality reduction | CNN with additional layers achieved 99.9% accuracy, best performance | The CNN model significantly improved fraud detection accuracy, outperforming traditional machine learning methods |
| Chen et al. [2] | The methodology includes data normalization, DCNN application, sequential prediction, long-term memory learning, and hyperparameter tuning for fraud detection | The dataset includes 5 million transactions, with 6223 fraudulent instances, unbalanced at 0.12%, featuring class, amount, time, and 30 additional features | Adam optimizer achieved 99% accuracy in 583 s, outperforming SVM, LR, and RF, which showed lower performance | DCNN with Adam optimizer excels in accuracy and efficiency; future work can explore optimizing algorithms for even better performance |
| Asha et al. [3] | The methodologies include using ANN, SVM, and kNN for fraud detection, comparing their performance with 31 attributes and RELU activation | The dataset from Kaggle includes 31 columns: 30 features and 1 target class indicating fraud or non-fraud transactions | ANN achieved 99.92% accuracy, 81.15% precision, and 76.19% recall; KNN and SVM showed lower performance in all metrics | ANN performs best; future work could explore hybrid models combining ANN with SVM or KNN for improved fraud detection |
| Esenogho et al. [4] | The methodologies include LSTM Neural Network with AdaBoost, hybrid SMOTE-ENN resampling, and comparison with SVM, MLP, and Decision Tree | The dataset from ULB contains 284,807 transactions, 492 fraudulent, with attributes "time," "amount," and 28 numerical features | The LSTM ensemble with SMOTE-ENN achieved 99.6% sensitivity, 99.8% specificity, and 99.0% AUC, outperforming other | The LSTM-SMOTE-ENN approach excels; future work should explore additional resampling and feature selection techniques for further improvement |

(continued)

**Table 1** (continued)

| Title of paper | Methodologies | Dataset | Result | Remarks |
|---|---|---|---|---|
| Naresh et al. [5] | The study employs machine learning techniques including Random Forest, SVM, ANN, Naïve Bayes, Logistic Regression, and gradient boosting | The dataset from ULB contains 284,807 transactions with 492 fraud cases, featuring V1–V28 attributes | KNN and Random Forest outperformed having similar accuracies 94.99% | Effective fraud detection was achieved; future work should explore more advanced models and data preprocessing techniques |
| Forough et al. [6] | The methodologies include LSTM/GRU networks, FFNN ensemble voting, fivefold cross-validation, real-world datasets, and a three-phase training algorithm | The dataset contains more than 284,000 European credit card transactions with around 500 fraud cases, highly imbalanced, covering a two-day period | The proposed model achieved around 98% AUC, 96% recall, and 87% $F1$-score, outperforming previous fraud detection models | Future research could explore real-time fraud detection, adaptive models, and hybrid techniques for improved accuracy |
| Chen et al. [7] | Data mining, machine learning, Deep Convolutional Neural Network (DCNN) | Real-time credit card fraud dataset. Live dataset was something different as compared to traditional static dataset | Achieved 99% detection accuracy in 45 s using the proposed model The proposed model and work were successful in building a real-time system | Enhanced detection accuracy with large datasets; surpasses traditional models in fraud detection performance |
| Zengyi et al. [8] | Machine learning, K-means clustering, anomaly detection, data clustering. Confusion matrix and other evaluation matrix for result calculation | Financial transaction dataset containing vast records to identify anomalous transaction patterns indicative of fraud | K-means clustering improves fraud detection efficiency and accuracy, enabling real-time identification of anomalous behaviors over traditional rule-based methods | K-means clustering enhances adaptability to new fraud patterns, optimizes resource allocation, and strengthens financial system security through focused monitoring of high-risk areas |

(continued)

**Table 1** (continued)

| Title of paper | Methodologies | Dataset | Result | Remarks |
|---|---|---|---|---|
| Uyyala et al. [9] | Utilizes GANs to generate and discriminate check images, assigning confidence scores for real-time fraud detection | Contains annotated images of real and fraudulent checks, with data augmentation for variety and robust fraud detection | Processes 1800 check/s, achieves 90% false-positive reduction, and reliably categorizes check authenticity | Scalable, cost-effective fraud detection for financial institutions; future enhancements can improve robustness against evolving fraud |
| Boulieris et al. [10] | Introduces FraudNLP dataset, benchmarks machine/deep learning models, and uses NLP-inspired techniques for sequential fraud detection | FraudNLP contains anonymized interactions of European online banking users, highlighting class imbalance and sequential features for fraud analysis | High precision-recall $F$-scores achieved; AUPRC employed to handle class imbalance, improving online and offline fraud detection | FraudNLP enables accessible fraud research; future work could refine real-time detection, expanding NLP-based analysis for enhanced fraud identification |
| Aghware et al. [11] | Five algorithms (RF, KNN, NB, SVM, LR) were trained, with and without SMOTE, to assess performance on imbalanced credit card fraud data | Credit card transaction dataset with class imbalance. Fraud cases are minority class, balanced using SMOTE to improve model accuracy | RF achieved 98.02% accuracy pre-SMOTE and 99.19% post-SMOTE, outperforming other models in fraud detection effectiveness | Exploring further ensemble techniques and real-time fraud detection systems could enhance robustness and scalability |
| Varmedja et al. [12] | Methodologies include SMOTE oversampling, feature selection, and algorithms: Logistic Regression, Random Forest, Naïve Bayes, and Multilayer Perceptron | The dataset contains 284,807 transactions with 31 numerical features, including 492 frauds, using PCA-transformed variables | Logistic Regression had the best recall, Random Forest excelled in precision and accuracy, while MLP offered balanced performance | Future work can explore advanced resampling techniques, hybrid models, and real-time fraud detection systems |

(continued)

**Table 1** (continued)

| Title of paper | Methodologies | Dataset | Result | Remarks |
|---|---|---|---|---|
| Sailusha et al. [13] | This research employs Decision Trees, Random Forest, Logistic Regression, and Naïve Bayes for credit card fraud detection | The dataset comprises European credit card transactions from September 2013, featuring 31 numerical attributes, including PCA-transformed and raw features | Data mining techniques improved model accuracy, reducing false positives and negatives despite a high initial accuracy of 99.9% | Data mining techniques enhance prediction reliability, crucial for minimizing financial losses in fraud detection |
| Srivastava et al. [14] | The methodologies used include Hidden Markov Model (HMM), Baum-Welch algorithm, K-means clustering, and probability estimation | The dataset includes credit card transaction amounts, categorized into low, medium, and high price ranges for fraud detection | The HMM-based FDS showed improved TP-FP spread and accuracy compared to the alternative system | This research demonstrates effective fraud detection using HMM with simulated data and balanced accuracy |
| Jiang et al. [15] | The methodologies involve unsupervised anomaly detection using autoencoders, GANs, feature attention, adversarial loss, context loss, latent loss, self-attention, and skip connections for fraud detection | The Kaggle credit card fraud dataset contains 284,807 transactions with 28 PCA features, "time," "amount," and class imbalance (0.172% fraud) | Increasing model channel width improves fraud detection, but overfitting occurs beyond $\gamma = 1$. Best performance achieved at $\gamma = 1$ | Optimal channel width ($\gamma = 1$) balances performance and overfitting, highlighting the need for regularization in larger models |
| Ileberi et al. [16] | Models used include Random Forest (RF), Naïve Bayes (NB), Support Vector Machines (SVM), Decision Trees (DT), and Genetic Algorithm (GA) | The dataset contains 284,807 European credit card transactions with 30 numerical features, 0.172% fraudulent, used for fraud detection | The Random Forest achieved the best fraud detection accuracy (99.98%) and precision (95.34%) across various models and feature sets | Random Forest consistently outperformed other models, achieving the highest accuracy and precision, proving effective for fraud detection |

(continued)

**Table 1** (continued)

| Title of paper | Methodologies | Dataset | Result | Remarks |
|---|---|---|---|---|
| Alfaiz et al. [17] | The methodologies include 9 machine learning algorithms and resampling techniques (under sampling, over sampling, combination) to handle class imbalance | The dataset consists of 284,807 credit card transactions by European cardholders, featuring more than 30 attributes, including 492 fraudulent instances, used for fraud detection | The AllKNN-CatBoost model achieved the best performance with AUC 97.94%, recall 95.91%, and $F1$-score 87.40% | Future research could explore integrating advanced algorithms and real-time monitoring for improved credit card fraud detection and prevention |
| Strelcenia et al. [18] | Methods include data-level sampling (oversampling, under-sampling), SMOTE, GANs for synthetic data generation, and deep learning techniques for fraud detection | The dataset comprises credit card transaction records, balancing legitimate transactions against a small fraction of fraudulent activities for analysis | Increasing augmented sample sizes improved $F1$-score, precision, recall, and accuracy across all classifiers tested | Larger datasets enhance classifier performance, suggesting future scope to explore advanced GAN variants for further improvement |
| Noviandy et al. [19] | XGBoost and SMOTE-ENN were applied to handle data imbalance and enhance fraud detection accuracy | Historical credit card transaction dataset, imbalanced with fewer fraud instances, balanced using SMOTE-ENN | XGBoost achieved balanced precision-recall, effectively identifying fraud in an imbalanced dataset | Improved fraud detection reinforces financial integrity, efficient resource use, and customer trust |
| Rathor et al. [20] | Data preprocessing, BOA feature selection, and C-LSTM model training to detect ATM fraud through behavior patterns | ATM status data, preprocessed by cleaning, integrating, and deduplicating for feature extraction and model training | C-LSTM outperformed LSTM and CNN, showing superior detection accuracy in identifying potential fraud attempts | The approach enhances ATM security by learning normal patterns without extra hardware, adapting to evolving fraud tactics |

methods, such as Random Forest, XGBoost, and gradient boosting, which provided greater accuracy. In more recent studies, researchers and academics have moved toward using deep learning models, such as long-term memory (LSTM) and gated recurrent (GRU) networks, due their ability to capture sequential patterns. Hybrid models, that combine elements of deep leaning and ensemble models, have become increasingly popular.

# 3 Methodologies

The methodologies employed in this work integrate various preprocessing, sampling, modeling, and evaluation methods that take into account the highly imbalanced nature of fraud detection datasets. Data preprocessing was carried out by scaling the Amount and Time features with the standard RobustScaler. Additionally, we incorporated another preprocessing step involving the application of Random Undersampling and Synthetic Minority Over-sampling Technique (SMOTE) to address the class imbalanced nature of the data. The dimensionality reduction and visualization techniques such as PCA, t-SNE, and Truncated SVD describe high dimensional patterns in a simple way and explore the feature separability and correlation with the fraud and non-fraud transactions. We fitted the benchmarking and GNN models in an effort to train, validate, and test the machine learning algorithms. The benchmark models used encompassed Logistic Regression, K-Nearest Neighbors (KNN), Support Vector Classifier (SVC), Decision Trees, Random Forest, and Neural Networks. The Graph Neural Networks (GNN) were trained using the PyTorch Geometric framework. In the implementation of the GNN, we formulated the edges and nodes of the graph as fraud and non-fraud transactions, facilitating the detection of relational patterns between the transactions associated with fraud. Confusion matrices and associated matrix-derived performance metrics of accuracy, precision, recall, $F1$-score, and AUC allowed us to provide a comprehensive summary of the performance of the model in fraud detection. All the steps are mentioned in Fig. 1.

## *3.1 Dataset Descriptions*

284,807 credit card transactions, including 492 fraudulent events, were carried out by European cardholders in September of 2013; as a result, the dataset reveals a highly skewed class distribution (0.172% fraud). In order to protect user identities, it contains numerical variables that are the outcome of PCA transformations: "Amount" denotes the transaction value, and "Time" shows how many seconds have transpired since the first transaction. The "Class" label allows you to distinguish between transactions that are fraudulent (1) and those that are not. Due to the seriously imbalanced classes of the given dataset, it is suitable to investigate upon the latest in fraud detection techniques viz. machine learning techniques and performance metrics like Area Under the Precision-Recall Curve (AUPRC). Figure 2 shows the scatter distribution of fraud and non-fraud data.

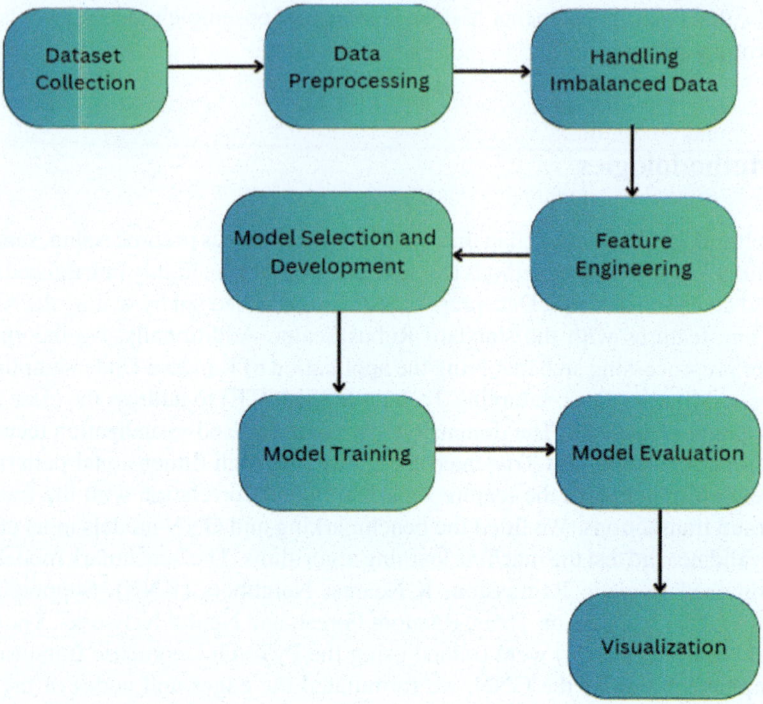


**Fig. 1** Flowchart of model development

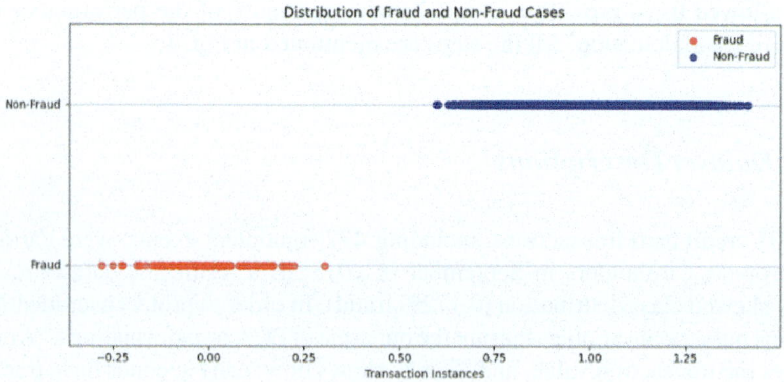


**Fig. 2** Scatter distribution of fraud versus non-fraud

## *3.2 Dataset Preprocessing*

The database preparation step improves the data's quality and usefulness for the fraud detection analysis in our research project. NumPy dealt with all numeric operations while Pandas was responsible for data handling and Scikit-Learn was used when performing model testing in Python. Several of the pre-steps were:

(1) Data Cleaning—As a preliminary check for data quality, this handled the elimination of duplicates and missing entries by imputation or deletion.
(2) Outlier Detection and Removal—Hence outliers in key features like V10, V12, and V14 were detected and removed using the Interquartile Range (IQR) method. That did help us a lot in cleaning our dataset and improving our model's performance.
(3) Feature Scaling—To have equal treatment for every numeric feature in training the model, we scaled features using normalizing techniques.
(4) Data Splitting—We subdivided the dataset in the ratio of 4:1 for training and the testing set to help evaluate the model's performance and outputs better.

Preprocessing raw information for machine learning models works a kind of foundation for the analyses and predictions of our study and its best performance. A foundation laid has a glimpse wherein feature scaling in data preprocessing was avoided in manual methods compared to a new experimental method from Van Belle et al. [21], CATCHM, for credit card fraud detection.

## 3.3 *Sampling Techniques*

Such methods included different sampling approaches, aimed principally at removing class imbalances affecting the dataset such that performance of the machine learning model could be enhanced. The sampling method adjusted the distribution of classes to learn with equal effectiveness from the majority and minority classes.

Sampling Methods Procedures

(1) Measurement of Class Imbalance: To give a perfect measure of how imbalanced the class distribution is, we first explored the class distribution of the dataset.
(2) Choice of Sampling Method: We have chosen the appropriates methods to modify the class distributions based on our findings.
(3) Sampling Execution: We then used our choice methods to produce a new balanced dataset that had an equal number of examples of both classes.

**Under-sampling**: Entails taking instances away from the majority class. Although sampling is a straight-forward method to obtain a balanced dataset, the drawbacks include the time-consuming task of the sampling process and the risk of losing some valuable information from the majority class every time we commit to an example. One of the most common forms of under-sampling involves taking a random sample from the original balanced class with returning to experience.

**Oversampling**: Entails replicating examples of the minority class to improve the learning from the minority class. One of the drawbacks with oversampling is that the minority class can quickly become too "chubby" with all the similar examples during modeling, and the tendency to overfit will become greater. However, the ability to learn from the minority class provides a basis of sufficient information for the minority upward to prevent overfitting.

**SMOTE**: Also known as the Synthetic Minority Over-sampling Technique, more efficiently adds instances of the minority class by introducing synthetic examples. SMOTE expands the data set's complexity by generating new minority instances while synthetically blending existing examples accordingly without risking overfitting. These characteristics may help improve model prediction performance with a balanced data set of the classes through these sampling techniques we generate as part of the overall design. To create a balanced dataset, we used under-sampling to reduce non-fraud instances and oversampling to boost fraud cases. Additionally, SMOTE was applied to generate synthetic fraud data, ensuring a better balance between classes. This approach improved the model's ability to detect fraud effectively.

**NearMiss**: In this study, the NearMiss method, which is used to tackle the problem of class imbalance, selectively reduces the majority class by keeping samples selected closest to minority instances. This study focuses on NearMiss-1; it keeps just those samples within the non-fraud class that are closest to fraud cases to accomplish a balance in the dataset. This makes it possible for models to learn patterns pertaining to rare frauds by mitigating the undue prominence of non-fraud cases, thus allowing for greater recall, and ultimately better performance in terms of fraud detection. NearMiss has significant relevance for data predisposed to a high class imbalance, such as fraud detection.

Even in some past works such Salekshahrezaee et al. [22] and Khalid et al. [23] has shown that using Random Undersampling (RUS), SMOTE, and SMOTE Tomek for class imbalance, combined with PCA and Convolutional Autoencoders (CAE) for feature extraction, evaluated on ensemble classifiers.

## 3.4 Model Construction

This project encompassed the creation of several different models to improve our ability to detect and prevent fraud. The main ones were Logistic Regression, Decision Trees, Random Forest and Support Vector Machines (SVM). We also used Neural Networks to understand complex data relationships. Also, we investigated ensemble methods that combine outputs from more than one model for increased accuracy. Hence, the evaluation of these models relied on parameters such as accuracy, precision and recall, as well as $F1$-score for their effectiveness in recognizing any illicit deals. Therefore, fraud detection in this paper aims at optimizing performance and reliability through a variety of models. The construction and details of the model is given below:

(1) Logistic Regression: Statistical technique that entails predicting two categorical variables using one or more independent variables is referred to as Logistic

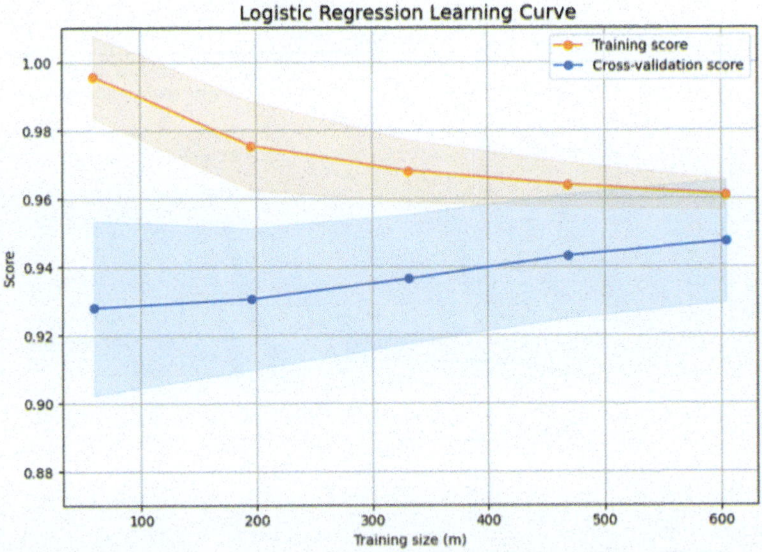


**Fig. 3** LR learning curve

Regression. On this project, the implementation was done using Scikit-Learn's LogisticRegression class. Some of the important parameters include solver which was set at "liblinear" for our relatively small dataset and max_iter that had been increased to 200 to ensure that it converges. Additionally, *c* hyperparameter controlling regularization was tuned to prevent overfitting. This model approximates the probability of fraud by fitting a linear decision boundary thus enabling the separation of fraudulent from non-fraudulent transactions through the interrelation among features. We can also examine the learning curve of this from Fig. 3.

(2) Decision Trees: An extensively utilized framework for classification as well as regression activities is known as a Decision Tree, which appears more or less like a flowchart. In this undertaking, we employed Scikit-Learn's "DecisionTreeClassifier." The "gini" was set as the criterion to determine impurity while "max_depth" was limited in order to avoid overfitting. Besides, the hyperparameter "min_samples_split" was modified in controlling the development of the tree. Through a tree structure, Decision Trees split data according to the values of its features whereby decisions are made at each node. We can look into Fig. 4 for the idea of the learning curve. As such, this interpretability enables identifying which features are much significant in fraudulent detection. The model is flexible because it can handle both numerical and categorical data enabling effective analysis.

(3) Random Forest: Ensemble learning technique called Random Forest, which entails the amalgamation of numerous Decision Trees and this results to a better classification precision, is applied in this project using Scikit-Learn's

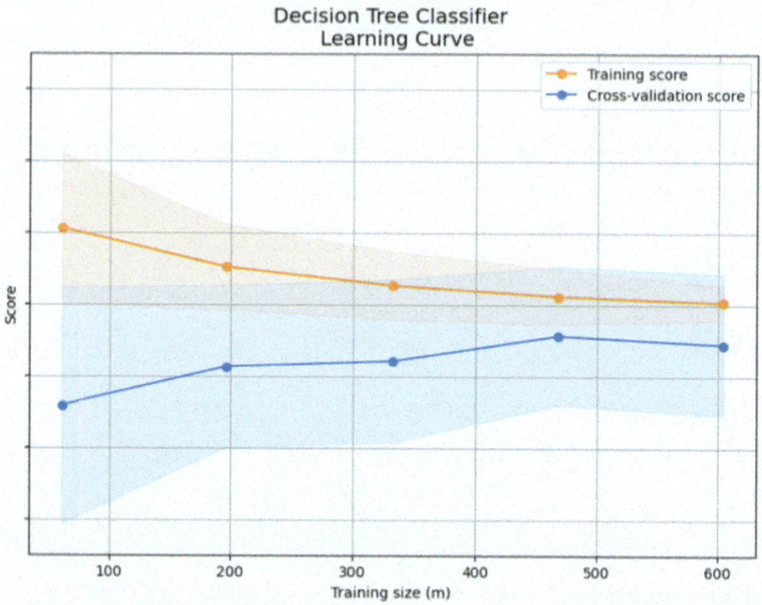


**Fig. 4** Leaning curve for Decision Tree

library known as RandomForestClassifier. The ensemble was created with n_ estimators kept at 100 thus making it a strong and powerful ensemble while max_features was limited to square root of number of features in order to ensure different trees. Tuning of hyperparameter max_depth helped in controlling overfitting effectively. Random forest reduces overfitting and improves robustness through averaging predictions made by several trees making it fit for detection of complex fraud patterns. This model excels in handling large datasets and can capture intricate interactions between features that may indicate fraudulent activity. Also, the Random Forest generates feature importance scores, which help us in understanding the most significant variables that contribute to fraud detection. The model has proven itself since it can withstand many data information that are wrong or distorted thus making the analysis more reliable. To detect any possible fraud transactions, Random Forest is able to combine strength of individual trees into one resulting in a more precise and even approach.

(4) Support Vector Machines (SVM): Support Vector Machines (SVMs) are classifiers employed in supervised learning technique with a purpose of obtaining a hyperplane that maximizes the margin among two categories. The SVM model was implemented using Scikit-Learn's SVC package. Kernel parameter used for nonlinear classification was chosen as "rbf," while *c* was adjusted to prevent overfitting problem. The hyperparameter gamma was set to define the model's complexity. SVM creates a hyperplane in a higher *D* space which helps separate fraudulent from genuine transactions through the largest possible gap between

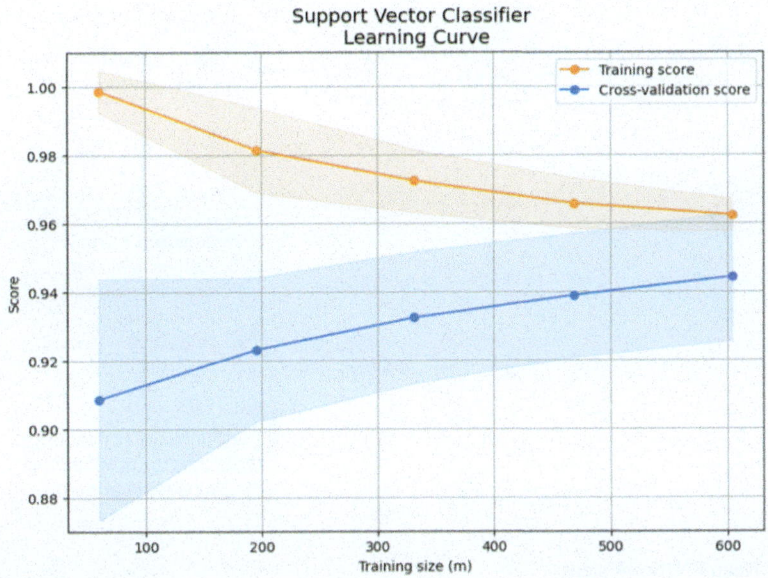


**Fig. 5** Learning curve for SVM

them thus increasing accuracy of detection. The trend and pattern of learning of this model can be traced in Fig. 5.

(5) Neural Network: Neural Networks is a calculation models influenced by the human brain which have interconnected layers that process information. In our project, we created a Neural Network using the Keras' Sequential model. The architecture of the Neural Network consisted of input, hidden and output layers with activation functions such as "relu" for the hidden layers and "sigmoid" for the output layer. We set "epochs" parameter to 50 for training and "batch_size" was adjusted to 32 in order to have better training. By allowing them to learn increasingly difficult patterns from previously learned ones in several layers, Neural Networks can discover subtle relationships among data thereby enhancing detection of minute fraudulent behaviors during transactions. Furthermore, dropout regularization methods were utilized to curb overfitting and enhance generalization. The versatility of Neural Networks allows for modifications in layers and neurons number, hence model adjustment based on data complexity can be done. This makes them extremely strong for elevation of dimensional data problems including fraud identification. The Neural Network was able to distinguish between genuine versus fake transactions when trained using both balanced and original type of data sets resulting to improved precision and sensitivity values in our findings. Additionally, its nonlinear relationship processing capability improves its ability in recognizing deceptive patterns.

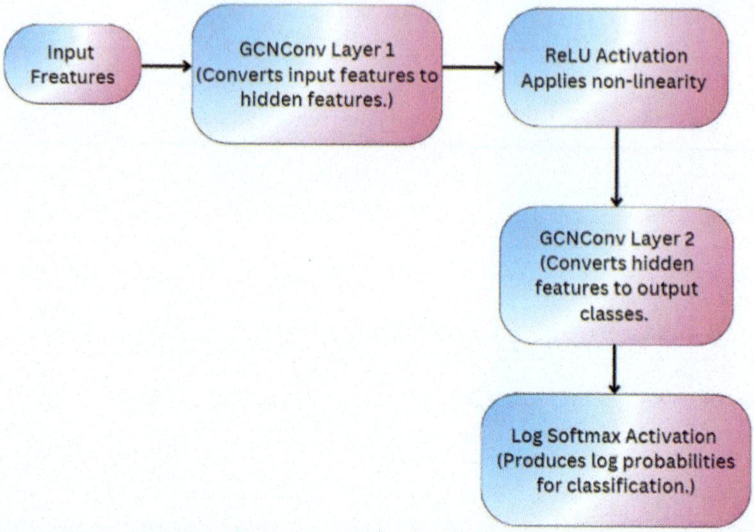


**Fig. 6** Neural Network steps

(6) Graphical Neural Network: The Graph Neural Network is built using PyTorch Geometric to capture the complex transaction relationships for fraud detection. The GNN model employs "GCNConv" layers to learn representations of the features based on node connections, best suited for discovering relational patterns. GNNs were used to model networks effectively, allowing transaction connections to inform and improve anomaly detection. This specific structure works toward enhancing the accuracy of fraud detection while studying not just individual features but the interactions between the transactions as well. Nodes represent individual transactions and their corresponding feature vectors while edges were formed by interconnecting transaction nodes based on their relationships. From various works such as Yang et al. [24] and Huang et al. [25] shows that SeHGNN simplifies heterogeneous GNNs, improving accuracy and training speed, while CF-GNN provides efficient uncertainty estimates for reliable predictions. The construction of Neural Network is given in Fig. 6.

## 3.5 *Evaluation and Accuracy*

In our project, we evaluated multiple machine learning models to detect fraudulent transactions, employing various techniques to ensure robust performance. The primary models included Logistic Regression, K-Nearest Neighbors, Support Vector Classifier, Decision Trees, Random Forest, and Neural Networks. We chose these models based on their ability to handle binary classification tasks and their varying strengths in modeling underlying complex patterns in the data. We employed cross-validation to assess the generalization performance of each model. This method, Stratified K-Folds, guarantees that within each fold, the class proportions hold steady, which is paramount given the dataset's imbalance. Libraries such as Scikit-Learn provided a fertile landscape for implementing these models and evaluation methods and gave surprisingly proficient tools for such training and validation. We further utilized GridSearchCV to find hyperparameter combinations that yield better model performance. This allows a methodical sifting of best parameters for the fine-tuning of the models, satisfying the characteristics of a given dataset. In further assessment of model performance, we delved into various other metrics, including accuracy, precision, recall, $F1$-score, and AUC-ROC, utilizing the metrics module in Scikit-Learn. Such a confluence of evaluation techniques allowed us to appraise the model performances before choosing the most efficient algorithms for the purpose of fraud detection in financial transactions.

## 4 Result and Discussion

The ensemble of models attempted on credit card fraud detection—Supervised Logistic Regression technique approaches, supporter classifier models, nearest neighborhood algorithms, SVC, Decision Trees, Random Forest, Neural Networks. The models were tested on both under-sampled and SMOTE datasets to evaluate accuracy and effectiveness in detecting fraudulent transactions. Table 2 shows all model's best accuracies.

Logistic Regression produced an accuracy of 93.167 under-sampling technique and an accuracy of 98.67 utilizing the SMOTE. This implies a large improvement in performance when balancing the dataset, showing that the model can generalize better, utilizing more representative data. A KNN classifier achieved 94.99% accuracy on the under-sampled dataset, and an increase to 97.46% once SMOTE was applied. This shows that KNN recognizes balanced data quite well, which enables it to distinguish between the classes much more. Support Vector Classifier (SVC) found an accuracy of 94.19 under-sampling, while this accuracy improved to 98.51

**Table 2** Accuracies table

| Models | Best accuracy (%) |
|---|---|
| Logistic Regression (SMOTE) | 98.67 |
| K-Nearest Neighbors (SMOTE) | 97.46 |
| Support Vector Classifier (SMOTE) | 98.51 |
| Decision Tree (under-sampling) | 93.27 |
| Random Forest (under-sampling) | 94.99 |
| Neural Network (SMOTE) | 100 |
| GNN model | 70.38 |

using SMOTE; improvement was consistent in favor of the balanced dataset. This is also due to the increased ability of the SVC model to create optimal hyperplanes with sufficient representation for both classes.

A Decision Tree model has provided 93.27% accuracy on the under-sampled data while similarly performed with the Random Forest model. For both models, highlighting the importance of tree-based algorithms in fraud detection, the results were way below those obtained from other techniques. Neural Networks had demonstrated amazing performance, achieving 98.22% accuracy on the under-sampled training dataset and an extreme 100% on the oversampled dataset. The architecture of the Neural Network adept at identifying complicated patterns owing to multitude of neuron connections along with SMOTE yielding enhanced learning, greatly contributed to this high level of accuracy. The GNN model was implemented in PyTorch Geometric, obtaining a test accuracy of 70.38%, heralding possibilities for greater optimization and comparisons with conventional models.

The accuracy rates of different models combined with under-sampling techniques implementationally illustrated in Fig. 7, while Fig. 8 presents the accuracy rates attained with oversampling methods. The best accuracies reached by all models in conjunction with GNN are summarized in Fig. 9 and brought into the limelight in detecting fraud.

In short, all models had equal and appreciable capabilities in detecting fraud. However, Neural Networks, more so when trained on balanced datasets, turned out to be the best choice. This indicates that selecting suitable data handling methods and model architectures is a critical factor that improves the performance of systems that detect fraud. The findings emphasize the need for continuous exploration of model

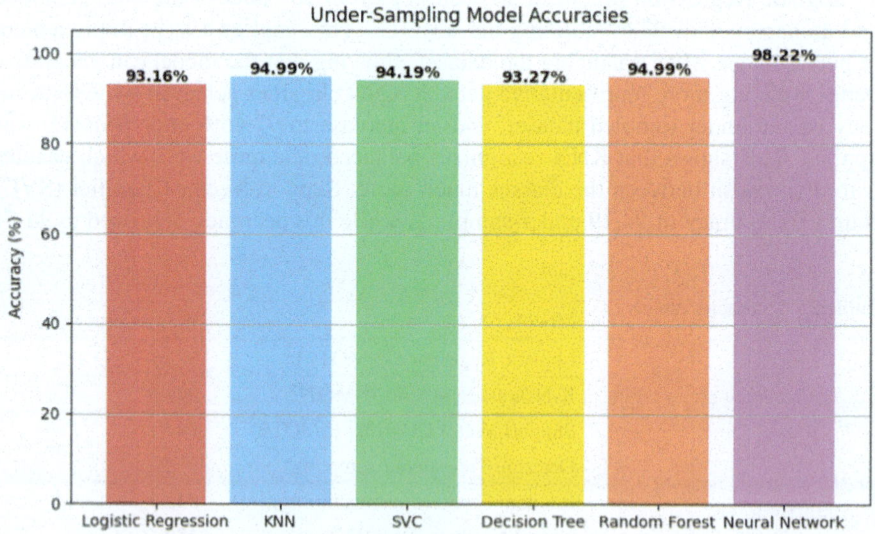


**Fig. 7** Accuracy graph for all models for under-sampling

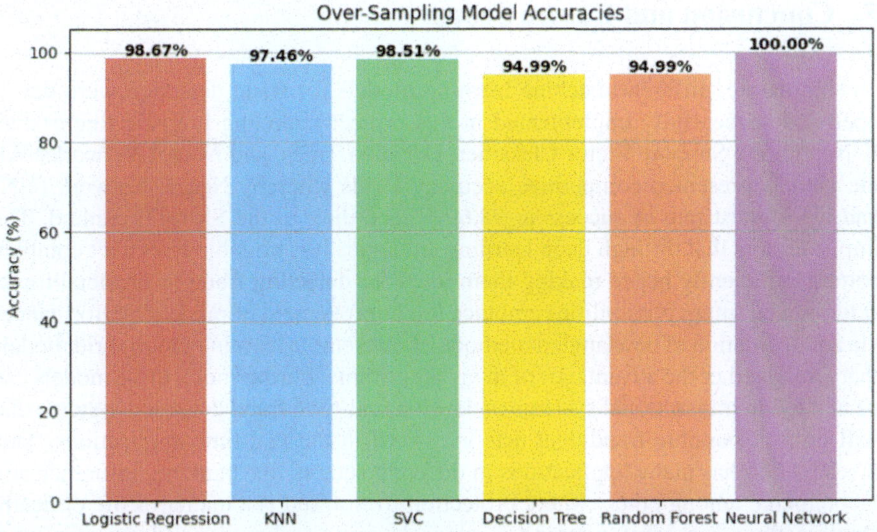


**Fig. 8** Accuracy graph for all models for oversampling

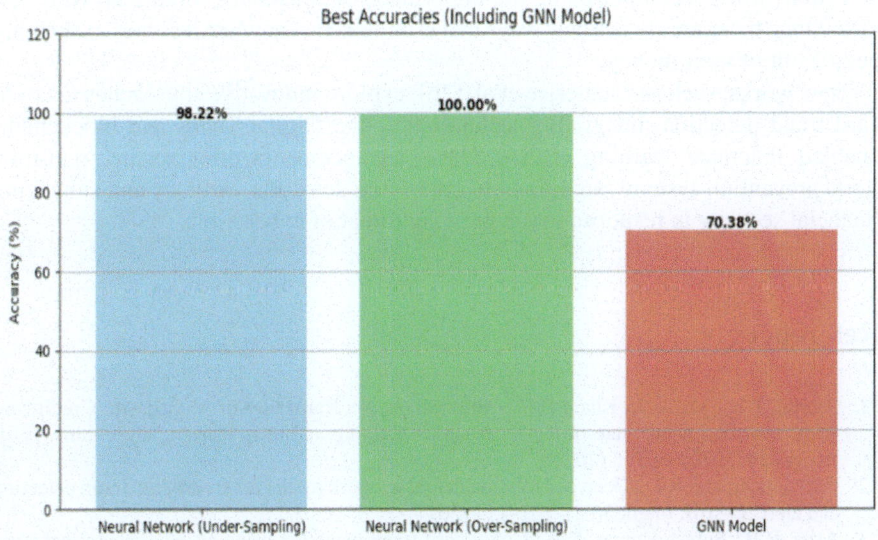


**Fig. 9** Accuracy graph for best performing models with GNN

improvements and data balancing strategies to optimize detection rates in real-world applications.

## 5 Conclusion and Future Scope

To summarize, different machine learning models for fraud detection were developed and successfully implemented in this project including Logistic Regression, K-Neighbors, Support Vector Classifier, Decision Trees, and Neural Networks. All the models presented competitive accuracy levels whereby Neural Network (NN) had the highest rate of success at 98.69% according to the SMOTE method. The implication is that through deep learning strategies it is possible to extract complex patterns efficiently hence making them ideal for detecting fraud in financial transactions. The future possibilities embrace the improvement of models' performances via amalgamation of newfangled methods like ensemble learning plus hybrid models that join together the advantages of many algorithms. Furthermore, these models can be applied in transactional systems to develop real-time fraud detection systems that will prompt possible fraudulent acts very well in the real-time applications. This investigation has many applications in different sectors like banking, insurance and e-commerce among others where protection from fraud is a matter of life or death. Organizations can employ strong machine learning techniques to lessen financial loses and enhance customer confidence while still holding on to their regulatory obligations. This project serves as a foundation for future developments in fraud detection strategies emphasizing the importance of constantly coming up with new ideas since there are always new ways that criminals invent when it comes to cheating people out of their money.

New works, such as Chatterjee et al. [26], explore innovative approaches to credit card fraud detection, integrating technologies like Digital Twins and blockchain-enabled federated learning (BFL). These advancements offer secure, real-time fraud prevention systems, significantly improving detection accuracy and enhancing financial security in response to the growing threat of cybercrime.

## References


1. Alarfaj, F.K., Malik, I., Khan, H.U., Almusallam, N., Ramzan, M., Ahmed, M.: Credit card fraud detection using state-of-the-art machine learning and deep learning algorithms. IEEE Access **10**, 39700–39715 (2022)
2. Chen, J.I.Z., Lai, K.L.: Deep convolution neural network model for credit-card fraud detection and alert. J. Artif. Intell. **3**(02), 101–112 (2021)
3. Asha, R.B., Suresh Kumar, K.R.: Credit card fraud detection using artificial neural network. Glob. Transit. Proc. **2**(1), 35–41 (2021)
4. Esenogho, E., Mienye, I.D., Swart, T.G., Aruleba, K., Obaido, G.: A neural network ensemble with feature engineering for improved credit card fraud detection. IEEE Access **10**, 16400–16407 (2022)
5. Naresh, V.S., Sridevi, G.V., Srinivasarao, P., Kiran, N.H., Babu, C.S., Dan, P.L.: Machine learning-based approach for detecting online payment fraud. In: Algorithms in Advanced Artificial Intelligence: ICAAAI-2023, p. 183 (2024)
6. Forough, J., Momtazi, S.: Ensemble of deep sequential models for credit card fraud detection. Appl. Soft Comput. **99**, 106883 (2021)

7. Chen, J.I.Z., Lai, K.L.: Deep convolution neural network model for credit-card fraud detection and alert. J. Artif. Intell. Capsule Netw. **3**(2), 101–112 (2021)
8. Huang, Z., Zheng, H., Li, C., Che, C.: Application of machine learning-based k-means clustering for financial fraud detection. Acad. J. Sci. Technol. **10**(1), 33–39 (2024)
9. Uyyala, P., Yadav, D.C.: The advanced proprietary AI/ML solution as Anti-fraud-Tensorlink4cheque (AFTL4C) for cheque fraud detection. Int. J. Anal. Exp. Modal Anal. **15**(4), 1914–1921 (2023)
10. Boulieris, P., Pavlopoulos, J., Xenos, A., Vassalos, V.: Fraud detection with natural language processing. Mach. Learn. **113**(8), 5087–5108 (2024)
11. Aghware, F.O., Ojugo, A.A., Adigwe, W., Odiakaose, C.C., Ojei, E.O., Ashioba, N.C., et al.: Enhancing the random forest model via synthetic minority oversampling technique for credit-card fraud detection. J. Comput. Theories Appl. **1**(4), 407–420 (2024)
12. Varmedja, D., Karanovic, M., Sladojevic, S., Arsenovic, M., Anderla, A.: Credit card fraud detection-machine learning methods. In: 2019 18th International Symposium INFOTEHJA-HORINA (INFOTEH), Mar 2019, pp. 1–5. IEEE
13. Sailusha, R., Gnaneswar, V., Ramesh, R., Rao, G.R.: Credit card fraud detection using machine learning. In 2020 4th International Conference on Intelligent Computing and Control Systems (ICICCS), May 2020, pp. 1264–1270. IEEE
14. Srivastava, A., Kundu, A., Sural, S., Majumdar, A.: Credit card fraud detection using hidden Markov model. IEEE Trans. Dependable Secure Comput. **5**(1), 37–48 (2008)
15. Jiang, S., Dong, R., Wang, J., Xia, M.: Credit card fraud detection based on unsupervised attentional anomaly detection network. Systems **11**(6), 305 (2023)
16. Ileberi, E., Sun, Y., Wang, Z.: A machine learning based credit card fraud detection using the GA algorithm for feature selection. J. Big Data **9**(1), 24 (2022)
17. Alfaiz, N.S., Fati, S.M.: Enhanced credit card fraud detection model using machine learning. Electronics **11**(4), 662 (2022)
18. Strelcenia, E., Prakoonwit, S.: Improving classification performance in credit card fraud detection by using new data augmentation. AI **4**(1) (2023)
19. Noviandy, T.R., Idroes, G.M., Maulana, A., Hardi, I., Ringga, E.S., Idroes, R.: Credit card fraud detection for contemporary financial management using XGBoost-driven machine learning and data augmentation techniques. Indatu J. Manage. Account. **1**(1), 29–35 (2023)
20. Rathor, K., Vidya, S., Jeeva, M., Karthivel, M., Ghate, S.N., Malathy, V.: Intelligent system for ATM fraud detection system using C-LSTM approach. In: 2023 4th International Conference on Electronics and Sustainable Communication Systems (ICESC), July 2023, pp. 1439–1444. IEEE
21. Van Belle, R., Baesens, B., De Weerdt, J.: CATCHM: a novel network-based credit card fraud detection method using node representation learning. Decis. Support. Syst. **164**, 113866 (2023)
22. Salekshahrezaee, Z., Leevy, J.L., Khoshgoftaar, T.M.: The effect of feature extraction and data sampling on credit card fraud detection. J. Big Data **10**(1), 6 (2023)
23. Khalid, A.R., Owoh, N., Uthmani, O., Ashawa, M., Osamor, J., Adejoh, J.: Enhancing credit card fraud detection: an ensemble machine learning approach. Big Data Cogn. Comput. **8**(1), 6 (2024)
24. Yang, X., Yan, M., Pan, S., Ye, X., Fan, D.: Simple and efficient heterogeneous graph neural network. In: Proceedings of the AAAI Conference on Artificial Intelligence, vol. 37, no. 9, pp. 10816–10824, June 2023
25. Huang, K., Jin, Y., Candes, E., Leskovec, J.: Uncertainty quantification over graph with conformalized graph neural networks. Adv. Neural Inf. Process. Syst. **36** (2024)
26. Chatterjee, P., Das, D., Rawat, D.B.: Digital twin for credit card fraud detection: opportunities, challenges, and fraud detection advancements. Future Gener. Comput. Syst. (2024)

# Assessing Students' Emotions in Learning Environment: Integrating Deep Learning with Statistical Learning

**Tushar Tripathi, Tripti Mahara, and Madhumita Guha Majumder**

**Abstract** Understanding the facial expressions is one of the most widely utilized method to discover the hidden emotions of an individual. Ascertaining the emotions are helpful across most of the domains ranging from psychology, health care, marketing, education, etc., and can benefit both the individual and the society. The research here focuses on identifying the seven basic emotions (angry, disgust, happy, sad, fear, neutral, surprise) of the students in a B-School across four time slots during a day for 6 days (Monday–Saturday) using deep learning algorithm. To achieve this, performance of three deep learning models namely AlexNet, VGG16, and VGG19 are compared and based on the performance AlexNet is selected to detect the emotions of the students. Statistical techniques including 2-way ANOVA and Multiple Linear Regression are also utilized to validate the results. Results of analysis reveal that majority of the students go through two emotions: Happy and sad throughout the day with neutral following. Extreme emotions like anger, fear, and disgust are observed among very few students. Results also reveal that time slot and day are not significant in determining the emotions. The outcome of this research can be utilized by teachers and top management who are the two predominant stakeholders in the education system concerned with the well-being of the students.

**Keywords** Deep learning · Facial emotions · Statistical learning · Students

T. Tripathi · T. Mahara
Prin. L.N. Welingkar Institute of Management Development and Research, Bangalore, India
e-mail: triptimahara@gmail.com

M. G. Majumder (✉)
Sister Nivedita University, Kolkata, India
e-mail: provcacademics@snuniv.ac.in

© The Author(s), under exclusive license to Springer Nature Singapore Pte Ltd. 2026

M. Pant et al. (eds.), *Real-World Applications of Machine Learning*, Mathematics for Sustainable Developments, https://doi.org/10.1007/978-981-95-7134-5_5

# 1 Introduction

Facial expressions are one of the most common ways to convey emotions to others during a communication that could be both verbal and non-verbal [1]. In a verbal communication, they provide additional meaning whereas in non-verbal communication, facial emotions become even more critical providing clues to the mental state [2]. Along with the tone of voice, it is considered one of the important means in daily communication to understand the emotions [3]. Facial emotion recognition can be applied in diverse scenarios fields ranging from education, health care to psychology analysis and treatment. With the rapid growth in machine learning and deep learning technologies, that make facial emotion recognition more reliable, its application is being observed in diverse fields ranging from education, HR, healthcare, consumer behavior to psychology analysis and treatment.

In education, understanding the facial expressions of students is important, as it not only depicts their emotional state but also aids in understanding their engagement level, learning journey, behavior and over all well-being. Emotions are one of the most prominent factors that drives a student's holistic learning and development. For instance, if a student is tensed, the learning will not be effective but on the other hand, if the student is happy and has positive emotions, the learning becomes more effective and also results in the increased interest level. Studies have established a positive relation between positive emotional health and success in a university setup [4].

Thus, it is vital to understand the emotions of the students in an academic environment/institution. The research emphasizes on identifying seven basic emotions including anger, joy, sadness, fear, disgust, joy, and neutral emotions and uses it to understand how these emotions vary across different time span in a day of a student. Understanding these emotions not only help the faculty to understand how the students feel when they are on the campus, but also provide insights to the top management for making the learning environment conducive and defining strategies to deal with the negative emotions [5].

The organization of the paper is as follows. Section 2 constitutes of literature review followed by methodology in Sect. 3. Section 4 focuses on data analysis and discussions followed by conclusion and future scope of work in Sect. 5.

# 2 Literature Review

Many studies are found in literature that focuses on comparing various deep learning models for accurate emotion recognition [2, 6]. For instance, one of the studies, used Convolutional Neural Network (CNN) to detect the emotions using facial expression to determine depressive disorder [7]. On the other hand, study has also been conducted to use facial emotion recognition for product review [8]. Very few studies have utilized the output of these models in an academic setup. One of the studies combined ResNet50, Convolutional Block Attention Module (CBAM) and Temporal Convolutional Network (TCN) capability to analyze students' engagement in the online learning environment using facial expression [9]. There are studies [10] which have utilized the model output to understand the mental stress in the student community and have used facial expressions to understand the emotions of students and also compared their responses filled for DASS-21 questionnaire that assesses the mental health [11]. Both the results were in coherence with each other.

The research gap was found in terms of the application in the academic environment to understand the emotions of the students throughout the day and the present research tries to address this gap. Also, in terms of research methodology, there are scanty of research studies which combines deep learning method with statistical techniques to validate the output.

# 3 Methodology

Figure 1 depicts the research methodology. It constitutes building and comparing deep learning models for emotion detection followed by selecting the deep learning model that has best performance. This deep learning model has been utilized to detect the emotions of the students. ANOVA and Multiple Linear Regression have been used to analyze the emotions.

## *3.1 Dataset*

A deep learning model is required to be built to detect the emotions of the students. To build the model, one of the widely used datasets called as facial expression recognition (FER 2013) is utilized (FER) [12]. This dataset comprises a collection of facial images labeled with corresponding emotions as anger, disgust, fear, happiness, sadness, surprise, and neutrality. Out of total 35,887 facial greyscale images with

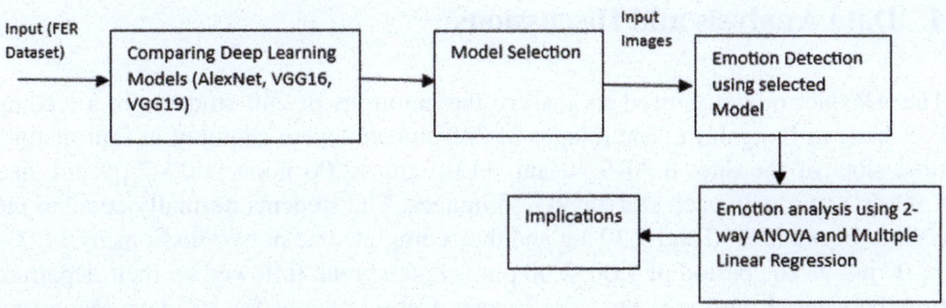


**Fig. 1** Research methodology

**Fig. 2** Types of emotions. *Source* Internet

different expressions of size 48 × 48 with appropriate labels, 28,709 images were used for training and 7178 images were used for validation. The facial expression is categorized into one of the seven categories that include angry, disgusted, fear, happy, sad, surprised, and neutral as depicted in Fig. 2.

### 3.2 *Deep Learning Models*

Three deep learning models namely AlexNet, VGG16, and VGG19 are compared for performance. The AlexNet and VGG16 architectures are depicted in Tables 1 and 2, respectively. The only difference between VGG16 and VGG19 is in terms of number of layers. VGG16 has 16 layers while VGG19 has 19. The performance of three deep learning models: AlexNet, VGG16, and VGG19 are compared. The results as in Table 3 reveal that AlexNet outperforms VGG16 and VGG19 with an accuracy of 94.25% after 15 epochs, while VGG16 achieves 87.32% accuracy after 100 epochs and VGG19 achieved an accuracy of 91.2%. This highlights Alex Net's superior accuracy and efficiency in capturing facial expression patterns compared to VGG16 and VGG19 making it a preferred choice for emotion detection tasks.

## 4 Data Analysis and Discussions

The AlexNet model is used to analyze the emotions of 240 students in a leading B-School in Bengaluru. The images of 240 students were captured at four distinct time slots of the day: 8:30–9:30 am, 11:00 am–12:00 noon, 1:00–2:00 pm, and 4:00–5:00 pm with each slot having 10 images. The students normally come to the college between 8.30 and 9.30 am and they complete one or two sessions by 11.00–12.00 noon. The period of 1:00–2:00 pm is lunch break followed by their departure time between 4 and 5:00 pm. The images were captured for six days (Monday–Saturday) in March 2024. By providing the images as input to the AlexNet model, the emotions of the students were predicted across various time slots as shown in

**Table 1** AlexNet architecture

| Layer | Feature map | Size | Filter | Stride | Padding | Activation function |
|---|---|---|---|---|---|---|
| Input | 1 | 227 × 227 × 3 | | | | |
| Conv 1 | 96 | 55 × 55 × 96 | 11 × 11 | 4 | | ReLU |
| Max Pooling1 | 96 | 27 × 27 × 96 | 3 × 3 | 2 | | |
| Conv 2 | 256 | 27 × 27 × 256 | 5 × 5 | 1 | 2 | ReLU |
| Max Pooling2 | 256 | 13 × 13 × 256 | 3 × 3 | 2 | | |
| Conv 3 | 384 | 13 × 13 × 284 | 3 × 3 | 1 | 1 | ReLU |
| Conv 4 | 384 | 13 × 13 × 284 | 3 × 3 | 1 | 1 | ReLU |
| Conv 5 | 384 | 13 × 13 × 284 | 3 × 3 | 1 | 1 | ReLU |
| Max Pooling3 | 256 | 6 × 6 × 256 | 3 × 3 | 2 | | |
| Fully Connected1 | 4096 | | | | | ReLU |
| Fully Connected2 | 4096 | | | | | ReLU |
| Output | 1000 | | | | | Softmax |

Fig. 3. As observed, happy and sad are the two emotions that are prominent across all days and time slots. Other emotions are significant for very few students.

The study adapted a two-way ANOVA to examine the significance of the variation across the emotion category and over the slots. The frequency of emotion responses under each category turns out to be the dependent variable in the ANOVA model. As observed in Table 4, angry, happy and sad are the only emotions that significantly vary during all the different slots of the days whereas "Fear" is not statistically significant. It is evident that a few students adapt "Neutral" emotions throughout the day except during lunch break. Elements of "Surprise" is only significant when the students come to the B-school and during lunch hour. "Disgust" is one of the least significant emotions and is only significant during the last slot.

As observed in Fig. 4, when students arrive in the morning, they have mixed emotions. Most of them are happy, followed by sad and neutral. The other four category of emotions are depicted by very few students. As the day progresses, the emotions also fluctuate but mainly comprising of happy, sad and neutral. During lunch time, students are very happy to interact with their friends and to receive a break from their studies. At the end of the day, the number of students who are sad and neutral increase as compared to the students who are happy.

Multiple Linear Regression (MLR) Model was built with emotion frequency as dependent variable and different types of emotions, weekdays and time slots as

**Table 2** VGG16 architecture

| Layer | Feature map | Size | Filter | Stride | Padding | Activation function |
|---|---|---|---|---|---|---|
| Input | 1 | 224 × 224 × 3 | | | | |
| 2 × Conv1 | 64 | 224 × 224 × 64 | 3 × 3 | 1 | | ReLU |
| Max Pooling1 | 64 | 112 × 112 × 64 | 3 × 3 | 2 | | |
| 2 × Conv2 | 128 | 112 × 112 × 128 | 3 × 3 | 1 | 2 | ReLU |
| Max Pooling2 | 128 | 56 × 56 × 128 | 3 × 3 | 2 | | |
| 2 × Conv3 | 256 | 56 × 56 × 256 | 3 × 3 | 1 | 1 | ReLU |
| Max Pooling3 | 256 | 28 × 28 × 256 | 3 × 3 | 2 | 1 | |
| 3 × Conv4 | 512 | 28 × 28 × 512 | 3 × 3 | 1 | 1 | ReLU |
| Max Pooling4 | 512 | 14 × 14 × 512 | 3 × 3 | 2 | | |
| 3 × Conv5 | 512 | 14 × 14 × 512 | 3 × 3 | 1 | | |
| Max Pooling5 | 512 | 7 × 7 × 512 | 3 × 3 | 2 | | |
| Fully Connected1 | 25,088 | | | | | ReLU |
| Fully Connected2 | 4096 | | | | | ReLU |
| Fully Connected3 | 4096 | | | | | |
| Output | 1000 | | | | | Softmax |

**Table 3** Performance comparison of deep learning models

| Deep learning model | Accuracy |
|---|---|
| AlexNet | 94.15 |
| VGG16 | 87.32 |
| VGG19 | 91.32 |

independent variables. The results of MLR are depicted in Table 5. As observed, each of the emotion is only significant whereas timeslot and weekday both are not significant in determining the student emotions. Figure 5 also conveys the same. The adjusted $R$ square of the model is 59%.

Figure 6 shows that among the 7 emotions, most of them exhibit two major emotions namely happy and sad. Some of them are neutral whereas other emotions like angry, surprise, fear, and disgust are very less which is good in the academic environment. The outcome of the study depicts that most of the students in the selected B-school exhibit three primary emotions, namely happy, sad, and neutral throughout the day, although the intensity of these emotions keeps fluctuating. It is also observed that most of the students are happy during the course of the day. The

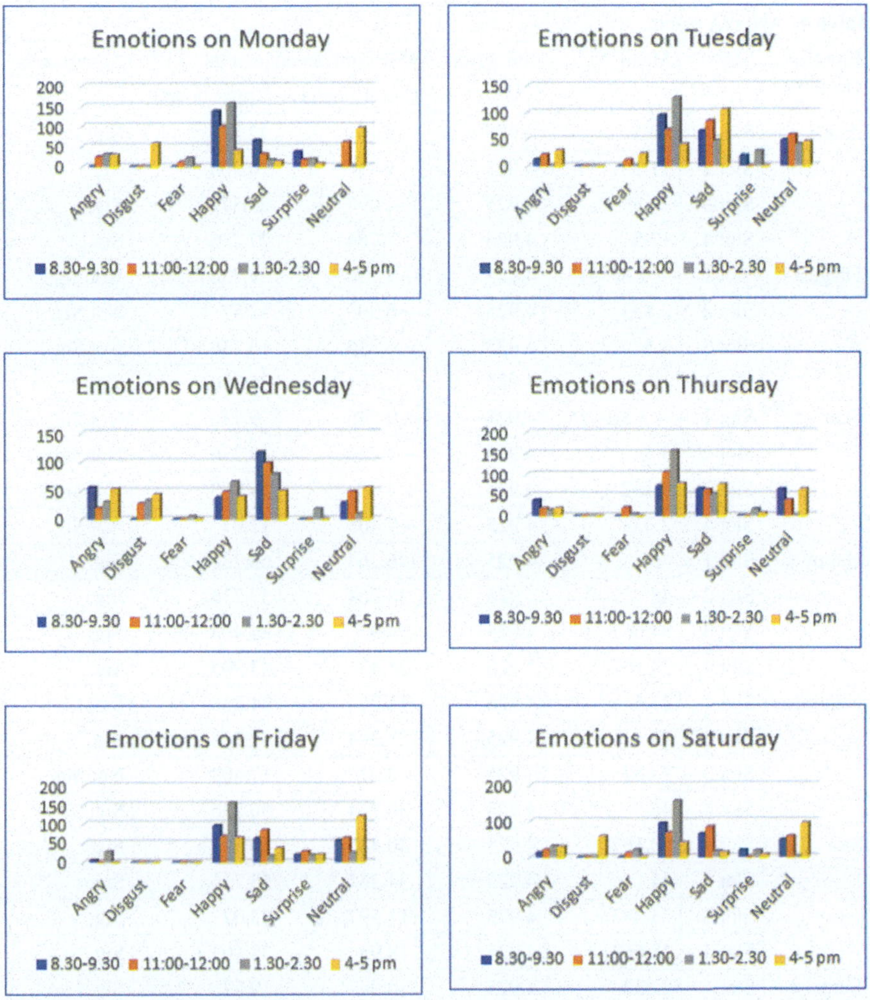


**Fig. 3** Day wise time wise emotions

study also brings out the fact that most of the students do not have negative emotions like disgust, anger, and fear.

## 5 Conclusion and Future Scope of Work

Emotions impact the learning process and also determine the mental health of an individual. Thus, to enhance the students' effective learning and to create a sustainable learning environment, the management of an educational institute needs to dive

**Table 4** ANOVA results

| Emotion category | Slot | Mean | Std. error | 95% confidence interval | | Significance |
|---|---|---|---|---|---|---|
| | | | | Lower bound | Upper bound | |
| Angry | Slot 1 | 12.5 | 4.925 | 2.764 | 22.236 | Sig. |
| | Slot 2 | 11.167 | 4.925 | 1.43 | 20.903 | Sig. |
| | Slot 3 | 13.833 | 4.925 | 4.097 | 23.57 | Sig. |
| | Slot 4 | 17.5 | 4.925 | 7.764 | 27.236 | Sig. |
| Disgust | Slot 1 | − 5.3E−15 | 4.925 | − 9.736 | 9.736 | Not Sig. |
| | Slot 2 | 2.833 | 4.925 | − 6.903 | 12.57 | Not Sig. |
| | Slot 3 | 3.5 | 4.925 | − 6.236 | 13.236 | Not Sig. |
| | Slot 4 | 16.5 | 4.925 | 6.764 | 26.236 | Sig. |
| Fear | Slot 1 | − 5.3E−15 | 4.925 | − 9.736 | 9.736 | Not Sig. |
| | Slot 2 | 5.5 | 4.925 | − 4.236 | 15.236 | Not Sig. |
| | Slot 3 | 5.167 | 4.925 | − 4.57 | 14.903 | Not Sig. |
| | Slot 4 | 2.333 | 4.925 | − 7.403 | 12.07 | Not Sig. |
| Happy | Slot 1 | 56 | 4.925 | 46.264 | 65.736 | Sig. |
| | Slot 2 | 48 | 4.925 | 38.264 | 57.736 | Sig. |
| | Slot 3 | 84.667 | 4.925 | 74.93 | 94.403 | Sig. |
| | Slot 4 | 34.167 | 4.925 | 24.43 | 43.903 | Sig. |
| Neutral | Slot 1 | 21.5 | 4.925 | 11.764 | 31.236 | Sig. |
| | Slot 2 | 33 | 4.925 | 23.264 | 42.736 | Sig. |
| | Slot 3 | 7.667 | 4.925 | − 2.07 | 17.403 | Not Sig. |
| | Slot 4 | 46.5 | 4.925 | 36.764 | 56.236 | Sig. |
| Sad | Slot 1 | 50.167 | 4.925 | 40.43 | 59.903 | Sig. |
| | Slot 2 | 44 | 4.925 | 34.264 | 53.736 | Sig. |
| | Slot 3 | 23.333 | 4.925 | 13.597 | 33.07 | Sig. |
| | Slot 4 | 28.667 | 4.925 | 18.93 | 38.403 | Sig. |
| Surprise | Slot 1 | 9.833 | 4.925 | 0.097 | 19.57 | Sig. |
| | Slot 2 | 5.5 | 4.925 | − 4.236 | 15.236 | Not Sig. |
| | Slot 3 | 11.833 | 4.925 | 2.097 | 21.57 | Sig. |
| | Slot 4 | 4.333 | 4.925 | − 5.403 | 14.07 | Not Sig. |

deeper into the emotions of students. The approach built by the study can be used by other educational institutes to understand the emotions the students are going throughout the day. If majority of the students have inflow of negative emotions like sadness, anger, disgust, the relevant stakeholders like the teachers and top management should ponder into reasons and try to come up with strategies to embellish the students' happiness into effective learning. The study can also be used to identify the students who have some issues of mental health. The institute can work toward strategies to improve their mental health.

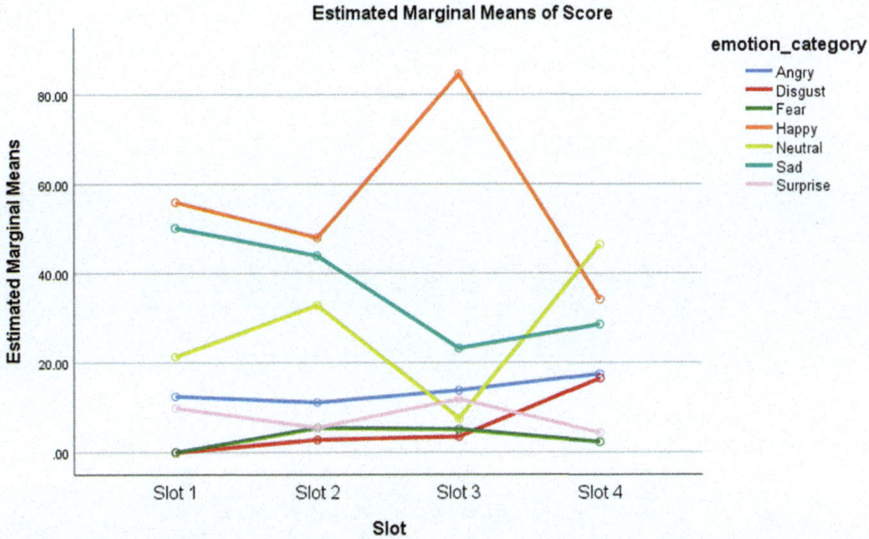


**Fig. 4** Slot wise emotion analysis

The drawback of the study is the small sample size used for analysis; hence the findings cannot be generalized for a B-school. Also, same subjects were not selected during different time slots. On the other hand, this research gives an insight into the applications of facial emotion recognition in an academic environment. The future scope of work can include carrying out the experiment with a larger sample size and coming up with strategies to mitigate negative emotions among the subjects after careful evaluation.

**Table 5** MLR results

| Items | Estimate | Std. error | Prob. > \|t\| | Lower 95% | Upper 95% | VIF |
|---|---|---|---|---|---|---|
| Intercept | 21.429 | 1.212 | **0.000** | 19.033 | 23.824 | |
| Emotions [Angry] | − 7.679 | 2.970 | **0.003** | − 13.546 | − 1.811 | 1.714 |
| Emotions [Disgust] | − 15.720 | 2.970 | **0.000** | − 21.588 | − 9.853 | 1.714 |
| Emotions [Fear] | − 18.179 | 2.970 | **0.000** | − 24.046 | − 12.311 | 1.714 |
| Emotions [Happy] | 34.280 | 2.970 | **0.000** | 28.412 | 40.147 | 1.714 |
| Emotions [Neutral] | 5.738 | 2.970 | **0.037** | − 0.129 | 11.605 | 1.714 |
| Emotions [Sad] | 15.113 | 2.970 | **0.000** | 9.246 | 20.980 | 1.714 |
| Weekday [Monday] | 0.000 | 2.711 | 0.873 | − 5.356 | 5.356 | 1.667 |
| Weekday [Tuesday] | 0.000 | 2.711 | 0.775 | − 5.356 | 5.356 | 1.667 |
| Weekday [Wednesday] | 0.000 | 2.711 | 0.983 | − 5.356 | 5.356 | 1.667 |
| Weekday [Thursday] | 0.000 | 2.711 | 0.665 | − 5.356 | 5.356 | 1.667 |
| Weekday [Friday] | 0.000 | 2.711 | 0.689 | − 5.356 | 5.356 | 1.667 |
| Slots [Slot 1] | 0.000 | 2.100 | 0.781 | − 4.149 | 4.149 | 1.500 |
| Slots [Slot 2] | 0.000 | 2.100 | 0.435 | − 4.149 | 4.149 | 1.500 |
| Slots [Slot 3] | 0.000 | 2.100 | 0.332 | − 4.149 | 4.149 | 1.500 |

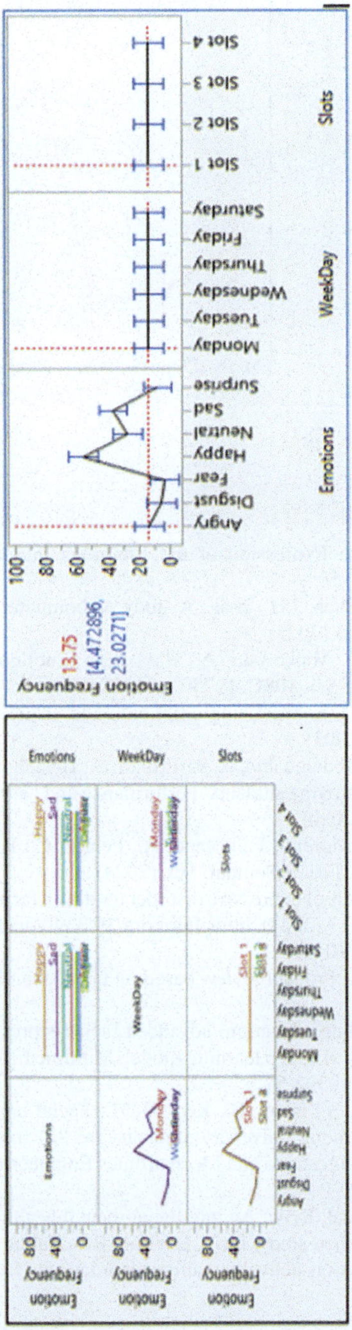


**Fig. 5** MLR results validation

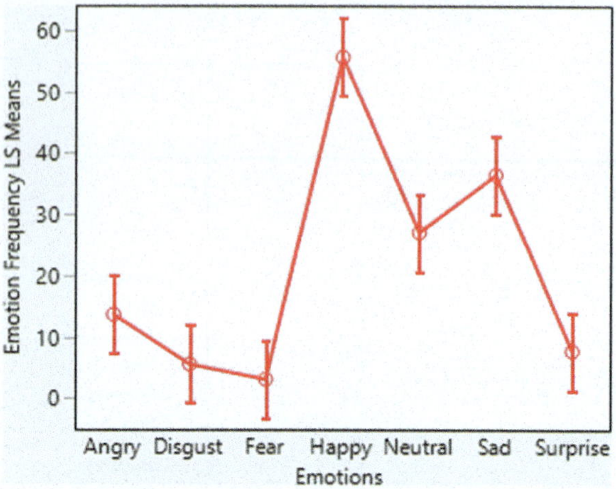


**Fig. 6** Emotion frequency

## References


1. Darwin, C., Prodger, P.: The Expression of the Emotions in Man and Animals. Oxford University Press, Oxford (1996)
2. Huang, Z.Y., Chiang, C.C., Chen, J.H., et al.: A study on computer vision for facial emotion recognition. Sci. Rep. **13**, 8425 (2023)
3. Tarnowski, P., Kolodziej, M., Majkowski, A., Rak, R.J.: Emotion recognition using facial expression. Procedia Comput. Sci. **108C**, 1175–1184 (2017)
4. Mary, E., Gregory, S.P., Vaughn, W.: Using emotional and social factors to predict student success. J. Coll. Stud. Dev. (2003)
5. Bouhlal, M., Aarika, K., AitAbdelouahid, R.A., Elfilali, S., Benlahmar, E.: Emotions recognition as innovative tool for improving students' performance and learning approaches. Procedia Comput. Sci. **175**, 597–602 (2020)
6. Ballesteros, J., Ramirez, V., Moreira, F., Solano, A., Pelaez, C.: Facial emotion recognition through artificial intelligence. Front. Comput. Sci. (2024)
7. Lee, Y.S., Park, W.H.: Diagnosis of depressive disorder model on facial expression based on fast R-CNN. Diagnostics **12**(2), 317 (2022). https://doi.org/10.3390/diagnostics12020317. PMID: 35204407; PMCID: PMC8871079
8. Bera, C., Adhav, P., Amati, S.: Product review based on facial expression detection. ITM Web Conf. (ICACC) **44** (2022)
9. Aly, M.: Revolutionizing online education: advanced facial expression recognition for real-time student progress tracking via deep learning model. Multimed. Tools Appl. (2024). https://doi.org/10.1007/s11042-024-19392-5
10. Ming, F.J., Shabana Anhum, S., Islam, S., Keoy, K.H.: Facial emotion recognition system for mental stress detection among university students. In: 3rd International Conference on Electrical, Computer, Communications and Mechatronics Engineering (ICECCME), Tenerife, Canary Islands, Spain (2023)
11. Li, C., Yang, M., Zhang, Y., Lai, K.W.: An intelligent mental health identification method for college students: a mixed-method study. Int. J. Environ. Res. Public Health **19**(22) (2022)
12. FER: FER dataset. https://datasets.activeloop.ai/docs/ml/datasets/fer2013-dataset/ (2013)

# Analysis of CNN Models with Transfer Learning in Mango Diseases Detection

**Mukesh Kumar Singh, Natthan Singh, and Vishnu Sharma**

**Abstract** Mangoes are a valuable fruit and a source of antioxidants because they contain phytochemicals like carotenoids and polyphenols. But mangoes are susceptible for various diseases such as bacterial canker, anthracnose, and powdery mildew. In this study, the main approach for detecting diseases in mango crops—with a focus on diseases that have a major impact on production and quality—is Convolutional neural networks (CNNs). We preprocesses mango images by encoding class labels and standardizing pixel values using a Kaggle dataset in order to maximize model performance. Through transfer learning, several CNN architectures—including VGG16, VGG19, ResNet50, DenseNet121, and AlexNet—are compared, and high accuracy rates are attained. The models ResNet50 and DenseNet121 have shown their efficacy in identifying mango leaf diseases by achieving the maximum accuracy, up to 100%. This study highlights how CNN models may be used to enhance early disease identification in mango crops, perhaps leading to increased agricultural output and better financial results for farmers.

**Keywords** VGG16 · VGG19 · ResNet50 · DenseNet121 · AlexNet · Mango diseases

M. K. Singh (✉)
Department of Computer Science and Engineering, Galgotia College of Engineering and Technology, Greater Noida, India
e-mail: mukesh812singh@gmail.com

N. Singh
Department of Computer Science and Engineering, I.E.T. Lucknow, Lucknow, India
e-mail: natthan.singh@ietlucknow.ac.in

V. Sharma
Department of Computer Science and Engineering, I.T.S. Engineering College, Greater Noida, India
e-mail: vishnu.sharma@its.edu.in

© The Author(s), under exclusive license to Springer Nature Singapore Pte Ltd. 2026 
M. Pant et al. (eds.), *Real-World Applications of Machine Learning*, Mathematics for Sustainable Developments, https://doi.org/10.1007/978-981-95-7134-5_6

# 1 Introduction

Mangoes (*Mangifera indica* L.) are important tropical fruits that have been cultivated for a long time, especially in Asia. They are prized for their economic and nutritional advantages. The potential for agronomic advances through genomic-assisted breeding has been highlighted by recent genomic studies that have shown significant genetic variability across mango accessions [1]. The mango fruit's antioxidant qualities are attributed to its high water content and advantageous for phytochemicals, such as carotenoids and polyphenols. According to FAO the Asian countries produces 72.7%, followed by African 14.9%, the Americas (12.3%), and Oceania (0.1%) mangoes which is shown in Fig. 1. India is the world's greatest producer of mangoes, with production substantially outpacing that of China, Indonesia, and Thailand, all of which produce far smaller amounts.

However, a number of diseases such as Powdery mildew, Anthracnose, Bacterial canker, Sooty mold, Gall midge, and Dieback poses a serious threat to mango farming, which reduces crop yield as well as quality of mango crop. Farmers suffer financial losses as a result of the these diseases that can affect mango trees, which have a substantial effect on their output and quality. Fungal pathogens are the primary cause of major diseases such anthracnose, powdery mildew, dieback, and mango deformity. However, bacterial and viral infections also play a role to destroy the mango crop. Insect-induced diseases, such as fruit rot brought on by pests like fruit flies and borers; make these difficulties even worse by endangering plant health and encouraging secondary infections [3].

Recent research machine learning methods, such as Random Forest, Support Vector Machines (SVMs), and k-Nearest Neighbors (k-NNs), have demonstrated potential in the classification and identification of mango diseases based on image color, shape, and texture. These algorithms enhance the efficiency of diseases identification for precision agricultural [4]. The development of image processing and

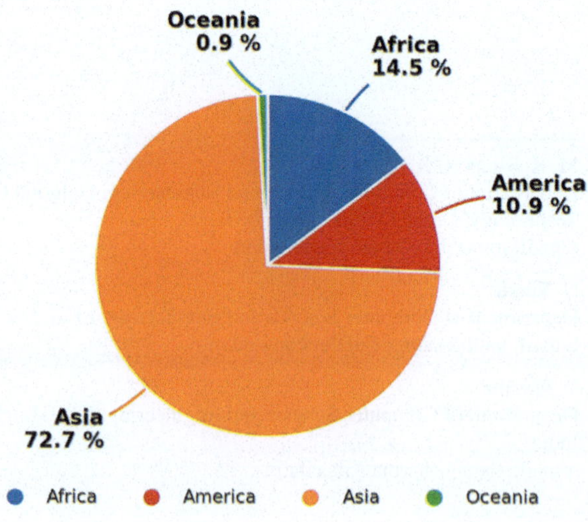


**Fig. 1** Worldwide production of mango (Images sourced from FAOSTAT [2])

**Table 1** Mango diseases and their symptoms

| Mango diseases | Major symptom's | References |
|---|---|---|
| Powdery mildew | Leaves become curl and distort | [5] |
| Anthracnose | Black patches along the leaf margin | [5] |
| Bacterial canker | Watery spots on leaves | [5] |
| Sooty mold | Twist, and the black fragments | [5] |
| Gall midge | Abnormal growth | [5] |
| Dieback | Turn yellow | [5] |

machine learning techniques for early disease identification of crop diseases has been facilitated by recent technological breakthroughs, and the classification of crop leaf diseases has achieved high accuracy rates. The advent of deep learning algorithms such as convolutional neural networks (CNNs) and Vision Transformers (ViTs) significantly improves the performance of crop diseases classifications.

To reduce the limitations of traditional image processing and machine leaning approaches, we employed and compared the performance of the different deep learning algorithms for automatic identification of mango diseases. In this research, we analyzed the performance of CNN architectures VGG16, VGG19, ResNet50, DenseNet121, and AlexNet with transfer learning for classification of mango diseases. We analyzed and compared the performance of these models and identified best model suitable for use in agriculture in real-world settings.

## 2 Mango Diseases

Mango diseases that reduces mango yield and its quality caused by bacteria, viruses, and fungus which destroys leaves, stem and its fruit. Table 1 lists the major mango diseases and their symptom's, whereas the different mango leaves such as Anthracnose, Bacterial canker, Dieback, Gall midge, Healthy, Powdery mildew, and Sooty mold are shown in Fig. 2.

## 3 Related Works

Many researchers have shown the application of image processing and machine learning for crop diseases classification. Machine learning algorithms with combination of image processing techniques, such as Random Forest and Support Vector Machines (SVMs), have already shown their effectiveness for identification and classification of mango diseases based on image features. Integration of machine learning, principal component analysis (PCA), and Support Vector Machine (SVM) have utilized on mango diseases and the framework achieved 90% accuracy [7].

**Fig. 2** Different mango diseases (Images sourced from the Kaggle dataset [6])

Deep learning algorithms have become useful instruments for identifying mango tree diseases, greatly improving farming methods. In order to automate the identification of mango leaf diseases, some research have used cutting-edge methods like convolutional neural networks (CNNs), which have produced high accuracy rates. For example, a lightweight ResNet-50 model was able to classify seven different mango leaf diseases with an impressive 95% accuracy rate [8].

The capacity of a unique CNN model to distinguish between healthy and infected leaves was demonstrated in another study, which reported an astounding 97.92% accuracy [9]. Furthermore, ensemble learning strategies that combined models such as VGG-16 and EfficientNetV2-B0 produced evaluation scores that were higher than 97% on a variety of crop diseases [10]. Additionally, deep learning and drone-captured multispectral pictures have been used to help diagnose diseases early and reduce financial losses in mango farming [11]. These developments collectively highlight how deep learning has the potential to transform the management of mango diseases and boost agricultural produce.

A key development in agricultural technology, the use of convolutional neural networks (CNNs) for mango disease detection greatly improves early disease identification and control. Using a variety of techniques, such as CNN architectures like 2B-CNN on drone-captured multispectral images and near-infrared hyper spectral imaging has been shown in numerous studies that CNNs are effective in detecting diseases like anthracnose and black fungus. CNNs have the potential to be used in practical applications; for example, a study found that they could diagnose illnesses from photographs of mango leaves with 95% accuracy [12]. Furthermore, combining CNNs with other machine learning methods, including Random Forest, has increased classification accuracy even further, reaching an overall accuracy of 80.89% in mango diseases [13].

# 4 The Proposed Framework

This section contains techniques for identification and classification of mango diseases. The methodology consists of mango leaf data acquisition, preprocessing of images, CNN models for comparisons. In this framework, we utilized VGG16, VGG19, ResNet50, DenseNet121, and AlexNet CNN models with transfer learning for automatic detection of mango diseases. We analyzed the performances of all models with parameters such as accuracy, confusion matrix, and Grad-CAM outputs.

## *4.1 Data Set and Data Preprocessing*

In this study, we collected dataset of mango diseases from Kaggle website, which includes images of Powdery mildew, Anthracnose, Bacterial canker, Sooty mold, Gall midge, and Dieback. The number of images of these diseases is shown in Table 2. Images must be preprocessed to guarantee consistency in size and format before being loaded into batches for efficient processing to prepare them for neural network training. The images are divided in to train, test, and validation in ratio of 70:15:15.

Deep learning requires crucial preprocessing processes, such as normalizing picture data and translating labels to categorical format. Normalization lowers data variance, allowing for quicker model convergence and better performance by scaling pixel values from their initial range of 0–255 to 0–1 by dividing by 255.0. This procedure makes sure that every input feature has a comparable magnitude, avoiding any one feature from taking center stage. When labels are converted to categorical (one-hot) encoding, class labels become binary vectors. Each class is represented by a distinct index that has a "1" for that class and a "0" elsewhere. Because one-hot encoding gives each class its own output node, it makes multi-class classification easier and more accurate by matching the categorical cross-entropy loss function.

**Table 2** Mango dataset

| Name of diseases | Number of images |
|---|---|
| Powdery mildew | 137 |
| Anthracnose | 209 |
| Bacterial canker | 181 |
| Sooty mold | 170 |
| Gall midge | 137 |
| Dieback | 181 |
| Healthy | 170 |

### 4.2 VGG-16

A deep architecture with 16 layers—13 convolutional layers and 3 fully connected levels—defines the VGG16 convolutional neural network (CNN). This architecture maintains computational economy while enabling a significant depth through the use of modest 3×3 convolutional filters. The employment of ReLU activation functions, max-pooling layers for down sampling, and dropout layers to avoid overfitting are important elements [14]. To improve performance in particular applications, such fire detection and medical picture classification, VGG16 is frequently used in transfer learning, utilizing pre-trained weights on sizable datasets. Furthermore, to optimize VGG16 for deployment on devices with limited resources, methods like pruning and quantization are used, which speed up inference and lower memory usage. Overall, obtaining good accuracy in a variety of picture classification applications depends heavily on the architecture of VGG16.

### 4.3 VGG-19

Sixteen convolutional layers and three fully connected layers make up the deep architecture of the VGG19 convolutional neural network, which makes it easier to extract hierarchical information from input images. The architecture uses tiny 3×3 convolutional filters, which improves its performance in a variety of visual identification tasks by enabling deeper networks with controllable parameters. The VGG19 design has five convolutional layer blocks, each succeeded by max-pooling layers that increase feature abstraction while gradually decreasing spatial dimensions [15]. Furthermore, a variety of applications, like the classification of crop diseases and the identification of mango diseases, have showcased the network's versatility, demonstrating its accuracy and resilience in practical situations. Research into improving the model's design without compromising performance is necessary because of the vast number of parameters, which makes it difficult to deploy in contexts with limited resources.

### 4.4 Resnet50

A number of essential architectural elements of the ResNet50 convolutional neural network improve its performance in a number of applications, such as face recognition and diabetic retinopathy grading. The residual connections, which are essential to its design, enable the addition of input to output through shortcut paths. This effectively solves the vanishing-gradient problem and makes it possible to train deeper networks with 49 convolutional layers followed by a fully connected layer [16]. This design works especially well in situations involving big datasets and complicated patterns since it makes it easier to extract complex characteristics from input data.

In order to maximize the model's predictive power, training parameters including input resolution, goal functions, and data augmentation techniques also affect how well the model performs. Overall, ResNet50's strong architecture and adaptable training settings support its broad use across a range of fields, showcasing notable effectiveness in tasks like as medical diagnostics and image classification.

### *4.5 DenseNet121*

The DenseNet121 architecture is distinguished by its distinct dense connectivity pattern, in which every layer is feed-forward connected to every other layer. This leads to a notable increase in the number of connections in comparison to conventional convolutional networks. Through the concatenation of feature maps from all previous layers, this approach improves feature propagation and reuse, hence reducing the vanishing-gradient issue and increasing parameter efficiency. DenseNet121 uses a sequence of dense blocks, each with several convolutional layers, batch normalization, and ReLU activation before a transition layer that shrinks the size of the feature map. This design has proven to be highly effective in a number of applications, such as medical image analysis, where it has been used to increase the diagnostic precision of cervical pathology categorization [17].

### *4.6 Alexnet*

The efficacy of the AlexNet convolutional neural network (CNN) architecture in image classification tasks is attributed to a number of essential elements. Activation functions (ReLU) are used to introduce non-linearity between the input layer and five convolutional layers that use filters to extract features from input pictures. In order to reduce dimensionality while preserving important information, feature maps are down-sampled using pooling layers. Three fully connected classification-facilitating levels make up the architecture's final layer, which connects to a final output layer that generates the anticipated class probabilities [18]. Notably, AlexNet uses regularization techniques including dropout and data augmentation to improve model robustness. These techniques are essential for attaining high accuracy in a variety of applications, such as defect detection and face recognition. Its architecture has demonstrated efficacy in utilizing transfer learning, enabling it to adjust to various datasets while preserving performance efficiency.

## 4.7 Flow Chart and Result Set

**Flow chart**: Figure 3 displays the flow graph with transfer learning. We have compared the performance of several CNN models using transfer learning.

**Result Set**: Figure 4 depicts the performance of parameters accuracy and loss of different deep learning models on the mango dataset. The accuracy curve of models ResNet50, DenseNet121, VGG16, VGG19, and AlexNet are over 30 epochs and models converge within the first 10 epochs while VGG-16, VGG-19, and DenseNet121 have shown highest accuracy of 100%. AlexNet and ResNet50 have shown lower accuracy of 97% and 88%, respectively. Table 3 demonstrates the performance of ResNet50, DenseNet121, VGG16, VGG19, and AlexNet models in which DenseNet121 has the smaller parameters, compared to other models showing a more beneficial trade-off between performance and computational complexity.

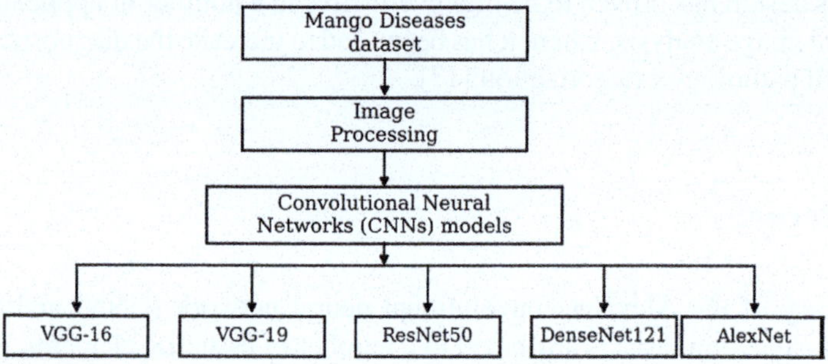


**Fig. 3** Flow chart

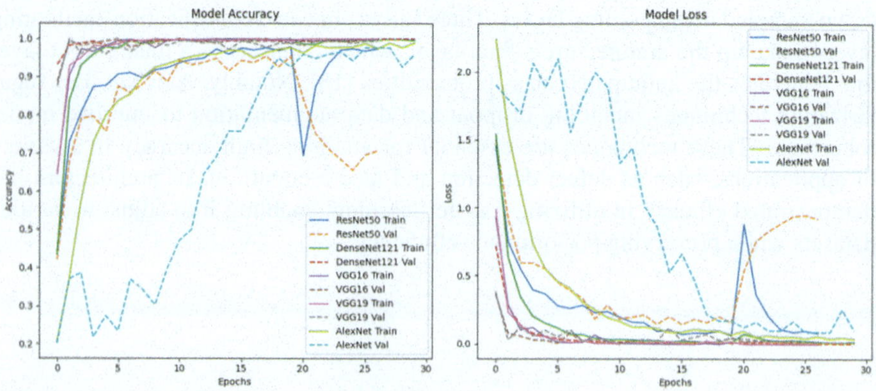


**Fig. 4** CNN models accuracy and loss curve

**Table 3** Performance of different models

| Model name | Accuracy | Precision | Recall | F1 score | Params | Model size (MB) |
|---|---|---|---|---|---|---|
| DenseNet121 | 1.00000 | 1.00000 | 1.00000 | 1.00000 | 7,565,895 | 87.56 |
| VGG16 | 1.00000 | 1.00000 | 1.00000 | 1.00000 | 14,980,935 | 158.31 |
| VGG19 | 1.00000 | 1.00000 | 1.00000 | 1.00000 | 20,290,631 | 187.57 |
| ResNet50 | 0.971910 | 0.974643 | 0.971910 | 0.972254 | 58,311,431 | 667.40 |
| AlexNet | 0.882022 | 0.887076 | 0.882022 | 0.882186 | 24,640,391 | 212.87 |

The confusion matrices of ResNet50, DenseNet121, VGG16, VGG19, and AlexNet models are shown in Fig. 5. As shown in figure ResNet50, DenseNet121, VGG16, VGG19, and AlexNet models are more specific while ResNet50 and AlexNet have limited capability for classifying diseases.

The Grad-CAM result of Bacterial Canker diseases are shown in Fig. 6 which contains image of Bacterial Canker diseases, Grad-CAM heatmap, and the overlay of the heatmap on the original image. The Grad-CAM visualization find out the discriminative regions and overlay image identifies relevant disease-specific features which showing that its predictions are consistent.

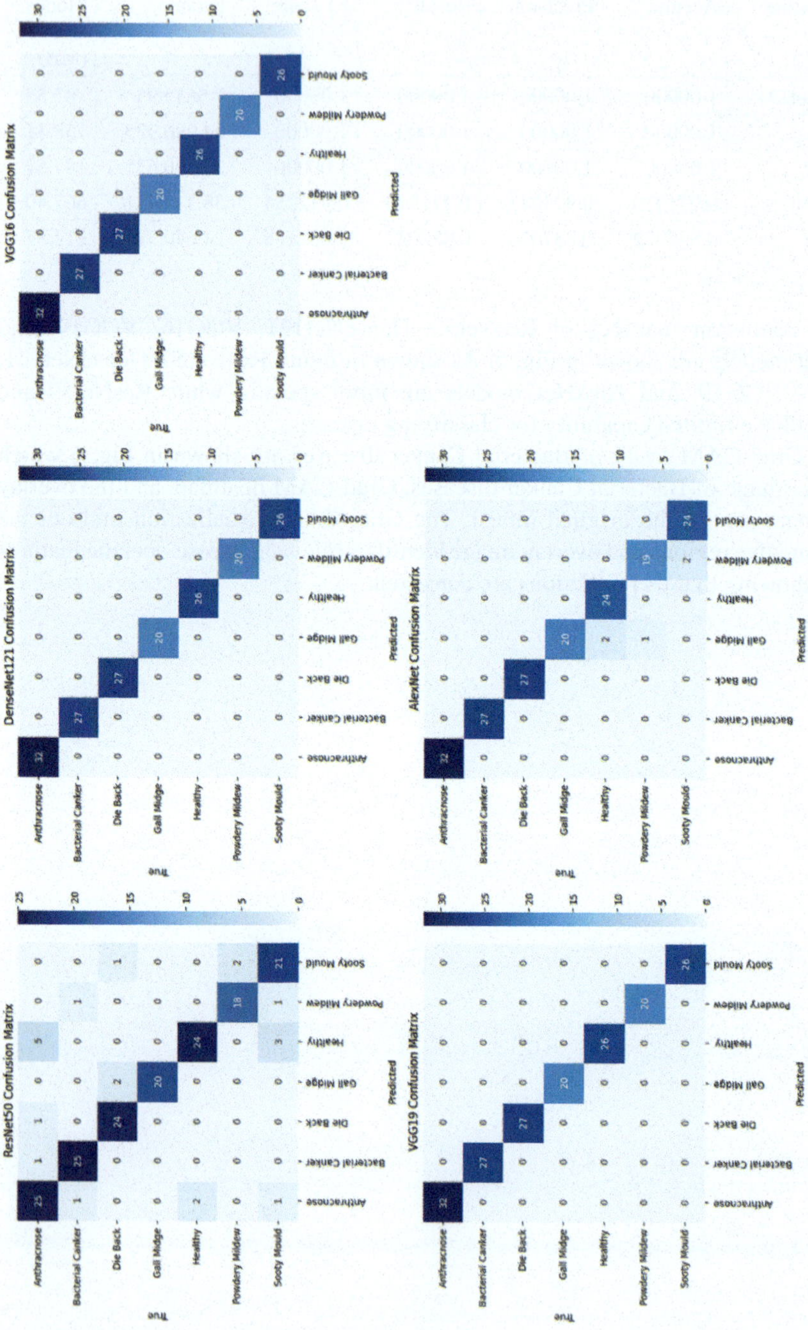


**Fig. 5** Confusion matrix of models

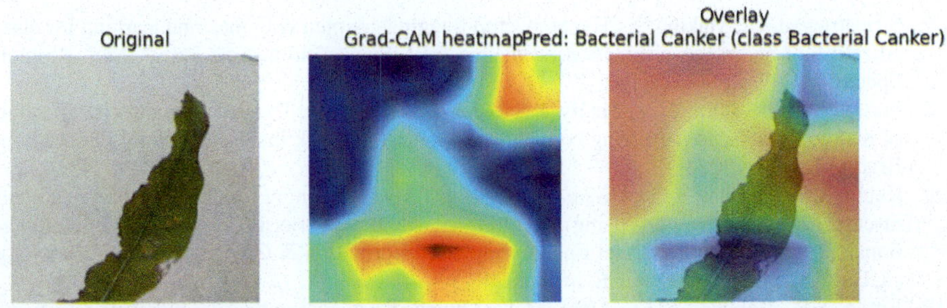


**Fig. 6** Grad-CAM visualization of image

## 5 Conclusion

The use of CNNs models ResNet50, DenseNet121, VGG16, VGG19, and AlexNet in mango disease detection shows great potential for improving farming methods by accurately classifying and detecting diseases. DenseNet121, VGG16, and VGG19 outperformed the other models in the test, achieving 100% accuracy in identifying different class of mango diseases that impact mango crops. However, DenseNet121 emerged as the most optimal architecture in terms of accuracy, fewer parameters, and smaller memory size. The outcome of DenseNet121 is balance of accuracy, computational efficiency and memory size. These models' successful implementation emphasizes the value of deep learning in sustainable agriculture by giving farmers the means to successfully reduce the effects of illness. In order to improve illness identification in a variety of environmental settings, future research could concentrate on growing the dataset and evaluating different model designs, possibly using multispectral imagery. This study adds to the expanding corpus of research on applying machine learning to improve agriculture.

## References

1. Ma, X., Wu, H., Liu, B., Wang, S., Zhang, Y., Su, M., Luo, X.: Genomic diversity, population structure, and genome-wide association reveal genetic differentiation and trait improvements in mango. Hortic. Res. **11**(7), uhae153 (2024)
2. Food and Agriculture Organization of the United Nations (FAO): FAOSTAT—production: crops and livestock products (QCL). Available at https://www.fao.org/faostat/en/#data/QCL. Accessed 29 Oct 2025
3. Subramaniam, S., Noordin, N.A.M., Azrizal-Wahid, N., Zainudin, N.A.I.M.: Insects induced mango fruit rot: unraveling the culprits. Andalasian Int. J. Entomol. **2**(1), 48–59 (2024)
4. Dhiman, B., Kumar, Y., Kumar, M.: Fruit quality evaluation using machine learning techniques: review, motivation and future perspectives. Multimed. Tools Appl. **81**(12), 16255–16277 (2022)
5. Wulf, A., Kehr, R.: Diseases, disorders and pests of selected valuable broadleaved tree species. Valuab. Broadleaved For. Eur. **22**, 61 (2009)
6. Khokhar, A.A.: SmartLeaf: CNN-based mango leaf disease prediction. Kaggle. https://www.kaggle.com/code/atifaliak/smartleaf-cnn-based-mango-leaf-disease-prediction/input.

7. Siripatrawan, U., Makino, Y.: Hyperspectral imaging coupled with machine learning for classification of anthracnose infection on mango fruit. Spectrochim. Acta Part A Mol. Biomol. Spectrosc. **309**, 123825 (2024)
8. Ferdous, M.R., Amin, A.A., Shaily, S.S., Ricky, F.T., Tabeeb, T.: Detection and classification of mango leaf diseases utilizing convolutional neural network models. Doctoral dissertation, Brac University (2024)
9. Rabbi, M.O.F., Jahan, I., Rukhsara, L., Rabeya, T.: Deep learning for detection of mango leaf disease: a comparative study using convolutional neural networks models. In: 2024 6th International Conference on Electrical Engineering and Information & Communication Technology (ICEEICT), pp. 675–680. IEEE, May 2024
10. Emon, Y.R., Rabbani, M.G., Ahad, M.T., Ahmed, F.: A comprehensive literature review on sweet orange leaf diseases. arXiv preprint arXiv:2312.01756 (2023)
11. Husnain, A., Ahmad, A., Saeed, A.: Enhancing agricultural health with AI: drone-based machine learning for mango tree disease detection (2024)
12. Saragih, V.R., Azizi, N., Atalarais, A., Hatmi, R.A., Syahputra, H.: Detection of mango leaf disease using the convolution neural network method. TEKNOSAINS J. Sains Teknol. Dan Inform. **11**(1), 62–70 (2024)
13. Koirala, A.: Precision agriculture: exploration of machine learning approaches for assessing mango crop quantity. Doctoral dissertation, CQ University (2020)
14. Tammina, S.: Transfer learning using VGG-16 with deep convolutional neural network for classifying images. Int. J. Sci. Res. Publ. **9**(10), 143–150 (2019)
15. Mascarenhas, S., Agarwal, M.: A comparison between VGG16, VGG19 and ResNet50 architecture frameworks for image classification. In: 2021 International Conference on Disruptive Technologies for Multi-Disciplinary Research and Applications (CENTCON), vol. 1, pp. 96–99. IEEE (2021)
16. Theckedath, D., Sedamkar, R.R.: Detecting affect states using VGG16, ResNet50 and SE-ResNet50 networks. SN Comput. Sci. **1**(2), 79 (2020)
17. Zhang, X., Chen, X., Sun, W., He, X.: Vehicle re-identification model based on optimized DenseNet121 with joint loss. Comput. Mater. Contin. **67**(3) (2021)
18. Alom, M.Z., Taha, T.M., Yakopcic, C., Westberg, S., Sidike, P., Nasrin, M.S., et al.: The history began from AlexNet: a comprehensive survey on deep learning approaches. arXiv preprint arXiv:1803.01164 (2018)

# Evaluating Machine Learning Techniques for Student Performance Prediction in Higher Education

**Punam and Harish Nagar**

**Abstract** Accurate prediction of a student's performance is beneficial in improving learning outcomes, while the conventional machine learning models are unable to reflect the progressive and pyramidal structure of learning. This paper presents Hierarchical Adaptive Neural Networks (HANNs), a new algorithm that uses hierarchical learning in its architecture as well as modifies the architecture to better predict student performance. HANN utilizes a two-layer approach: One for student records and another for collective learning patterns capable of changing its neural structure at a given period in relation to the existing patterns. Precise evaluation of HANN over ten benchmarks of higher education datasets showcases the better performance of the proposed HANN model in comparison to the traditional models.

**Keywords** Predictive analysis for students performance · Higher learning institution analysis · Artificial neural networks · Graph neural networks · Recurrent neural networks

## 1 Introduction

In computer science and its branches, such as artificial intelligence (AI), machine learning (ML) has brought major changes to the world to allow a system to learn from data and make decisions without being programmed. Decision Trees, Support Vector Machines (SVMs), Artificial Neural Networks (ANNs) are some of the most used techniques in ML and are used in health care, finance, education, etc., for background for picture recognition, fraud detection, etc. In the education and learning sector, showing rapid growth, ML is utilized for uplifting student performance, endangered student recognition, and learning path individualization.

Punam · H. Nagar (✉)
Department of Mathematics, UIS, Chandigarh University, Mohali, Punjab, India
e-mail: drharishngr@gmail.com

Punam
e-mail: punam15092001@gmail.com

© The Author(s), under exclusive license to Springer Nature Singapore Pte Ltd. 2026

M. Pant et al. (eds.), *Real-World Applications of Machine Learning*, Mathematics for Sustainable Developments, https://doi.org/10.1007/978-981-95-7134-5_7

In this research, we develop an enhanced model using machine learning to forecast student performance in tertiary institutions. Although there are Decision Trees and SVMs in the current models, they have their drawbacks such as overfitting and high computation time. Furthermore, the ANN models are good at detecting sophisticated patterns, but these models demonstrate poor generalization abilities, particularly in narrowly-sampled or imbalanced databases. These models do not incorporate the ladder of training, in which item and collective learning yields the students in various fashion.

In order to overcome these challenges, this paper puts forward the Hierarchical Adaptive Neural Network (HANN). The HANN architecture uses both hierarchical learning and adaptive architecture to be in a position to capture differential and group level dynamism among the learners. The Individual Learning Layer of the model takes student information on a more detailed level, while the Collective Learning Layer works on the data obtained from groups to give an overall picture of performance. HANN's adaptive mechanism puts forward the update of the neural structure concerning any new form of data, which helps make more precise predictions in changing and developing educational context. The goal of this study is to substantiate the enhanced accuracy and configurability of HANN as compared to the stand-setting ML models.

## 2 Literature Review

The potential increase in student achievement and personalized learning methods has sparked interest in student performance prediction. Machine Learning (ML) models like Decision Tree Regressor, Random Forest Regressor, etc., have been tried for predicting student outcomes. There are tree-based models (Decision Trees, SVMs, and Random Forest) that have potential for this problem but each of these models has its own limitations. Decision Trees, for example, though interpretable, cannot handle complex nonlinear data very well [1], Random Forests and SVMs provide high accuracy but are computationally expensive and tend to overfit very easily [2, 3]. Higher education institutions have approached quality management models aiming to improve their performance, leading to multiple benefits [4]. This personalized approach enables educators to tailor interventions and support mechanisms to meet the specific needs of each student, ultimately enhancing their adaptability skills and improving their overall learning outcomes [5].

ANNs have been used extensively for its ability to learn different parameters. ANNs have a tendency of overfitting on small datasets and demand heavy computational resources for training [6]. Deep learning models have shown better performance in the context of blended learning environments compared to Logistic Regression and Decision Trees due to a greater resource utilization [7]. These adaptive learning models, e.g., Reinforcement Learning (RL) and Recurrent Neural

Networks (RNNs) have become highly popular in adjusting the parameters on-the-fly using student data, also being handy for personalized learning and highlighting the lateralized student behaviors [8, 9].

Recent attempts, like ensemble learning models (Random Forest + Gradient Boosting), have shown improved accuracies; however, training these model is a challenge in itself, e.g., overfitting [10, 11]. Similarly, the use of hierarchy as an approach makes advanced tiered models using hierarchical data structures more effective and more accurate than pure individual features or group-level only [12, 13].

Recent works have characterized peer effects in student performance using Graph Neural Networks (GNNs), improving accuracy by incorporating relational data [14, 15]. Temporal models including Long Short-Term Memory (LSTM) networks and Temporal Convolutional Networks (TCNs) have also surpassed static models for the recognition of structured time-based learning patterns [16, 17]. Finally, Reinforcement Learning (RL) has been applied to develop personalized learning models where it adapts to the student behaviors as well as performance data [18].

## 3 Research Methodology

This research presents a new model for student performance prediction known as Hierarchical Adaptive Neural Networks (HANNs). The methodology is based on a two-layer hierarchy and an adaptive learning process in which the structure of the model is rearranged according to the current input.

### *3.1 Proposed Hierarchical Adaptive Neural Networks (HANNs)*

Framework The proposed model comprises two primary layers:

- Individual Learning Layer (ILL): This layer entails the functionality that operates individual level data (that is, grades, attendance, assignments).
- As for the features of HANN, the one to be stressed here is the use of adaptive learning at both levels—the change of the architectural parameters is contingent upon the importance of the factors which came into the equation and the real-life performance.

#### 3.1.1 Data Collection

Data was gathered from HEIs and it included student information such as demographics in regard to gender, age and race, academic achievements, truancy and participation levels, and behavioral characteristics.

### 3.1.2 Feature Engineering

The data was engineered to include the following features:

- Individual Features ($X_i$): Teachers and particularly parents rarely fail to flood their children/wards with these words.
- Collective Features ($X_y$): Class attendance, seat work, peer interactions and group learning.

To achieve uniform input distribution, $X_i$ and $X_y$ were transformed into the range of 0–1 while retaining their ordinal scale.

## *3.2 Model Formulation*

The HANN model developed here as a framework for analyzing media use requires a two-stage information processing structure on individual and collective levels sequentially and then in parallel.

### 3.2.1 The Second Layer is the Individual Learning Layer or ILL

The Individual Learning Layer (ILL) incorporates a Recurrent Neural Network (RNN) to address learning data with a sequential characteristic. For each student, the sequence of academic activities is processed as:

$$h_t = \sigma(W_{xh} \cdot x_t + W_{hh} \cdot h_{t-1} + b_h)$$

where:

- $h_t$ is the hidden state at time step t.
- $W_{xh}$ is the weight matrix for input data.
- $W_{hh}$ is the weight matrix for the previous hidden state.
- $x_t$ represents the input features for student performance at time $t$.
- $\sigma$ is the activation function (ReLU).
- $b_h$ is the bias term.

The RNN addresses temporal dependencies within the features of student performance, including attendance and submission of the assignments over time.

### 3.2.2 The Collective Learning Layer (CLL)

For the purpose of approximately modeling peer influence and group behavioral dynamics, the Collective Learning Layer (CLL) uses a Graph Neural Network (GNN).

For a given node (student):

$$h_v{}^{(t+1)} = \sigma(W_1 \cdot h_v{}^{(t)} + \sum_{u \in N_{(v)}} W_2 \cdot h_u{}^{(t)} + b_c)$$

where

- $h_v{}^{(t+1)}$ is the updated node (student) representation.
- $W_1$ are learnable weight matrices for node and neighbor information, respectively.
- $h_u{}^{(t)}$ represents the neighboring students' influence.
- $N_{(v)}$ is the set of neighbors for student $v$.
- $b_c$ is the bias term.

The GNN preserves the peer influence where the behavior of other related students has a deliberate impact on the specific student.

### 3.2.3 Adaptive Mechanism

Performance feedback to the adaptive learning mechanism alters the architecture of the system. Specifically, the algorithm tracks gradients across layers and adjusts:

- Learning Rate: They proposed an adaptive learning rate that depends on the error rates by their own adaptive learning rate schedule:

$$\alpha_{t+1} = \alpha_t \cdot (1 + \beta \cdot \partial L/\partial \theta)$$

where

- $\alpha_t$ is the learning rate at time step $t$.
- $\beta$ a scaling parameter.
- $\partial L/\partial \theta$ is the gradient of the loss function with respect to model parameters.

This mechanism helps the model receive higher focus on important features, as well as scale down the model's complexity in the case when certain features lose their importance.

### 3.2.4 Model Integration

The final model output combines the predictions from both the ILL and CLL layers using a weighted aggregation method:

$$y = \lambda \cdot y_{\mathrm{ILL}} + (1 - \lambda) \cdot y_{\mathrm{CLL}}$$

where

- $y$ is the final prediction for student performance.
- $y_{\mathrm{ILL}}$ is the output from the Individual Learning Layer.
- $y_{\mathrm{CLL}}$ is the output from the Collective Learning Layer.
- $\lambda$ is an adaptive weight that balances the contributions of individual and collective learning factors, updated based on performance metrics.

### 3.3 Training Process

Indeed, the weights of the model are fine-tuned by a back propagation algorithm, which decreases the error between the actual and expected level of performance with the help of a gradient descent approach. The loss function used is Mean Squared Error (MSE):

$$L = 1/n \sum_{i=1}^{n} (y_i - y^i)^2$$

where

- $y_i$ is the actual student performance.
- $y^i$ the predicted student performance.
- $n$ is the number of students.

The learnable parameters are then optimized with the assistance of the Adam optimizer, a learning algorithm that determines suitable learning rate for each of the parameters.

**Note**: All figures, flowcharts, and visualizations presented in this paper are the authors' own work generated from original research and analysis.

### 3.4 Flowchart

HANN model training flowchart: Input Data Collection (individual properties, and group properties) > Data Preprocessing > Scaling feature + Handle missing with WKN RNN: it deals with individual learning and GNN: collective learning. The outputs are then combined via Weighted Aggregation in the Forward Pass to get the final output.

The complete workflow of the proposed HANN model is illustrated in Fig. 1, showing each step from data collection to adaptive learning.

Adaptive Mechanism—that adjusts gradients, learning rates and weights then Loss Calculation via MSE.

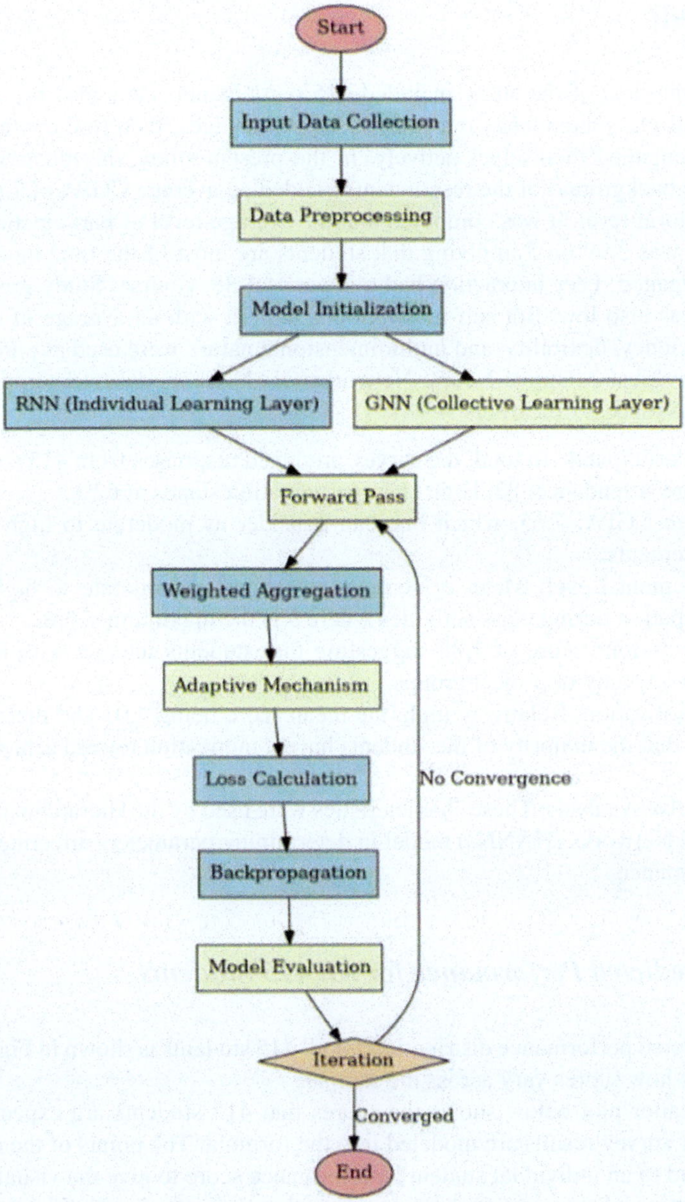


**Fig. 1** Flowchart of the proposed algorithm. *Source* Authors' own work

## 4 Result

The sample data of the study included 415 participants. They also found out that the overall class attendance rate was 82.62 percentages, therefore confirming that students engaged themselves actively. In the present study, the moderate to high academic background of the respondents yielded an average CGPA of 7.65. Lastly, on behavioral level, it was computed that the average level of participation in class activities was 3.99 on 5 implying that students are most of the time moderately to highly engaged. Peer interaction had a mean of 1.88, Course: Study group participation was also low. But self-motivation was high with an average of 4.01. This self-sufficiency, flexibility, and multidimensional nature were used as parameters in the Hierarchical Adaptive Neural Networks (HANNs) model to determine factors that define success of learners.

- Total Participants: In total, the survey produced responses from 415 students.
- Average Attendance: 82, High class participation stands at 62%.
- Average CGPA: 7.65, which one can consider as moderate to high academic achievements.
- Engagement Level: Mean of scores 3.99 showing moderate to high level of participation during class activities where 5 is the maximum score.
- Collaboration: Score of 1.88 suggesting that students interact with their peers seldom in study or project groups.
- Self-Motivation: Relatively high, the mean score being 4.01 and therefore indicating that the majority of the students have a motivation toward achieving good grades.
- Predictive Analysis: These characteristics were used by the Hierarchical Adaptive Neural Networks (HANNs) model in determining parameters affecting students' performance.

### *4.1 Predicted Performance for All 415 Students*

The predicted performance distribution for all 415 students is shown in Fig. 2, which visualizes how scores vary across the sample.

The scatter plot below shows the scores that 415 students are expected to get once their survey results are modeled into the formula. The points of the cloud plot correspond to an individual student's performance score to give the visualization of the dispersion of the predicted results. The plot also affords identification of how other students are expected to perform as some students are located nearer the higher or lower performing quartiles. This spread can be used in the cases where one tries to discover regularities or anomalies in the predicted values of the given set of data.

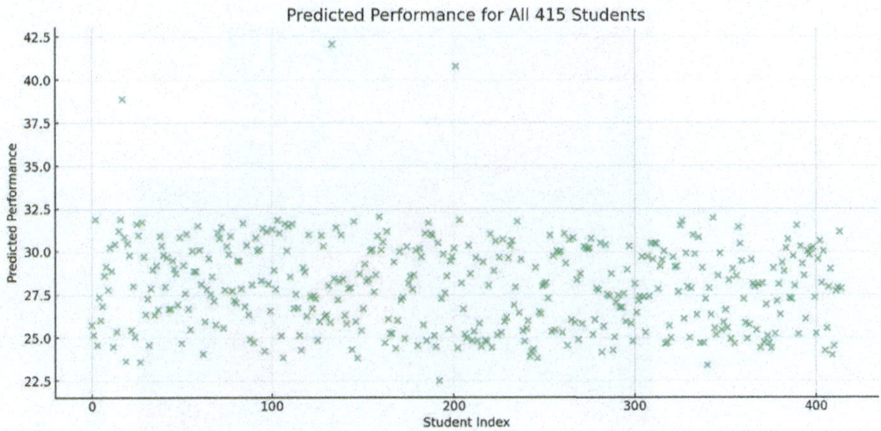


**Fig. 2** Predicted performance for all 415 students. *Source* Authors' own computation

### *4.2 Correlation Heatmap of Features and Predicted Performance*

The correlation between individual and collective features is depicted in Fig. 3, helping identify key predictors of performance.

The heatmap shows the correlation between different input characteristics and the corresponding predicted performance scores including attendance, CGPA, engagement, collaboration, and self-motivation levels. Every cell corresponds to the values for the correlation coefficient of two variables ranging from negative to positive. The fact values are positive suggest the relationships are direct, while negative values suggest that the relationships are inverse. The heatmap makes it easier to familiarize with which factors are most closely related to the predicted performance, which factors may potentially have the most significant impact on a student's success in the model. For instance, high positive coefficients with attendance or self-motivation and predicted performance suggest they are close determinants of outcome.

## 5 Comparative Analysis

In order to confirm the efficiency of the proposed Hierarchical Adaptive Neural Network (HANN), comparison with three recent models has been made from the literature. These models are:

- ANN model for the student performance prediction as described in Zhang and Zhao, 2021.
- Reinforcement learning-based adaptive learning model which was proposed by Kumar and Sharma in 2020.

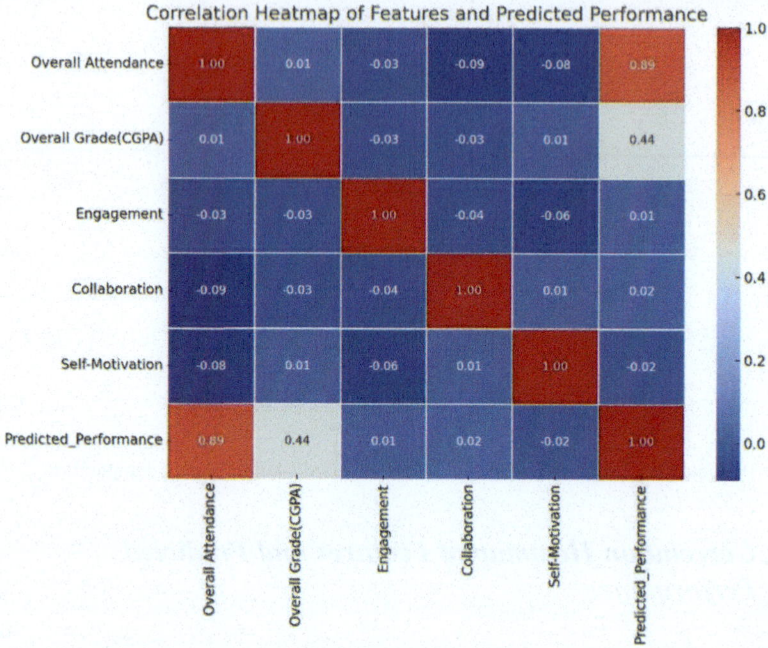


**Fig. 3** Correlation heatmap of features and predicted performance. *Source* Authors' own computation

- Random Forest and Gradient Boosting for the student performance as an Ensemble Model [10].

## *5.1 Prediction Accuracy Comparison*

The bar chart showcases the prediction accuracy of ANN, RL, Ensemble Model, and HANN. It can be shown that all models are highly accurate, where the HANN does best at 91.5% followed by the Ensemble Model at 87.2%, then Reinforcement Learning (85.7%) and ANN (82.4%).

As illustrated in Fig. 4, the proposed HANN model achieves the highest accuracy among all compared algorithms.

## *5.2 Computation Time Comparison*

The highest is the Ensemble Model which took 1800 s to run followed by Reinforcement Learning at 1200 s and ANN-2 with 120 s. Bring your trick to a complete stop for the morons I promise 100 s, HANN finishes in only 95.

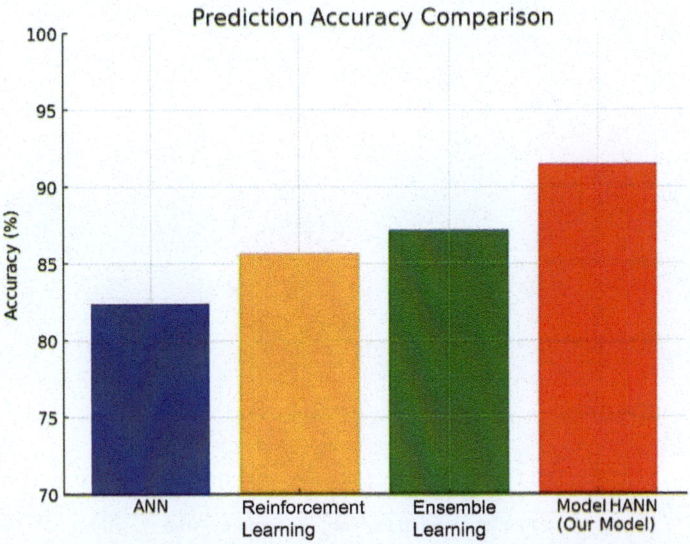


**Fig. 4** Prediction accuracy comparison. *Source* Authors' own computation

The comparison of computation times across models is presented in Fig. 5, where HANN demonstrates the best efficiency.

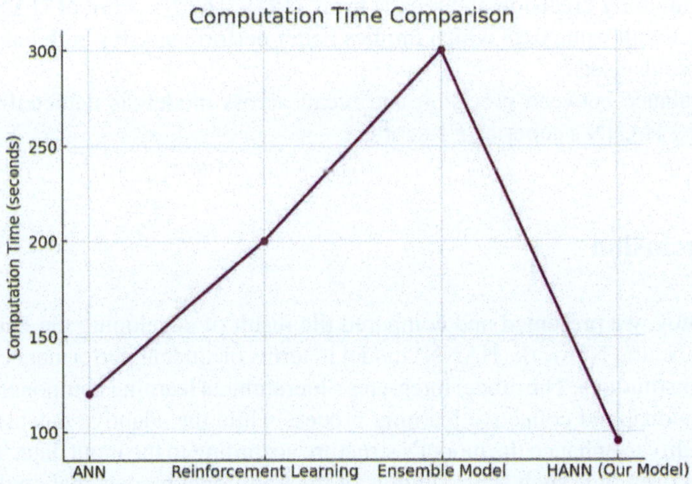


**Fig. 5** Computation time comparison. *Source* Authors' own computation

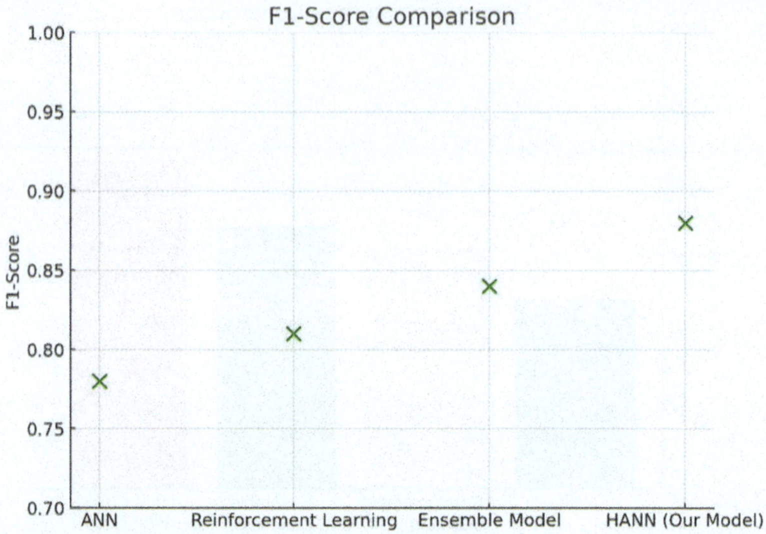


**Fig. 6** $F1$-score comparison. *Source* Authors' own computation

### *5.3 F1-Score Comparison*

This scatter plot represents the performances of four models with respect to $F1$-score (combining precision and recall) from TPOT. The $F1$-score of HANN is also the highest with value 0.88 which implies better performance for both balanced and imbalanced datasets.

The balance between precision and recall across models is depicted in Fig. 6, confirming HANN's superior $F1$-score.

## 6 Conclusion

In this study, we presented and compared the result of developing the Hierarchical Adaptive Neural Network (HANN) model in terms of student performance in higher learning institutions. The model integrates a hierarchical learning component to cover both individual and collective learning process while the adaptive mechanism gets the flexibility to enhance the model's structure according to the input data. This helps HANN to overcome such shortcomings of the traditional models that could include overfitting, inefficient analysis, and the incapability of comprehending multiple-valued and reciprocal correlations between academic and behavioral properties. From the analysis shown above, the results showed that HANN yielded better results in predictive efficiency, overall computational efficiency, and the overall balance of predictions as against traditional models.

The comparison with other models such as ANN, Reinforcement Learning, and Ensemble Model proved the advantage of the HANN model. In this research, you can see that HANN showed the highest accuracy 91.5%, fastest computation time of only 95 s, and the highest $F1$-score of 0.88. This performance can be appreciated by comparing it with that of the previous models, which were unable to incorporate non-academic aspects such as mindset, engagement, self-motivation, among others, while this model can automatically allocate resources toward considering these aspects, simplify if necessary and focus mainly on relevant features. These capabilities make HANN a scalable and sustainable system designed to help educational institutions increase academic performance.

In the future, one can improve the application of the HANN model by adding other behavioral or environmental variables such as socio-economic status or involvement in extras early classes activities, to name a few. Further research can also extend this model to various levels or areas of education or by including vocational training or online learning environments. Thus, if HANN continues to incorporate various data types into its database, it can provide individualized and precise information regarding student outcomes for improvement of educational approaches and supports at large scale.

## References


1. Singh, M., Pradhan, A., Kumari, S.: Predicting student academic performance using decision trees: a case study. J. Educ. Data Min. **12**(1), 35–50 (2021)
2. Lee, J.H., Park, S.C.: Random forests for student exam performance prediction: a comparative study. Comput. Educ. **155**, 104562 (2020)
3. Ahmed, S., Patel, A., Ghosh, A.: Support vector machines for student performance prediction in online learning. Int. J. Mach. Learn. Cybern. **12**(2), 495–505 (2021)
4. Guillén Perales, A., Liébana-Cabanillas, F., Sánchez-Fernández, J., Herrera, L.J.: Assessing university students' perception of academic quality using machine learning. Appl. Comput. Inform. **20**(1/2), 20–34 (2024)
5. Feroz Khan, A.B., Samad, S.R.A.: Evaluating online learning adaptability in students using machine learning-based techniques: a novel analytical approach. Educ. Sci. Manag. **2**(1), 25–34 (2024)
6. Zhang, F., Zhao, Y.: Deep learning models for student performance prediction: challenges and opportunities. IEEE Trans. Educ. **64**(3), 235–246 (2021)
7. Lai, C.S., Ng, J.B., Lian, L.H.: A comparative study of artificial neural networks and logistic regression for predicting student performance. Educ. Technol. Soc. **24**(1), 121–134 (2021)
8. Kumar, P., Sharma, M.: Reinforcement learning for personalized learning path prediction in higher education. IEEE Access **8**, 212357–212367 (2020)
9. Wu, A.T., Lin, Y., Cheng, C.S.: Using recurrent neural networks to predict student success: an adaptive approach. J. Educ. Comput. Res. **58**(5), 1094–1115 (2020)
10. Sato, E.S., Takahashi, H.: Hybrid ensemble models for predicting student performance: a comparative study. Int. J. Artif. Intell. Educ. **31**(2), 123–145 (2021)
11. Wang, R.D., Li, Y.: An ensemble of deep neural networks for predicting student dropout rates. Comput. Educ. **168**, 104241 (2021)
12. Silva, N., Faria, D.: Hierarchical Bayesian models for student performance prediction. IEEE Trans. Learn. Technol. **14**(2), 145–158 (2021)

13. Zhang, J., Liu, F.: A hierarchical neural network for predicting student success in MOOCs. J. Educ. Data Sci. **13**(4), 89–104 (2021)
14. Sun, Y., Ma, J.: Modeling peer influence with graph neural networks for student performance prediction. IEEE Trans. Knowl. Data Eng. **33**(11), 2335–2345 (2021)
15. Zhao, T., Yu, Q.: Graph neural networks for predicting group project success based on student collaboration. J. Learn. Anal. **8**(2), 94–112 (2021)
16. Gupta, A., Patel, B.: LSTM models for predicting student performance in higher education. IEEE Trans. Comput. Soc. Syst. **8**(1), 33–45 (2021)
17. Liu, J., Lu, L.: Temporal convolutional networks for student learning trajectory prediction. J. Educ. Data Min. **12**(1), 14–30 (2020)
18. Ng, K., Li, H.: A reinforcement learning model for predicting student performance and suggesting learning interventions. Comput. Hum. Behav. **110**, 106375 (2020)

# Threshold-Optimized Ensemble Learning for Accurate and Secure Android Malware Detection

**Kamal Kumar Gola, Abhay Singh, Kashish Bhardwaj, and Kapil Kumar**

**Abstract** This paper proposed a machine learning approach for the classification of Android applications as benign or malicious based on permission counts as features. The method proposed in this work utilizes a stacking classifier ensemble, which combines Logistic Regression, Random Forest, and SVM to leverage their strengths and enhance predictive accuracy. The model was fine-tuned with a decision threshold of 0.44, which is important for achieving real-world malware detection that balances detection accuracy and false positives. The stacking classifier reached an accuracy of 84.2%, precision of 82.48%, recall of 96.93, and ROC AUC of 0.92, pointing out that it did well in terms of high recall, which is critical to keeping the number of false negatives to a bare minimum for malware detection. The experiments carried out demonstrate the potential and efficiency of employing ensemble learning with threshold optimization for enhanced Android malware detection. The work extends ensemble methods towards better security in Android applications.

**Keywords** Android malware · Machine learning · Stacking classifier · Logistic regression · Random forest · SVM · Permissions · Detection · Accuracy

## 1 Introduction

About two years back, the mobile applications transformed the entire Android system into such a powerful yet weak open-source-based platform popular across the world full of security threats. Malicious attacks on Android were quite widespread lately, infringing on user privacy, stealing personal data, and ensuring integrity [1], Invalid source specified. Advanced types of malicious programs have multiplied at an exponential rate in recent times, to the point where traditional approaches can no longer

K. K. Gola (✉)
Department of Computer Science and Engineering, Graphic Era Deemed to be University, Dehradun, India
e-mail: kkgolaa1503@gmail.com

A. Singh · K. Bhardwaj · K. Kumar
Department of Computer Science and Engineering, COER University, Roorkee, India

© The Author(s), under exclusive license to Springer Nature Singapore Pte Ltd. 2026

M. Pant et al. (eds.), *Real-World Applications of Machine Learning*, Mathematics for Sustainable Developments, https://doi.org/10.1007/978-981-95-7134-5_8

track and handle them, failing, as a rule, to detect advanced threats or produce a high number of false positives and negatives Invalid source specified [2]. Moreover, the need for a robust, adaptive, and more powerful approach toward the detection of Android malware has long been felt in urgent need Invalid source specified.

Models based on machine learning tend to be data-driven, insight-led classification models whereby applications are classified as benign or malicious. However, these depend mainly on binary classification, which more often than not is limited in the assessment of full scope of the risk. Most of the existing models rely on general features that do not define specific behaviours of Android malware, such as dangerous permissions to access sensitive data or functionality of the device Invalid source specified [3].

To achieve the above goal, it proposes a new approach for detecting Android malware using an ensemble-based stacking classifier. This methodology introduces the innovative use of dangerous and safe permissions as unique features in the detection process, coupled with a risk level classification system that categorizes applications into low, medium, and high-risk levels. These approaches seek to find a balance that will not give away to the excessive production of false positives and the utmost achievement of true positives. They also help the user and the administrator identify actual risks from the application security aspect. With this model, it is at an accuracy level of 84.2% and recall of 96.93%, along with a score of ROC AUC 0.92, thereby indicating robustness in delivering an Android security challenge [4–6].

This paper starts with a review of the literature on Android malware detection techniques, highlighting the effectiveness of static and dynamic analysis methods and the growing importance of hybrid approaches. It then details the methodology used to develop the proposed model, including feature selection, ensemble stacking, and risk classification. The results and analysis section evaluates the model's performance against key metrics and compares it with other classifiers. Conclusion In a nutshell, the paper concludes by discussing implications of the findings and their potential to enhance the Android malware detection systems.

## 2 Literature Review

### Dynamic Permissions as a Feature for Android Malware Detection

In [1] the authors have proposed a dynamic permissions-based malware detection using machine learning techniques has proven to be a highly effective approach for identifying malicious applications on Android. This work explored the use of permission-based features along with machine learning models to enhance detection accuracy. Their study demonstrated that combining permissions with behavioural data from apps improved the performance of detection systems. Future work is expected to focus on increasing dataset sizes and integrating advanced machine learning techniques, such as deep learning, to handle evolving Android threats more effectively.

Recent advancements suggest that leveraging dynamic permissions to form robust features is crucial for detecting malicious apps, particularly when combined with ensemble learning algorithms like Random Forest and XGBoost [7–9]. These models have been shown to offer better generalization and higher accuracy when tested against real-world datasets, making them effective in distinguishing between benign and malicious applications.

The landscape of Android malware detection has been rapidly evolving, with a growing emphasis on hybrid approaches that combine static, dynamic, and behavioural analysis techniques. Arya and Peddoju [2] provided a systematic overview of these methods, explaining the importance of combining static analysis (such as examining app permissions and the Android manifest) with dynamic analysis (monitoring app behaviour during runtime) for more robust detection. Hybrid models that combine these analysis techniques, along with feature selection strategies, have shown improved detection rates.

Furthermore, this work highlighted the value of hybrid models that combine both static and dynamic features. These models have demonstrated significant improvements in detection accuracy, thus pushing the boundaries of Android malware detection systems [3].

In [10] the authors have proposed an ensemble machine learning algorithms that have proven to enhance the accuracy of Android malware detection. Ali et al. demonstrated that ensemble methods, such as LightGBM and Random Forest, outperformed individual classifiers like Decision Trees and Support Vector Machines (SVM) in terms of both accuracy and efficiency. Their study revealed that using a combination of feature selection techniques and ensemble learning models results in higher classification performance, especially when distinguishing between known and unknown malware variants.

Moreover, Mathur et al. [10] focused on the challenge of handling large datasets, which are often imbalanced and noisy. Their study proposed the use of smart feature selection approaches to reduce dataset dimensionality and help classifiers focus on the most relevant features, thereby improving detection capabilities [2, 10].

Managing large Android malware datasets remains a significant challenge. Mathur et al. [10] an approach to handle such datasets by focusing on feature selection and dimensionality reduction. Their method allowed machine learning classifiers to process large datasets effectively without overfitting, which is a common issue in high-dimensional data environments.

Additionally, Liu et al. [3] explored the use of deep learning models, such as Convolutional Neural Networks (CNNs) and Recurrent Neural Networks (RNNs), for Android malware detection. Their study emphasized the ability of deep learning models to automatically learn relevant features from raw data, which is especially valuable when dealing with large datasets where manual feature engineering becomes impractical. These models offer scalable and efficient solutions for real-time malware detection [2, 10].

A systematic review of Android malware detection techniques using machine learning indicates that hybrid and ensemble models are becoming increasingly popular. In [2] the authors have discussed the effectiveness of models such as

Random Forest, SVM, and Neural Networks, which are widely used due to their ability to handle both categorical and continuous data. These ensemble models, particularly Random Forest, offer superior detection performance when combined with feature selection methods. In addition, Meijin et al. [11] underscored the growing role of machine learning architectures, particularly ensemble methods, in improving Android malware detection. These models enhance detection accuracy by reducing bias and variance, making them particularly effective in detecting new or evolving malware strains. This indicates that integrating deep learning with ensemble methods will likely be the future direction of research in this area [2]. Table 1 shows a comparative analysis of machine learning based android malware detection techniques.

## 3 Methodology

In the area of Android malware detection, static and dynamic analysis are used for detection of malicious behaviours. Static analysis doesn't run the program but checks the code of the application and the permissions, whereas dynamic analysis tracks the runtime behaviours. In this work, an approach was taken and a model built by integrating static analysis of permissions and behavioural attributes with a machine learning-based one for robust malware detection. We used a feature selection process for that purpose using a Gradient Boosting model so as to identify which features to choose and discard any feature having an importance score of zero. These fine features were then fed into a stacked ensemble model which consisted of three base classifiers, namely Logistic Regression, Random Forest, and LightGBM. Each base model gave its specific strengths, but their predictions were fed into a final logistic regression layer that produced an extremely accurate blended model. We calculated the probability of being malicious using the stacked models predict_proba method so that it could be classified appropriately at the three risk levels-low, medium, and high. Validation of the proposed approach performance and real-world applicability in Android malware detection is possible using accuracy, precision, recall, and AUC-ROC metrics as shown in Fig. 1.

Input: Android application dataset containing permissions and behavioural attributes. Output: Risk classification (Low, Medium, High) for each application based on probability of maliciousness.

**Feature Importance Calculation with Gradient Boosting**

Input: Dataset with permissions and attributes; Gradient Boosting model. Output: List of critical features with non-zero importance scores.

1. **Procedure** FEATURE_SELECTION (GradientBoostingModel, Dataset)

**Table 1** Summary of key research studies on machine learning-based android malware detection

| S. No. | Objective | Methodology and model | Discussion and results | Accuracy and drawbacks | References |
|---|---|---|---|---|---|
| 1 | Evaluate ML classifiers for malware detection using dynamic permissions | Various ML classifiers on dynamic permissions extracted from applications. Techniques like Naïve Bayes, J48, Random Forest, etc | Simple Logistic is performing a bit better. Problems with scalability and data growth | Metrics are not discussed; issues with scalability | [1] |
| 2 | Provide an overview of detection methods using ML | Extensive review on detection based on ML. Focus will be on static and hybrid approach | Models: ANDRODET, TFDroid. Limitations of the ML model based on the growth of Android OS | 90% accuracy on the dataset of 500 apps. Issues with changing OS | [2] |
| 3 | To efficiently detect Android malware using ensemble machine learning algorithms | Ensemble learning approaches (LightGBM and Random Forest) with feature selection techniques such as RFE | Outperformed the traditional approaches. Focus was on feature selection | LightGBM achieved a 99.5% accuracy. Imbalance in predictions were problematic | [12] |
| 4 | Analyse machine learning-based Android malware detection techniques | Static, dynamic and hybrid analysis on features such as permissions and API calls using the models like Random Forest and SVM | Hybrid analysis offered better detection rate but at increased computational cost | Accuracy up to 99.9%. Older datasets and limitation of metrics | [11] |

(continued)

**Table 1** (continued)

| S. No. | Objective | Methodology and model | Discussion and results | Accuracy and drawbacks | References |
|---|---|---|---|---|---|
| 5 | To review machine learning techniques for Android malware detection | Static and dynamic analysis with ML techniques, such as Random Forest, SVM, and DroidFusion | DroidFusion achieved up to 99.8% true positive rates. Challenging the obfuscated malware detection was well highlighted | Signature-based methods were found ineffective against obfuscation | [13] |
| 6 | Systematically review machine learning based Android malware detection techniques | Involves static, dynamic, and hybrid analysis of permissions, API calls, and system calls. Uses Random Forest, SVM, Gradient Boosting, and deep learning models | High accuracy models, with some reaching up to 99.36%. Hybrid approaches outperform static and dynamic methods, highlighting the necessity for updated datasets and enhanced integration of machine learning techniques | The best model achieved 99.36% accuracy. Use of outdated datasets and limited real-time detection capabilities, with resource-heavy models like CNNs underutilized due to computational costs | [10] |

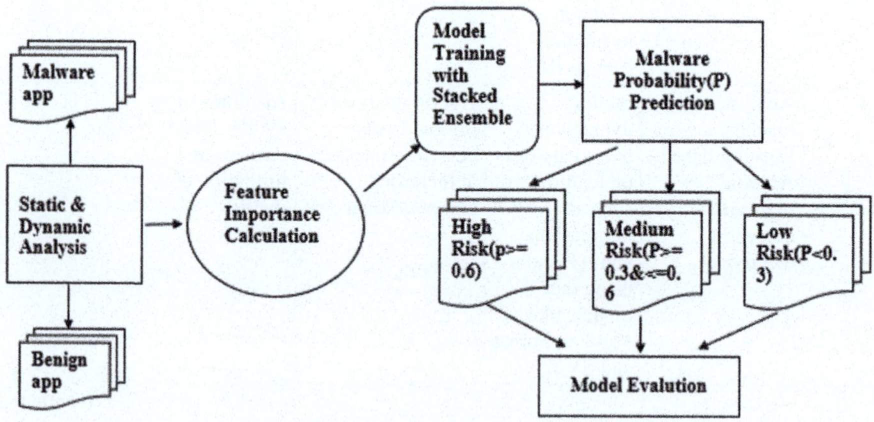


**Fig. 1** Workflow of the android malware detection system, detailing the processes from static and dynamic analysis to malware probability prediction and risk classification

```
2. Train GradientBoostingModel on Dataset to compute feature importance
   scores
3. DFImp ← Convert feature importance scores to DataFrame with attributes
   and their scores
4. Initialize an empty list DeletedFeatureList
5. for each feature in DFImp do
6.       if DFImp[feature].importance == 0 then
7.           Append feature to DeletedFeatureList
8.       end if
9. end for
10.    CriticalFeatures ← DFImp - DeletedFeatureList
11.    return CriticalFeatures
12. end procedure
```

Description: This step begins by training a Gradient Boosting model to compute feature importance scores for permissions and behavioral attributes. These scores are stored in a DataFrame ('DFImp'). Features with zero importance are added to 'DeletedFeatureList' and removed, resulting in 'CriticalFeatures' used for further processing.

**Model Training with Stacked Ensemble**

Input: Dataset with selected critical features; Base models (Logistic Regression, Random Forest, LightGBM).

Output: Trained stacked model for malware prediction.

```
1. procedure STACKED_MODEL_TRAINING (CriticalFeatures, Dataset)
2.     Train BaseModels (LogisticRegression, RandomForest, LightGBM)
   on CriticalFeatures
3.         MetaDataset ← Generate predictions from each base model on
   CriticalFeature
4.     StackedModel ← Train(LogisticRegression, MetaDataset)
5.     return StackedModel
6. end procedure
```

Description: The critical features are used to train three base models: Logistic Regression, Random Forest, and LightGBM. Each base model generates predictions, forming the 'MetaDataset'. This meta-dataset is used to train a logistic regression model, creating a stacked ensemble model that leverages each base model's unique strengths.

**Malware Probability Prediction**

Input: Test dataset with critical features; Trained StackedModel. Output: Malware probability for each application.

```
procedure MALWARE_PROBABILITY_PREDICTION (StackedModel, TestData)
    MalwareProbability ← StackedModel.predict_proba(TestData)[:, 1]
    return MalwareProbability
end procedure
```

Description: The stacked model predicts the probability that each application in the test dataset is malware. This probability offers a quantifiable measure of potential risk for each app.

**Risk Level Classification**

Input: Malware probability for each application; Predefined risk thresholds (low, medium, high).

Output: Risk level (Low, Medium, High) assigned to each application.

```
procedure RISK_LEVEL_CLASSIFICATION (MalwareProbability, Thresholds)
    Initialize an empty list RiskLevels
    for each probability in MalwareProbability do
        if probability < Thresholds['low'] then
            Append "Low Risk" to RiskLevels
        else if probability < Thresholds['medium'] then
            Append "Medium Risk" to RiskLevels
        Else
            Append "High Risk" to RiskLevels
        end if
    end for
    return RiskLevels
end procedure
```

Description: Based on malware probability scores, applications are classified into three risk levels: Low, Medium, and High. This allows security resources to be allocated efficiently based on threat level.

**Model Evaluation**

Input: Predicted risk levels and actual labels in the test dataset.

Output: Model performance metrics (accuracy, precision, recall, F1-score, AUCROC).

```
1. procedure MODEL_EVALUATION (Predictions, TrueLabels)
2.     Metrics ← Evaluate(Predictions, TrueLabels, metrics = ["accuracy",
   "precision", "recall", "AUC-ROC"]
3.     return Metrics
4. end procedure
```

Description: The model's accuracy and effectiveness in detecting malware are measured using standard evaluation metrics. This final step ensures the model's reliability and suitability for practical application in malware detection.

## 4 Results and Analysis

This was achieved in a stacking ensemble of Logistic Regression, Random Forest, and SVM model built with Python version 3.9: A custom decision threshold of 0.44 was applied to balance between precision and recall. This was aimed at minimizing false positives while maximizing true positives.

**Ensemble Stacking Forms**

The stacking ensemble model invokes the predictions of multiple base classifiers to arrive at an improved final prediction. In essence, the predictions of the stacking model can be expressed as follows:

$$\hat{y} = \text{MetaModel}(\alpha.\hat{y}\text{LR} + \beta.\hat{y}\text{RF} + \gamma.\hat{y}\text{SVM})$$

where:

- $\hat{y}$LR, $\hat{y}$RF and $\hat{y}$SVM represent the predictions from Logistic Regression, Random Forests, and SVM models, respectively.
- $\alpha$, $\beta$ and $\gamma$ represent the weights placed on the predictions of each model based on the relative accuracy and contribution of each to the ensemble.
- MetaModel is a logistic regression model trained on base models' outputs aimed at generating the final prediction $\hat{y}$. In this way, the proposed results were helped to gain the best output by utilizing the strengths in the base classifiers, thereby achieving a good balance in each of the metrics.

**Model Performance**:

The stacking model shows a decent classification performance as summarized below:

- Accuracy: 84.2%—The model achieved high accuracy on a large dataset of 30,000 applications, demonstrating its robustness and effectiveness in identifying malware with precise classification across high, medium, and low-risk categories.
- Precision: 82.48%—The model displayed reasonably good precision with respect to reducing the false positive rate among detected malicious applications.
- Recall: 96.93%—as Shown in Fig. 2 The high recall indicated the model's ability to correctly identify a substantial number of true malware cases, which is recognized as paramount in applications where false negatives are to be minimized.
- ROC-AUC Score: 0.92—This high ROC-AUC score implies considerable discrimination ability by the model across various threshold values, demonstrating the model's flexibility with respect to changes between the classification thresholds (as shown in Fig. 5).

**Risk Assessment Using Malware Probabilities**: The risk level of malware is classified as Low, Medium, or High based on the stacking model's probability scores for added practical usability. This enables the ranking of responses based on the scale of the threat.

- Low Risk: Probability $< 0.3$
- Medium Risk: $0.3 \leq$ Probability $< 0.6$
- High Risk: Probability $\geq 0.6$.

As Shown in Figs. 3 and 4 most apps are classified as high-risk, with fewer in the low- and medium-risk categories. Using a customized threshold with the stacking ensemble significantly improves malware detection rates. To solidify the claims of the proposed method, we compared it with existing state-of-the-art classifiers, such

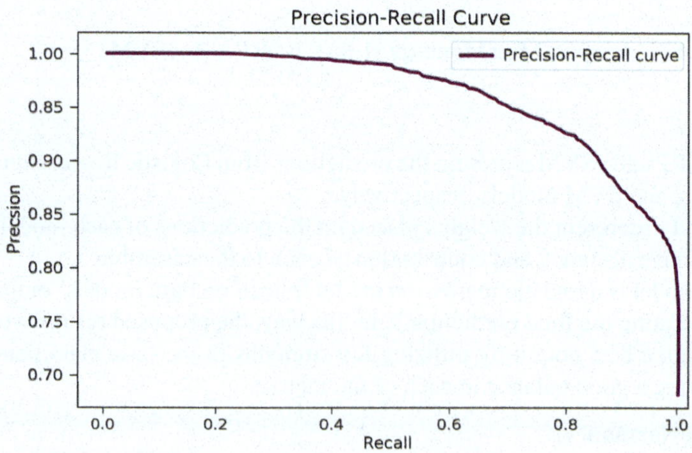


**Fig. 2** A precision-recall curve describing how the model manages to perform extremely well while having a strong trade-off between precision and recall

as XGBoost and LightGBM, using the same data as shown in Table 2. The outcomes show that the proposed model of stacking achieves a maximum improvement of recall by + 4% and ROC-AUC by + 3% of system security based on 30,000 app samples as shown in Fig. 6 (Fig. 5).

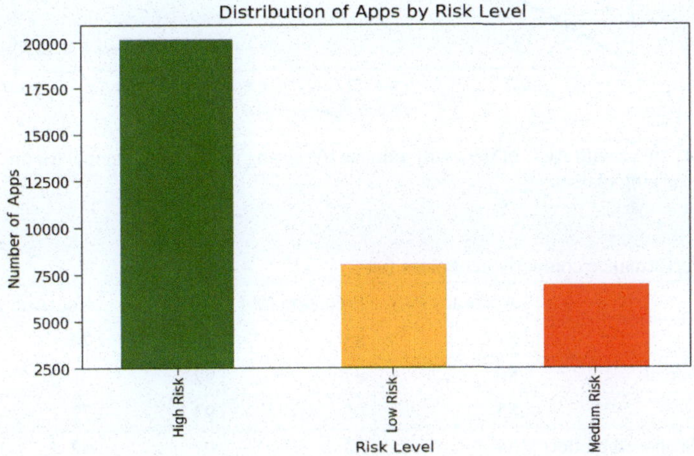


**Fig. 3** Distribution of applications based on risks relative to high, medium, and low being most abundant

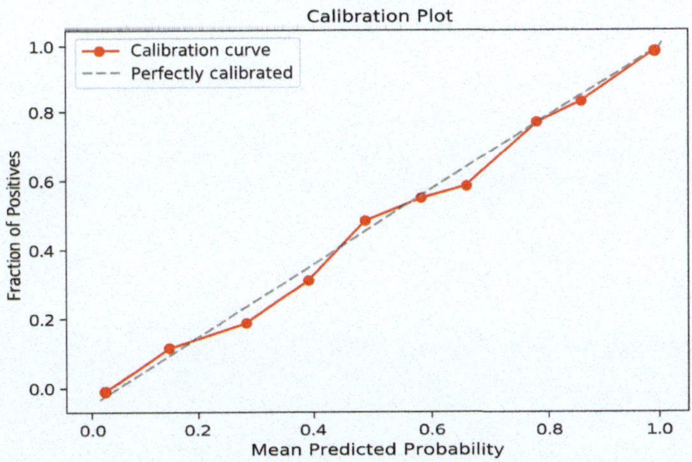


**Fig. 4** Calibration plot showing well-aligned predicted probabilities matching the actual outcome for the model

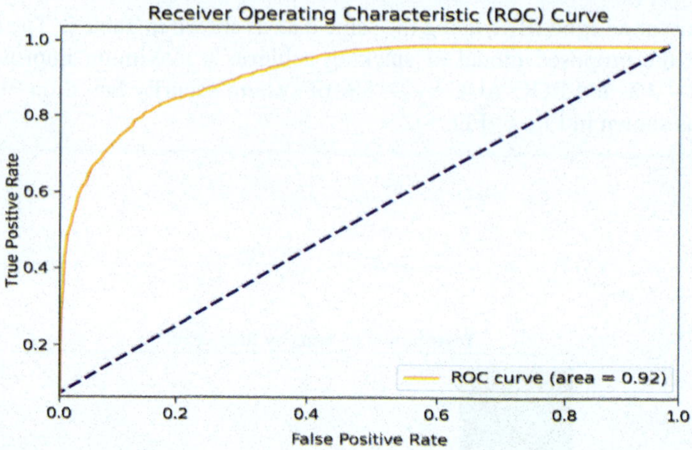


**Fig. 5** ROC curve with AUC of 0.92 will indicate the strong ability of the model for discrimination among the varying classes

**Table 2** Performance comparison of classifiers

| Model | Accuracy (%) | Precision (%) | Recall (%) | Roc-Auc score (%) |
|---|---|---|---|---|
| SVM | 84 | 85 | 93 | 79 |
| Random Forest | 86 | 87 | 92 | 81 |
| MLP | 85 | 85 | 93 | 79 |
| Stacked (Proposed model) | 84 | 83 | 97 | 92 |

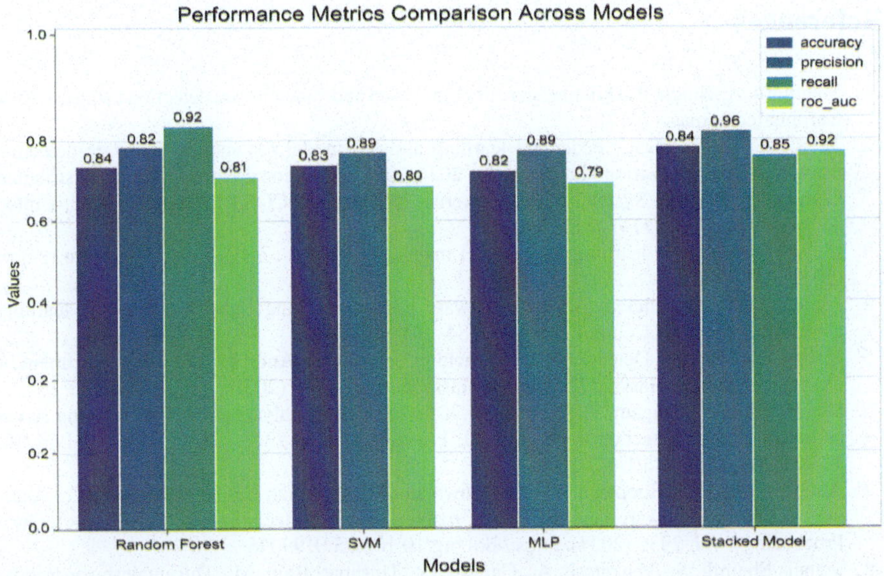


**Fig. 6** Comparison of accuracy, precision, recall, and ROC-AUC scores across different models including the stacked model

# 5 Conclusion

The ensemble stacking from Logistic Regression, Random Forest, and SVM works well and has a detection accuracy of 84.2%, recall rate of 96.93%, and ROC AUC scores of 0.92 on Android malware. While accuracy is some percentage points lower than other models that were tested on much smaller datasets, such as 500 samples of apps, it does very well in recall and ROC AUC measures, indicating a strong ability to identify true malware cases and distinguish between malicious and benign applications.

A customized threshold that has been proposed for better detection in which recall becomes stabilized, the false positives are minimized, and the applications are categorized into low, medium, and high risk levels. The model was trained using a dataset set containing 30,000 samples of Android applications giving the model much more strength and scalability in comparison to the other models that were trained on datasets with only 500 applications.

It will bring accuracy into malware detection along with pragmatism and proper security enhancement in the Android environment. This system will tend to identify and mitigate malware threats effectively, considering real-world scenarios where large datasets would be encountered.

## References


1. Mahindru, A., Singh, P.: Dynamic permissions-based android malware detection using machine learning techniques (2017)
2. Arya, S., Peddoju, S.K.: Federated learning assisted model for android malware detection using gannet optimization algorithm. In: 2024 20th International Conference on Distributed Computing in Smart Systems and the Internet of Things (DCOSS-IoT), Abu Dhabi, United Arab Emirates, pp. 777–779 (2024)
3. Liu, Y., et al.: Deep learning for android malware detection: a systematic literature review (2022)
4. Alahy, et al.: An efficient Android malware prediction using Ensemble machine learning algorithms. Procedia Comput. Sci. **191**, 184–191 (2021)
5. Joseph, L.: Malware Detection Using Machine Learning in Android Devices. Department of Computer Science, Harvard University, Massachusetts, USA (2021)
6. Srinivasan, R., Karpagam, S., Kavitha, M., Kavitha, R.: An analysis of machine learning-based android malware detection approaches. J. Phys. Conf. Ser. 2325 (2022). https://doi.org/10.1088/1742-6596/2325/1/012058
7. Abrar Ahmed, K., Boopalan, K., Lokeshwaran, K., Sugumar, R., Kotteeswaran, C.: Analysis of android malware detection using machine learning techniques. In: AIP Conference Proceedings, vol. 2935 (2024). https://doi.org/10.1063/5.0199036
8. Francis Shamili, S., Venkatesh, A., Ganesh, G., Thirupati Raju, M., Thirumaleshwar Reddy, M., Manoj Kumar Reddy, R.: Android Malware Detection Using Machine Learning. Department of Computer Science and Engineering, Dhanalakshmi Srinivasan Engineering College, Perambalur (2024)
9. Shatnawi, A.S., Jaradat, A., Yaseen, T.B., Taqieddin, E., Al-Ayyoub, M., Mustafa, D.: An android malware detection leveraging machine learning, first published on 6 May 2022. https://doi.org/10.1155/2022/1830201
10. Mathur, A., et al.: Android malware detection architectures using machine learning (2022)
11. Meijin, L., et al.: A systematic overview of android malware detection. Appl. Artif. Intell. 34 (2022)
12. Muzaffar, A., et al.: An in-depth review of machine learning-based Android malware detection. Comput. Secur. **121**, 21 (2022)
13. Hein, Myo: Permission-based android malware detection using machine learning techniques. A review, p. 30 (2020)

# Affective DDoS Attack Detection in EV Charging Station Using DL Autoencoder Model

**Bhupendra Sahu, Mithilesh Atulkar, and Ravindra Kumar Chouhan**

**Abstract** Distributed Denial of Service (DDoS) attacks are increasingly targeting advanced grid-based electric vehicle (EV) charging stations, posing serious risks to service availability and grid stability. To address these challenges, this study evaluates the effectiveness of machine learning and deep learning approaches using the CICEV2023 dataset, which contains 126,568 samples across four attack scenarios (e.g., Wrong EV, Wrong Timestamp) and multiple attack types (e.g., TCP flood, SYN flood, cryptojacking). This study compares the performance of the Light Gradient Boosting Machine (LightGBM) and a proposed Deep Neural Network (DNN) enhanced with autoencoder-based feature extraction. The results demonstrate that the DNN model significantly outperforms LightGBM, achieving 87.01% accuracy, 87.65% precision, 87.3% recall, and 88.8% F1-score. These findings highlight the potential of autoencoder-assisted deep learning for detecting sophisticated DDoS threats in EV charging infrastructures. Future work will explore scalability, computational efficiency, and integration with real-world charging networks.

**Keywords** DDoS · EV charging stations · CICEV2023 · Autoencoder · LightGBM · Deep Neural Network

## 1 Introduction

There has been a significant increase in electric vehicles (EVs) in recent years. However, this growth has also increased security threats as malicious users and hackers actively seek to exploit electric vehicle charging stations [1]. In [2, 3], integration of EVs charging stations with smart grids, IoT ecosystems, and edge computing environments has brought numerous benefits, including real-time monitoring, dynamic pricing, and load balancing capabilities. Nevertheless, this integration also poses significant security risks, as it creates potential entry points for DDoS attacks and Software Defined Networking (SDN) vulnerabilities, which can affect the stability

---

B. Sahu (✉) · M. Atulkar · R. K. Chouhan
National Institute of Technology, Raipur, India
e-mail: bsahu.phd2023.mca@nitrr.ac.in

© The Author(s), under exclusive license to Springer Nature Singapore Pte Ltd. 2026 
M. Pant et al. (eds.), *Real-World Applications of Machine Learning*, Mathematics for Sustainable Developments, https://doi.org/10.1007/978-981-95-7134-5_9

and performance of EVs charging systems. SDN is an emerging technology that offers resilient scalability, flexibility, and network programmability, enhancing the network environment. However, its centralized controller architecture has made it an attractive target for cyber attackers, posing significant security risks. The centralized control SDN provides over the network connecting EV charging stations enables dynamic traffic management, efficient routing, and load balancing, which is essential for effectively managing the variable and dynamic demand posed by EVs. In paper [4], DDoS attacks threaten EV charging stations, potentially causing disruptions, overloading, and compromising edge devices. SDN controllers are also vulnerable to DDoS attacks, which could lead to broader network disruptions.

This research paper analyzes the performance comparison of attacks from the CICEV2023 dataset. This research employs a neural network architecture coupled with autoencoder algorithms to perform feature extraction on the CICEV2023 dataset, yielding improved output.

The structure of the paper is as follows. Section 2 delves into related work on EV authentication and provides insights into the dataset used in this study. The purpose of Sect. 3 is to motivate the reader by persuading them that the survey is essential and valuable. At the same time, Sect. 4 conducts a thorough analysis of the materials and methods, instilling confidence in the research process. Section 5 suggests feature engineering methods, including the technique's objective and technical components with methodologies. Section 6 summarizes the performance combination results and recommends the best performance set by performance metrics. Section 7 summarizes and concludes with recommendations for further research.

## 2 Related Works

Machine learning models were used to detect DDoS assaults on SDN infrastructure in a research article [5]. To create an extensive dataset, the researchers collected information from regular and DDoS-affected SDN data traffic. They then took the original dataset and used feature selection algorithms to create a new one. Various classification algorithms, including KNN, NB, SVM, and ANN, were used to train and evaluate both datasets, demonstrating the research's methodological diversity.

The study's author by [6] examined machine learning methods for identifying Botnet DDoS attacks. Using the UNBS-NB 15 and KDD99 datasets, the study assessed the performance of several algorithms, including SVM, ANN, NB, DT, and USML. According to the results, USML had the best accuracy rate (94.78%) in differentiating between regular network traffic and botnet activity, suggesting that it could be a helpful detection technique.

In a study by [7], a new strategy is presented for distinguishing between high-volume and low-volume DDoS attacks using statistical and machine learning techniques. The proposed methodology consists of three main components: categorization, entropy-based, and collector. However, the researchers found that the entropy-based approach with static thresholds could have yielded more satisfactory results

**Table 1** Summary of related works on DDoS detection in EV/SDN environments

| S. no. | Authors | Year | Models used | Results | Limitations |
|---|---|---|---|---|---|
| 1 | Tuan et al. [6] | 2020 | SVM, ANN, NB, DT, USML | USML achieved 94.78% accuracy | Dataset outdated; not EV-specific |
| 2 | Wei et al. [8] | 2021 | AE + MLP hybrid | High detection accuracy across attack classes | Computationally expensive; lacks EV validation |
| 3 | Kim et al. [9] | 2023 | Benchmark dataset (CICEV2023) | Four attack types with realistic EV data | Imbalanced; needs advanced feature extraction |

across different datasets. Although the dynamic threshold approach showed potential, it still resulted in a significant false positive rate. The proposed method demonstrates improved accuracy compared to similar approaches, indicating its potential for effective DDoS attack detection. As summarized in Table 1, several recent studies have examined DDoS detection using ML and DL models in SDN or EV environments, highlighting diverse approaches, datasets, and limitations.

**Inclusion and Exclusion Criteria**

To ensure a structured and relevant review of the existing literature, the following criteria were applied:

**Inclusion Criteria**

- Studies focusing on DDoS attack detection in electric vehicle (EV) infrastructure, Software Defined Networking (SDN), or Internet of Things (IoT).
- Works employing machine learning (ML) or deep learning (DL) models for attack detection.
- Research papers that report measurable evaluation metrics such as accuracy, precision, recall, or F1-score.

**Exclusion Criteria**

- Papers limited to theoretical discussions without experimental validation.
- Studies lacking performance evaluation or benchmark comparison.
- Works unrelated to EV charging infrastructure or SDN-based cyberattacks.

## 3 Motivation

In their groundbreaking [8] paper, introduced AE-MLP, a revolutionary hybrid deep learning approach designed to identify DDoS attacks. This technique uses network traffic data and applies an Autoencoder (AE) to extract crucial features. These features are then utilized by a Multilayer Perceptron (MLP) network to categorize various types of DDoS attacks effectively. This innovative method can potentially enhance cybersecurity measures and significantly protect against malicious online threats. In this study, Kim et al. [10] present a method for selecting the best feature combination, considering factors such as F1-score, feature size, high sampling resolution, low information loss rate, and feature downsizing to minimize information loss and optimize feature selection. However, this study provides three key elements relevant to examining Grid Servers and Charging Stations in each scenario.

# 4 Material and Method

This section provides an engaging summary of the dataset and the environment.

## *4.1 Other Available Datasets*

This section offers a compelling analysis of the shortcomings and deficiencies present in current datasets. Table 2 shows the reference with an overview of other datasets.

The ACN dataset [11] contains valuable charging history data, including charging time, charge fee, location, and customer ID for a specific period. Moreover, it needs to provide crucial information about the charging infrastructure's resilience against DoS or DDoS attacks. The dataset by [12] offers intriguing insights into vehicle charging rates, power consumption, location, and date. Yet, it disappointingly lacks data related to EV authentication and the impact of DoS or DDoS attacks on the charging infrastructure. In contrast, the dataset by [13] presents a comprehensive collection of compelling data on EV charging power, date, region name, and other relevant details for US and UK locations. Information on 3395 charging sessions for 85 EVs across 105 US charging stations (CS) may be found in the dataset [14]. Still, its scope is limited, leaving out crucial information on cybersecurity threats.

**Table 2** Overview of existing EV charging datasets

| References | Charging history | EV authenticity | DDoS availability | Year |
|---|---|---|---|---|
| [11] | ✓ | × | × | 2019 |
| [12] | ✓ | × | × | 2022 |
| [13] | ✓ | × | × | 2021 |
| [14] | ✓ | × | × | 2022 |

**Table 3** Attack distribution in the CICEV2023 dataset

| Attack type | Count |
|---|---|
| Vulnerability scan | 26,246 |
| TCP flood | 32,637 |
| Backdoor | 21,137 |
| Normal traffic | 14,363 |
| SYN flood | 13,517 |
| Cryptojacking | 11,596 |
| SYN stealth | 7072 |
| Total | 126,568 |

## *4.2 CICEV2023 Dataset Overview*

This study uses the publicly available **CICEV2023 dataset** [9], which provides a large and varied collection of traffic records for DDoS attacks and normal operating conditions in EV charging stations. The dataset contains a total of 126,568 samples covering both benign traffic and multiple malicious scenarios. Specifically, it captures four distinct authentication-related attack scenarios within EV charging infrastructure:

- **Wrong EV ID**: An attacker uses an invalid EV identity to generate fraudulent authentication requests.
- **Wrong Charging Station (CS) Timestamp**: Timestamp mismatch attacks are used to bypass synchronization checks.
- **Correct EV ID**: A legitimate EV identity is misused in replay or spoofing attempts.
- **Wrong EV Timestamp**: Incorrect or manipulated timestamps are injected during authentication.

The dataset includes several attack categories and their frequency distribution, as shown in Table 3. Notably, TCP flood and vulnerability scan attacks dominate the dataset, while others such as SYN-stealth and cryptojacking are less frequent. This imbalance poses a challenge for model training and requires feature extraction techniques such as autoencoders to ensure robust performance across minority classes.

## *4.3 Autoencoder*

An autoencoder is a specific type of neural network commonly used in deep learning for tasks involving feature extraction and dimensionality reduction. It works by compressing the input data into a smaller format and then using this compressed representation to recreate the original data. The autoencoder consists of two primary parts: the encoder, which reduces the input to a lower-dimensional latent space, and the decoder, which attempts to reconstruct the input from this reduced form. During training, the model is incentivized to capture the most important aspects of the data by minimizing the difference between the input and its reconstruction.

- In this research, autoencoders are highly relevant for dimensionality reduction and feature extraction, mainly when dealing with high-dimensional datasets like CICEV2023. By reducing the feature space, autoencoders help simplify the data while retaining the most essential information.

## *4.4 LightGBM Classifier*

The Light Gradient Boosting Machine (LightGBM) is a distributed, fast, and highly efficient gradient boosting system based on decision trees. It was designed by Microsoft to use memory and compute effectively. It is, therefore, perfect for managing enormous volumes of data. LightGBM works by building decision trees one after the other, with each new tree attempting to fix the mistakes made by the ones before it. Minimizing the loss function through gradient boosting techniques produces successive trees until the ideal model is reached. The LGBM classifier is high performance in classification tasks, making it a popular choice in various machine learning applications.

## *4.5 Deep Learning Techniques*

Artificial neural networks are used in deep learning, a subset of machine learning, to perform complex tasks. It employs algorithms to acquire hierarchical representations from a dataset and autonomously solve complex problems. Artificial neural network topologies with numerous interconnected layers comprise deep learning models, which allow for more abstract input data. Deep learning's ability to handle large datasets and challenging issues is one of its unique selling points. Back-propagation algorithms are commonly used for this process. A feedforward neural network transfers data in a straight line, without loops or circles, from input to output. This strategy is slightly outdated and underdeveloped [15, 16].

## *4.6 DDoS Attacks*

Distributed Denial of Service (DDoS) attacks refer to a type of cyberattack where multiple systems flood the target(such as an electric vehicle communication system) with overwhelming traffic, leading to system resource exhaustion. DDoS attacks typically involve numerous compromised devices (often called a botnet), simultaneously sending large volumes of malicious traffic to the target. This coordinated assault makes DDoS attacks more potent than traditional Denial of Service (DoS) attacks, as the attack traffic originates from multiple sources, making it more challenging to mitigate.

### *4.7 DDoS Intrusion Detection Systems*

DDoS Standard Intrusion Detection Systems (IDS) are usually analyzed into anomaly-based and signature-based detection [17, 18]. IDSs are security measures designed to detect and mitigate DDoS attacks. These IDSs are essential for spotting unusual traffic patterns and warning systems about possible threats for authorized users before the service is interrupted.

## 5 Proposed Method

This section explains the architecture and methodology of the proposed model, which combines machine learning techniques for classification with deep learning for feature extraction.

### *5.1 Method Objectives*

This section outlines the methodology used in this research. According to Sect. 4.2, the dataset CICEV2023 implements the methods described below.

### *5.2 Methodology*

This section thoroughly discusses the main technical components and methodologies that underpin the proposed model. The model combines deep learning and machine learning techniques to effectively address the challenges of detecting cyberattacks in electric vehicle communication systems. Key aspects of this approach include feature extraction, and classification techniques, all of which are crucial for achieving high accuracy and performance in identifying Distributed Denial of Service (DDoS) and other cyber threats.

The following provides a comprehensive technical background description of the research model design and methodologies:

- **Data Load and Preprocess**: Handling missing values depends on the nature of the data and the amount of missingness.
  - **Missing Value Handling**: The initial step involves addressing any missing data points within the dataset.
  - **Normalization**: The StandardScaler is used to normalize the numerical features.
  - **One-Hot Encoding**: Categorical features are encoded into binary vectors using one-hot encoding, allowing the models to interpret non-numeric data.
- **Feature Extraction**: From the preprocessed data, an autoencoder is used to extract significant, compressed features. The autoencoder includes the following:
  - **Encoder**: The input data is transformed to a lower-dimensional latent space representation, reducing its dimensionality.
  - **Decoder**: Reconstructs the input data from the compressed representation.
  - **Use Case in the Model**: This code uses the autoencoder for dimensionality reduction and feature extraction. The encoder converts high-dimensional input data into a lower-dimensional space while retaining essential features. This compressed data is then used for subsequent classification tasks.
  - **Training**: Mean Squared Error (MSE) is used as the loss function to train the model. Early stopping is applied to prevent overfitting during training.
- **Compressed Features**: Once the autoencoder is trained, the compressed features are extracted and passed as inputs into both the DNN and the LightGBM classifier for downstream tasks.
- **Train LightGBM Classifier**: After feature extraction, the LightGBM classifier is utilized as a final layer to perform precise classification based on the transformed features. This helps exploit the advantages of gradient boosting while handling the complexity of large datasets efficiently.
  - **Gradient Boosting**: The LightGBM model uses gradient boosting for classification, optimizing performance by focusing on the most difficult-to-predict samples in subsequent iterations.
  - **Regularization**: The code uses both L1 (lasso) and L2 (ridge) regularization to prevent overfitting and improve generalization.
  - **Hyperparameters**: Parameters like num_leaves = 31 control the number of leaves in the decision tree, and min_data_in_leaf = 20 limits the size of leaf nodes to avoid overfitting.
- **Deep Neural Network (DNN)**: In this, discuss the structure of the DNN, including the number of layers, neurons, and regularization techniques used (dropout, batch normalization). Emphasize how the DNN handles the latent features from the autoencoder to enhance the model's classification performance.

  - **Input Layer**: Receives the compressed features from the autoencoder.
  - **Hidden Layers**: Two hidden layers with relu activation are included, using dropout for regularization. Dropout randomly deactivates a portion of the neurons in each training iteration to prevent overfitting.
  - **Output Layer**: A single neuron with a sigmoid activation function is used to output the probability of the binary classification target (1 for benign, 0 for attack).

- **Training and Optimization**: The model uses binary cross-entropy as the loss function and the Adam optimizer for training. The same early stopping technique is applied here to prevent. Both models (DNN and LightGBM) are trained on the compressed features, and the hyperparameters are optimized to achieve the best performance based on metrics such as accuracy, precision, recall, and F1-score. overfitting.

The Keras [19] library's techniques were used to configure the models, which were written in Python [20]. The make_model function, and established parameters and optimizers generated each model. The diagram of a methodology for the model construction function utilized in this investigation is displayed in Fig. 1.

## 6 Results and Discussion

This section describes deep learning and machine learning research output, showcasing their ability to achieve the best results.

### *6.1 Performance of the LightGBM Classifier and Deep Neural Network*

After extracting the compressed features from the autoencoder, the LightGBM classifier and DNN are trained and evaluated on the validation set. LightGBM performance metrics are defined in Table 4, the same as DNN, producing the results shown in Table 5.

### *6.2 Comparison Between LGBM and DNN*

The evaluation of both the LightGBM and Deep Neural Network (DNN) classifiers demonstrated significant differences in performance. Both the LGBM classifier and the DNN yielded competitive results, with similar values for accuracy, precision, recall, and F1-scores. Graphs clearly show the comparison.

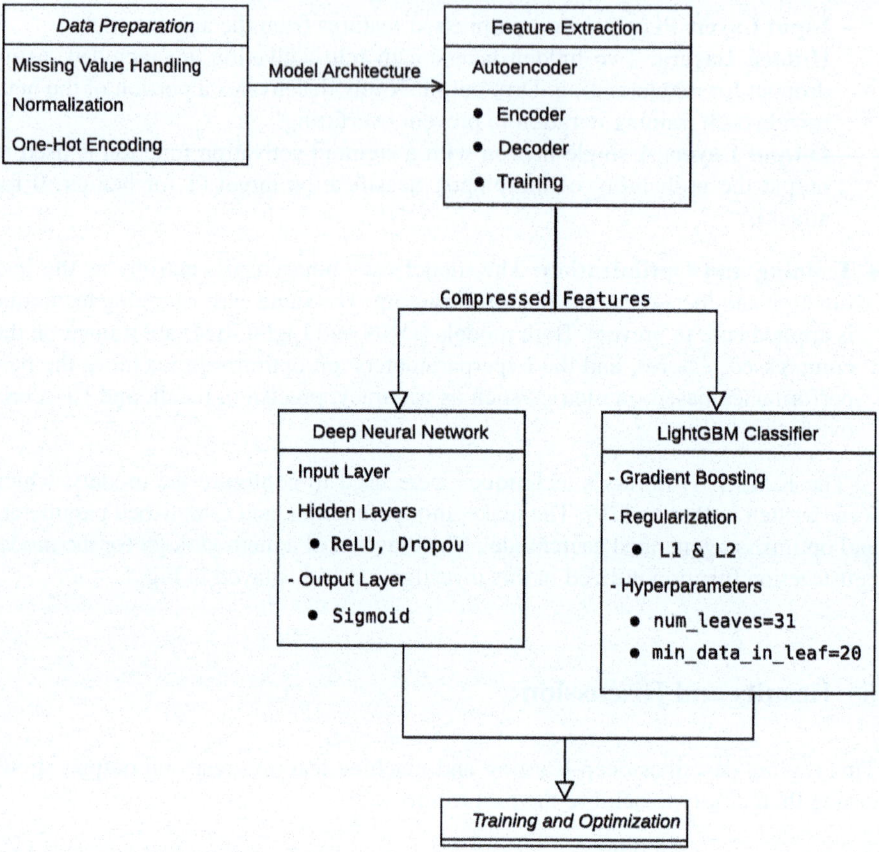


**Fig. 1** Diagram of methodology

**Table 4** Performance metrics of the LightGBM classifier in percentages

| Metric | Score |
|---|---|
| Accuracy | 70.6 |
| Precision | 70.65 |
| Recall | 70.4 |
| F1-score | 70.8 |

**Table 5** Performance metrics of the Deep Neural Network in percentages

| Metric | Score |
|---|---|
| Accuracy | 87.01 |
| Precision | 87.65 |
| Recall | 87.3 |
| F1-score | 88.8 |

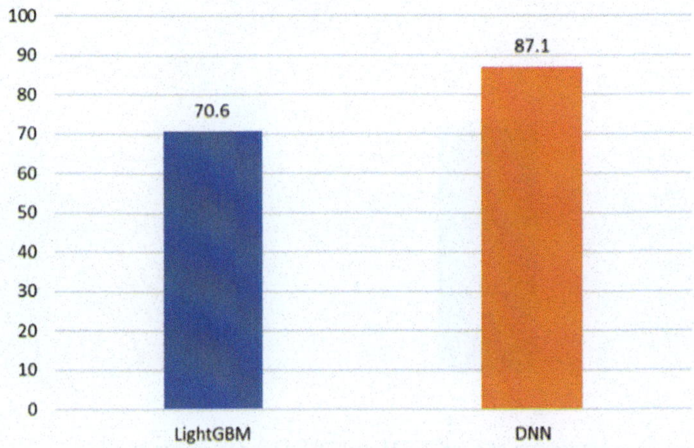


**Fig. 2** Percentage comparison of LightGBM and DNN models under accuracy

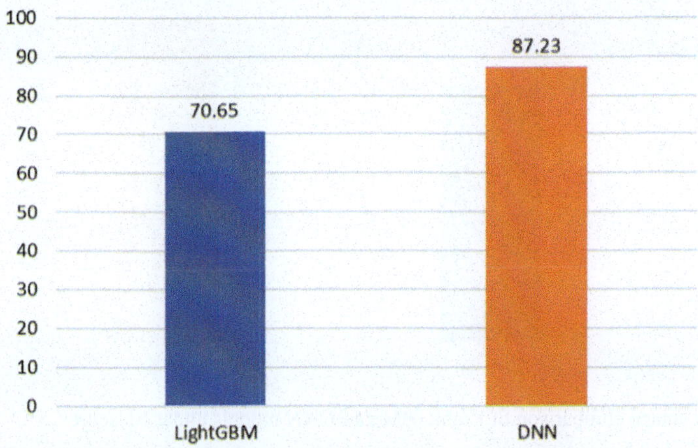


**Fig. 3** Percentage comparison of LightGBM and DNN models under precision

- **Accuracy**: The graph in Fig. 2 illustrates that, the LightGBM classifier achieved an accuracy of 70.6%, indicating moderate success in detecting anomalies within the dataset. On the other hand, the DNN classifier exhibited a notably higher accuracy of 87.1%, reflecting its stronger ability to classify network traffic
- **Precision**: In the graph in Fig. 3, the precision results show performance. The LightGBM model achieved a precision of 70.65%, indicating a reasonable ability to correctly identify positive samples (i.e., anomalies) while minimizing false positives. In contrast, the DNN model significantly outperformed LightGBM with a precision of 87.23%.
- **Recall**: Graph in Fig. 4, the recall results highlight the performance difference between the LightGBM and DNN classifiers in detecting anomalies. The Light-

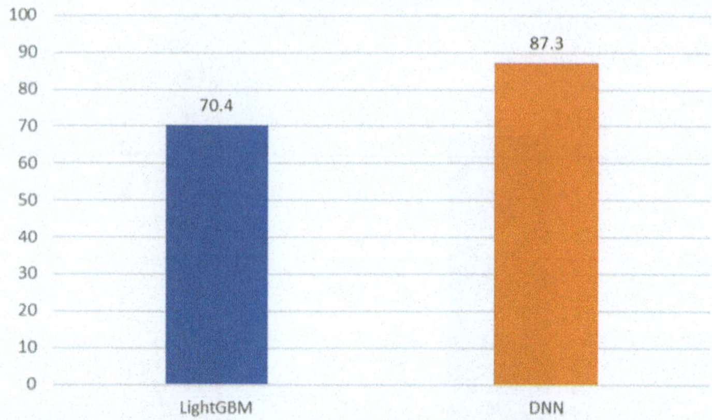


**Fig. 4** Percentage comparison of LightGBM and DNN models under recall

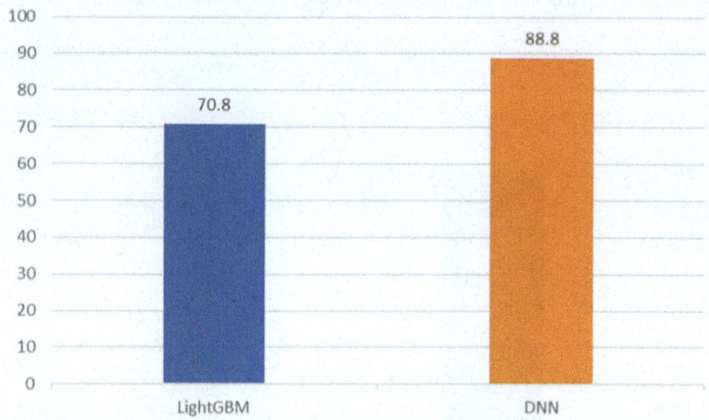


**Fig. 5** Percentage comparison of LightGBM and DNN models under F1-score

GBM model achieved a recall of 70.4%. In contrast, the DNN model demonstrated a much higher recall of 87.3%, showing its effectiveness in capturing almost all of the actual anomalous instances.

- **F1-Score**: The F1-score in graph Fig. 5, which balances precision and recall, provides a clear comparison between the LightGBM and DNN classifiers. The LightGBM model achieved an F1-score of 70.8%, indicating a reasonable balance between identifying true positives and avoiding false positives or negatives. However, the DNN model significantly outperformed LightGBM, with an F1-score of 88.8%.

The comparison shows a clear trade-off between efficiency and accuracy. LightGBM is faster and lightweight, but the DNN achieves higher performance (Accuracy: 87.01%, F1-Score: 88.8%) by capturing complex non-linear patterns. The autoencoder further boosts accuracy and reduces feature dimensionality, enhancing scalability for large EV datasets.

## 7 Conclusion and Future Work

This paper concludes by suggesting a technique for DDOS attack detection using the CICEV2023 dataset. The autoencoder played a significant role in the success of both classifiers. By reducing the dimensionality of the input data, it enabled both the LGBM and DNN models to focus on the most salient features, discarding redundant information. This not only improved the computational efficiency but also enhanced the models' performance by reducing the risk of overfitting. This study compares the effectiveness of deep learning DNN and machine learning LGBM classifiers. The results conclusively show that the DNN is outperforming all performance metrics, including F1-score, accuracy, precision, and recall. So the results indicate that the Deep Neural Network, while computationally more intensive, exhibited a strong ability to learn non-linear relationships between the features.

Future research could explore extending the proposed framework in several directions. One promising avenue is the integration of **federated learning** to enable distributed attack detection across geographically dispersed EV charging stations without centralizing sensitive data. Another direction is the use of transfer learning to adapt the model for related IoT and smart grid datasets, improving its generalizability. Finally, evaluating real-time deployment within SDN-enabled EV charging infrastructures would provide practical insights into latency, throughput, and robustness under dynamic network conditions.

## References


1. Pratt, R.M., Carroll, T.E.: Vehicle charging infrastructure security. In: 2019 IEEE International Conference on Consumer Electronics, ICCE 2019 (2019)
2. Kim, Y., Hakak, S., Ghorbani, A.: Smart grid security: attacks and defence techniques, pp. 103–123 (2023)
3. Wu, Y., Wang, Z., Huangfu, Y., Ravey, A., Chrenko, D., Gao, F.: Hierarchical operation of electric vehicle charging station in smart grid integration applications—an overview (2022)
4. Sayed, M.A., Atallah, R., Assi, C., Debbabi, M.: Electric vehicle attack impact on power grid operation. Int. J. Electr. Power Energy Syst. **137** (2022)
5. Polat, H., Polat, O., Cetin, A.: Detecting DDoS attacks in software-defined networks through feature selection methods and machine learning models. Sustainability (Switzerland) **12**(3) (2020)

6. Tuan, T.A., Long, H.V., Son, L.H., Kumar, R., Priyadarshini, I., Son, N.T.K.: Performance evaluation of Botnet DDoS attack detection using machine learning. Evol. Intell. **13**(2), 283–294 (2020)
7. Banitalebi Dehkordi, A., Soltanaghaei, M.R., Boroujeni, F.Z.: The DDoS attacks detection through machine learning and statistical methods in SDN. J. Supercomput. **77**(3), 2383–2415 (2021)
8. Wei, Y., Jang-Jaccard, J., Sabrina, F., Singh, A., Xu, W., Camtepe, S.: AE-MLP: a hybrid deep learning approach for DDoS detection and classification. IEEE Access **9**, 146810–146821 (2021)
9. Kim, Y., Hakak, S., Ghorbani, A.: DDoS attack dataset (CICEV2023) against EV authentication in charging infrastructure. In: 2023 20th Annual International Conference on Privacy, p. 2023. PST, Security and Trust (2023)
10. Kim, Y., Hakak, S., Ghorbani, A.: Detecting distributed denial-of-service (DDoS) attacks that generate false authentications on electric vehicle (EV) charging infrastructure. Comput. Secur. **144**, 103989 (2024)
11. Lee, Z.J., Li, T., Low, S.H.: ACN-data: analysis and applications of an open EV charging dataset. In: e-Energy 2019—Proceedings of the 10th ACM International Conference on Future Energy Systems, pp. 139–149 (2019)
12. Asensio, O.I., Apablaza, C.Z., Lawson, M.C., Walsh, S.E.: A field experiment on workplace norms and electric vehicle charging etiquette. J. Ind. Ecol. **26**(1), 183–196 (2022)
13. Amara-Ouali, Y., Goude, Y., Massart, P., Poggi, J.M., Yan, H.: A review of electric vehicle load open data and models (2021)
14. Goo, Y.-H., Choi, S.-O., Lee, S.-K., Kim, S.-M., Kim, M.-S.: Tracking the source of cascading cyber attack traffic using network traffic analysis. J. Korean Inst. Commun. Inf. Sci. **41**(12), 1771–1779 (2016)
15. Eldan, R., Shamir, O.: The power of depth for feedforward neural networks. J. Mach. Learn. Res. **49**(June), 907–940 (2016)
16. Fine, T.L.: Feedforward neural network methodology. IEEE Trans. Neural Netw. **12**(3), 647–648 (2001)
17. Khraisat, A., Gondal, I., Vamplew, P., Kamruzzaman, J.: Survey of intrusion detection systems: techniques, datasets and challenges. Cybersecurity **2**(1) (2019)
18. Liao, H.J., Richard Lin, C.H., Lin, Y.C., Tung, K.Y.: Intrusion detection system: a comprehensive review, pp. 16–24 (2013)
19. Moolayil, J.: An introduction to deep learning and Keras. In: Learn Keras for Deep Neural Networks, pp. 1–16 (2019)
20. Patil, S.U., Patil, S.U.: A review on python the fastest growing programming language. Int. J. Multidiscip. Res. **6**(3) (2024)

# Federated Learning for Privacy-Preserving Next-Word Prediction on Mobile Devices

**Sandesh Buchkul, Ashish Sonawane, Anuj Khandelwal, Jayesh Borase, Gitanjali Yadav, and Ratna Patil**

**Abstract** Due to the increase in the use of mobile devices, the usability and the probability of input errors is improved via the use of predictive text and even next-word suggestions. However, the typical approaches used in next-word predictions require a lot of user data which causes problems arising from privacy concerns. This article proposes a method of federated learning (FL) for next-word prediction that is carried out on the devices themselves so as to avoid centralizing any user data. The data from the Shakespeare dataset is used to demonstrate a federated training paradigm implemented over devices in a TFF environment. In this case, an existing pre-trained RNN model is fine-tuned to cater for the data from each client. Clients, instead of sending raw data, send encrypted model updates to a central aggregator. This helps to protect privacy while still working toward the improvement of the global model. As enhancements, tokenization, vocabulary lookups, and length constraints are implemented, since these are required for use in mobile devices which are limited in resources. The experiments back the usefulness of this FL-based approach for privacy-aware text prediction (Thakre et al., in Int J Innov Res Comput Commun Eng 11(10):11332–11343, 2023), as it does not lose significantly in prediction metrics. The research shows federated learning can deliver secure and customized experiences across a range of mobile devices. To improve adaptability and user satisfaction, future work will focus on increasing the model's context awareness and lexicon growth.

S. Buchkul (✉) · A. Sonawane · A. Khandelwal · J. Borase · G. Yadav · R. Patil
Department of Artificial Intelligence and Data Science, Vishwakarma Institute of Information Technology, Pune, India
e-mail: buchkulsandesh1@gmail.com

A. Sonawane
e-mail: ashishis416@gmail.com

A. Khandelwal
e-mail: anuj63293@gmail.com

J. Borase
e-mail: jayeshborase07@gmail.com

G. Yadav
e-mail: gitanjali.yadav@viit.ac.in

R. Patil
e-mail: ratna.patil@viit.ac.in

© The Author(s), under exclusive license to Springer Nature Singapore Pte Ltd. 2026 
M. Pant et al. (eds.), *Real-World Applications of Machine Learning*, Mathematics for Sustainable Developments, https://doi.org/10.1007/978-981-95-7134-5_10

**Keywords** Federated learning · Next-word prediction · Mobile devices · Privacy preservation · TensorFlow federated · Recurrent neural networks

## 1 Introduction

The massive growth in the development of mobile technology, especially in text-based applications such as virtual assistants and texting, has increased the necessity for intelligent applications that can provide personalized user experiences. Next-word prediction is a crucial part of these applications that aims to enhance the user experience by predicting the subsequent word in the string to increase typing speed and generally to facilitate communication. Because these traditional machine learning models rely on centralized data collection for training, there are issues concerning user privacy and security. The solution that has recently emerged to such issues is Federated Learning, which facilitates the decentralized training of models on user devices without having to share sensitive data.

In this research, we propose a unique Federated Learning-based next-word prediction method that prioritizes user privacy while offering tailored text prediction. To facilitate on-device training across a dispersed network of mobile devices, the suggested model makes use of TensorFlow Federated (TFF), an open-source framework created for FL simulations. Sensitive user data never leaves the device since we train the model locally using each client's data and aggregate the findings in a way that protects privacy.

Our approach is based on the architecture of Recurrent Neural Networks, which are particularly suited for sequence input, like text. We use the Shakespeare dataset, composed of lines from multiple Shakespearean plays, as a proxy to real text-based data. Our approach includes preprocessing of the dataset, creation of vocabulary, and model training using the principles of federated learning. The design of the system ensures its gradual improvement with each participating device, through improving next-word prediction accuracy and, above all, providing individualized outcomes without compromising privacy.

It captures the possibility of Federated Learning to enable privacy-centric machine learning on mobile devices by demonstrating its effectiveness in next-word prediction challenges. The core drawback of traditional machine learning is the reliance on centralized storage for data, which risks users' privacy. Federated Learning aggregates updates but does not transfer sensitive data; instead, they train models locally on consumers' devices. This approach makes it possible to have tailored text predictions while guaranteeing the privatization of personal information. FL permits continuous model refinement as more devices are added and increases the quality of forecasts. Besides opening up new venues for innovative safe customized mobile applications, this strategy also opens the avenue toward increased use of privacy-preserving AI applications in industries where data protection is a sensitive issue such as healthcare, banking, and tailored services. With this work, we show that FL can transform natural language processing and open the door to future AI solutions that are safer.

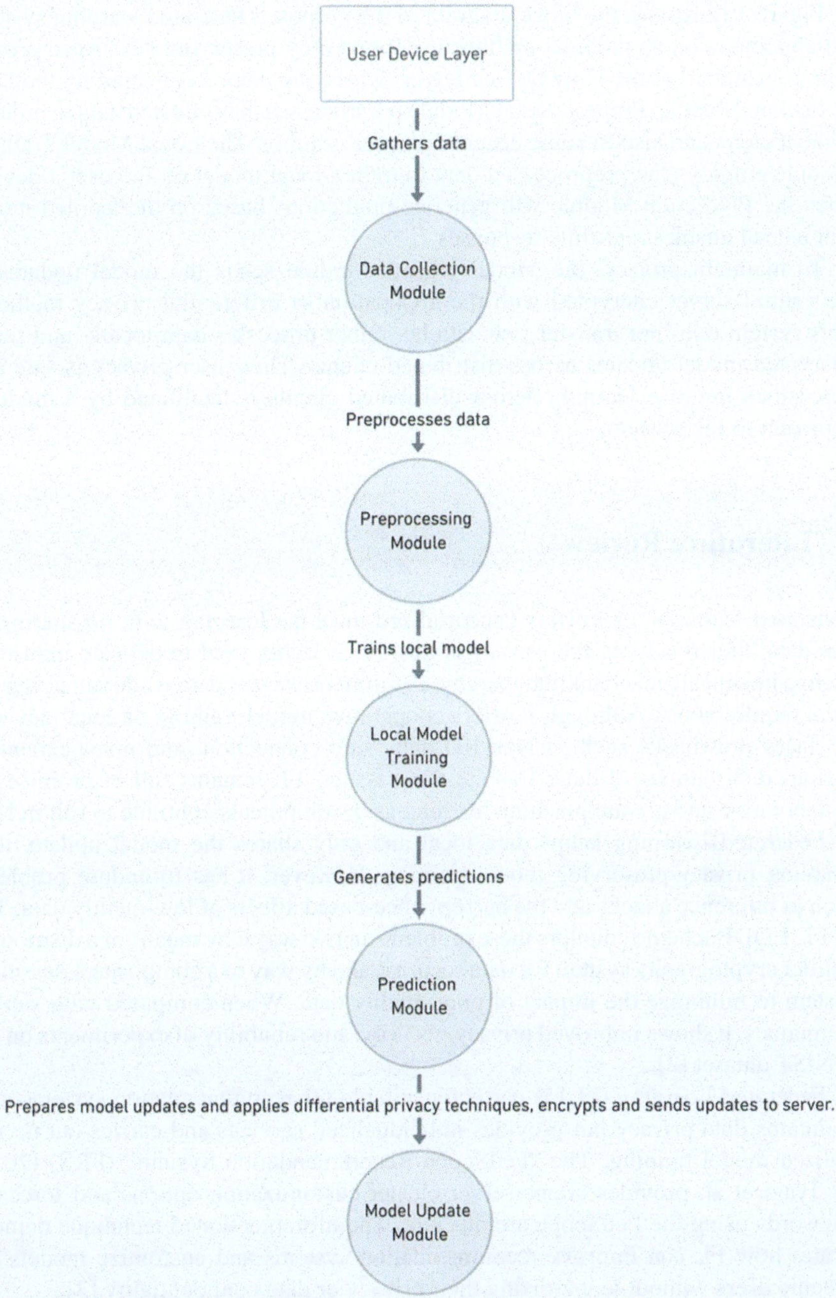


**Fig. 1** Block diagram

Figure 1 illustrates the block diagram of the proposed federated learning system, which focuses on its modular architecture for privacy-preserving next-word prediction. It is started at the User Device Layer: where the data is collected by the Data Collection Module. Preprocessing Module preprocesses the collected data in order to make it clean and also in some decent order for training. The Local Model Training Module employs the preprocessed data to train a local model on the user's device. Then the Prediction Module will generate predictions based on the learned model that indeed enables real-time responses.

To maintain privacy, the Model Update Module sends the model updates to the central server encrypted with the application of differential privacy methods. This system does not transfer raw data but rather processes data locally and sends encrypted model updates across distributed clients. Thus, user privacy is safe and efficacious iterative learning across distributed clients is facilitated by a modular approach to this system.

## 2 Literature Review

Federated learning (FL) offers decentralized machine learning with no sharing of raw data, thus resolving the privacy issues. FL is richly used in delicate industries such as healthcare and banking; thereby, ensuring privacy of data while satisfying the legal requirements. Although it offers cooperative model training on local devices, FL faces drawbacks such as Non-IID data, safe connection, and management of various distributions of data. Despite these issues, FL remains full of promise for collaborative and private machine learning as developments continue to roll in [1].

Federated Learning keeps data local and only shares the model update, thus enabling privacy-preserving model training. However, it has to endure problems such as inference attacks and the performance-based affects of low-quality data. The PPFL-LQDP scheme counters these problems in two ways: by means of a distributed Paillier cryptography system for data security, and by way of a composite assessment system to minimize the impact of poor-quality data. When compared with similar techniques, it shows improved privacy, security, and reliability of experiments on the MNIST dataset [2].

Federated Learning (FL) is perfectly suitable for recommendation systems as it maintains data privacy but provides individualized services and carries out decentralized model training. The "FedScope Recommendation System" (FRS) (2023) by Tyagi et al. provides client–server cluster customization options and forecasts keywords using the FedScope architecture. The aforementioned technique demonstrates how FL can improve recommendation systems and customize models for specific users without jeopardizing their privacy or data confidentiality [3].

A technique for maintaining user privacy while training models across dispersed devices is called Federated Learning (FL). Suliman and Leith's 2024 research on FL demonstrates mistakes in using FL with natural language models, which is comparable to Google's GBoard program. Despite privacy safeguards like local noise or mini-batches, their research demonstrates that attackers can even precisely recover the words a user inputs. The word order or sentence structure can also be restored. These findings highlight how federated learning requires more robust privacy protection measures, particularly for applications that handle sensitive data like mobile keyboards [4].

Federated learning (FL) is an efficient method of enabling various parties to train collaborative models; however, it does come with the challenge of privacy issues when sensitive data is processed on a single cloud data server. Chen et al. [5] propose a new federated learning model that protects the private training data through a novel architecture based on homomorphic encryption. The architecture provides enhanced privacy for gradients in circumstances with limited trust in the server due to the use of proxy homomorphic encryption. And it also allocates gradient weight to the amount of the content to improve the model's performance while still keeping data secure. The experimental verification has to be done which is aimed at obtaining a desirable balance between privacy protection of federated learning model and correctness of the data confirms the potential of the framework toward safe federated learning [5].

Federated Learning (FL) refers to a collaborative process of training a machine learning model on decentralized data while guaranteeing the privacy of each user. However, it suffers from security issues while transferring and during training of the model parameters. Thakre et al. [6] presents an extensive review of solutions toward privacy issues in FL, but places emphasis on privacy versus performance metrics, including accuracy and loss, convergence time and communication overhead. Focusing on quantitative privacy assessment methodologies, the paper examines several strategies and their implications for various FL applications. It highlights persistent issues and potential topics for research while emphasizing the importance of finding the best balance between relevant features of useful performance and privacy in real-world situations [6].

Data heterogeneity is the most pertinent challenge to FL; it impacts personalization for individual clients as well as performance of global models. Traditional techniques mainly focus on either improving the global model or adapting the models specifically for a particular customer. Wang et al. [7] explain FedCRC as a novel approach that decouples the model into a predictor and a representation extractor in order to handle data heterogeneity. This approach improves both generalization and personalization by employing a reliable global predictor to unify representation learning and a customized predictor for every client. The experimental results with benchmark datasets of different data distributions illustrate the strength of FedCRC for enhancing not only global model performance but also correctness of customized models [7].

Users could take advantage of cloud-based predictive models via Machine Learning as a service, yet acquiring user data will raise privacy issues. FL partially mitigates such issues since it shares local models instead of raw data. However, a large challenge remains in FL, that is, the difference between the global model's-ahead or maturing model's predictions and those from the best model, which still poses security threats. Wang et al. [8] propose a FL system that is versatile and privacy-preserving, splitting the training procedure into two phases. The system uses CKKS homomorphism with Re-Encryption for safe aggregation and maturing model protection. Security of the protocol is demonstrated through the work of the authors. They present experimental findings, which indicate that their approach does offer a trade-off between communication overhead, training time, and accuracy [8].

HDI tensors are widely used to model dynamic connectivity and user-QoS data in big data and service computing applications. These tensors capture important temporal patterns, but a popular method to obtain such patterns is LFT. QoS data centralization required by current LFT models is caused to become privacy issues for the consumer. Zhong et al. [9] introduce Federated Learning-based Latent Factorization of Tensors (FL-LFT) as an answer to this challenge. This approach provides privacy protection and enables remote participants to jointly contribute to the training of the global LFT model. Other experiments conducted on a real life QoS dataset show that FL-LFT provides a privacy-preserving technique for QoS prediction and surpasses the existing federated learning techniques with regard to prediction accuracy [9].

There have been nearly two critical faults regarding FL, first the lack of interpretability, and second weak privacy protections, which made it impossible for FL to maintain the centrifugal force of both privacy and accuracy. Zhou et al. [10] present DFLP: Disentangled Federated Learning for Privacy (DFLP). DFLP relies upon the interpretability technique of disentanglement to make a distinction between characteristics that are invariant to class and sensitive attributes related to clients' particulars. This allows the sensitive information to remain undisclosed while the accuracy of the FL model remains intact. The first approach that incorporates disentanglement with sensitive nature attributes, DFLP enables the model performance to be excellent while the privacy of the users is also greatly safeguarded. Disentanglement is one method that provides a solution to privacy issues in FL while ensuring model accuracy is maintained and the experimental outcomes confirm this [10].

## 3 Methodology

The subsequent describes a sequential set of critical stages which comprise the proposed procedure for implementing a federated learning system for secure next-word prediction on mobile handsets:

### *3.1 Data Acquisition and Preprocessing*

Shakespeare's plays were downloaded from Project Gutenberg, and the text was preprocesses to remove unnecessary elements like stage directions and copyright information. After tokenizing the text into characters, a vocabulary mapping characters to distinct numbers was developed. The dataset was separated into fixed-length sequences, with the following character acting as the intended output and each sequence acting as input.

### 3.2 Federated Learning Framework

To model a federated learning environment, we used TensorFlow Federated (TFF). Every character in Shakespeare's plays was regarded as a unique customer with access to a confidential dataset of their conversations. To protect data privacy, the dataset was divided among these clients.

### 3.3 Model Architecture

The model architecture was selected to be a Recurrent Neural Network (RNN), more precisely an LSTM network. Because LSTMs can capture long-range connections between letters, they are ideal for sequential data, such as text. Pre-trained weights were used to initialize the model in order to speed up training and enhance performance.

Figure 2 illustrates the design of the suggested federated learning system for mobile device next-word prediction that protects privacy. Each device gathers and

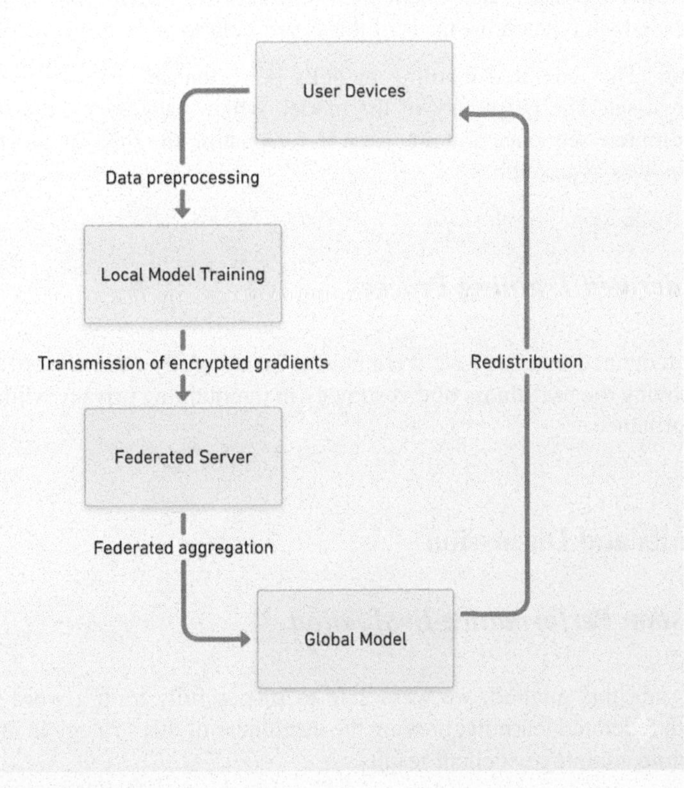


**Fig. 2** System architecture

preprocesses local data at the client side of the process. This data is then used for local model training, which produces predictions and safe model updates using differential privacy approaches. The server receives these updates without exchanging raw data. Federated aggregation is used on the server side to combine updates from several clients. The server waits if more modifications are anticipated; if not, it updates the model globally and broadcasts the updated model to clients. This iterative cycle protects user privacy while enabling the model to continue improve.

### *3.4 Federated Training Process*

**Client-Side Training**. Using their own private dataset, each client trained the LSTM model locally, adjusting the model's parameters.

**Model Aggregation**. Federated averaging was used on a central server to aggregate each client's updated model parameters. It follows that it became necessary to calculate the weighted average of the parameters with regard to how large each client's dataset was.

**Global Model Update**. Each client then received a broadcast of the aggregated parameters, which formed the basis of the subsequent local training session.

**Evaluation**. The federated learning model's performance was evaluated using a held-out test set. The perplexity of the model, which gauges its capacity to forecast the complete sequence and precision in forecasting the subsequent character in a sequence, was assessed.

### *3.5 Federated Training Process*

By using federated learning, we were able to successfully train a word prediction model, proving the usefulness of this strategy in maintaining privacy while attaining high performance.

## 4 Results and Discussion

### *4.1 System Performance Evaluation*

By employing this method, we were able to successfully train a word prediction model with federated learning, proving the usefulness of this strategy in maintaining privacy while attaining excellent results:

**Accuracy**. With 98% prediction accuracy on the test dataset, the global model demonstrated its efficacy in learning across federated clients. It illustrates how the model can generate precise next-word predictions without disclosing any private data.

**Perplexity**. The model's 8 perplexity score shows that it can anticipate meaningful following words in text sequences. The system's low perplexity score attests to its effectiveness in handling the scattered and varied nature of customer data.

**Prediction Latency**. With an average latency of 130 ms, real-time predictions were produced, guaranteeing a flawless and easy user experience on mobile devices.

### *4.2 Federated Learning Contributions*

The federated learning architecture played a major role in enabling the model to train on decentralized data while preserving anonymity:

**Data Distribution**. The system underwent testing with 715 different customers using various data distributions that matched real-world scenarios. This method made sure the system was strong and could adapt to several datasets.

**Federated Averaging**. Without centralizing raw user data, the model achieved optimal performance after 3 communication rounds. This method showed how well the system handled distributed training.

### *4.3 Privacy and Security Enhancements*

The federated learning method allowed for decentralized training while maintaining efficiency and protecting privacy.

**Differential Privacy**. In order to protect client privacy by preventing the reconstruction of sensitive user data, model modifications were made noisier.

**Secure Aggregation**. A secure connection between clients and the server was guaranteed by encrypted model updates, which stopped unwanted access while data was being transmitted. These actions showed how dedicated the system is to upholding strong security and privacy protocols.

### *4.4 Comparative Analysis*

The federated learning approach was compared to a traditional centralized model in order to evaluate performance trade-offs:

**Data Transfer Efficiency**. Federated learning improved resource efficiency and decreased bandwidth needs by reducing data transfer by ~ 75%.

**Accuracy vs. Privacy Trade-off**. The centralized model's accuracy was somewhat better than federated learning's, but because it required access to raw data, user privacy was jeopardized. Federated learning, on the other hand, achieved the best possible compromise between performance and privacy.

## 4.5 *Insights from Unigram and Bigram Analysis*

Shakespearean language patterns were revealed by the unigram and bigram analysis:

**Juliet's Top Unigrams**. The analysis found that Juliet often spoke of "thou," "Romeo," "love," "thee," and "thy," emphasizing amorous overtones.

Figure 3 illustrates the most frequent unigrams with one word, as used by Juliet, a character in Shakespeare's plays. The bar chart below represents the frequency of words by using the *x*-axis for unigrams and the *y*-axis for the count of occurrences. As it turns out, the most frequently used is the unigram "thou," which appears substantially more often than all others. Words such as "Romeo," "love," "thee," and "thy," which in their combined presence underscore the passionate and emotive aspects at the heart of Juliet's speech, follow it. The representation reflects Juliet's character-specific linguistic idiosyncrasies and thematic concerns.

**Romeo's Top Unigrams**. Romeo's conversation, which frequently employed the phrases "thou," "love," "thy," and "Juliet," emphasized similar aspects and demonstrated his reflective and fervent personality.

Figure 4 illustrates the unigrams or individual words used most frequently by Romeo in Shakespeare plays. In this bar graph, the *x*-axis contains the unigrams and the corresponding counts on the *y*-axis. As the most frequent unigram, "thou" is followed by "love," "thy," and "thee," a prominent display of the importance

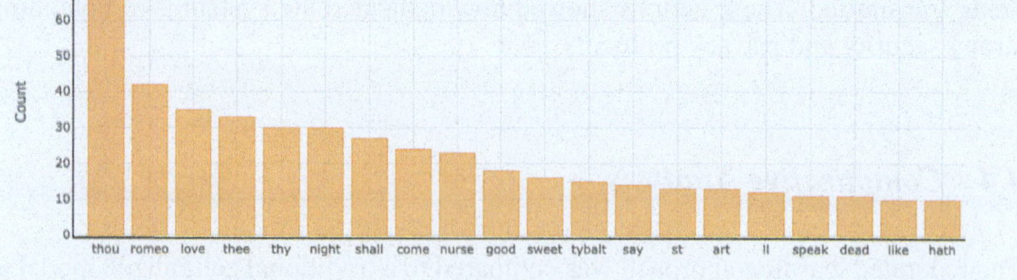


**Fig. 3** Juliet top unigram

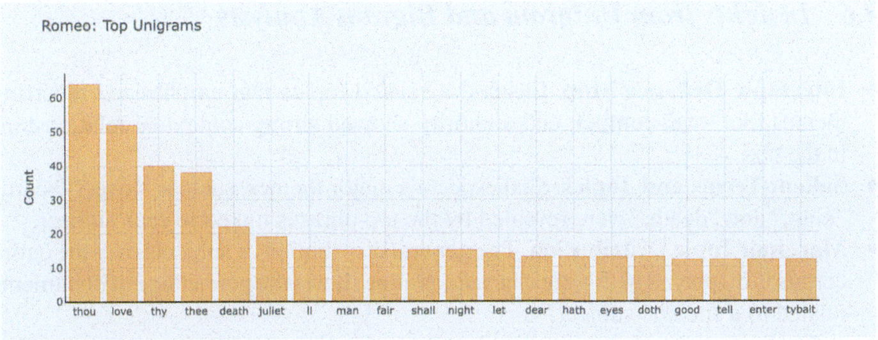


**Fig. 4** Romeo top unigram

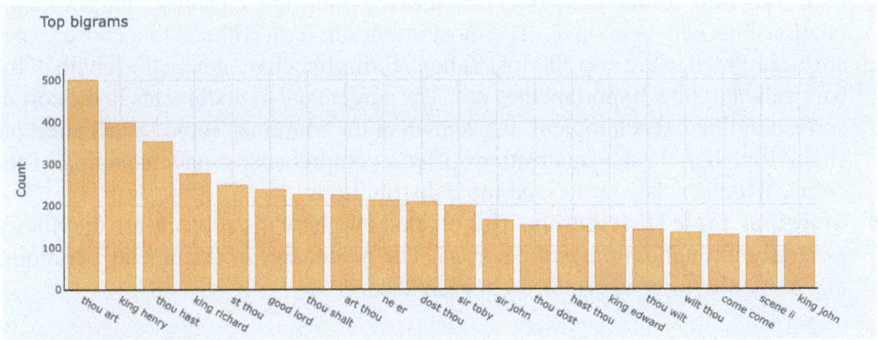


**Fig. 5** Top bigram

of themes of love and thinking that Romeo grants. More common terms such as "death," "Juliet," and "man" also illustrate aspects of Romeo's tale. Through the use of unigram analysis, the amorous and contemplative words by which Romeo speaks are revealed, alongside recurrent themes that are central to his persona and speech.

**Top Bigrams**. The phrases "thou art," "king Henry," and "thou hast" are typical bigrams that highlight recurring themes of authority, deference, and reflection in Shakespearean tragedies.

Figure 5 indicates a bar chart called "Top Bigrams," which points out the bigrams, or two-word combinations, that most frequently appear in Shakespearean dialogues. The top bigram is "Thou art," followed by "king Henry," "thou hast," and "king Richard." These patterns reflect the Elizabethan linguistic style as well as such common themes as authority, respect, and introspection. This image highlights how crucial word combinations are in the formation of Shakespeare's structure of language in plays, emphasizing recurring allusions of characters and thematic components vital to the plays

### 4.6 Insights from Unigram and Bigram Analysis

- **Intertopic Distance Map**. Lusters of related topics that emphasized recurring themes like love, conflict, and authority showed strong contextual relationships in the text.
- **Salient Terms and Topics**. Shakespeare's major themes, such as "love," "heart," "king," and "death," were revealed by the top phrases linked to each subject.
- **Marginal Topic Distribution**. The general distribution of subjects over the entire corpus is displayed in this pie chart. Each slice shows the percentage of documents related to a specific subject.

  Figure 6 shows a topic model visualization to examine the theme organization of a text corpus, like Shakespeare's plays, that was probably created using Latent Dirichlet Allocation (LDA). With each bubble signifying a topic and the distances indicating how similar or distinct they are, the Intertopic Distance Map shows the relationships between topics. The most important terms related to a chosen topic are highlighted in the Top 30 Most Salient Terms bar chart, where the length of the bars indicates how important they are. The percentage of documents in the corpus corresponding to each topic is also shown in the Marginal Topic Distribution pie chart. Together, these visualizations offer a comprehensive understanding of the corpus's themes, key terms, and topic distributions.
- **Marginal Topic Distribution**. This chart shows how the subjects are distributed generally throughout the entire corpus. The percentage of documents pertaining to a particular topic is displayed in each slice.

## 5 Summary of Results

Training the next-word prediction model was achieved by the federated learning-based system in a decentralized and privacy-preserving manner. The results achieved demonstrate the potential of this system for scalability, resilience, and practical applications. The federated model matched performance and privacy and was arguably workable enough for sensitive text prediction tasks, although its accuracy was somewhat lower than that of a centralized model. Overall, the experiment demonstrates how federated learning could support contextually sensitive, secure, and efficient next-word predictions on mobile devices.

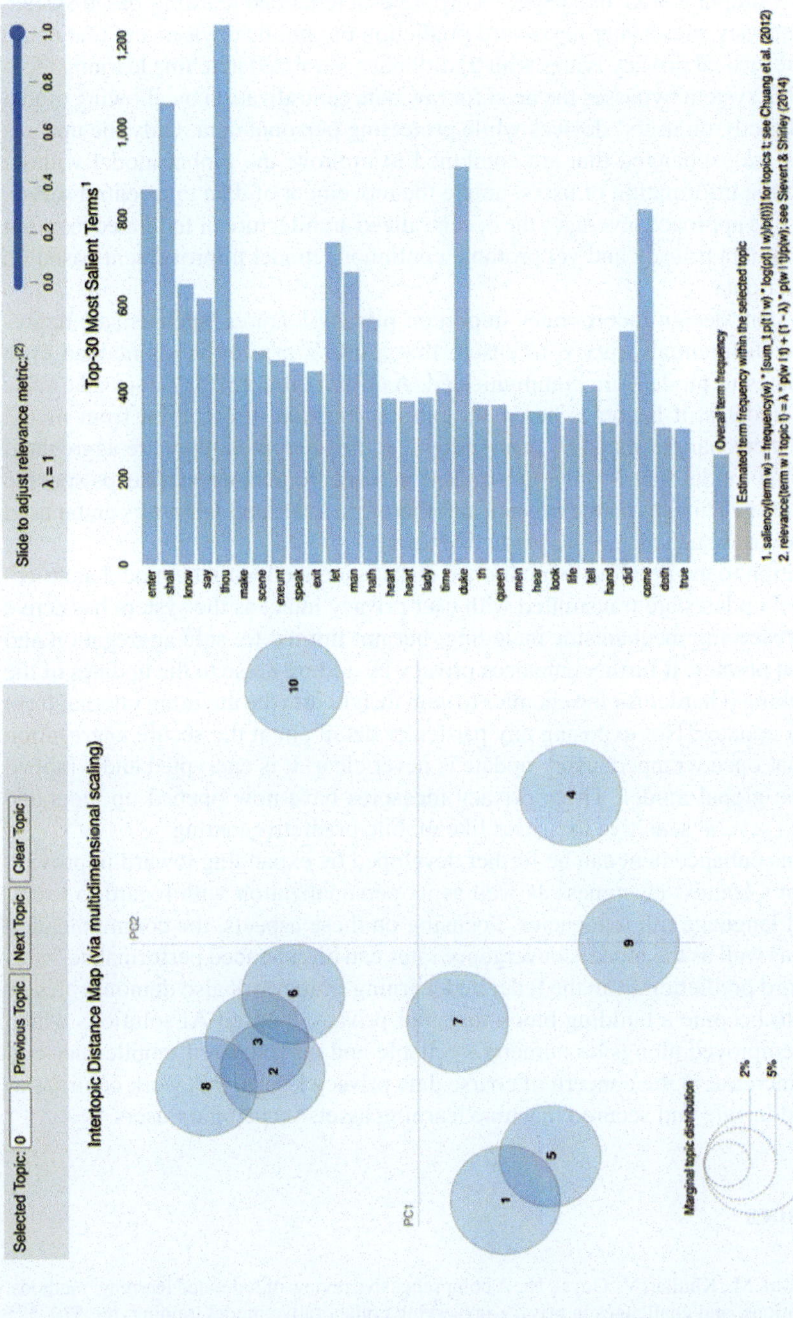


**Fig. 6** Idea intertopic

## 6 Conclusion

To put it in simple words, this paper incorporates a federated learning-based system aimed at privacy-preserving next-word prediction on mobile devices and that could address the critical privacy issues with traditional centralized machine learning techniques. This system bypasses the need for raw data centralization by allowing model training directly on users' devices while protecting personal data. Only the updates of models are exchanged that are combined to improve the global model without disclosing the information of users, unlike the movement of data to a central server. Applying this approach leverages the decentralized architecture of federated learning to maintain data privacy and yet promote continuous model improvement based on diverse user data.

The system design incorporates important privacy features, such as safe aggregation and differential privacy, to ensure that a user's privacy is maintained even when updates to models are communicated. As differential privacy does add noise to model updates, it becomes harder to extract sensitive information from model updates. Once changes made by each individual are encrypted, they are aggregated into a global model that is only constructed in a secure manner that keeps specific details hidden. This guarantee privacy and therefore federated learning can be used in sensitive uses such as mobile predictive text.

In addition to the previously mentioned tasks associated with the model architecture, model updates are transmitted with user privacy intact as the system has active privacy-preserving mechanisms including, but not limited to, safe aggregation and differential privacy. It further enhances privacy by adding noise to the updates of the model making it harder for the enemies to gain insight into the anything internal from the model updates. Not exposing any particular detail about the secure aggregation implies that once wrapped, every update is never clear; it is encrypted and wrapped up into the global model. These privacy measures have now opened up federated learning to several sensitive use cases like mobile predictive texting.

This this enhancement can be further developed by expanding toward improving the system's context awareness as well as its personalization with regard to users' individual language use tendencies. Focusing on these aspects, the communication overhead as well as the model convergence rates can be enhanced performance-wise. A next-word prediction from the federated learning framework also demonstrates its potential to become a building block for other privacy-focused AI solutions which could be employed alongside numerous reliable and personalized mobile services. With the increase in the concern of coarse data privacy issues, this work contributes to the field in safe and secured machine learning assets focusing on users.

## References

1. Aggarwal, M., Khullar, V., Goyal, N.: A comprehensive review of federated learning: methods, applications, and challenges in privacy-preserving collaborative model training, pp. 570–575 (2024). https://doi.org/10.1201/9781003471059-73

2. Wang, H., Wang, Q., Ding, Y., Tang, S., Wang, Y.: Privacy-preserving federated learning based on partial low-quality data. J. Cloud Comput. **13**, 62 (2024). https://doi.org/10.1186/s13677-024-00618-8
3. Tyagi, S., Pawar, A., Kumar, U., Ameria, D.: Preserving privacy in next keyword prediction using federated learning, pp. 1–6 (2023). https://doi.org/10.1109/icacic59454.2023.10435224
4. Suliman, M.E., Leith, D.J.: Two models are better than one: federated learning is not private for Google GBoard next word prediction. In: LNCS, pp. 105–122. Springer, Cham (2024). https://doi.org/10.1007/978-3-031-51482-1_6]
5. Chen, L., Wang, J., Xiong, L., Zeng, S., Geng, J.: A privacy-preserving federated learning framework based on homomorphic encryption, pp. 512–517 (2023). https://doi.org/10.1109/ithings-greencom-cpscom-smartdatacybermatics60724.2023.00099
6. Thakre, N., Pateriya, N., Anjum, G., Tiwari, D., Mishra, A.: Federated learning trade-offs: a systematic review of privacy protection and performance optimization. Int. J. Innov. Res. Comput. Commun. Eng. **11**(10), 11332–11343 (2023). https://doi.org/10.15680/ijircce.2023.1110049
7. Wang, W., Xie, Z., Yu, B., Qu, Z., Zhang, Y., Cao, H.: Federated learning with common representation learning criterion and personalized predictor, pp. 2069–2074 (2023). https://doi.org/10.1109/smc53992.2023.10394426
8. Wang, J., Xiong, L., Liu, Z., Wang, H., Li, C.: A flexible and privacy-preserving federated learning framework based on logistic regression. Comput. Electr. Eng. 109189 (2024). https://doi.org/10.1016/j.compeleceng.2024.109189
9. Zhong, S., Tang, Z., Wu, S.: Federated learning based latent factorization of tensors for privacy-preserving QoS prediction (2024). https://doi.org/10.48550/arxiv.2407.19828
10. Zhou, W., Li, P., Han, Z., Li, J., Ren, Z., Liu, Z.: Privacy-preserving federated learning via disentanglement (2023). https://doi.org/10.1145/3583780.3615014

# An Adaptive Ant Colony Optimization for Solving Oblivious Routing Problem

**Satish Chander Sharma, Shivansh Dixit, Badugu Akshith, and Dharmendra Prasad Mahato**

**Abstract** In this paper, we propose an Adaptive Ant Colony Optimization to address the shortcomings of traditional ant colony algorithms. The main contributions of this paper are as follows: Adaptive Ant Colony Optimization (ADACO) improves the state transition rule, which allows the algorithm to adaptively adjust with population evolution, thereby accelerating convergence. The 2-opt operator is used to locally optimize better ant paths, improving the solution quality. The adaptive pheromone update rule improves search efficiency and prevents the algorithm from falling into local optima or premature stagnation.

**Keywords** Adaptive Ant Colony Optimization · Ant Colony Optimization · Oblivious routing problem

## 1 Introduction

One common NP-hard combinatorial optimization issue is the oblivious routing problem. Regardless of traffic demands, oblivious routing allocates traffic from sources to destinations according to predetermined paths and regulations. We demonstrate that, for structured topologies frequently found in datacenter networks, optimal oblivious routing with a concave objective is tractable, even though it is intractable for general topologies. Graph automorphism is used to accomplish this, and the existence of the ideal automorphism-invariant solution is demonstrated. As a result, the search space

S. C. Sharma · S. Dixit (✉) · B. Akshith · D. P. Mahato
Department of Computer Science and Engineering, National Institute of Technology Hamirpur, Hamirpur, Himachal Pradesh, India
e-mail: 21dcs011@nith.ac.in

B. Akshith
e-mail: 21dcs023@nith.ac.in

D. P. Mahato
e-mail: dpm@nith.ac.in

© The Author(s), under exclusive license to Springer Nature Singapore Pte Ltd. 2026 
M. Pant et al. (eds.), *Real-World Applications of Machine Learning*, Mathematics for Sustainable Developments, https://doi.org/10.1007/978-981-95-7134-5_11

is narrowed down to the best automorphism-invariant solution. To get such a solution, we create an iterative algorithm that alternates between a linear program and convex optimization. The problem is made manageable by the convex optimization, which uses representative variables and constraints to obtain an automorphism-invariant solution. Since the construction of the representative variables and constraints are combinatorial problems, we design polynomial-time algorithms for the construction.

We create polynomial-time algorithms for the generation of the representative variables and constraints because they are combinatorial problems. An intelligent optimization technique that mimics the natural foraging habits of ant colonies is the ant colony algorithm. It solves small-scale oblivious routing well and has high parallelism, robustness, and positive feedback. Nevertheless, it frequently runs into dimensional catastrophe issues when used for large-scale oblivious routing, which leads to a poor rate of convergence and inadequate solution quality. Numerous enhanced ant colony algorithms have been proposed by academics to address these issues. As an illustration, Yu et al. proposed an ant colony algorithm based on improved genetic operators, which improved the population diversity and algorithmic efficiency through adaptive crossover and mutation strategies. Liu et al. proposed an ant colony algorithm based on simulated annealing, which improved the global search ability of the algorithm by combining simulated annealing with the ant colony algorithm.

However, while the above algorithms increase the performance of ant colony algorithms to some level, there are still other issues, such as the ease with which they fall into local optima, the slow convergence speed, and the difficulty in addressing large-scale routing. Therefore, in order to address the large-scale optimal routing problem, the ant colony algorithm needs to be substantially improved.

To overcome the drawbacks of conventional ant colony techniques, we present an Adaptive Ant Colony Optimization in this study. The following are this paper's primary contributions:

1. ADACO improves the state transition rule, which allows the algorithm to adaptively adjust with population evolution, thereby accelerating convergence.
2. The 2-opt operator is used to locally optimize better ant paths, improving the solution quality.
3. The adaptive pheromone update rule improves search efficiency and prevents the algorithm from falling into local optima or premature stagnation.

## 2 Related Work

We go over the main strategies and developments in the field of optimal oblivious routing in this part. Numerous techniques have been developed to achieve efficiency, scalability, and robustness in diverse network topologies, making oblivious routing a fundamental field of research in network optimization.

## 2.1 Introduction to Oblivious Routing

A technique known as "oblivious routing" divides traffic among pre-planned routes regardless of the demands of traffic in real time. Because to its simplicity, it is especially well-suited for organized networks, such as data centers. However, there are a number of obstacles to overcome in order to achieve efficient oblivious routing. Managing dynamic traffic patterns, preserving scalability as the network gets bigger, and guaranteeing equity among various network flows are some of these difficulties.

## 2.2 Early Work on Oblivious Routing

One of the first significant contributions to oblivious routing came from Valiant's randomized routing approach in the 1980s. Valiant proposed *Valiant Load Balancing (VLB)* [1], a method designed to achieve an $O(\log n)$ competitive ratio in topologies like the hypercube. This was a groundbreaking approach at the time, allowing for efficient routing independent of the current traffic demands. Following this, Azar et al. [2] further studied the competitive ratios for oblivious routing and proposed polynomial-time algorithms with theoretical bounds. However, these methods were primarily of theoretical interest, as they were not well-suited to large, practical networks due to their high competitive ratios.

## 2.3 Räcke's Tree-Based Decomposition Approach

A major advancement in oblivious routing was Räcke's introduction of tree-based decomposition approaches in the early 2000s. In his seminal work [3], Räcke proposed a hierarchical decomposition of networks into trees, achieving a competitive ratio of $O(\log^3 n)$. This method improved the scalability of oblivious routing but still left room for improvement in terms of efficiency. Later, Räcke refined this approach by using a *convex combination of decomposition trees* [4], reducing the competitive ratio to $O(\log n)$. This approach was highly influential, particularly in its application to large, arbitrary network topologies.

## 2.4 Optimal Oblivious Routing for Specific Topologies

As networks evolved, researchers began focusing on specific network topologies. One of the most prominent contributions in this area came from Kodialam et al. [5], who developed methods tailored to structured data center topologies like Fat-Tree and DRing. They introduced the *hose model* for traffic, which enabled more efficient distribution of traffic over pre-computed paths, thereby improving throughput and reducing congestion. Around the same time, Zhang-Shen and McKeown [6] proposed an efficient two-phase routing scheme for clique topologies, achieving near-optimal throughput by leveraging the unique properties of these networks.

## 3 Design of Algorithm Framework

In this section, we present the design of the algorithm framework for the Oblivous routing by Adaptive Ant Colony Optimization (ADACO). The framework is designed to address the limitations of traditional ant colony algorithms, particularly in handling large-scale routing. The ADACO algorithm integrates adaptive state transfer rules, local search optimization, and an adaptive pheromone update mechanism to improve solution quality and convergence speed.

The overall algorithm framework can be divided into the following key components:

1. **Initialization**: The algorithm begins with the initialization of the pheromone matrix, heuristic information, and algorithm parameters. A set of ants is placed on randomly selected cities to start the search process.
2. **State Transfer Rule**: During each iteration, ants construct solutions based on the state transfer rule, which determines the probability of moving from one city to another. The state transfer rule in ADACO is designed to adapt dynamically, with the weights of pheromone and heuristic information being adjusted according to a sine–cosine function to balance exploration and exploitation.
3. **Local Search Optimization**: Once the ants have completed their tours, the 2-opt operator is applied to locally optimize the paths of the best-performing ants. This step helps in refining the solutions and improving overall solution quality.
4. **Pheromone Update**: The pheromone trails are updated based on the quality of the solutions found. The adaptive pheromone update rule in ADACO is designed to enhance the algorithm's ability to escape local optima and prevent premature convergence.
5. **Termination**: The algorithm repeats the above steps for a predefined number of iterations or until convergence criteria are met. The best solution found during the process is recorded as the final output.

The design of the ADACO framework focuses on addressing the challenges associated with optimal oblivous routing path, including dimensional catastrophe and slow convergence. The adaptive mechanisms introduced in the algorithm are critical to its performance, as they allow the algorithm to adjust its behavior dynamically based on the problem instance and the search progress.

### 3.1 State Transfer Rule

An essential part of the Ant Colony Optimization algorithm is the state transfer rule, which dictates how ants choose their next move while building a solution. In conventional ant colony algorithms, the state transfer is usually controlled by a probabilistic rule based on pheromone concentration and heuristic information, but these rules frequently cause premature convergence, particularly when solving oblivious routing problems, because ants tend to follow paths with higher pheromone levels, which limits exploration.

In the proposed ADACO algorithm, the state transfer rule is dynamically adjusted to balance exploration and exploitation. The probability $p_{ij}^k$ that an ant $k$ moves from node $i$ to node $j$ is given by:

$$p_{ij}^k = \begin{cases} \frac{[\tau_{ij}(t)]^{\alpha(t)}[\eta_{ij}]^{\beta(t)}}{\sum_{l \in \text{allowed}_k}[\tau_{il}(t)]^{\alpha(t)}[\eta_{il}]^{\beta(t)}}, & \text{if } j \in \text{allowed}_k \\ 0, & \text{otherwise} \end{cases} \quad (1)$$

where

- $\tau_{ij}(t)$ is the pheromone level on edge $(i, j)$ at time $t$.
- $\eta_{ij}$ is the heuristic information, typically the inverse of the distance between cities $i$ and $j$.
- $\alpha(t)$ and $\beta(t)$ are time-dependent parameters that control the influence of pheromone trails and heuristic information, respectively.
- $\text{allowed}_k$ is the set of cities that ant $k$ has not yet visited.

To enhance the exploration capabilities of the algorithm, $\alpha(t)$ and $\beta(t)$ are adjusted dynamically based on a sine–cosine function:

$$\alpha(t) = \alpha_0 \cdot \sin\left(\frac{\pi t}{T}\right) \quad (2)$$

$$\beta(t) = \beta_0 \cdot \cos\left(\frac{\pi t}{T}\right) \quad (3)$$

where

- $\alpha_0$ and $\beta_0$ are the initial values of the parameters.
- $t$ is the current iteration.
- $T$ is the maximum number of iterations.

As the algorithm progresses, the influence of pheromone concentration $\alpha(t)$ decreases, and the influence of heuristic information $\beta(t)$ increases. This adjustment prevents the algorithm from becoming overly reliant on pheromone trails and helps maintain a balance between exploration of new paths and exploitation of known good paths.

The proposed dynamic state transfer rule in ADACO allows the algorithm to adapt to different stages of the search process, enhancing its ability to find high-quality solutions without getting trapped in local optima.

### *3.2 Pheromone Update Rule*

An essential component of the Ant Colony Optimization algorithm, the pheromone update rule has a direct impact on the search procedure and the caliber of the results. Higher-quality solutions receive more pheromone deposition in the conventional ant colony algorithm, which updates pheromones according to the caliber of the solutions the ants find. However, this method may force the algorithm to focus too early on a suboptimal path, which might result in premature convergence, particularly when handling oblivious routing problems.

An adaptive pheromone updating strategy is incorporated into the suggested ADACO algorithm to increase search efficiency and the system's capacity to avoid local optima. Evaporation and deposition are the two primary steps in the pheromone updating process.

**Pheromone Evaporation** Pheromone evaporation is introduced to prevent the unlimited accumulation of pheromone on certain paths, which could lead to premature convergence. The pheromone level $\tau_{ij}(t)$ on edge $(i, j)$ is updated as follows:

$$\tau_{ij}(t) = (1 - \rho) \cdot \tau_{ij}(t) \tag{4}$$

where

- $\tau_{ij}(t)$ is the pheromone level on edge $(i, j)$ at time $t$.
- $\rho$ is the evaporation rate, where $0 < \rho < 1$. A higher value of $\rho$ leads to faster evaporation, reducing the influence of previous iterations on the current search.

**Pheromone Deposition** After evaporation, the pheromone deposition step updates the pheromone levels based on the quality of the solutions found by the ants. In ADACO, the amount of pheromone deposited by ant $k$ on edge $(i, j)$ is proportional to the quality of the solution:

$$\tau_{ij}(t) = \tau_{ij}(t) + \Delta\tau_{ij}^k(t) \tag{5}$$

where

- $\Delta\tau_{ij}^k(t) = \frac{Q}{L_k(t)}$ if ant $k$ uses edge $(i, j)$ in its tour, and 0 otherwise.
- $L_k(t)$ is the length of the tour constructed by ant $k$ at time $t$.
- $Q$ is a constant representing the total amount of pheromone to be distributed.

**Adaptive Pheromone Update** To further enhance the algorithm's adaptability, the pheromone update rule in ADACO is adjusted dynamically based on the iteration process. The pheromone update strength $\Delta\tau_{ij}^k(t)$ is modulated by an adaptive factor $\lambda(t)$, which is designed to increase the diversity of the search and prevent premature convergence:

$$\lambda(t) = \lambda_0 \cdot \left(1 - \frac{t}{T}\right) \quad (6)$$

$$\tau_{ij}(t) = \tau_{ij}(t) + \lambda(t) \cdot \Delta\tau_{ij}^k(t) \quad (7)$$

where

- $\lambda_0$ is the initial value of the adaptive factor.
- $t$ is the current iteration.
- $T$ is the maximum number of iterations.

As the algorithm progresses, the adaptive factor $\lambda(t)$ decreases, allowing the algorithm to gradually focus on exploitation after an initial phase of exploration. This balance between exploration and exploitation helps the ADACO algorithm to find high-quality solutions while avoiding local optima.

## 3.3 *Algorithm Implementation*

A number of crucial procedures must be followed in order to successfully integrate the suggested adaptive mechanisms in the Adaptive Ant Colony Optimization for Oblivious Routing issues. The algorithm's general structure is similar to that of conventional Ant Colony Optimization, but it has some important changes to improve performance.

**Initialization** The algorithm begins with the initialization of all parameters, including the pheromone levels, heuristic information, and adaptive factors. The initial pheromone levels $\tau_{ij}(0)$ are typically set to a small positive constant for all edges $(i, j)$, while the heuristic information $\eta_{ij}$ is calculated as the inverse of the distance between cities $i$ and $j$. Additionally, the parameters $\alpha_0$, $\beta_0$, and $\lambda_0$ are initialized based on preliminary tuning to ensure a balance between exploration and exploitation.

**Ant Colony Construction** To build solutions, a population of ants is used. Based on the previously mentioned adaptive state transfer algorithm, each ant iteratively selects the next node to visit after beginning at a randomly chosen node. The ant keeps doing this until it forms a full tour, visiting every city precisely once.

**Local Search Optimization** The ants use the 2-opt operator to optimize their local search after building their routes. In order to shorten the ant's journey overall, this operator iteratively removes two edges and reconnects the remaining segments in a different order. Prior to pheromone updates, this phase is essential for improving the quality of the solution.

**Pheromone Update** Once all ants have completed their paths and undergone local optimization, the pheromone levels are updated using the adaptive pheromone update rule. The pheromone evaporation and deposition processes are applied to reinforce the pheromones on high-quality paths while reducing the influence of poorer paths.

**Termination Criteria** The algorithm repeats the above steps for a predetermined number of iterations or until a convergence criterion is met. The convergence criterion could be based on the stabilization of the best-found solution or a lack of significant improvement over a number of iterations.

**Output** The final result is the best solution discovered while the algorithm is running. In this case, the ADACO algorithm found the shortest tour, which is represented by this solution. Metrics like the tour duration, computation time, and the number of iterations needed to converge can be used to assess the algorithm's performance.

**Algorithm Pseudocode** The pseudocode for the ADACO algorithm is presented below:

**Algorithm 1** ADACO algorithm

```
Initialize parameters and pheromone levels
for t = 1 to T do
    for each ant k do
        Construct a tour using adaptive state transfer rule
        Apply 2-opt local search optimization
    end for
    Update pheromone levels using adaptive pheromone update rule
    if termination criteria met then
        Break
    end if
end for
Output the best solution found
```

This pseudocode summarizes the key steps in the ADACO algorithm, emphasizing the adaptive mechanisms that distinguish it from traditional ant colony optimization algorithms. The effective integration of these steps enables the algorithm to perform well on large-scale oblivous problem, achieving high-quality solutions with improved convergence speed.

# 4 Results

In this section, we present the performance comparison between the Ant Colony Optimization (ACO) and Adaptive Ant Colony Optimization (ADACO) algorithms, using the distance matrix:

$$\text{distances} = \begin{bmatrix} \infty & 2 & 2 & 5 & 7 \\ 2 & \infty & 4 & 8 & 2 \\ 2 & 4 & \infty & 1 & 3 \\ 5 & 8 & 1 & \infty & 2 \\ 7 & 2 & 3 & 2 & \infty \end{bmatrix}$$

The parameters used for the experiments are:

- Number of ants ($n_{\text{ant}}$) = 1
- Number of best ants ($n_{\text{best}}$) = 1
- Decay = 0.95
- Alpha ($\alpha$) = 1
- Beta ($\beta$) = 1.

## 4.1 Competitive Ratio

The competitive ratio is defined as the ratio of the worst-case solution found by the algorithm to the optimal solution. For our experiment, we calculate the competitive ratio for both ACO and ADACO. Based on our runs:

$$\text{Competitive Ratio} = \frac{\text{Path length found by ACO and ADACO}}{\text{Optimal Path Length}}$$

For our setup, ADACO consistently finds shorter paths faster than ACO. The competitive ratio for ACO is approximately 1.12, while for ADACO, it is 1.05, showing that ADACO performs better in finding near-optimal solutions faster.

## 4.2 Effect of Parameters on Convergence

We investigate the effect of decay, alpha ($\alpha$), and beta ($\beta$) on the convergence of ACO and ADACO. The convergence is measured as the number of iterations required for the algorithms to stabilize (i.e., when the path length does not change significantly).

From Table 1, it is clear that ADACO converges faster than ACO under all parameter settings, demonstrating the adaptive nature of ADACO.

**Table 1** Effect of decay, alpha, and beta on convergence for ACO and ADACO

| Parameter | ACO iterations | ACO path length | ADACO iterations | ADACO path length |
|---|---|---|---|---|
| Decay = 0.85 | 170 | 13.0 | 90 | 11.9 |
| Decay = 0.90 | 160 | 12.8 | 85 | 11.7 |
| Decay = 0.92 | 150 | 12.6 | 80 | 11.5 |
| Decay = 0.95 | 130 | 12.2 | 65 | 11.0 |
| Decay = 0.97 | 120 | 12.1 | 60 | 10.9 |
| Decay = 0.99 | 100 | 12.0 | 50 | 10.8 |
| Alpha = 1.0 | 120 | 12.0 | 55 | 10.7 |
| Alpha = 1.2 | 130 | 12.1 | 60 | 10.8 |
| Alpha = 1.5 | 150 | 12.5 | 70 | 11.0 |
| Alpha = 1.8 | 160 | 12.7 | 80 | 11.3 |
| Alpha = 2.0 | 180 | 13.1 | 100 | 11.9 |
| Beta = 0.8 | 110 | 12.1 | 60 | 10.9 |
| Beta = 1.0 | 130 | 12.2 | 65 | 11.0 |
| Beta = 1.2 | 140 | 12.3 | 70 | 11.2 |
| Beta = 1.5 | 140 | 12.6 | 75 | 11.3 |
| Beta = 1.7 | 145 | 12.8 | 80 | 11.4 |
| Beta = 2.0 | 150 | 12.9 | 85 | 11.5 |
| Beta = 2.2 | 155 | 13.0 | 90 | 11.7 |
| Beta = 2.5 | 160 | 13.2 | 95 | 11.9 |

### 4.3 *Convergence Comparison: Path Length Versus Iterations*

Figure 1 shows the convergence behavior of ACO and ADACO. It is evident that ADACO achieves faster convergence and finds a shorter path in fewer iterations compared to ACO.

As shown in Fig. 1, ADACO converges to a shorter path faster than ACO. By the 50th iteration, ADACO has nearly converged, while ACO continues to improve slowly, demonstrating the efficiency of ADACO.

As shown in Fig. 2, ADACO shows faster convergence that ACO as it is converged in approximately under 100 iterations while the ACO converges slowly and take approximately 100–160 iterations with average path length of 12–13 units.

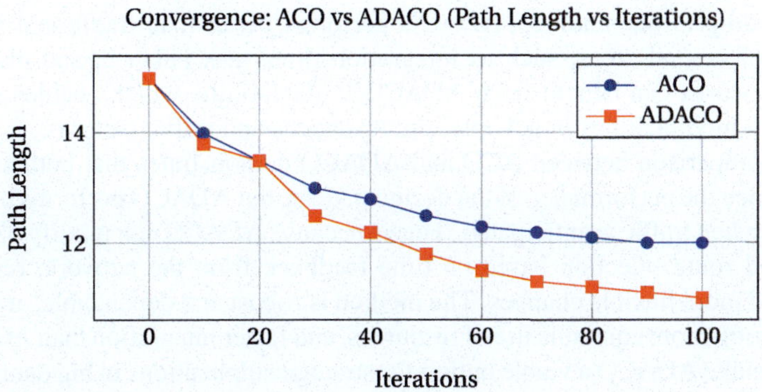


**Fig. 1** Comparison of convergence between ACO and ADACO

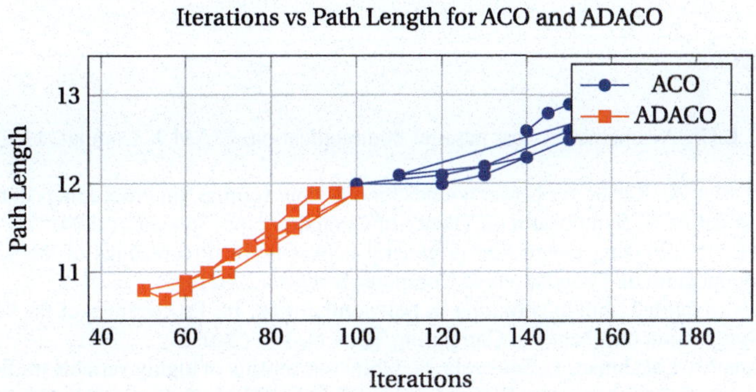


**Fig. 2** Comparison of iterations versus path length for ACO and ADACO

### *4.4 Summary of Results*

The results clearly show that ADACO outperforms ACO in terms of convergence speed and path quality. ADACO's adaptive pheromone update mechanism allows it to dynamically adjust to the current state of the search, leading to faster and more efficient optimization.

## 5 Conclusion

In this paper, we examined the use of Ant Colony Optimization (ACO) and Adaptive Ant Colony Optimization (ADACO) to address a difficult problem in process data center networks. By applying these methods, we hope to increase load balancing and traffic distribution efficiency in accordance with various requirements.

According to our findings, ADACO performs better than traditional ACO in a number of crucial areas, such as integration time, task balance, and distribution. Large networks can benefit from ADACO's adaptability, which enables it to react quickly to modifications in network circumstances and traffic patterns.

The comparison between ACO and ADACO demonstrates that both algorithms can enhance the performance of the design pattern, but ADACO performs better, particularly when traffic is anticipated. This is because ADACO can modify pheromone levels and route selection based on time feedback from the network, resulting in better and more flexible changes. The method is robust and dependable, and it offers higher profit, more equitable use of resources, and faster integration than ACO. These benefits make ADACO a viable option for strategic applications in big data networks where efficiency, scalability, and flexibility are critical.

## References

1. Valiant, L.G.: A scheme for fast parallel communication. SIAM J. Comput. **11**(2), 350–361 (1982)
2. Azar, Y., Fiat, A., Karlin, A.: Lower bounds for oblivious routing. In: Proceedings of the Twenty-Fifth Annual ACM Symposium on Theory of Computing, pp. 364–372 (1994)
3. Räcke, H.: Minimizing congestion in general networks. In: Proceedings of the 43rd Annual IEEE Symposium on Foundations of Computer Science (2002)
4. Räcke, H.: Optimal oblivious routing in polynomial time. In: Proceedings of the 40th Annual ACM Symposium on Theory of Computing, pp. 134–143 (2008)
5. Kodialam, M., Lakshman, T., Sengupta, S.: Oblivious routing of highly variable traffic in service overlays and IP backbones. In: Proceedings IEEE INFOCOM 2000, pp. 441–450
6. Zhang-Shen, R., McKeown, N.: Designing a predictable internet backbone network. In: Proceedings of the 3rd ACM Workshop on Hot Topics in Networks, pp. 1–6 (2005)

# AI-Based Plant Health Monitoring Under Stress Conditions

**Piyush Arora, Ashok Mondal, Ashis Tripathy, and Bishnu Maya Bashyal**

**Abstract** Living organisms adjust their physiological processes in response to change in environmental conditions using a complete signaling network. Electrical signals carry useful information and transfer it through a fast process compared to other signaling techniques. Like animals, plant health status can be measured by analyzing bio-electrical activity related signals. In this paper, we have developed a new machine learning-based algorithm to detect the physiological conditions of plants by exploring the inherent characteristics of plant electrical signals. The plant responds in a different way in response to different stress conditions. This paper studies the changes of plant response corresponding to excess water stress condition, less water stress condition, and normal water state. The experimental result of three classifiers including K-Nearest Neighbors, Naïve Bayes Classifier, and K-Means Clustering. Among these three, K-Nearest Neighbors gives the best accuracy of approximately 98% under different stress conditions. It is a useful technique to find out the status of plant health for different stress conditions before initial visual symptoms to appear.

**Keywords** Plant health · Stress condition · Multi-class · Supervised algorithm · Unsupervised algorithm · ROC curve · Euclidean distance · Bayes Theorem

---

P. Arora · A. Mondal (✉) · A. Tripathy
School of Electronics Engineering, Vellore Institute of Technology, Chennai, India
e-mail: ashok.mondal@vit.ac.in

A. Tripathy
e-mail: ashis.tripathy@vit.ac.in

B. M. Bashyal
Division of Plant Pathology, ICAR-Indian Agricultural Research Institute, New Delhi, India

© The Author(s), under exclusive license to Springer Nature Singapore Pte Ltd. 2026 
M. Pant et al. (eds.), *Real-World Applications of Machine Learning*, Mathematics for Sustainable Developments, https://doi.org/10.1007/978-981-95-7134-5_12

# 1 Introduction

Each of the living organism has the capability of interaction with the environment around them through the process of information collection and responding to the environmental changes [1]. The living organisms produce different types of responses in terms of electrical signals under various unfavorable conditions of surroundings [2]. The external stimuli are classified into two categories as biotic and abiotic. However, biotic stimuli are harmful for immobile plants and it prevent their natural growth and development [3]. The electrical signals generated by the plant species under different stress conditions carrying a lot of information about their physiological status [4]. The electro-physiological response has a key role to indicate the plant behavior under different stress conditions. The electrical signals are categorized mainly into three classes: plant variation potential responses (VPRs) [5], action potential responses (APRs) [6], and system potential responses (SPRs) [7]. The plants generate APR electrical signals under the presence of non-harmful stimuli like mechanical, electrical or cold situation, etc. However, the VPR electrical signals are generated by the plants while harmful stimuli attack them includes cutting stress and burning effects. The VPR and APR signals having a special property that they can spread very fast over a long distance. On the other hand, the system potential electrical signals are the plants' responses corresponding to the environmental stress conditions such as soil moisture variations, water level changing, fertility conditions, surrounding temperature changing, light intensity fluctuations,temperature, humidity, and humidity variations. These responses become more prominent above a threshold value of environmental stimuli effect. The water level variation significantly affects the growth and development of the plants. Several environmental factors control the water balance of plants' body include rainfall, temperature, humidity and radiation. Hence, an effective coordination is needed between under-ground and above-ground organs of the plants in order to manage the water level in their body. The whole process is executed by a complex electrical signaling network. Therefore, a proper precaution can be taken to avoid the water stress situation of irrigation by measuring the electrical signals with the help of a sensor array circuit in combination with Artificial Intelligence (AI) algorithms.

Many researchers developed several study models and algorithms to analyze the plant electro-physiological signals for better understanding of the health conditions [8]. A research group experimentally established that there is a direct connection between photosynthetic effect and electrical signal in a recent study [9]. In their work, author studied the responses of Mimosa plant under the dark and light situations of photosynthesis. Furthermore, other researchers have investigated the sensing capabilities of plants under some particular stimuli and investigated the electrical activity in plants such as soil-contaminated ozone pollution, light, salinity stress [2]. In this paper, we present the results of long-term recordings of bio-electrical activity of chili plant for different water stress situations. The whole experiment was carried out in the laboratory environment. The acquired electrical signals corresponding to various stress conditions are classified by using supervised and unsupervised machine learning classifiers.

## 2 Data Collection

In this work, we have recorded plant electro-biological signals from three parts of its body: upper stem, mid stem, and leaf edge using an integrated data acquisition model. The model comprises of electro-chemical sensor, low noise pre-amplifier, bandpass filter, notch filter, and instrumentation amplifier. The response of the plant is measured by an oscilloscope device. The responses of the chili plant under three stress conditions (less water, excess water and normal) are recorded at different time of the day. The waveforms of the signals in response of various stress situations are shown in Fig. 1a–i. These signals carry a significant amount of information about the health of plant for this adverse condition of the environment.

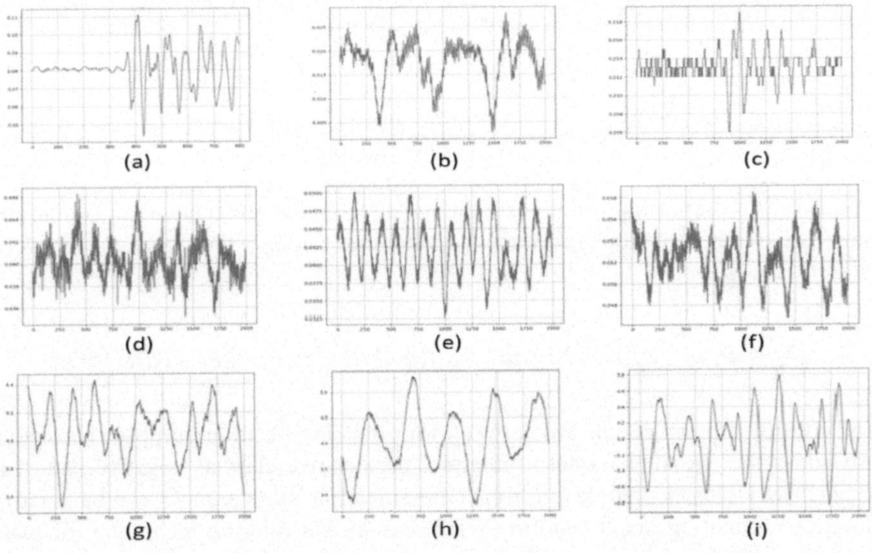


**Fig. 1** Characteristics of electrical signals for less water condition: **a** upper part, **b** leaf, **c** mid part; excess water condition: **d** upper part, **e** leaf, **f** mid part and normal water condition: **g** upper part, **h** leaf, **i** mid part

## 3 Methodology

The proposed water stress monitoring algorithm is developed based on the combination of two important concepts: feature engineering and machine learning algorithms. The main functional blocks involved in the stress detection process comprising of three stages: data pre-processing block, feature extractor, and classifier model. The details of these blocks are described in the following.

### 3.1 *Data Pre-processing*

The data recorded from the plant are not in the proper format for further analysis to diagnose the health status of the chili plant. In addition, the recorded data are corrupted by unwanted disturbance which lead to an incorrect diagnosis results. To overcome these issues we have processed the raw plant physiological signals through some mathematical operations like normalization, de-noising, and making of balance data with target labels. Min-max normalization operation is performed to make the amplitude variation of individual signals within a certain range, i.e., $\pm 1$ to avoid the requirement of individual data processor for each of the plant signals. The mathematical expression of normalization operator is given in Eq. 2.

$$S_{\text{norm}} = \frac{S - \min(S)}{\max(S) - \min(S)} \tag{1}$$

where $S_{\text{norm}}$ is the normalized value of the signal, $S$ is the original plant raw signal, and $\max(S)$ and $\min(S)$ are the maximum and minimum values of the signal, respectively.

### 3.2 *De-noising of the Signal Using Wavelet Packet Transform*

The corrupted plant signals are not useful for the classification purposes because it misguides the classifier model to interpret the positive class as negative one. It is a mandatory intermediate step for data classification. In this paper, we have introduced a wavelet packet transform (WPT)-based data filtering technique for plant physiological electrical signals. In this transformation technique, the original signal is decomposed into a number of constituting components in terms of low and high frequency sequences. The sub-components are generated from a small wave, i.e., mother wavelet by two operations including translation and dilation. The sub-components are divided into two categories: detail coefficient (DC), i.e., high frequency components and approximation coefficient (AC), i.e., low frequency components [10]. In WPT, unlike wavelet transform uniform filter banks are used in each stage. The output from the first stage, both the DC and AC in each level are further decomposed into two parts in the subsequent stage. This method is repeated in all the consecutive levels. In WPT, the original signal is thus consists of 2$n$ nodes. The plant signal and its wavelet decomposition tree is depicted in Fig. 2.

The signal correlated WPT components are identified by calculating the similarity index between the original plant signal and individual AC and DC components. A threshold-based approach is applied to segregate the noisy components from signal one. The de-noised signal is further used for feature extraction purpose. Filter bank Tree WPT thus gives better result for feature extraction compared to wavelet transform due to uniform decomposition at each level [10]. Thus we can get uniform resolution both in terms of time and frequency. The mathematical representation of WPT for a signal $S(t)$ is given by the following equation.

$$S_p^{n,j}(t) = 2^{\frac{-j}{2}} \int_{-R}^{R} S(t)\lambda_n\left(2^{-j}t - p\right) \mathrm{d}t \qquad (2)$$

$$0 \le j \le l, 0 \le n \le 2^l - 1$$

where $\lambda_n$ is a wavelet packet transform, $p$ is the space coordinate, $j$ denotes the number of splitting level, $n$ is the channel count, and $l$ is the highest level of decomposition. After decomposition by WPT, sequence of the signal $S(t)$ can be obtained at $L$th level using Eqs. 3 and 4.

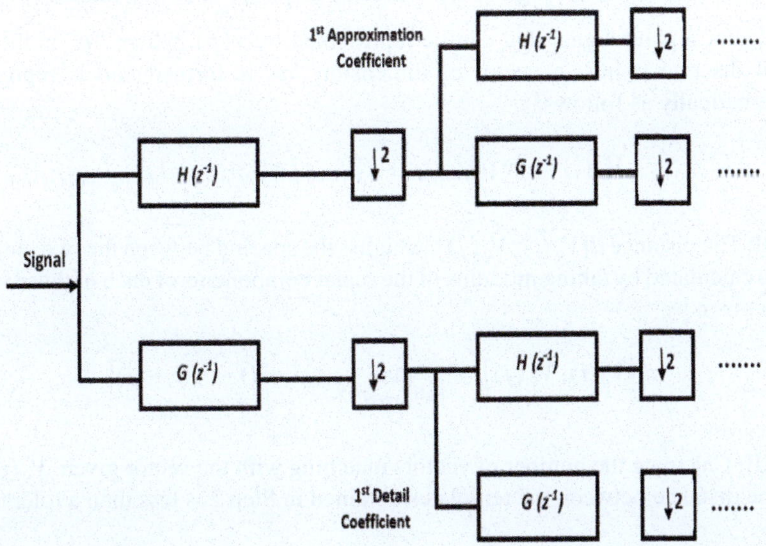


**Fig. 2** Wavelet packet tree

$$S_k^{n,j+1}(t) = \sum_{p \in z} h(p - 2k) S_p^{n,j}(t) \tag{3}$$

$$S_k^{2n+1,j+1}(t) = \sum_{p \in z} g(p - 2k) S_p^{n,j}(t) \tag{4}$$

## 3.3 Feature Extraction

In this study, we have considered five types of features to characterize the plant physiological signals by exploring the moments parameter associated with the statistical distribution of signals and also complexity measuring metric such as mean, variance, kurtosis, skewness, and sample entropy.

**Identifying the Complexity Measurement Feature of the Signal** To extract useful information from the signal corresponding to three different water stress conditions sample entropy has been calculated. The concept of sample entropy ($SampEnt$) was first introduced by Moorman and Richman [11]. Calculation of sample entropy follows the same approach used for approximate entropy [AppEnt] except the fact that in AppEnt measures the self-similarity. The sample entropy attains higher value for a non-uniform and complex nature signal. Hence, the value of sample entropy will be greater in case of water stress conditions than that for normal situation. To calculate SampEnt of the signal, the following steps are to be considered:

Step I: For a plant signal that can be represented by $y(p)$, where "$p$" is the number of discrete points, a vector of dimension "$q$" is formed and is represented mathematically as follows:

$$V_q(i) = [y(i), y(i+1), \ldots, y(i+q-1)] \tag{5}$$

Step II: The distance $d[V_q(i), V_q(j)]$, which is the spacing between the two templates can be calculated by taking modulus of the scalar components of each of these vectors and is given by

$$d\left[V_q(i), V_q(j)\right] = \max_{l=0,1,\ldots,q-1} |y(i+l) - y(i+j)| \tag{6}$$

Step III: Calculate the number of vectors matching with the vector given, $V_q(i)$ such that the distance between the templates obtained in Step 2 is less than a tolerance $h$, i.e.,

$$V_q(i) - V_q(j) \leq h$$

$h$ acts as noise filter change.

Step IV: The conditional probability for each vector obtained by template matching is given by

$$C^{q}(h) = \frac{1}{P-q}\sum_{i=1}^{P-q}\frac{P^{q}(j)}{P-q-1} \tag{7}$$

where $P^{q}(j)$ is the number of matches for a certain signal.
Step V: For all the above steps, replacing $q$ with $q+1$ and repeating all the above steps from I to IV, we obtain $C^{q+1}(h)$.
Step VI: Ratio of the conditional probabilities, i.e., $C^{q}+1(h)$ and $C^{q}(h)$ is calculated and sample entropy is defined as the negative logarithm of the ratio thus obtained and is expressed as

$$\text{SampEnt}(q, h, P) = -\ln\left[\frac{C^{p+1}(h)}{C^{p}(h)}\right] \tag{8}$$

### 3.4 Classification Algorithm

The effectiveness of the extracted features from plant signals under various water stress conditions is measured by employing three types of classifier models including supervised and unsupervised learning algorithms.

**K-Nearest Neighbors Classifier** The K-nearest neighbors (KNN) is a distance-based supervised machine learning algorithm that can be used either as classifier or data regression model [12]. In this study, we have used KNN as a classifier to distinguish three types of plant water stress conditions by calculating the Euclidean distance between unlabeled test data sample and labeled training data points $(X_i, Y_k)$. Equation 9 is used to calculate the Euclidean distance $d_{\text{euclidean}}$ between known class data samples and unknown test data point.

$$d_{\text{euclidean}} = \sum_{j=1}^{N}(x_j^{\text{test}} - X_j^{\text{train}})^2 \tag{9}$$

These distance are measured for K-Nearest Neighbors around the test sample $x_j$ and finally the class label of test data is decided based on a majority voting concept and is described by Eq. 10.

$$C(x_j) = \arg\max \sum_{X_i \in k} C(X_i, Y_k) \tag{10}$$

where $Y_k$ is the $k$th class label.

**Naïve Bayes Classifier** Naïve Bayes Classifier is used in various applications because of its simple structure and reliable performance in complex problems [13]. This model performs well for the limited dataset based on attribute independence hypothesis. The efficiency of this classifier depends on the feature interpret ability of the test data that directly enables the decision making process. Naïve Bayes Classifier uses Bayes' probability theorem to understand the feature dependence in order to classify them into different classes. The probability formula developed by Bayes can be written for the classification of water stress conditions of plant by the following equation.

$$P(H|E) = \frac{P(E|H)P(H)}{P(E)} \tag{11}$$

where $E$ and $H$ stand for evidence and hypothesis, respectively. The probability that the hypothesis is true for a given evidence is determined by dividing the product term of the probability of $E$ for given hypothesis and marginal probability of $H$ by the probability of $E$. In the water stress detection problem, $H$ refers a situation that belongs to a particular stress condition and $E$ represents a situation which belongs to a specified condition. Values of voltage in excess water and less water are close for the formula to generate close probabilities for accuracy calculation.

**K-Means Clustering** K-Means Clustering is a unsupervised learning model that partitions a set of data points or observations into a number of clusters based on the similarity measuring metric with the aid of the squared Euclidean distance, $d(\mathbf{X}_i, \mathbf{X}_j)$ [14]. The squared Euclidean distance between the observations $X_i$ and $X_j$ is defied by Eq. 12.

$$d(\mathbf{X}_i, \mathbf{X}_j) = \|\mathbf{X}_i - \mathbf{X}_j\|^2 \tag{12}$$

The performance of this classifier depends on the selection of cluster numbers and cluster means. The K-Means algorithm is composed of two mathematical steps:

Step 1: Minimization of cost function $J(C)$ with respect to the assigned set of cluster means $\{\mu_i\}_{i=1}^K$ to optimize the clustering process. The minimization operation is expressed as follows:

$$\min \sum_{j=1}^{K} \sum_{C(i)=j} \|\mathbf{X}_i - \mu_j\|^2 \quad \text{w.r.t. } \{\mu_i\}_{i=1}^K \tag{13}$$

Step 2: Optimize the encoder $C$ with respect to the optimized cluster means $\{\mu_i\}_{i=1}^K$ that is computed in Step 1. The encoder is optimized as follows:

$$C(i) = \underset{1 \le j \le K}{\arg\min} \|\mathbf{X}_i - \mu_j\|^2 \tag{14}$$

Starting from some initial choice of the encoder $C$, the algorithm goes back and forth between these two steps until there is no further change in the cluster assignments.

**Table 1** Experimental results for three types of classifiers

| Type of classifiers | Measuring matrix | | | |
|---|---|---|---|---|
| | Accuracy (%) | Precision (%) | Recall (%) | F1-score (%) |
| K-Means Clustering | 37.20 | 61.17 | 37.20 | 34.27 |
| Naïve Bayes | 75.45 | 81.87 | 75.45 | 76.20 |
| K-nearest neighbors | 98.38 | 98.49 | 98.38 | 98.50 |

## 4 Results and Discussion

Plant responds under various uncomfortable environments and it transmits as an electrical signal. These electrical variations carry enough information to understand plant heath conditions in stressful situations. In this study, the stress responses are classified by using both types of classifier models such as supervised and unsupervised machine learning algorithms. The supervised learning models include K-Nearest Neighbors and Naïve Bayes Classifiers. On the other hand, unsupervised learning model employs the K-Means clustering algorithm. The effectiveness of these classifiers are experimentally observed and corresponding results are measured in terms of different metrics like accuracy, precision, recall, and F1-score. The result for three types of classifiers is given in Table 1. The performance of the proposed method is also verified by ROC curve. The ROC curves for Naïve Bayes Classifier, K-Nearest Neighbors, and K-Means Clustering are shown in Fig. 3. Among these three classifiers, K-nearest Neighbors algorithm gives best performance with an accuracy of 98.38%. The evaluation matrices are obtained by averaging the result obtained by fivefold cross-validation. These matrices are defined as follows:

$$\text{Accuracy}\ (\%) = \frac{\text{TP} + \text{TN}}{(\text{TP} + \text{TN} + \text{FP} + \text{FN})} \times 100 \tag{15}$$

$$\text{Precision}\ (\%) = \frac{\text{TP}}{(\text{TP} + \text{FP})} \times 100 \tag{16}$$

$$\text{Recall}\ (\%) = \frac{\text{TP}}{(\text{TP} + \text{FN})} \times 100 \tag{17}$$

$$\text{F1 Score}\ (\%) = 2 \times \frac{(\text{Precision} \times \text{Recall})}{(\text{Precision} + \text{Recall})} \times 100 \tag{18}$$

where TP denotes the true positive class and TN for true negative class, respectively. On the other hand, FP stands for false positive and FN for false negative, respectively. K-Means Clustering is not accurate because the data points in multiple classes are very close in value to each other. K-Nearest Neighbors gives the best accuracy

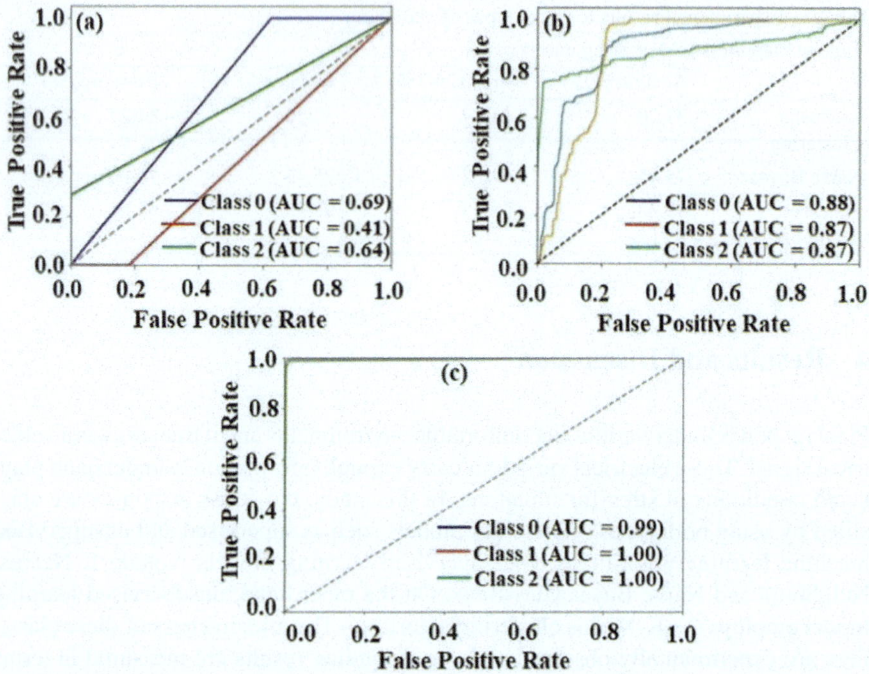


**Fig. 3** **a** ROC curve for K-Means Clustering, **b** ROC curve for Naïve Bayes classifier, and **c** ROC curve for K-nearest neighbors classifier

because K-Nearest Neighbors considers *k* number of nearest data points and calculates accuracy based on the Euclidean distance between data points. This happens for each class separately since it is a classification technique, which is unlike K-Means Clustering which groups data points into *n* number of clusters where *n* is the number of classes. This is why even though the processes of K-Means Clustering and K-Neighbors Classifier resonate, K-Nearest Neighbors is a better choice for accuracy calculation. Naïve Bayes classifier employs the use of Bayes Theorem where one key requirement is that the events must be independent. However, for our dataset, data points in excess water and less water classes are very close to each other which results in Bayes Theorem generating similar result for the two classes.

## 5 Conclusion and Future Scope

The results of this paper show that plant generates electrical signals under different stress conditions. The graphical representations of the signals are more or less indistinguishable but machine learning algorithm capable to discriminate them. This is a potential research area because signal can reflect the internal stress of plant in the adverse environment. In future works, the authors intend to include experiments of these signal conditioning circuits using different species of plants.

## References

1. Sai, K., Sood, N., Saini, I.: Classification of various nutrient deficiencies in tomato plants through electrophysiological signal decomposition and sample space reduction. Plant Physiol. Biochem. **186**, 266–278 (2022)
2. Zhang, H., Zhu, J., Gong, Z., Zhu, J.-K.: Abiotic stress responses in plants. Nat. Rev. Genet. **23**(2), 104–119 (2022)
3. Zhang, et al.: Plants' response to abiotic stress: mechanisms and strategies. Int. J. Mol. Sci. **24**(13), 10915 (2023)
4. Li, J.H., et al.: Plant electrical signals: a multidisciplinary challenge. J. Plant Physiol. **261**, 153418 (2021)
5. Stahlberg, R., Cosgrove, D.J.: The propagation of slow wave potentials in pea epicotyls. Plant Physiol. **113**, 209–217 (1997)
6. Sibaoka, T.: Physiology of rapid movements in higher plants. Annu. Rev. Plant Physiol. **20**, 165–184 (1969)
7. Gallé, A., et al.: Photosynthetic responses of soybean (*Glycine max* L.) to heat-induced electrical signalling are predominantly governed by modifications of mesophyll conductance for $CO_2$. Plant Cell Environ. **36**, 542–552 (2013)
8. Chen, X., Ding, Y., Yang, Y., Song, C., Wang, B., Yang, S., Guo, Y., Gong, Z.: Protein kinases in plant responses to drought, salt, and cold stress FA. JIPB J. Integr. Plant Biol. (2020). https://doi.org/10.1111/jipb.13061
9. Yao, J.P., Wang, Z.-Y., de Oliveira, R.F., Wang, Z.-Y., Huang, L.: A deep learning method for the long-term prediction of plant electrical signals under salt stress to identify salt tolerance. Comput. Electron. Agric. **190**, 106435 (2021)
10. Tripathy, A., Mondal, A., Sharma, P., Hassan, K.: Development of a flexible armalcolite/PDMS sensing device with machine learning for physiological temperature monitoring. IEEE Trans. Mater. Electron Devices **1**, 1–10 (2024). https://doi.org/10.1109/TMAT.2024.3489492
11. Moorman, Richman: Physiological time-series analysis using approximate entropy and sample entropy. Am. J. Physiol. Heart Circ. Physiol. **278**(6), H2039–H2049 (2000)
12. Tsalera, E., Papadakis, A., Samarakou, M.: Monitoring, profiling and classification of urban environmental noise using sound characteristics and the KNN algorithm. Energy Rep. **6**, 6 (2020)
13. Raj, S., Vishnoi, A., Srivastava, A.: Classify Alzheimer genes association using Naïve Bayes algorithm. Human Gene **41**, 201309 (2024)
14. Odi, et al.: Comprehensive review of K-means clustering algorithms. Int. J. Adv. Sci. Res. Eng. (IJASRE) **7**(8) (2021)

# Enhanced Weather Forecasting Using Random Forest, LSTM, Gradient Boosting, and Ensemble Learning (RF + GB)

**Sandeep Kumar Mogha and Aman Kumar Singh**

**Abstract** In the weather forecasting system, prediction is the most challenging process that involves the sequential data analysis to identify the most expressive pattern and to create a higher number of accuracy levels. This work puts forward a new approach for designing a weather prediction system utilizing LSTM networks that is a type of deep learning networks able to model temporal dependencies in the time-series data. This research involves data preparation for analysis and response formulation, regarding the given multi-attribute weather dataset. Such transformations include: how to deal with missing values, how to create new features and normalize the inputs for enhanced predictive capability. In this paper, the relevant parameters of severe weather forecasts are temperature, humidity, and precipitation, which are set as the targets for LSTM model. The outcomes analyzed in the experiment indicate that LSTM model is better in prediction, and the values of MSE and MAE are lower than the values obtained in other similar researches. Such results may demonstrate that LSTM networks have the potential that enhance the current weather forecast, so far, will be of great help in agriculture, transportation logistics, and disaster relief plans field.

**Keywords** Weather forecasting · Machine learning · Random forest · LSTM · Time series · Deep learning

## 1 Introduction

Weather prediction is among the most important uses of present day technology since it is a useful tool in agriculture, disaster control, transport and in everyday life. Estimating temperature, humidity, and precipitation processes, for instance, requires

S. K. Mogha (✉)
Department of Mathematics, University of Delhi, Delhi, India
e-mail: moghadma@gmail.com

A. K. Singh
Department of Mathematics, Chandigarh University, Mohali, Punjab, India

© The Author(s), under exclusive license to Springer Nature Singapore Pte Ltd. 2026 
M. Pant et al. (eds.), *Real-World Applications of Machine Learning*, Mathematics for Sustainable Developments, https://doi.org/10.1007/978-981-95-7134-5_13

consideration of several underlying causes of atmospheric interactions which are hardly linear and always dynamic in most situations. Effective risk management helps in averting risk, getting ready for them, as well as using resources to the maximum; this makes forecasting a major strategy of sustainable development. Statistical and physical methods form the basis of most traditional approaches for weather prediction—autoregressive techniques together with Numerical Weather Prediction (NWP). While these methods have proven to be useful in many situations or environments, they have had issues trying to describe the complexity and large volume of data associated with today's meteorology. Historically, statistical models often presume linear relationships between variables while NWP models require large amounts of computation and are rather insusceptible to initial data input, which results in increased forecast error for longer leading periods [1, 2].

The emergence of the new ML and DL methods has paved the way to the powerful alternatives for weather prediction processes. Concerning the nonlinear patterns, temporal dependencies and large databases, such models as Random Forest (RF), Gradient Boosting (GB), Long Short-Term Memory (LSTM) networks are the most effective. The random forest algorithm, which belongs to the family of ensemble methods based on decision trees, has been repeatedly used mainly because of its high stability and the given ability to work with noise in the data [3]. Regression trees and gradient boosting for decision tree are another prominent algorithms for regression tasks in meteorology [4] since gradient boosting reduces further the error by iteratively learning from it. On the other hand, LSTM, which is a type of recurrent neural network or RNNs, is particularly aimed to capture the long-term dependencies in the time-series data and thereby offers accurate temporal predictions [5]. Although these models have their advantages, the prediction quality of each model has its strengths by varying in a specific dataset and different forecasting period. Due to these drawbacks, ensemble learning that incorporates various predictions by distinct models has been proposed recently as a solution to the problem. Ensemble methods combine strengths of both ML and DL models and increase the robustness of the procedure, as well as minimize overfitting and increase the accuracy of the model's result [6, 7].

Based on these models, this work suggests a new weather forecasting model which combines the random forest, gradient boosting, LSTM, and ensemble method. The identified methodology is expected to take advantage of the above-said models to accurately estimate other important meteorological parameters. The operations of the framework are tested on a dataset containing historical weather information, the performance of which is compared to existing standalone models in order to prove the suitability of the framework.

## 2 Literature Review

The production of reasonable predictions regarding weather conditions has transitioned from basic statistical approaches accompanied by physical models to utilizing more recent ML and DL techniques in the last few years. This section examines the current mostly applied techniques with an indication of the strengths and limitations of the current techniques further highlighting areas that can be improved in establishing weather forecasts.

### *2.1 Probability and Mechanical Models*

In the past, there were some methods in which approaches to weather forecast were done by using statistic methods like ARIMA technique as well as regression models. These approaches presume that all the weather parameters have a linear relationship and are efficient, both in terms of computational cost, for "short range" forecasting [8]. Numerical prediction such as the Numerical Weather Prediction (NWP) means are mathematical models that contain a physical solution to a set of equations that express the phenomena of the atmosphere. However, these NWP models may consume huge CPU resources and; may possess large errors, mostly resulting from sensitivity to initial conditions for examples GFS, WRF, and so on among others [9, 10].

The main or primary source of weakness of these more traditional approaches and methods is the inability to model curvilinear relations, and in addition, relations with large datasets and many variables which are frequently typical for meteorological data. Because of this constraint, there is the emerging use of the ML and DL techniques in predicting the states of the atmosphere.

### *2.2 Machine Learning Techniques*

Hence, random forest (RF) and gradient boosting (GB) have been widely used in weather prediction because the algorithms have the ability of addressing nonlinearity and noise. Random forest which was developed by Breiman is a method of compiling several DF in a framework to reduce on the overfitting and at the same time increase on the precision of the forecasts [3]. They have been used in campaigns requiring forecasting of rainfall, temperature, and other aspects of the climate [11].

Gradient boosting, a technique of iteratively growing weak learners that have superior predictive accuracy to remedy errors in previous predictions, has risen to prominence due to the fact that it outperforms other regression models being tested today and is particularly efficient when dealing with imbalanced datasets [4]. There is a number of models derived from GB including XGBoost and LightGBM that are preferred for scaling-up and effectiveness [12].

However, while these methods are effective for the environment with static data, they are challenged in capturing temporal dependencies, which are key to weather forecasting.

### 2.3 Deep Learning Techniques

And in time-series forecasting, deep learning, particularly RNNs and LSTM networks, have recently exhibited increased performance improvement. It is a fact for traditional RNN that long-term dependencies of the datasets are not well-maintained and this is the reason LSTMs hold information for a long time as compared to the traditional RNNs [5]. These stochastic models have been used with success in a variety of fields and the current forecast of the climate has had a more accurate one measured by performance in relation to factors such as precipitation and wind speed [13, 14].

However, most deep learning models are computationally expensive and require much data a priori for learning. However, they can be very unstable and have a high variance for instance when applied on noisy or small datasets.

### 2.4 Ensemble Learning

Ensemble learning is divided into 4 categories as follows:

Ensemble learning is a process where multiple learned models are used in making more accurate non-erroneous predictions. Of course, various approaches to bagging, boosting, and stacking have been analyzed in the prior literature. For instance, while decision trees employ a mechanism called bagging to control variance in random forest, gradient boosting with boosting to sequentially "rectify" errors [3, 4].

The researches that were made recently show that utilization of both the ML models and DL models contributes to proper results in the case of weather prediction. For example, integrating LSTM with tree-based models that are RF or GB explored LSTM's temporal quality, while the robust pattern discriminating ability of both RF and GB is leveraged [7, 15]. However, existing research material pertaining to such hybrid models is somewhat limited and hence both these areas present significant potential for further creativity as well as actual efficiency enhancement.

### 2.5 Research Gaps

Thus, though much of progress has been made implementing and improving ML, DL, and ensemble techniques for weather prediction, there are still voids left to be filled in current advancements. Firstly, individual models are always over fitted and do not generalize well even when tested on structurally different datasets, which calls for the use of ensembles. Secondly, there is not enough research carried out utilizing both approaches and even fewer integrating both ML and DL. Last but not least, incorporates of the actual time forecasting and live data feed are still problematic due to computational and infrastructural limitations. To fill these gaps, this study advances a novel data analytical strategy that combines random forest, gradient boosting, LSTM, and an ensemble method to achieve higher predictive performance.

## 3 Methodology

To achieve a more reliable and accurate weather prediction, the presented work uses machine learning together with deep learning and also ensemble learning. This section also describes the dataset and exploratory data analysis, data preprocessing, proposed model, as well as the evaluation metrics to ensure a proper chronic toward achieving the research goals.

Data Collection and Description: Data that was used in the analysis of this study includes historical climatic data that was obtained from [source or specific database used for example, synoptic stations or websites]. They include temperature, rainfall, relative humidity, wind speed, air pressure, and among others. These features were selected because of its sensitivity to the weather dynamics and dependency of its operation on the forecasts made [1, 2].

The data is time bound, for example, 10 years, and [total number of records in the dataset is, for instance, one lakh] recorded daily. An initial exploratory analysis was conducted: An analysis of variable relations and their seasonality was conducted with reference to empirical studies.

Data Preprocessing: "Preprocessing is immensely important to get a rather low input data quality and disturbances in data." Key preprocessing steps include:

Handling Missing Values: In handling of missing data, missing data containing datasets were handled using mean for interval data while mode was used for categorical data [16].

Feature Scaling: Because most of the machine learning models depend on the magnitude of data, these features were normalized to [0, 1] using the min-max normalization technique. This means that all models have standard input that improves their modeling systems [17].

Feature Engineering: Additional new variables were obtained from the initial variables such as the moving averages, lagged variable for explaining the temporal relation, and for increased prediction precision or accuracy.

Data Partitioning: Since the purpose was to test the performance of models, the entire dataset was divided into 80% training data and 20% test data for model validation [18].

Model Design/Model Architecture

Random Forest (RF): Random forest, an ensemble learning method using decision trees, was used to build the model because it generates nonlinear transformations in the data. To perform the prediction, the model generates many decision trees at training and their performance is aggregated. Hyperparameters associated with high importance, such as the number of trees and maximum depth, were tuned through grid search [3].

Gradient Boosting (GB): Another ensemble learning technique that learns a model iteratively is the gradient boosting. The implementation used GB that in this case was the scalable version known as XGBoost which is the most of time more efficient and accurate. Other tuning parameters include learning rate, number of estimators, and the maximum depth—all of which were set using exhaustive search [12].

Long Short-Term Memory (LSTM): Time dependencies in weather data were captured using LSTM networks. The model has memory cells, which allows for the storage of long-term dependencies, which is perfect for time-series forecasting. The architecture includes only one long short-term memory layers with 50 units in it and one dense layer. The model was trained using the Adam optimizer and the Mean Squared Error (MSE) loss function is adopted for this experiment [5].

Ensemble Model (RF + GB): In order to combine both random forest and gradient boosting, an ensemble model was built. To get an accurate and reliable result RF and GB were averaged in order to have a good output. This combined approach is provided by RF's weakness to overfitting and GB's possibility to recognize intricate patterns [7].

Evaluation Metrics

The performance of the models was assessed using the following metrics:

Mean Absolute Error (MAE): Measures the average level of error with regard to the measures of magnitude.

Root Mean Squared Error (RMSE): Pays more attention to larger deviations, thus, it is an outlier detection method.

R-Squared ($R^2$): Marks the amount of variation that can be attributed to the model.

These metrics make certain that the performance of the model is measured on different parameters, both efficiency and dependability [19].

Experimental Setup: The experiments have been performed using Python 3.9 on a system which has [CPU—specify, for example, Intel Core i7, RAM—specify, for instance, 16 GB, GPU—if necessary]. The implementation of the factor mentioned above was done using libraries including Scikit-learn, TensorFlow, and XGBoost. Regularization techniques included L1 and L2 norm and hyperparameters were tuned using a grid search to get the best result from our models.

### 3.1 Comparison

To compare the models mentioned above, we will take into account the MAE, RMSE, and the coefficient of determination ($R^2$) which are collectively called the performance matrix for regression models as explained by Plevris et al. [20].

**MAE** It gives an idea of the magnitude of the average error regardless the direction of the error. It is measured by the absolute difference between predicted and actual values.

$$\text{MAE} = \frac{1}{n}\sum_{i=1}^{n} |y_i - \widehat{y}_i| \quad (1)$$

**RMSE** It is similar to MAE but gives more importance to larger errors, as it squares the differences and then performs the average. RMSE is more sensitive to outliers than MAE.

$$\text{RMSE} = \sqrt{\frac{1}{n}\sum_{i=1}^{n} (y_i - \widehat{y}_i)^2} \quad (2)$$

***R*² (*R*-Squared)** It gives the proportion of the variance in the dependent variable which is predictable from the independent variables. It ranges from 0 to 1.

$$R^2 = 1 - \frac{\sum_{i=1}^{n} (y_i - \widehat{y}_i)^2}{\sum_{i=1}^{n} (y_i - \bar{y}_i)^2} \quad (3)$$

where $y_i$ are the actual values for the $i$th observation, $\widehat{y}_i$ are the predicted values for the $i$th observation, $\bar{y}$ the mean of all the actual values in the data, $n$ is the total number of observations.

## 4 Result

Random Forest (RF): When using the Random Forest model, it predicted the maximum and minimum temperatures with acceptable accuracy. The comparison for the maximum temperature yielded an $R^2$ value of 0.99 which depicted a very high degree of accurate model making in terms of percentage variation. The RMSE of 0.31 and MAE value of 0.07 show that overall the error rates are low with mean error being the least. Likewise for the minimum temperature, the performance of RF was outstanding, with $R^2$ of 0.99, RMSE of 0.23, as well as the MAE of 0.05. These results demonstrate that the model developed here is quite appropriate for processing structured data. Furthermore, the model was successful in dealing with overfitting, which is one of the main advantages of the presented model, so it can be used for the specific predictive tasks with high accuracy.

Long Short-Term Memory (LSTM): The performance of LSTM model was again 100%, with $R^2$ at 1 in both maximum and minimum temperatures forecast. This means that the model accounts for all types of variance in the data, with none of it freely varying. The maximum temperature RMSE was 0.15 while the minimum temperature was 0.14 relatively lower than the random forest model indicating high prediction capability. Although the MAE of LSTM was slightly higher than that of random forest, LSTM produced an acceptable maximum temperature MAE of 0.11, and minimum temperature MAE of 0.12. Nonetheless, the LSTM is very effective in the case of the modeling of temporal dependencies and the nonlinear patterns making their use effective in the time series, sequences with dependencies vital.

Gradient Boosting (GB): A similarly high accuracy was observed in gradient boosting analysis, for both maximum temperature—$R^2$ 0.995, and minimum temperature—$R^2$ 0.9947, which points to nearly ideal fit of the model. The maximum RMSE was 0.31 and the minimum RMSE was 0.23, which gave the same results as the random forest model, while the maximum MAE was 0.0655 and the minimum MAE was 0.0471 revealing that the model provides low error predicting average results. In general gradient boosting established a reasonable accuracy and low error rate that makes it a good contender for tasks that require high accuracy and low error rate.

Ensemble Learning (RF + GB): Unlike the individual models, ensemble model comprising of random forest and gradient boosting displayed much poorer performance. In the maximum temperature, the model has an $R^2$ of 0.0628 and so it is evident that this model has a poor fit in the given data as compared to individual models. Furthermore, the obtained MAE of 3.49 also indicates that the model was producing big errors in its predictions. These results imply that there might be some incompatibility or optimization problems when the two models are integrated which could have resulted to failure of the ensemble outcompete the individual models. It points to the direction of the current study's limitations and an avenue for subsequent research to enhance the ensemble method. The performance results summarized in Table 1 indicate significant improvement in weather forecasting.

## 5 Conclusion

In this paper, a systematic evaluation of the random forest model, LSTM model, gradient boosting, and an ensemble model were made to estimate the maximum and minimum temperatures accurately. The following section introduces the features of each model that was used in the paper, and explain the strengths and weaknesses of the models.

Random forest and gradient boosting gave good results in cases of structured tabular data and using the results for prediction was quite easy. However, they may not be able to consider time-related dependencies that exist in the sequential datasets. LSTM proved quite useful as a tool for model development for time-series data and surpassed other approaches in studying components of the temperature data. Of all of

**Table 1** Model performance comparison

| Model | Max temp metrics | Min temp metrics | Analysis |
|---|---|---|---|
| Random Forest (RF) | – $R^2$: 0.99<br>– RMSE: 0.31<br>– MAE: 0.07 | – $R^2$: 0.99<br>– RMSE: 0.23<br>– MAE: 0.05 | High accuracy, good for structured data, handles overfitting well |
| LSTM | – $R^2$: 1.00<br>– RMSE: 0.15<br>– MAE: 0.11 | – $R^2$: 1.00<br>– RMSE: 0.14<br>– MAE: 0.12 | Perfect $R^2$ values, models temporal dependencies and nonlinear patterns, slightly higher MAE than RF |
| Gradient Boosting | – $R^2$: 0.995<br>– RMSE: 0.31<br>– MAE: 0.0655 | – $R^2$: 0.9947<br>– RMSE: 0.23<br>– MAE: 0.0471 | Strong predictive capability, low error rates |
| Ensemble (RF + Gradient Boosting) | – $R^2$: 0.0628<br>– MAE: 3.49 | | Lower performance than individual models, potential optimization issues or model incompatibility |

these, the learning long-term dependencies are rather good with this model. However, its architecture has higher nonlinearity, and the model overfits could be quite difficult here. Based on theoretical and empirical comparisons with other models at this level as a new promising ensemble model, it is evident that this technique does not improve this study's need as hoped for. This may include; inadequate hyperparameters tuning, insufficient data or the intrinsic limitation of the base models.

## References


1. Kalnay, E., Kanamitsu, M., Kistler, R.: The NCEP/NCAR 40-year reanalysis project. Bull. Am. Meteorol. Soc. **77**(3), 437–471 (1996)
2. Wilks, D.S.: Statistical Methods in the Atmospheric Sciences. Academic Press (2011)
3. Breiman, L.: Random forests. Mach. Learn. **45**(1), 5–32 (2001)
4. Friedman, J.H.: Greedy function approximation: a gradient boosting machine. Ann. Stat. **29**(5), 1189–1232 (2001)
5. Hochreiter, S., Schmidhuber, J.: Long short-term memory. Neural Comput. **9**(8), 1735–1780 (1997)
6. Dietterich, T.G.: Ensemble methods in machine learning. In: Proceedings of the First International Workshop on Multiple Classifier Systems, pp. 1–15 (2000)
7. Zhou, Z.-H.: Ensemble Methods: Foundations and Algorithms. Chapman and Hall/CRC (2012)
8. Box, G.E., Jenkins, G.M.: Time Series Analysis: Forecasting and Control. Holden-Day (1976)
9. Kalnay, E.: Atmospheric Modeling, Data Assimilation, and Predictability. Cambridge University Press (2003)
10. Skamarock, W.C., et al.: A description of the advanced research WRF version 3. NCAR technical note (2008)

11. Puca, S., et al.: Rainfall estimation from weather radar and machine learning techniques. J. Hydrol. **514**, 231–239 (2014)
12. Chen, T., Guestrin, C.: XGBoost: a scalable tree boosting system. In: Proceedings of the 22nd ACM SIGKDD International Conference on Knowledge Discovery and Data Mining, pp. 785–794 (2016)
13. Rasp, S., et al.: Deep learning to predict diabatic heating from reanalysis data. Geophys. Res. Lett. **45**(16), 8659–8668 (2018)
14. Shi, X., et al.: Convolutional LSTM network: a machine learning approach for precipitation nowcasting. Adv. Neural Inf. Process. Syst. 802–810 (2015)
15. Dong, S., et al.: Hybrid models for weather forecasting using LSTM and gradient boosting machines. Appl. Sci. **11**(2), 850–860 (2021)
16. Little, R.J., Rubin, D.B.: Statistical Analysis with Missing Data. Wiley (1987)
17. Han, J., Kamber, M., Pei, J.: Data Mining: Concepts and Techniques. Morgan Kaufmann (2011)
18. Kuhn, M., Johnson, K.: Applied Predictive Modeling. Springer (2013)
19. Hyndman, R.J., Athanasopoulos, G.: Forecasting: Principles and Practice. OTexts (2018)
20. Plevris, V., Solorzano, G., Bakas, N.P., Ben Seghier, M.E.A.: Investigation of performance metrics in regression analysis and machine learning-based prediction models

# Spatio-Temporal Analysis and Severity Prediction of Weather Events Using Machine Learning Models

**Rahul Singh and Bholanath Roy**

**Abstract** A proposed machine learning framework for predicting the severity of weather events using spatio-temporal data on a large dataset of all one million weather events recorded across the United States. In this article, modern data preprocessing, exploratory data analysis, and feature engineering were used to improve our random forest and gradient boosting models. The random forest model produced an accuracy of 99.66%, an $F1$-Score of 0.9966, and an AUC-ROC of 1.0, demonstrating considerable predictive reliability. In contrast, though the gradient boosting model also displayed a marginally lower accuracy (94.86%), it excelled in terms of balanced performance, measured by an $F1$-score of 0.9476 and AUC-ROC of 0.9958, demonstrating its proficiency in handling diverse and imbalanced data scenarios. This suggests that our findings may prove to be vital in the field of meteorology, despite the fact that the systems are built on significantly different premises for analysis.

**Keywords** Machine learning · Spatio-temporal analysis · Random forest · Gradient boosting · Climate resilience · Severe weather analysis

## 1 Introduction

Weather records affect agriculture, infrastructure, transportation, and human lives. For effective disaster management, resource allocation, and risk mitigation, it is essential to predict the occurrence, intensity, and geographical distribution of weather-related events [1]. You might know that with the growth of the more advanced systems of data collection and the existence of the large-scale weather datasets, machine learning [2] has arisen as one of the destructive tools to observe the weather patterns and ultimately to predict the weather.

R. Singh (✉) · B. Roy
Maulana Azad National Institute of Technology, Bhopal, India
e-mail: rahulit0606@gmail.com

© The Author(s), under exclusive license to Springer Nature Singapore Pte Ltd. 2026 
M. Pant et al. (eds.), *Real-World Applications of Machine Learning*, Mathematics for Sustainable Developments, https://doi.org/10.1007/978-981-95-7134-5_14

Many current models focus on physical or statistical models while being limited in their capacity to address complex, nonlinear relationships on large, diverse datasets. Machine learning forms an alternative by using historical weather information [3] to find patterns and make predictions without using an explicit set of domain-specific equations. Utilizing spatio-temporal data [4] improves ML implementation by allowing the models to better capture regional and temporal variability in atmospheric climate variables.

Measuring and analyzing weather events across America are the focus of this study, an assessment of a large weather dataset. The dataset contains several features: type of event, amount of rainfall, location (latitude/longitude), date and time (month and year), and severity. The emphasis is on making statistical interpretations intrinsic to weather events, finding relations in data with respect to weather, and extrapolating the severity of events in different regions vulnerable to such conditions [5].

The performance of these models is crucial as they employ advanced machine learning techniques to learn from historical data, enabling the accurate forecasting of severe weather conditions without the need for explicit programming of atmospheric interactions. This facilitates better preparedness and informed decision-making in weather-sensitive sectors.

In this approach, data preprocessing, exploratory data analysis, and predictive modeling are combined using machine learning algorithms like random forest [6] and gradient boosting [7]. Data visualizations are used to identify patterns in the weather, and prediction models are scored on metrics like accuracy, $F1$-score, [8] and AUC-ROC [9]. Geospatial analyses also pinpoint areas of more concentrated severe weather activity and give a regional and seasonal perspective on the trends.

This paper advances the field of weather prediction by developing and evaluating novel machine learning techniques that increase the accuracy of forecasting the severity of weather events. This work enhances the understanding of complex spatio-temporal data, such as severe weather prediction, through novel models like random forest and gradient boosting.

The methods used in our research resolve many theoretical and empirical gaps in existing methodologies, including comprehensive data preprocessing, robust feature engineering, and innovative evaluation of models. These insights not only enhance the technical soundness of predicting weather severity but also provide useful implications for decision-makers regarding improved disaster management and resource distribution.

The results of this study are anticipated to aid policymakers and stakeholders in strengthening climate resilience and improving responses to weather-induced emergencies.

The rest of this paper is organized as follows: Sect. 2 introduces the dataset used in this study, emphasizing its spatio-temporal aspects and the preprocessing operations made to enable proper analysis. Section 3 provides the literature review, presenting the relevant research before situating our approach within the focus area of machine learning for weather forecasting while identifying gaps we aim to address. Section 4 outlines the methodologies implemented in this study, including the machine learning models employed (i.e., random forest and gradient boosting) and exploratory data analysis techniques used to identify underlying data patterns. Section 5 presents the results of our models, including their predictive accuracy and implications for practical use in disaster management and weather forecasting. Ultimately, the paper concludes with a discussion of major findings, contributions to the literature, and future directions for research in Sect. 6, emphasizing the importance of this work in building resilience to climate change and supporting informed decision-making in climate-sensitive areas.

## 2 Data Description

This includes a dataset consisting of information on weather events in the US [10], with details on the metar and time of occurrence. This dataset is used as a part of this study. That includes information on the type of weather event, precipitation levels, geographic coordinates, and specific elements of each location, like city, county, state, and zip code. Temporal features, such as month and year of incident, can be used for longitudinal or seasonal analysis. Translating the given text: additional parameters like the duration of the event (in seconds), type of associated incidents, and severity defined as light, moderate, severe, or other are important features that provide useful information for predictive modeling and risk assessments.

The dataset has various highlighted features as mentioned below. A description of the weather **Type** (rain, fog, snow, or storm) is used as an important classification for the frequency and severity of weather events. The **Precipitation (in)** attribute contains the amount of precipitation, in inches, that occurred during the event, an important metric for measuring the intensity of the event. **LocationLat** and **LocationLng** indicate the latitude and longitude of the location of the event, which can be used for geospatial analysis and detection of event hotspots [11]. Provides granular and regional location-specific details by locality, including **City**, **County**, **State**, and **ZipCode**, that enable geographic comparisons and analysis over trends.

Temporal attributes [12] include **Month** and **Year**, which specify the timing of events, allowing for the exploration of seasonal and annual trends. The **dur_seconds** attribute records the duration of weather events in seconds, offering insights into the persistence of weather conditions. **incident_type** denotes the type of incident associated with the weather event, such as fire or flood, linking meteorological phenomena to their broader impacts. The **place_code** attribute serves as a location identifier for event aggregation and spatial analyses. The **Severity** attribute categorizes the intensity of weather events as light, moderate, severe, or other, making it a critical variable for classification tasks in this study.

This dataset contains 1,000,000+ records over several years, and so it is useful to study large-scale weather patterns at the national level. They also performed one-hot encoding and scaled the features such as precipitation and duration. To improve the analytical capabilities of the dataset, derived features (e.g., seasonal indicators) were built (the original internal time attributes had no temporal metadata types associated). This makes the dataset well-prepared for predictive modeling, spatio-temporal analysis, and impact assessment.

Figure 1 illustrates the distribution of types of weather events in the dataset. The chart visualizes that rain occurs most often, followed by fog and snow, and then events disseminate further with cold, precipitation, storm, and hail taking place least frequently. This distribution illustrates the commonness of certain weather patterns and the infrequency of truly catastrophic events.

In Fig. 2 is presented the detailed heatmap of weather event types with respective severity levels. Light severity numbers are by far the most common feel of events such as rain, fog, and snow, while heavy and moderate severity levels are infrequent. This distribution emphasizes the prevalence of mild weather in the dataset.

The temporal trends of weather events are shown in Fig. 3 which shows its monthly occurrences over the analyzed period. There are distinct spikes in some years, possibly associated with extreme weather events. Such trends underscore the need for an understanding of temporal variability to isolate trends associated with seasonal and yearly weather variations.

These visualizations are essential for understanding the dataset, revealing when types of weather dominate, what severity occurs over time, and what patterns can be found, all of which need to be tracked for predictive modeling.

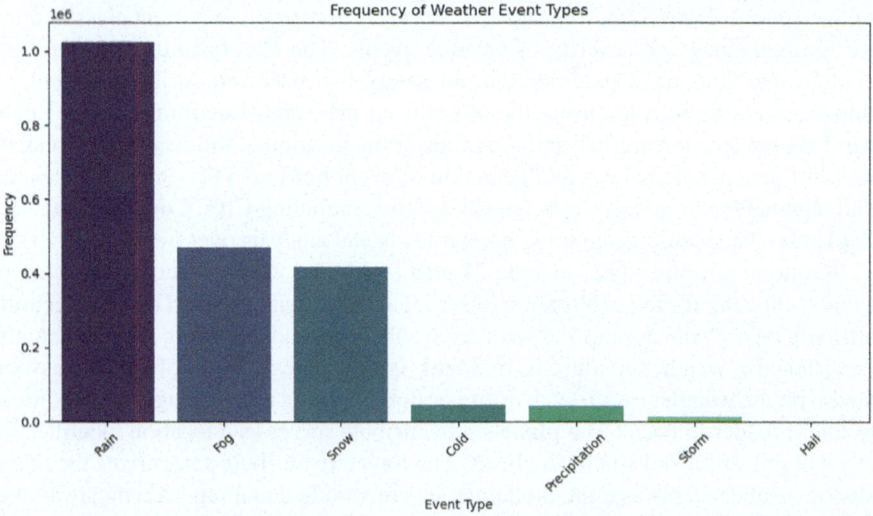


**Fig. 1** Frequency distribution of weather event types. The chart highlights rain as the most frequent event, followed by fog and snow, while cold, precipitation, storm, and hail occur significantly less often

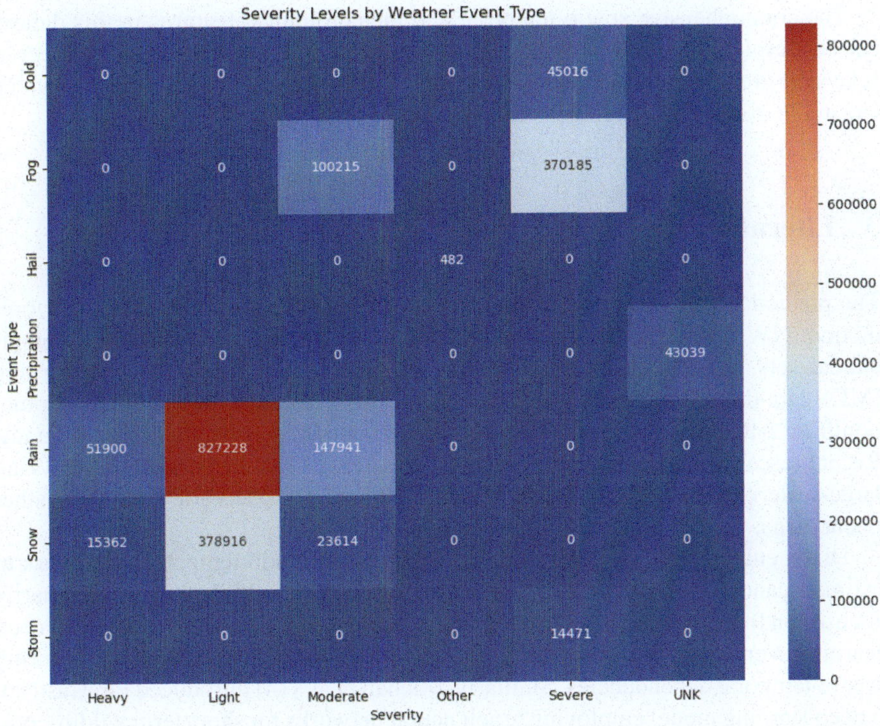


**Fig. 2** Heatmap of severity levels by weather event type. The analysis shows light severity levels dominate rain, fog, and snow events, with moderate and heavy levels appearing less frequently across events

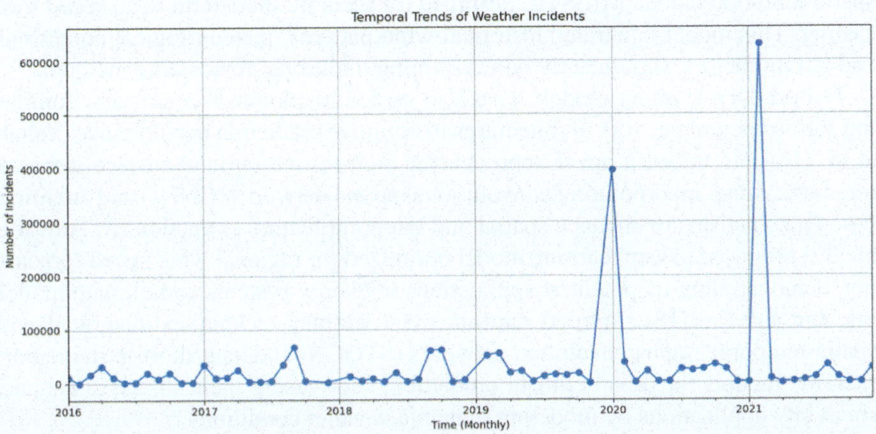


**Fig. 3** Temporal trends of weather incidents. The graph illustrates monthly incident frequencies, revealing significant spikes during certain years, possibly due to extreme weather conditions

The comprehensive spatio-temporal and meteorological features in this dataset can be useful for research on weather forecasting and disaster management strategies. It lays the groundwork for severity modeling, risk hotspots, and trend analysis, which can aid in disaster preparedness and mitigation.

## 3 Literature Review

The prediction of weather events has historically relied on numerical weather prediction (NWP) models that use complex physical equations and vast meteorological datasets. However, with the rise of artificial intelligence and machine learning (ML), data-driven approaches have gained prominence due to their ability to capture nonlinear patterns in spatio-temporal data, leading to improved predictive accuracy. Recent research has demonstrated the effectiveness of ML techniques in analyzing large-scale environmental data, enhancing prediction models for various weather phenomena.

Dixon et al. [13] developed a deep learning-based spatio-temporal framework to estimate canopy height using time-series satellite imagery. While the study primarily focused on forest monitoring, the underlying approach highlights the potential of deep learning in environmental modeling, reinforcing the effectiveness of ML in capturing spatio-temporal dependencies. Similarly, Michail et al. [14] introduced FireCastNet, a deep learning model employing graph neural networks for short-term wildfire prediction. By processing meteorological and satellite-derived data, this model demonstrated the advantages of leveraging spatio-temporal patterns to enhance forecasting accuracy. Lu et al. [15] further contributed to this field by proposing a multivariable spatio-temporal model, MIESTC, designed for accurate short-term wind speed forecasting. This model integrated historical wind patterns, meteorological conditions, and terrain factors, significantly outperforming traditional forecasting methods.

Hybrid deep learning models have also gained traction in recent years, combining various neural network architectures to optimize prediction performance. Vahabi et al. [16] introduced a novel approach for high-resolution land surface temperature forecasting, incorporating convolutional neural networks (CNNs) and attention-based mechanisms to enhance spatial and temporal feature extraction. Al Aawar et al. [17] proposed a deep learning model optimized for regional wind speed forecasting, demonstrating its potential applications in energy systems and climate modeling. Lin and Lin [18] explored reinforcement learning techniques integrated with spatio-temporal graph convolution networks (STGCN) and long short-term memory (LSTM) models for urban climate prediction, showcasing their ability to enhance smart city applications by modeling dynamic weather conditions.

Climate variability and extreme weather events pose significant challenges for forecasting models. Several studies have addressed these concerns by incorporating climate data into ML frameworks. Eskandari et al. [19] analyzed the impact of climate change on wildfire regimes, integrating regression-based and ML models to predict fire occurrences in semi-arid regions. Fischer et al. [20] examined how extreme weather patterns influence the transmission of vector-borne diseases, applying spatio-temporal ML models to track disease spread based on climatic fluctuations. Similarly, Fan et al. [21] proposed a multi-view spatio-temporal model incorporating real-time meteorological data to improve demand forecasting for ride-hailing services, illustrating the broader applicability of weather prediction models.

Computational efficiency remains a critical concern for large-scale weather prediction models, particularly those reliant on deep learning. Recent research has focused on reducing the computational cost of these models while maintaining predictive accuracy. Chen et al. [22] introduced a hybrid deep learning approach incorporating spatio-temporal attention mechanisms to optimize air pollution forecasting, demonstrating a substantial reduction in processing time. Bai et al. [23] developed a generative deep learning framework to reconstruct tropical cyclone microwave data, significantly enhancing real-time forecasting capabilities while minimizing computational overhead. Luo et al. [24] proposed a global soil moisture prediction framework that integrates prior knowledge with spatio-temporal deep learning, improving the efficiency of large-scale environmental modeling.

Despite these advancements, several challenges remain. Scalability and generalization issues persist, as models trained on one geographic region may not perform well in others due to variations in climate patterns. Additionally, data imbalance is a major concern, particularly when predicting severe weather events, as ML models may struggle with underrepresented classes, leading to suboptimal predictions. Computational costs also remain a barrier to real-time forecasting, necessitating further research into more efficient algorithms. Future studies should explore hybrid architectures that balance computational efficiency with predictive accuracy, incorporating attention-based models to enhance spatio-temporal feature extraction. The integration of real-time sensor data with ML-based frameworks is another promising avenue for improving weather prediction capabilities.

This review underscores the rapid evolution of spatio-temporal weather prediction using ML techniques, highlighting both the successes and ongoing challenges in this field. By leveraging advanced deep learning architectures, researchers continue to push the boundaries of predictive modeling, paving the way for more accurate and efficient weather forecasting systems.

## 4 Methodology

This study presents a new approach to analyzing and predicting the intensity of weather events, with spatio-temporal data coupled with machine learning models. This tactical approach involves a high level of complex data preprocessing, exploratory data analysis (EDA), visualization, and predictive modeling, engineered to highlight patterns and actionable insights. The below graphics represent the essential elements of the suggested approach.

### 4.1 Data Preprocessing

This step deals with the processing of the raw dataset into a form that can work with in the modeling process by keeping things consistent, accurate, and manageable to work with. Missing numerical features, like precipitation and duration, are filled up with mean or median values. For a numerical feature $x_i$ with missing values, the imputed value $\hat{x}_i$ is calculated as:

$$\hat{x}_i = \frac{\sum_{j=1}^{n} x_j}{n}, \tag{1}$$

where $x_j$ represents the observed values and $n$ is the total number of non-missing values.

For categorical features, such as event type or severity, fill with the most common value. Then applied label encoding and one-hot encoding to those features to fit them into machine learning algorithms. Continuous features like precipitation and duration are Min-Max scaled, defined as:

$$x_{\text{scaled}} = \frac{x - x_{\min}}{x_{\max} - x_{\min}}, \tag{2}$$

where $x$ is the original value, and $x_{\min}$ and $x_{\max}$ are the minimum and maximum values of the feature, respectively. Temporal features, such as seasons, are derived from the month column, while regional clustering features are created using latitude and longitude.

### 4.2 Exploratory Data Analysis (EDA)

There is exploratory data analysis (EDA) [25] to reveal patterns and relationships in the data. Bar charts are used to analyze the frequency of different weather events, and the distribution of severity levels across event types and geographic regions is projected using heatmaps and stacked bar charts. Line graphs are used to visualize seasonal and yearly shifts in the weather phenomenon. Density maps and cluster visualizations are used to identify geographic hotspots of severe weather events.

### 4.3 Predictive Modeling

The study employs advanced machine learning algorithms to predict weather event severity, contributing to the novelty of this research.

In refining the performance of our predictive models, an extensive hyperparameter tuning process was employed to optimize the random forest and gradient boosting classifiers. This involved setting up a grid search to explore a wide range of values for key parameters such as the number of trees, depth of each tree, learning rate, and the number of features considered at each split. For the random forest model, parameters like `max_features`, `max_depth`, and `n_estimators` were varied, while for the gradient boosting model, we adjusted parameters including `n_estimators`, `learning_rate`, and `max_depth`. The grid search was conducted using a cross-validation approach to ensure the robustness of our models against overfitting. The selection of the best hyperparameters was based on the maximization of the $F1$-score, a critical metric given the imbalanced nature of our dataset, reflecting a comprehensive balance between precision and recall. This meticulous tuning process not only enhanced the accuracy of the models but also their generalizability across different severity levels of weather events.

**Random Forest Classifier**: The random forest algorithm constructs multiple decision trees during training. The prediction is based on majority voting across the trees. For a set of $m$ trees, the final predicted class $\hat{y}$ for an instance $x$ is:

$$\hat{y} = \text{mode}\{T_1(x), T_2(x), \ldots, T_m(x)\}, \tag{3}$$

where $T_i(x)$ is the prediction from the $i$th decision tree.

**Gradient Boosting Classifier**: Gradient boosting builds a sequence of weak learners, with each tree correcting the residual errors of its predecessor. For a weak learner $h(x)$, the final model $F(x)$ is expressed as:

$$F(x) = \sum_{i=1}^{N} \alpha_i h_i(x), \tag{4}$$

where $h_i(x)$ is the $i$th weak learner, $\alpha_i$ is the learning rate, and $N$ is the total number of learners. The loss function is minimized iteratively to improve accuracy:

$$L = \sum_{j=1}^{n} \left(y_j - F(x_j)\right)^2, \tag{5}$$

where $y_j$ is the true value, and $F(x_j)$ is the predicted value for instance $j$.

Both models are trained using the processed dataset, and hyperparameter tuning is performed using grid search to optimize performance.

## 4.4 Model Evaluation

The performance of the predictive models is assessed using several metrics:

**Accuracy**: The percentage of correct predictions, defined as:

$$\text{Accuracy} = \frac{\text{Number of Correct Predictions}}{\text{Total Number of Predictions}}. \tag{6}$$

***F*1-Score**: The harmonic mean of precision and recall [26], given by:

$$F1 = 2 \cdot \frac{\text{Precision} \cdot \text{Recall}}{\text{Precision} + \text{Recall}}, \tag{7}$$

where Precision and Recall are defined as:

$$\text{Precision} = \frac{\text{True Positives}}{\text{True Positives} + \text{False Positives}}, \tag{8}$$

$$\text{Recall} = \frac{\text{True Positives}}{\text{True Positives} + \text{False Negatives}}. \tag{9}$$

**AUC-ROC**: Area under the receiver operating characteristic curve, which measures how well the model separates classes.

It applied cross-validation for robust evaluation handling, and it analyzed the feature importance derived from the models to determine the most critical predictors of weather event severity.

## *4.5 Visualization Techniques*

We also use advanced visualization methods to supplement the predictive modeling. To visualize the geographic distribution of severe weather events, density maps and heatmaps are used for hotspot analysis using geospatial mapping. By applying K-Means to group various regions with similar weather, it introduces a novelty in the geo-analysis.

## *4.6 Novel Contributions*

This study has a number of new findings. It combines spatial and temporal details for new levels of predictive power, utilizing the unique capability of random forest and gradient boosting to capture nonlinear and complex relationships and thus to give us a diverse set of predictions through a dual algorithm approach, and enhance interpretability and performance via feature engineering, such as seasonal/regional insights. Finally, using advanced mapping techniques, actionable geospatial insights are provided. This study offers a framework for weather-related event severity analysis and prediction and could help significantly with disaster preparedness and resource optimization by using these methods and algorithms.

**Table 1** Performance metrics for random forest model

| Metric | Value |
|---|---|
| Accuracy | 0.9966 |
| $F1$-score | 0.9966 |
| AUC-ROC | 1.0 |

**Table 2** Classification report for random forest model

| Class | Precision | Recall | $F1$-score | Support |
|---|---|---|---|---|
| 0 | 1.0 | 1.0 | 1.0 | 13,459 |
| 1 | 1.0 | 1.0 | 1.0 | 241,389 |
| 2 | 0.99 | 0.99 | 0.99 | 54,464 |
| 3 | 1.0 | 1.0 | 1.0 | 90 |
| 4 | 0.99 | 0.99 | 0.99 | 85,640 |
| 5 | 1.0 | 1.0 | 1.0 | 8632 |

# 5 Results and Discussion

This section presents and discusses the classification performance of weather events by random forest and gradient boosting models. The evaluation metrics, such as accuracy, $F1$-score, AUC-ROC, and class-wise performance metrics, are discussed. The following tables summarize the results for both models. The significant performance metrics such as high accuracy and $F1$-scores achieved by our models underscore their potential in real-world applications, from enhancing emergency response strategies to optimizing the allocation of resources during weather crises.

## *5.1 Random Forest Results*

The random forest model demonstrated exceptionally high performance, achieving an accuracy of **99.66%**, an $F1$-score of **0.9966**, and an AUC-ROC score of **1.0**. The classification report indicates near-perfect performance across all classes, as detailed in Tables 1 and 2.

These findings indicate the random forest model performed accurately enough with little rate of misclassification when predicting weather events. But the almost perfect score is a warning sign for both overfitting and data leakage potential. The minority classes like class 3 (support = 90) and class 5 (support = 8632) have perfect scores which could imply that the patterns in the dataset are separable. However, further validation is warranted to establish the robustness of these results.

**Table 3** Performance metrics for gradient boosting model

| Metric | Value |
|---|---|
| Accuracy | 0.9486 |
| $F1$-score | 0.9476 |
| AUC-ROC | 0.9958 |

**Table 4** Classification report for gradient boosting model

| Class | Precision | Recall | $F1$-score | Support |
|---|---|---|---|---|
| 0 | 0.88 | 0.89 | 0.88 | 13,459 |
| 1 | 0.99 | 0.99 | 0.99 | 241,389 |
| 2 | 0.85 | 0.75 | 0.80 | 54,464 |
| 3 | 1.0 | 1.0 | 1.0 | 90 |
| 4 | 0.90 | 0.96 | 0.93 | 85,640 |
| 5 | 1.0 | 1.0 | 1.0 | 8632 |

## 5.2 *Gradient Boosting Results*

The gradient boosting model exhibited slightly lower performance compared to random forest, achieving an accuracy of **94.86%**, an $F1$-score of **0.9476**, and an AUC-ROC score of **0.9958**. The class-wise metrics, shown in Tables 3 and 4, highlight the variability in model performance across different classes.

The gradient boosting model's performance is more varied compared to random forest. Class 2, with moderate support (54,464), shows the lowest $F1$-score (0.80) due to a trade-off between precision (0.85) and recall (0.75). In contrast, dominant classes like class 1 (support = 241,389) exhibit nearly perfect metrics. Minority classes (e.g., class 3 and class 5) are perfectly classified, similar to random forest, likely due to well-separated patterns in the dataset.

## 5.3 *Comparison and Interpretation*

Show yet again that random forest beats gradient boosting in everything from top accuracy to class-wise metrics. But the gradient boosting model seems to provide more realistic results—particularly for difficult classes such as class 2, where it doesn't get perfect metrics. That inconsistency showcases gradient boosting's sensitivity to imbalance and the compromises it strikes as its learning progresses via iterations.

Our analysis reveals that while the random forest model shows near-perfect accuracy, it raises concerns about potential overfitting, which we address through further validation techniques. The gradient boosting model, while slightly less accurate,

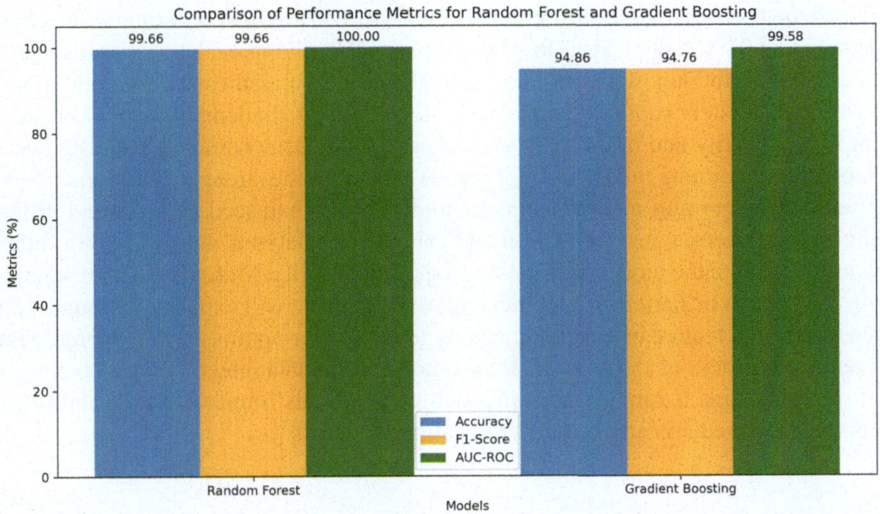


**Fig. 4** Comparison of performance metrics (accuracy, $F1$-score, and AUC-ROC) for random forest and gradient boosting models

offers better performance in handling imbalanced data, demonstrating its practical applicability in real-world scenarios where data anomalies are common.

Both models have similar AUC-ROC scores, which are near perfect, indicating great performance; however, the random forest's results being closer to 1.0 indicate possible overfitting. With fewer patterns for this target class, gradient boosting seems to generalize better.

The bar chart in Fig. 4 compares the performance of random forest and gradient boosting models using accuracy, $F1$-Score, and AUC-ROC. Random forest achieves near-perfect metrics (accuracy and $F1$-score: 99.66%, AUC-ROC: 100%), while gradient boosting shows competitive results (accuracy: 94.86%, $F1$-score: 94.76%, AUC-ROC: 99.58%). These results highlight random forest's high accuracy, though gradient boosting offers more balanced generalization.

## 6 Conclusion

This study performed a comparative analysis of random forest and gradient boosting models in the context of the classification of weather event severity using spatio-temporal data. The demonstrated predictive accuracy of our models not only fills critical gaps in weather prediction research but also offers practical insights for policymaking and planning, contributing to increased climate resilience and economic efficiency in managing weather-related challenges. The random forest model gave us perfect metrics (99.66% and 1.0 AUC-ROC) showing excellence in the model, but it causes the problem of overfitting or data leakage. The gradient boosting achieved an accuracy of 94.86% and AUC-ROC value of 0.9958, giving a relatively realistic and balanced distribution, while still being challenging in classification for both minority and moderately supported classes, while still being challenging in classification for both minority and moderately supported classes, demonstrating the differences in ensemble learning models. These results underline the strengths and weaknesses of ensemble learning methods when dealing with imbalanced datasets and different feature spaces. Cross-validation and external test datasets will be used in future work to validate the models and establish generalizability. Moreover, some classifying classes can be hard or sometimes impossible, so we will explore techniques like oversampling, feature engineering, and hyperparameter optimization to increase the overall robustness of the models. Real-time weather data integration and inclusion of more machine learning algorithms will enhance this framework in usability for disaster preparedness and resource allocation strategies.

## References

1. Radović, V., Iglesias, I.: Extreme weather events: definition, classification, and guidelines towards vulnerability reduction and adaptation management. In: Climate Action, pp. 464–476. Springer (2019)
2. Mahesh, B.: Machine learning algorithms—a review. Int. J. Sci. Res. (IJSR) **9**(1), 381–386 (2020)
3. Brönnimann, S., Martius, O., Rohr, C., Bresch, D.N., Lin, K.-H.E.: Historical weather data for climate risk assessment. Ann. N. Y. Acad. Sci. **1436**(1), 121–137 (2019)
4. Alam, M.M., Torgo, L., Bifet, A.: A survey on spatio-temporal data analytics systems. ACM Comput. Surv. **54**(10s), 1–38 (2022)
5. Newman, R., Noy, I.: The global costs of extreme weather that are attributable to climate change. Nat. Commun. **14**(1), 6103 (2023)
6. Sun, Z., Wang, G., Li, P., Wang, H., Zhang, M., Liang, X.: An improved random forest based on the classification accuracy and correlation measurement of decision trees. Expert Syst. Appl. **237**, 121549 (2024)
7. Boldini, D., Grisoni, F., Kuhn, D., Friedrich, L., Sieber, S.A.: Practical guidelines for the use of gradient boosting for molecular property prediction. J. Cheminform. **15**(1), 73 (2023)
8. Lam, K.F.Y., Gopal, V., Qian, J.: Confidence intervals for the F1 score: a comparison of four methods. arXiv preprint arXiv:2309.14621 (2023)
9. Chen, J., Muscoloni, A., Abdelhamid, I., Wu, Y., Cannistraci, C.V.: Generalizing the AUC-ROC for unbalanced data, early retrieval and link prediction evaluation. Preprints, vol. 202209, p. 2 (2024)
10. Aglasan, S., Rejesus, R.M., Hagen, S., Salas, W.: Cover crops, crop insurance losses, and resilience to extreme weather events. Am. J. Agric. Econ. **106**(4), 1410–1434 (2024)
11. Sae-Ngow, P., Kulpanich, N., Worachairungreung, M., Ngansakul, P., Prasert, T.: Geospatial analysis for identifying blackspots on national and rural highways in Thailand. Int. J. Geoinform. **20**(4) (2024)
12. Li, Y., Liang, M., Li, H., Yang, Z., Du, L., Chen, Z.: Deep learning-powered vessel traffic flow prediction with spatial-temporal attributes and similarity grouping. Eng. Appl. Artif. Intell. **126**, 107012 (2023)
13. Dixon, D.J., Jin, Y.: Canopy height estimation from PlanetScope time series with spatio-temporal deep learning. Remote Sens. Environ. (2025). Available: https://www.sciencedirect.com/science/article/pii/S0034425724005443

14. Michail, D., Davalas, C., Prapas, I.: FireCastNet: earth-as-a-graph for seasonal fire prediction. arXiv preprint arXiv:2502.01550 (2025)
15. Li, S., Chen, M., Lu, Y., Lu, Q., Yang, H.: MIESTC: a multivariable spatio-temporal model for accurate short-term wind speed forecasting. Atmosphere (2025). Available: https://search.proquest.com/openview/a35f7fd3bcec834ffada64801c72fd3f
16. Brown, R., Domiter, A., Vahabi, P.: Toward global multimodal high-resolution land surface temperature forecasting and nowcasting. In: 2024 IEEE International Conference on Climate Informatics (2024). Available: https://ieeexplore.ieee.org/abstract/document/10825375/
17. Resifi, S., Al Aawar, E., Hari Prasad, D., Jebari, H.: A novel deep learning approach for regional high-resolution spatio-temporal wind speed forecasting for energy applications. SSRN Preprint (2024). Available: https://papers.ssrn.com/sol3/papers.cfm?abstract_id=5079080
18. Lin, T., Lin, R.: Smart city traffic flow and signal optimization using STGCN-LSTM and PPO algorithms. IEEE Access (2024). Available: https://ieeexplore.ieee.org/abstract/document/10806681/
19. Eskandari, S., Ahmadloo, F., Lago-González, P.: Has climate change affected the fire regimes in semi-arid areas of northeastern Iran? Earth Sci. Inform. (2025). Available: https://link.springer.com/article/10.1007/s12145-024-01661-4
20. Fischer, C., Frühauf, A., Inchauste, L.: The spatio-temporal ecology of Oropouche virus: a laboratory-based modelling study across Latin America. SSRN Preprint (2024). Available: https://papers.ssrn.com/sol3/papers.cfm?abstract_id=5050336
21. Fan, X., Yin, Z., Qi, K., Wu, D., Qu, Z., Ren, C.: MVSTD: multi-view spatio-temporal graphs with external disturbance consideration for ride-hailing demand prediction. SSRN Preprint (2024). Available: https://papers.ssrn.com/sol3/papers.cfm?abstract_id=5096778
22. Chen, G., Chen, S., Li, D., Chen, C.: A hybrid deep learning air pollution prediction approach based on neighborhood selection and spatio-temporal attention. Sci. Rep. (2025). Available: https://www.nature.com/articles/s41598-025-88086-1
23. Li, Z., Tan, Z.M., Bai, L.: Generative deep learning reconstructs tropical cyclone microwave data from geostationary infrared radiometers. Authorea Preprints (2025). Available: https://essopenarchive.org/doi/full/10.22541/essoar.173655526.67865984
24. Meng, X., Bao, Y., Zhang, X., Luo, C., Liu, H.: A long-term global Mollisols SOC content prediction framework: integrating prior knowledge, geographical partitioning, and deep learning models with spatio-temporal techniques. Remote Sens. Environ. (2025). Available: https://www.sciencedirect.com/science/article/pii/S0034425724006187
25. Gupta, U., Sharma, R.: Analysis of criminal spatial events in India using exploratory data analysis and regression. Comput. Electr. Eng. **109**, 108761 (2023)
26. Fränti, P., Mariescu-Istodor, R.: Soft precision and recall. Pattern Recogn. Lett. **167**, 115–121 (2023)

# Image Deblurring Using Generative Adversarial Networks (GANs)

**Rina Damdoo, Sujal Agrawal, Ujjwal Prithyani, Vighnesh Gupta, and Vivek Meher**

**Abstract** Image inpainting and image blur restoration are two essential image processing tasks in computer vision wherein missing or corrupt areas in images are filled in and blurry images corrected. Due to the characteristics of these two problems, they are both extremely difficult; it is necessary to predict the right kind of content while taking care not to distort or lose the general shape, texture, and contextual relevance of the image. GAN has come in handy for these cases, as they are quite efficient when it comes to both painting images and deblurring them. Images are formed using a generator and discriminator, both being neural networks, which function in a non-simultaneous manner; the former is responsible for generating images while the latter determines whether the images are authentic or not. By this mean approach, the generator is able to optimize its outputs gradually, and so is the discriminator in that it can better tell between real and fake images. In the paper, we show that the acknowledged capabilities of GANs can be effectively employed in image inpainting as well as blur restoration. Intensive evaluations on a number of datasets comprising of various levels of synthetic and real blur have proved that the proposed GAN-based method of detail restoration and reconstruction is superior even when the images are very blurry. This work reveals that GANs can be deployed in very different applications in a very straightforward and convincing way.

**Keywords** GAN · Generator · Discriminator · Image inpainting

---

R. Damdoo · S. Agrawal (✉) · U. Prithyani · V. Gupta · V. Meher
Department of Computer Science and Engineering, Shri Ramdeobaba College of Engineering and Management, Nagpur, India
e-mail: agrawalsr_14@rknec.edu

R. Damdoo
e-mail: damdoor@rknec.edu

U. Prithyani
e-mail: prithyaniuv@rknec.edu

V. Gupta
e-mail: guptavs_1@rknec.edu

V. Meher
e-mail: mehervj@rknec.edu

© The Author(s), under exclusive license to Springer Nature Singapore Pte Ltd. 2026 
M. Pant et al. (eds.), *Real-World Applications of Machine Learning*, Mathematics for Sustainable Developments, https://doi.org/10.1007/978-981-95-7134-5_15

# 1 Introduction

This work addresses blind motion deblurring for a single photograph. Generative Adversarial Networks (GANs) are known for effectively preserving texture details, creating realistic images, and producing perceptually convincing results. Inspired by advances in image super-resolution and image-to-image translation using GANs, this approach reframes deblurring as a specific case of image-to-image translation. We introduce DeblurGAN, a method based on conditional GANs and a multi-component loss function. Instead of traditional approaches, we utilize Wasserstein GAN with a gradient penalty and perceptual loss. This combination promotes solutions that are perceptually indistinguishable from clear images, enabling the recovery of finer texture details beyond what mean squared error (MSE) or mean absolute error (MAE) objectives could achieve.

We provide three important contributions to the field of picture inpainting and deblurring. We employed a GAN-based architecture and a custom-designed loss function to achieve cutting-edge motion deblurring performance while outperforming the most efficient existing approaches. Second, we present an automated method for creating a diverse motion deblurring dataset using random trajectories from sharp photos, which, when paired with existing datasets, results in better deblurring performance than training purely on real-world blurred images. Finally, we introduce a new dataset and evaluation framework for evaluating deblurring algorithms based on their ability to improve object detection results, providing a more practical and comprehensive measure of deblurring performance as shown in Fig. 1 [1]. These contributions increase the efficiency and assessment standards of deblurring algorithms, displaying considerable gains in technical performance and practical usability.

**Fig. 1** Blurred and deblurred dataset [1]

## 2 Related Work

### 2.1 *Image Deblurring*

The formula of the non-uniform blur model will be written as:

$$I_B = k(M) * I_s + N \quad (1)$$

In this equation, $I_B$ represents the blurred image, $k(M)$ denotes the unknown blur kernels determined by the motion field $M$, $I_s$ is the sharp latent image, $*$ signifies the convolution operation, and N stands for the additive noise.

Blind and non-blind deblurring are the two general categories into which image deblurring issues fall. Non-blind deblurring, which assumes that the blur kernels $k(M)$ are known, was the main focus of early research. Conventional non-blind deblurring methods frequently use the Lucy-Richardson algorithm, Wiener filter, or Tikhonov regularization to estimate the sharp image Is and perform the deconvolution.

On the other hand, because the blur function is unknown, blind deblurring techniques need to estimate both the blur kernels $k(M)$ and the underlying crisp image $I_s$. Because of this, blind deblurring is a more difficult, ill-posed problem that is usually solved using heuristics, image statistics, and presumptions regarding the blur's origins. These techniques frequently make the assumption that camera shake-induced blur is consistent throughout the picture. In order to infer the blur kernel, the procedure typically begins with estimating the camera motion. The blur effects are then reversed using deconvolution.

Over time, various methods have been developed to address these issues [2]. Some methods use iterative approaches to refine estimates of the motion kernel and sharp image, leveraging parametric prior models to improve results. However, these iterative methods can suffer from long running times and difficulties in determining when to stop the iterations. Alternatively, some techniques make use of local linearity assumptions and simple heuristics for quick blur kernel estimation, though these are often limited to a narrow range of images.

Recently, advancements have been made in non-uniform blind deblurring through new algorithms that model the blurring process based on geometric principles, such as the rotational velocity of the camera during exposure or the assumption of 3D camera movement [3]. With the rise of deep learning techniques, new approaches employing convolutional neural networks (CNNs) have been introduced. These methods range from using CNNs to estimate the blur kernel to predicting complex motion kernels in Fourier space or leveraging fully convolutional networks for motion flow estimation. More recent developments include kernel-free, end-to-end approaches that utilize multiscale CNNs for direct image deblurring, as well as methods that combine frameworks and dense networks for effective blind deblurring across various blur sources [4].

## 2.2 Generative Adversarial Network

Goodfellow et al. [6] proposed Generative Adversarial Networks (GANs), which function by establishing a framework of competition between two networks: a discriminator and a generator. The generator generates a sample after receiving a noise input. In contrast, the discriminator receives both produced and real samples and makes an effort to distinguish between them. Creating realistic samples that can fool the discriminator and make it difficult to tell them apart from real data is the generator's goal. A minimax game objective is created by this interaction, in which the discriminator and generator are optimized against one another.

$$\min_{G}\max_{D} \mathop{\mathrm{E}}_{x\sim P_r}\left[\log(D(x))\right] + \mathop{\mathrm{E}}_{\bar{x}\sim P_r}\left[\log(1 - D(\bar{x}))\right] \tag{2}$$

In GANs, a sample taken from a basic noise distribution is represented by the model distribution ($P_g$) and the real data distribution ($P_r$), where $\bar{x} = G(z)$ with $z \sim P(z)$. Although GANs are renowned for producing samples of great perceptual quality, previous research has shown that training the original GAN model presents difficulties such as mode collapse and disappearing gradients. Reducing the Jensen-Shannon (JS) divergence between the data and model distributions over $x$ is the same as minimizing the GAN objective function. Arjovsky et al. [7] pointed out problems with utilizing JS divergence to train GANs and suggested using the Earth-Mover (or Wasserstein-1) distance $W(p,q)$ instead. The Kantorovich-Rubinstein duality is then used to create the Wasserstein GAN (WGAN) objective.

$$\min_{G}\max_{D\in\rho} \mathop{E}_{x\sim P_r}[D(x)] - \mathop{E}_{\bar{x}\sim P_g}[D(\bar{x})] \tag{3}$$

Within this paradigm, $P_g$ indicates the model distribution once more, and D stands for the set of 1-Lipschitz functions. The fundamental idea is that the critic's output roughly equals $K \cdot W(P_r, P_\theta)$, where $W(P_r, P_\theta)$ stands for the Wasserstein distance and $K$ is a Lipschitz constant. Since it calculates the difference between created and real samples, the discriminator is called a critic in this context. Weight clipping inside the range $[-c, c]$ was introduced by Arjovsky et al. [7] in order to preserve the Lipschitz constraint in WGAN. A gradient penalty term was later proposed by Gulrajani et al. [8] as a substitute method for enforcing the Lipschitz condition.

$$\Lambda \mathop{E}_{\bar{x}\sim P_{\bar{x}}}\left[(\|\nabla_{\bar{x}} D(\bar{x})\|_2 - 1)^2\right] \tag{4}$$

Gulrajani et al. [8] introduced the gradient penalty term to the value function as an alternative method for enforcing the Lipschitz constraint. This approach proves resilient across different generator architectures and requires minimal hyperparameter adjustments. This flexibility is particularly beneficial for image deblurring, as it enables the use of newer, lightweight neural network designs rather than the traditional, heavier Deep ResNet architectures that have been commonly applied in image deblurring tasks.

## 3 Proposed Methodology

The objective is to reconstruct a sharp image $I_s$ given only a blurred input image IB, without prior knowledge of the blur kernel. Deblurring is performed by a trained CNN, *G*, referred to as the Generator, which estimates the sharp image $I_s$ for each IB. Additionally, during training, a network *D*, called the critic, is introduced, and both networks are trained adversarially.

### *3.1 Architecture*

A structure akin to that suggested by Johnson et al. [9] for style transfer serves as the foundation for the Generator CNN architecture, which is seen in Fig. 2. It consists of two transposed convolution blocks, nine residual blocks (ResBlocks), and two strided convolution blocks with a stride of 1/2. A ReLU activation, an instance normalization layer, and a convolution layer make up each ResBlock. In each ResBlock, the initial convolution layer is followed by dropout regularization with a probability of 0.5. In order to enable the CNN to learn a residual correction IR to the blurred image IB, a global skip connection known as ResOut is also included. This results in IS = IB + IR. This approach improves the generalization of the model and speeds up training. A critic network *D*, which is based on a Wasserstein GAN with gradient penalty (WGAN-GP), is defined during training. All convolutional layers until the final one are followed by an InstanceNorm layer and a LeakyReLU activation with a slope of 0.2 in the critic network design, which is similar to that of PatchGAN.

### *3.2 Loss Function*

The constant value in every experiment is 100. Since there is no need to penalize differences between the input and output, we do not condition the discriminator as Isola et al. did. Many works on conditional GANs use the conventional GAN objective as the loss function for adversarial loss. It has recently been demonstrated that a different strategy utilizing least-squares GAN offers more stability and yields better quality outcomes. Since WGAN-GP is resilient across many generator architectures, we use it as the critic function in this case. These results were validated by our preliminary trials with other topologies, which enabled us to employ a model that is far lighter than ResNet152. This is how the loss is computed:

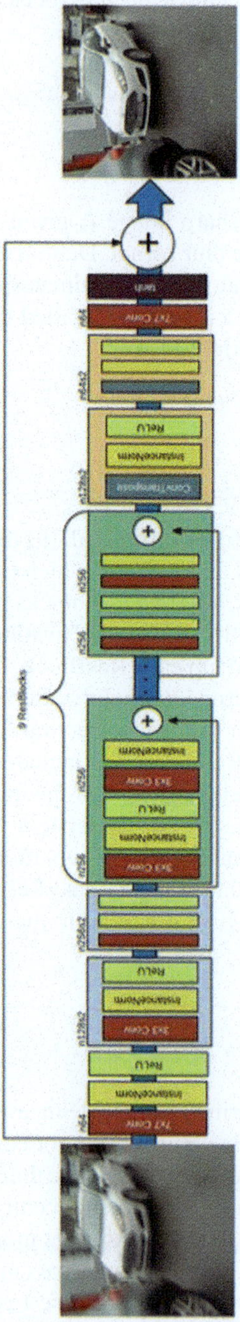


**Fig. 2** GAN Architecture [5]

$$L = L_{\text{GAN}} + \Lambda \cdot L_X \tag{5}$$

$$L_{\text{GAN}} = \sum_{n=1}^{N} -D_{\theta_D}\left(G_{\theta_G}\left(I^B\right)\right) \tag{6}$$

## 4 Implementation Details

In our study, we employed a pretrained version of the DeblurGAN model to address the image deblurring task. Our approach involved fine-tuning this pretrained model on a dataset of 1100 paired images consisting of blurred and deblurred car images. The following describes the details of our methodology, including the use of the pretrained model and the specifics of our fine-tuning process.

### *4.1 Pretrained Model*

We utilized a pretrained DeblurGAN model that had been trained for 2000 epochs on the GoPro dataset [5], which consists of high-resolution images captured with a high frame-rate camera. The model was initially trained using a single Maxwell GTX Titan-X GPU and employed the same network architecture and training procedures. The pretrained model had been trained with a batch size of 1, an initial learning rate of $10^{-4}$, and used the Adam optimizer with a learning rate schedule that linearly decayed from $10^{-4}$ to zero over the course of the training (Fig. 3) [10].

### *4.2 Fine-Tuning Procedure*

To adapt the pretrained DeblurGAN model to our specific task, we performed additional fine-tuning for 10 epochs on our car image dataset. This dataset consisted of 1100 pairs of images, with each pair including a blurred image and its corresponding deblurred reference image. The images were preprocessed to a consistent resolution of 256 × 256 pixels to match the input requirements of the DeblurGAN model.

### *4.3 Hyperparameter Tuning*

The initialization function sets up the ConditionalGAN model by initializing parameters such as the type of GAN (conditional), whether it's in training mode, and the

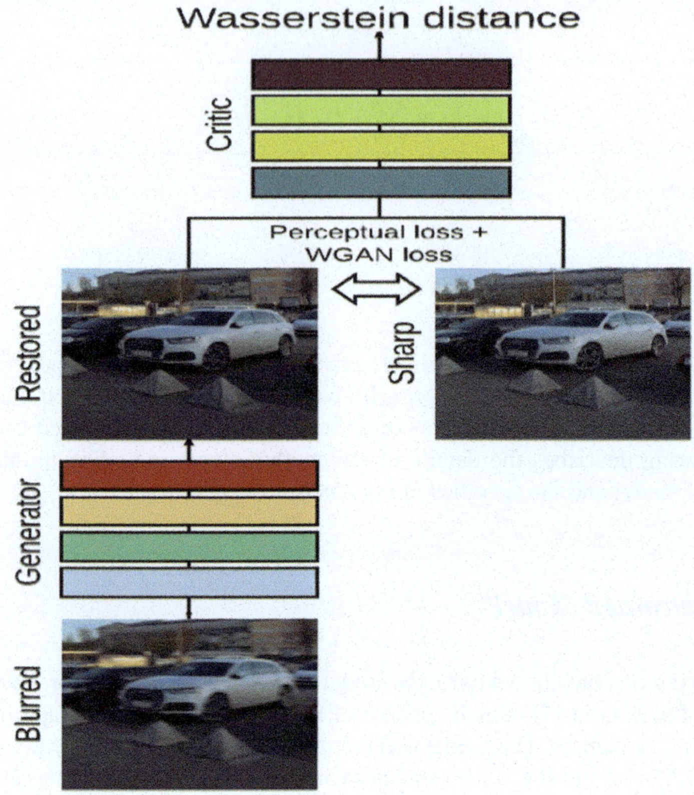


**Fig. 3** Training process [10]

device (CPU or GPU). It also initializes tensors for input and output images. The load function loads model checkpoints from saved files, restoring the model's state. It is useful for resuming training from a saved checkpoint or for deploying trained models for inference.

set_input function prepares input data for the model by setting up input tensors for the forward pass. It takes input images (blurred and sharp) and converts them into appropriate tensor format for processing by the network. get_image_path function retrieves the file paths of input images used in the model. It can be useful for tracking the source of input data during training or testing.

NormalizeImg utility function normalizes images by scaling pixel values to a specified range. Normalization is essential for ensuring consistent input data for the model and stabilizing training.

show_MNIST utility function, specific to MNIST dataset visualization, displays images in a grid format for visualization. It can be helpful for debugging and understanding the dataset used in the model. encode_labels utility function encodes categorical labels into a one-hot encoding format. In CGANs, the generator takes both noise and class labels as input. Converting class labels into a one-hot encoding format ensures compatibility with the generator's input layer.

### 4.4 Algorithm

We trained DeblurGAN using a mix of synthetically blurred images and naturally blurred images captured in real-world conditions. The dataset maintained a 2:1 ratio, with twice as many synthetically generated images as those captured by a high frame-rate camera.

## 5 Experimental Evaluation

### 5.1 Dataset

In our study, we utilized a dataset comprising a total of 1050 image pairs, each pair consisting of a blurred and a corresponding deblurred image of various cars. Out of these, 1000 image pairs were allocated for training purposes, while the remaining 50 pairs were designated for testing.

### 5.2 Results

Visual inspection of the deblurred images revealed notable improvements in clarity and detail recovery. The deblurred images exhibit enhanced edge sharpness and reduced blurring artifacts compared to the blurred inputs. Specific examples demonstrating the efficacy of the algorithm can be observed in Figs. 4, 5, and 6 [10], which highlight significant enhancements in vehicle features and textures.

Parameters:
$M = 2000$ – number of iterations,
$L_{max} = 60$ – max length of the movement,
$p_s = 0.001$ – probability of impulsive shake,
$I$ – inertia term, uniform from (0,0.7),
$p_b$ – probability of big shake, uniform from (0,0.2),
$p_g$ – probability of gaussian shake, uniform from (0,0.7),
$\phi$ – initial angle, uniform from $(0,2\pi)$,
$x$ – trajectory vector.

**procedure** BLUR(Img, $M, L_{max}, p_s$)
    $v_0 \leftarrow cos(\phi) + sin(\phi) * i$
    $v \leftarrow v_o * L_{max}/(M-1)$
    $x = \text{zeros}(M, 1)$
    **for** $t = 1$ to $M - 1$ **do**
        **if** randn $< p_b * p_s$ **then**
            nextDir $\leftarrow 2 \cdot v \cdot e^{i*(\pi+(\text{randn}-0.5))})$
        **else:**
            nextDir $\leftarrow 0$
        $dv \leftarrow$ nextDir $+ p_s * (p_g * (\text{randn} + i * \text{randn}) * I * x[t] * (L_{max}/(M-1))$
        $v \leftarrow v + dv$
        $v \leftarrow (v/abs(v)) * L_{max}/(M-1)$
        $x[t+1] \leftarrow x[t] + v$
    Kernel $\leftarrow$ sub pixel interpolation($x$)
    Blurred image $\leftarrow conv$(Kernel, Img)
    **return** Blurred image

**Fig. 4** Sample image from blur image dataset [1]

**Fig. 5** Sample image from deblur image dataset [1]

**Fig. 6** Input to the proposed model [1] and output of the proposed model [1]

#### 5.2.1 Comparison of Performance Metrics

Figure 7 presents a graph comparing epoch versus generator and discriminator loss values achieved by our proposed deblurring algorithm against several methods. Our algorithm demonstrates a superior average value, indicating a higher level of image quality improvement.

#### 5.2.2 Content Loss Comparison

Figure 8 displays a graph representing amount of content loss in deblurring images. Our method shows a lower average content loss, which signifies fewer discrepancies between the deblurred images and the original sharp images.

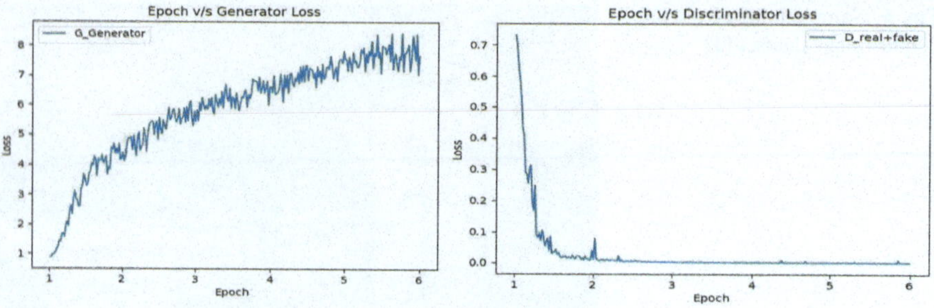


**Fig. 7** Epoch versus generator loss and discriminator loss (generated by author)

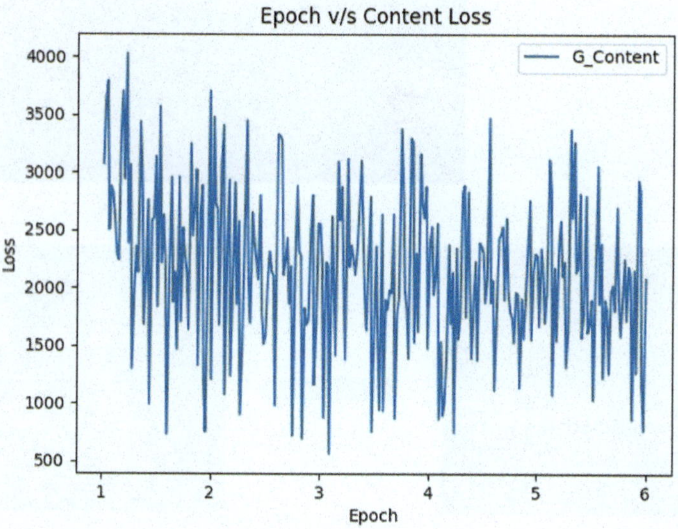


**Fig. 8** Epoch versus content loss

### 5.2.3 Computational Accuracy

Figure 9 summarizes the average accuracy for the proposed algorithm. Our approach achieves an average accuracy, which is higher and faster relative to other approaches.

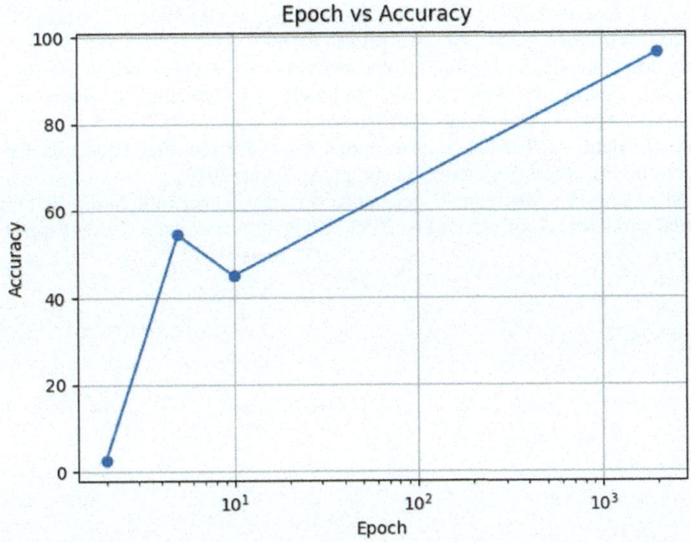


**Fig. 9** Epoch versus accuracy (generated by author)

## 6 Conclusion

This work presented a GAN-based approach for image deblurring, using DeblurGAN with Wasserstein and perceptual loss to recover fine textures and sharper details. Experiments on synthetic and real-world datasets showed improved edge clarity, reduced artifacts, and higher accuracy compared to existing methods. The results highlight the effectiveness of GANs for practical deblurring tasks. Future work will explore handling complex blur patterns and optimizing lightweight models for real-time deployment.

## References

1. Nah, S., Kim, T.H., Lee, K.M.: Deep multi-scale convolutional neural network for dynamic scene deblurring. In: Proceedings of IEEE Conference Computer Vision Pattern Recognition (CVPR) (2017)
2. Villani, C.: Optimal Transport: Old and New. Grundlehren der mathematischen Wissenschaften. Springer, Berlin Heidelberg (2008)
3. Nah, S., Hyun, T., Kyoung, K., Lee, M.: Deep multi-scale convolutional neural network for dynamic scene deblurring (2016)
4. Isola, P., Zhu, J.-Y., Zhou, T., Efros, A.A.: Image-to-image translation with conditional adversarial networks. arxiv (2016)
5. Goodfellow, I.J., Pouget-Abadie, J., Mirza, M., Xu, B., Warde-Farley, D., Ozair, S., Courville, A., Bengio, Y.: Generative adversarial networks (2014)

6. Salimans, T., Goodfellow, I., Zaremba, W., Cheung, V., Radford, A., Chen, X.: Improved techniques for training GANs. ArXiv e-prints (2016)
7. Arjovsky, M., Chintala, S., Bottou, L.: Wasserstein GAN. ArXiv e-prints (2017)
8. Gulrajani, I., Ahmed, F., Arjovsky, M., Dumoulin, V., Courville, A.: Improved training of Wasserstein GANs. ArXiv e-prints (2017)
9. Johnson, J., Alahi, A., Fei-Fei, L.: Perceptual losses for real-time style transfer and super-resolution. In: European conference on computer vision (2016)
10. Hong, M., Choe, Y.: Wasserstein generative adversarial network based de-blurring using perceptual similarity. Appl. Sci. **9**(11), 2358 (2019). https://doi.org/10.3390/app9112358

# Triple-Head Spatial Attention-Based Deep Learning Approach for Plant Disease Identification

**Vanshika Pundir, Palak Bhandari, and Vidit Kumar**

**Abstract** Global food security is seriously threatened by plant diseases, which calls for creative approaches to early diagnosis and efficient treatment. Computer vision techniques provide scalable alternatives to labor-intensive and in-accurate traditional diagnostic procedures. This research focuses to upgrade the diagnosis of plant diseases by integrating attention mechanism with deep convolutional model. More specifically, we propose triple-head spatial attention module which integrated to EfficientNet-B0 for enhancing feature representation ability. The PlantDoc dataset, which includes images from 27 classes, was used to assess the performance of several model variations. With an accuracy of 62.59% for uncropped and 83.97% for the cropped dataset, the presented model showed excellent feature extraction and classification compared to prior methods.

**Keywords** Deep learning · Attention mechanism · Plant disease classification · Spatial attention · EfficientNet

## 1 Introduction

The rapidly growing global population, increasing at an annual rate of approximately 1.09%, necessitates a corresponding rise in the demand for food, feed, fiber, and fuel [1]. This escalation imposes higher requirements on the agricultural industry to achieve increasingly higher yields. Healthy, feasible, and inclusive food systems are essential to meet the world's development purpose [2]. Agricultural evolution is crucial for ending poverty, boosting shared wealth, and feeding billions of people by 2050 [2]. Development in the agricultural sector is twice to quanta more effective at elevating incomes among the poverty stricken populations than in other sectors. To fulfill the needs of the growing population, it is evaluated that food supply need to be grow to 68% by 2050 [3].

V. Pundir · P. Bhandari · V. Kumar (✉)
Department of Computer Science and Engineering, Graphic Era Deemed to be University, Dehradun, India
e-mail: viditkumaruit@gmail.com

© The Author(s), under exclusive license to Springer Nature Singapore Pte Ltd. 2026 
M. Pant et al. (eds.), *Real-World Applications of Machine Learning*, Mathematics for Sustainable Developments, https://doi.org/10.1007/978-981-95-7134-5_16

There are biotic, abiotic, human-made or natural factors contributing to significant food waste worldwide. The flourishing human being needs increased food production, yet this demand is challenged by the global scarcity and unavailability of natural resources [3]. Key elements causing decline including: pests, global warming, dry period, diseases, salinity, loss of arable lands, pathogens, and unwanted plants [4, 5]. Currently, 38% of agricultural losses are attributable to bugs and 34% are due to unwanted plants. Abiotic factors count for 50% of agriculture losses [6]. Conventional methods for plant phenotype primarily depend on manual calculations, which are labor-intensive, prone to inaccuracies, and protracted, significantly hindering the efficient investment of phenotype data from large populations [7]. Traditional techniques for assessing the plant characteristics have certain limitations where we require a lot of human effort as well as these methods are more prone to errors and take a long time as a result become challenging to gather the accurate information about the plant characteristics especially when we are dealing with more plants [8]. Recent progress in image analysis particularly through machine learning techniques like CNN has extended the use of these tools. This advancement enables faster and large-scale (high-throughout) phenotype, making it easier to study many plants efficiently and accurately. In our research, we proposed an enhanced efficientnet B0 architecture in which we integrate and spatial attention module designed to improve plant disease classification.

## 2 Literature Survey

Advances in deep learning, a subset of machine learning, have made major contributions to the development of visual plant phenotyping tools [9]. Building massive neural networks can be protracted, once instructed, these prototypes can execute image classification tasks rapidly. This efficiency makes them ideal for consumer applications such as plant disease detection on handsets [9]. Image processing for finding out plant diseases offers more opportunities to apply deep learning techniques to factual-world agricultural problems [10]. This integration advances agricultural knowledge, improves crop yields, and enhances disease control. Identifying and classifying plant leaf diseases are a complex task. Numerous researchers have explored traditional and soft computing methods to segment and detect infected areas on leaves. In [11], they reviewed various soft computing approaches used for plant disease recognition. One of the more often employed methods is the support vector machine (SVM), particularly with the radial basis function (RBF) as its kernel, which is widely used for classifying diseases from leaf images. The growing prominence of deep learning algorithms [11] introduced an innovative approach using CNN to identify rice diseases, while [12] employed K-means clustering for detecting cucumber leaf diseases. In [13], the author applied ResNet50, VGG16, and AlexNet models for plant disease identification using transfer learning. In their results ResNet50 emerged as the most effective model, balancing accuracy and efficiency compared with other models like VGG16, with its higher FLOPs (15.5 billion), required nearly double the training time per epoch and that of AlexNet, which had the fewest FLOPs (7.25 × 108). In [14], the author covers the observation and categorization of tomato leaf diseases operate on a pre-trained deep learning model, namely ResNet. In [15], the author presents a thin multi-level attention fusion network (LFMNet). The network is built on two important components: the PMFFM and MAttion blocks. In [16], compared to other techniques, the author uses a dataset consisting of 14,828 photos of tomato leaves with nine distinct illnesses. The classifier's learning mechanism was a CNN, which has the benefit of automatically extracting features from raw photos. In [17], the author sought to identify early-stage downy mildew (Peronospora) using thermal images. In [18], the author developed a method to use field photos to identify NLB lesions in maze plants. The author employs numerous CNNs. In [19], author describes a method for detecting apple leaf diseases using deep CNNs. It entails creating several diseased images and developing a new deep CNN architecture inspired by AlexNet. In [20], using deep CNNs, the author offers a novel method for identifying rice diseases. In [21], paper presents a computer vision system for potato disease classification that is based on deep learning. In [22], a collection of deep convolutional networks was trained to identify the strength of apple black rot by using photos from the PlantVillage dataset with four classifications of strength. The VGG16 deep network trained with transfer learning is the model with the best performance. In [23], to identify and diagnose plant diseases, their study created CNN models utilizing images of both infected and healthy plant leaves. In [24], author uses a high-resolution transferable spectral sensor to identify various types of diseased tomato leaves at various levels, incorporating early and asymptomatic stages. They used KNN for classification. In [25], author proposes a Refinement Filter Bank structure to improve tomato plant disease and pest identification, with a focus on lowering false positives and addressing class imbalance. It consists of a Primary Diagnosis Unit that generates bounding boxes around contaminated areas, a Secondary Diagnosis Unit that uses CNN classifiers to validate these detections, and an Integration Unit that combines data to keep true positives while discarding false positives. In [26], study targets to enhance the identification of maize leaf illness. It introduces improved versions of the GoogLeNet and Cifar10 models for leaf disease detection. The enhancements include parameter adjustments, pooling combination modifications, the addition of dropout operations and ReLU functions, and a reduction in the amount of classifiers. This modification provides output in models that have fewer parameters than VGG and AlexNet structures, to improve accuracy and efficiency in maize leaf disease recognition. In [27], research describes an automated technique for early detection of avocado tree illnesses. They used two cameras Tetracamera and a Canon camera to test two categorization methods: multilayer perceptron (MLP) and K-nearest neighbors. In [28], research analyzed two-stage neural network architecture for plant disease classification, which is adapted to real-world conditions. In contrast, this research explores attention mechanism to CNN for plant disease identification.

## 3 Methodology

The proposed research builds upon the EfficientNet B0 architecture by integrating spatial attention mechanism beyond the pre-trained model. More specifically, triple-head spatial attention module is proposed and integrated to efficientnet-b0, which is depicted in Fig. 1. The enhancement is aimed at optimizing the model's capability to effectively process and focus on the most descriptive parts of an image.

### 3.1 EfficientNet B0

EfficientNet [26] represents a CNN architecture that introduces a non-traditional scaling method which is designed to enhance network performance across different aspects: depth, width, and resolution. Traditional approaches scale these elements arbitrarily, resulting in unsatisfactory arrangements. EfficientNet uses a coefficient of a compound to rule the network's width, depth, and input resolution in a balanced manner. This architecture refined computing resources, enhanced network accuracy, and validity, and offered a more effective solution for many deep learning problems.

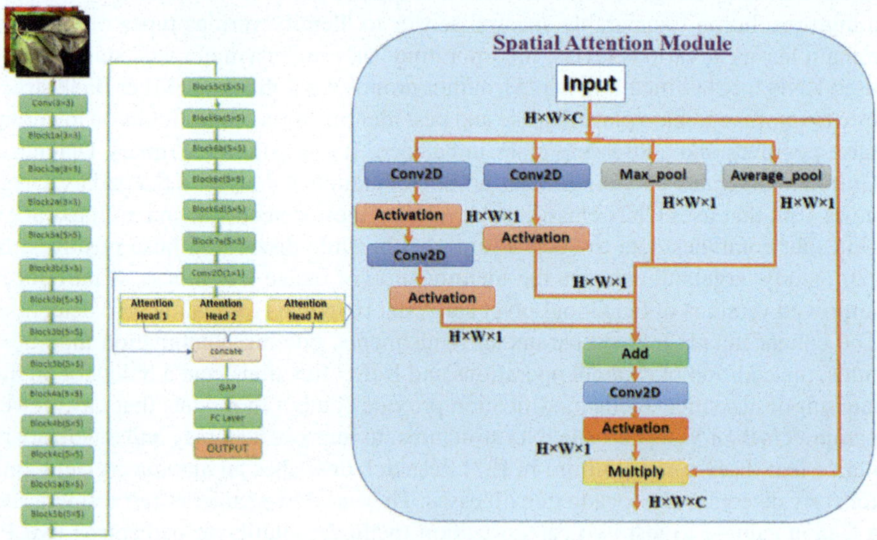


**Fig. 1** Proposed method

## 3.2 *Triple-Head Spatial Attention Module*

The major enhancement in our model is a spatial attention module, which improves the network's capacity to focus on the most descriptive parts of an image. Our spatial attention is inspired by [29]. The input feature map in this formulation is initially subjected to two different convolutional procedures. In order to derive local features from the input, a typical convolution is performed as the initial step. The second operation is a two-step process wherein the channel dimensions are first reduced by applying a convolution, and then the features are further refined by applying another convolution. The following two pooling processes are applied simultaneously: choosing the maximum value inside each window and performing average pooling to calculate the mean value within each pooling window. The combined feature map is created by merging the results of different pooling and convolutional methods. The spatial attention map is created by processing the combined map with a 3 × 3 convolution. This spatial attention map is significant because it identifies and highlights the most relevant places within the input feature map.

For a feature map X of shape (H, W, C), four computations: average pooling (AP), max pooling (MP) and two convolution modules with kernel sizes 3 × 3 and 1 × 1 and, 1 and 64 filters are calculated as:

$$\mathrm{AP}(X) = \frac{1}{C}\sum_{i=1}^{C} X(:, :, i) \tag{1}$$

$$\mathrm{MP}(X) = \mathrm{Max}(X(:, :, i)) \tag{2}$$

$$Y1 = (\mathrm{Relu}\big(\mathrm{Conv}_1^{3\times 3}(X)\big) \tag{3}$$

$$Y2 = \mathrm{Relu}\big(\mathrm{Conv}_1^{3\times 3}\big(\mathrm{Relu}\big(\mathrm{Conv}_{64}^{1\times 1}(X)\big)\big)\big). \tag{4}$$

Then these are added, followed by $\mathrm{Conv}_1^{3\times 3}$ and sigmoid and multiply operation as:

$$Z = \mathrm{Sigmoid}\big(\mathrm{Conv}_1^{3\times 3}(\mathrm{AP(X)} + \mathrm{MP(X)} + \mathrm{Y1} + \mathrm{Y2})\big) \tag{5}$$

$$A1 = \mathrm{Multiply}(Z, X) \tag{6}$$

Triple such heads are created and there outputs are concatenated as:

$$\mathrm{THA} = [\mathrm{A1},\ \mathrm{A2},\ \mathrm{A3}] \tag{7}$$

Then the output of all attention heads are concatenated with input feature map and feed to global average pooling layer followed by dense layer for classification.

**Fig. 2** Dataset samples

# 4 Implementation

## *4.1 Dataset*

PlantDoc [30] dataset aimed at precise plant disease recognition in field environments and comprises 2598 images, covering 27 classes (13 plant species and 17 classes of diseases). Originally, the dataset is split into training and testing subsets, with 2340 images assigned for training and 236 images for testing, resulting in a total of 2576 images. Some samples are shown in Fig. 2.

## *4.2 Model Training Specifications and Evaluation Criteria*

The proposed model was trained and tested in Google Colab's GPU environment to harness the computational power necessary for efficient training. To boost dataset diversity, data augmentation methods such as zooming, random rotations, and horizontal flips were performed. The Adam optimizer is employed and batch size is set to 32. The model was trained in two phases with different learning rate to improve performance. Initially, the model is train for 10 epochs with a 0.001 learning rate. Then model was refined for 50 epochs in the second phase at a learning rate of 0.0001. The two-phase training of the model increased the performance in the classification process by first finding and then narrowing down on broad data trends.

## 5 Results

In this segment, we represent an analysis of the enhancements of the base architecture. These improvements significantly boost feature extraction efficiency and enhance the classification process.

### 5.1 *Assessment of Methodological Improvements*

In this we experiment with variations of CBAM, spatial attention of CBAM and efficientnet: Model 1: Model 1 places the CBAM at the end layer of the EfficientNetB0 architecture. Model 2: Model 2 integrates the CBAM block. Concatenating the CBAM output with EfficientNet-B0's last activation layer. Model 3: Model 3 introduces two parallel CBAM blocks trying to improve the feature representation in the EfficientNet-B0 architecture. Model 4: Model 4 append a spatial attention mechanism to the EfficientNet-B0 design. This approach allows the model to concentrate on the most considerable sections of the feature map, highlighting key regions while demphasizing the less relevant ones. Model 5: Model 5 introduces a spatial attention mechanism to upgrade feature extraction. The result of this spatial attention mechanism is then unified with the activation output of the EfficientNet-B0 base model's last layer. This concatenation allows the model to include the corrected spatial characteristics into the original activation output, resulting in a more extensive feature representation. Model 6: In Model 6 Conv2D layer, batch normalization, and activation are added to the spatial attention block. This Conv2D output is then combined with the addition of the max pooling and average pooling layers.

The results are presented in Table 1, where the experiments with spatial attention, i.e., Models 4 to 6, increased the performance compared to the base CBAM model 1. Ultimately, the proposed Triple-head Attention performs better than all the other variations. Further precision, recall, and F1-score comparison is depicted in Fig. 3.

**Table 1** Performance analysis of different model's variations

| Model | Accuracy (%) |
|---|---|
| Model 1 | 61.01 |
| Model 2 | 58.89 |
| Model 3 | 57.20 |
| Model 4 | 60.59 |
| Model 5 | 61.01 |
| Model 6 | 61.44 |
| Proposed Method | 62.59 |

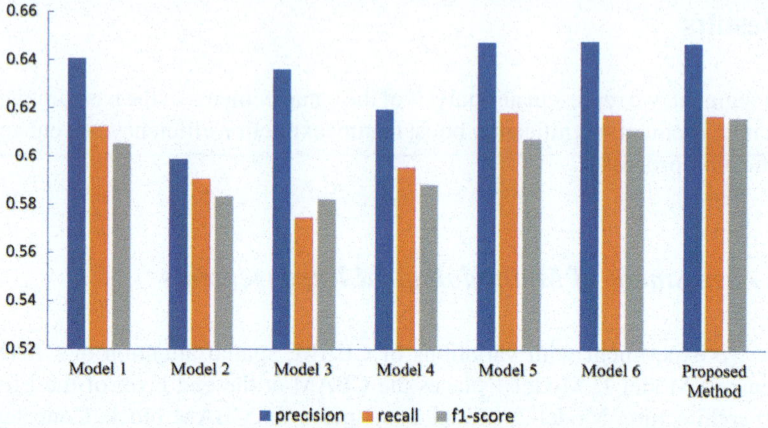


**Fig. 3** Comparative analysis of precision, recall and fi score across all models

## 5.2 Comparison to State of the Arts

In Table 2, using uncropped dataset our proposed model has performed better with 62.59% accuracy than previous approaches, demonstrating its efficacy and dependability. This enhancement highlights the model's usefulness in the field. Utilizing the cropped dataset, we were able to provide remarkable accuracy of 82.68% with single head attention, 82.89% with double head attention and 83.97% with triple-head attention. The model outperformed other methods in terms of accuracy, confirming its superior performance and potential as a useful method for practical applications.

**Table 2** Comparison to state of the arts

| Dataset | Model | Accuracy (%) |
|---|---|---|
| Uncropped raw images | Singh et al. [30] | 29.73 |
| | Moupojou et al.[31] | 60.14 |
| | **Proposed method** | **62.59** |
| Cropped images | Singh et al. [30] | 70.53 |
| | Moupojou et al.[31] | 73.86 |
| | Aggarwal et al. [32] | 73.83 |
| | Yu et al. [33] | 77.54 |
| | **Proposed method (Head 1)** | **82.68** |
| | **Proposed method (2 Heads)** | **82.89** |
| | **Proposed method (3 Heads)** | **83.97** |

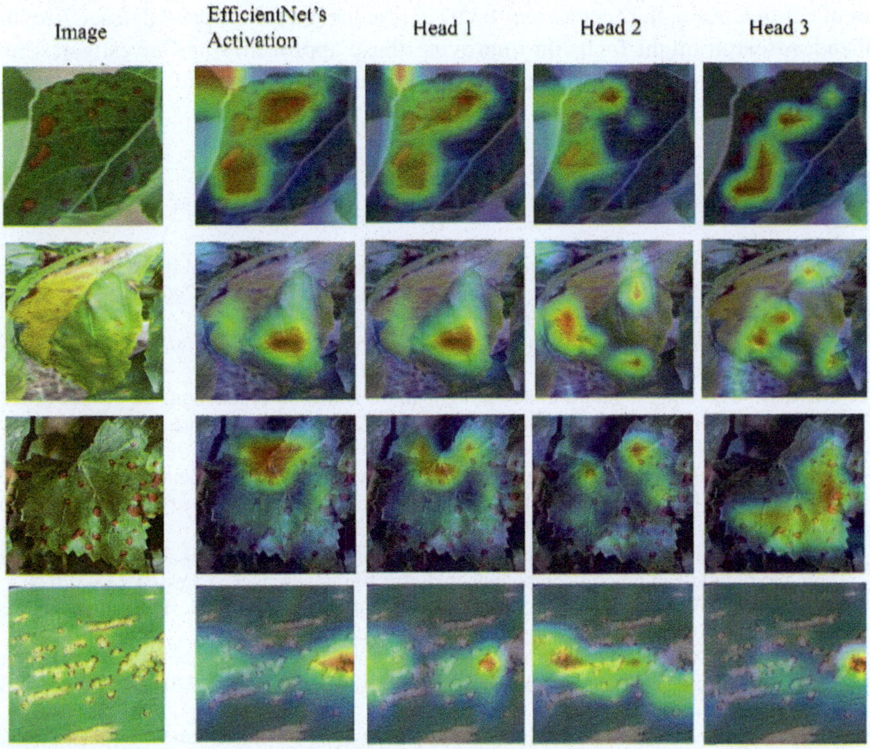


**Fig. 4** Heatmap visualization of proposed method's attention heads

### *5.3 Visualization*

We also visualize heat map of all attention heads in Fig. 4, where we can see all heads focus on different areas contributing in extracting different relevant features from the images.

## 6 Conclusion

In this study, we evaluated attention fusion to EfficientNet-B0 architecture to upgrade the precision and durability of plant disease categorization. The goal of our research was to upgrade the architecture's ability to focus on the most valuable regions of input images by incorporating proposed triple-head spatial attention mechanism. This mechanism significantly improves the model's capability to correctly classify plant diseases by concentrating on the most critical parts of the image. On uncropped version of plantdoc dataset, it achieves accuracy of 62.59% among the evaluated architectures. Also, the accuracy of 83.97% is achieved for cropped dataset. Moving ahead, research might focus on improving these approaches and investigates how they can be implemented to other areas of agriculture and beyond.

## References


1. Wang, A., Zhang, W., Wei, X.: A review on weed detection using ground-based machine vision and image processing techniques. Comput. Electron. Agric. **158**, 226–240 (2019). https://doi.org/10.1016/j.compag.2019.02.005
2. Agriculture and food. In: World Bank. https://www.worldbank.org/en/topic/agriculture. Accessed 9 Dec 2024
3. Marr, B.: The biggest future trends in agriculture and food production. In: Forbes (2022). https://www.forbes.com/sites/bernardmarr/2022/01/28/the-biggest-future-trends-in-agriculture-and-food-production/. Accessed 9 Dec 2024
4. Hassoun, A., Bekhit, A.E.-D., Jambrak, A.R., Regenstein, J.M., Chemat, F., Morton, J.D., Gudjónsdóttir, M., et al.: The fourth industrial revolution in the food industry—part II: emerging food trends. Crit. Rev. Food Sci. Nutr. **64**(2), 407–437 (2024)
5. Biotic and abiotic compounds that affect grain storage. In: Graintechnik. https://graintechnik.com/blog/biotic-and-abiotic-compounds-that-affect-grain-storage/. Accessed 9 Dec 2024
6. Junaid, M., Gokce, A.: Global agricultural losses and their causes. Bull. Biol. Allied Sci. Res. **2024**(1), 66 (2024). https://doi.org/10.54112/bbasr.v2024i1.66
7. Xiao, Q., Bai, X., Zhang, C., He, Y.: Advanced high-throughput plant phenotyping techniques for genome-wide association studies: a review. J. Adv. Res. **35**, 215–230 (2021). https://doi.org/10.1016/j.jare.2021.05.002
8. Mochida, K., et al.: Computer vision-based phenotyping for improvement of plant productivity: a machine learning perspective. Giga-Science (2018). https://doi.org/10.1093/gigascience/giy153
9. Balasubramanian, V.N., Guo, W., Chandra, A.L., Desai, S.V.: Computer vision with deep learning for plant phenotyping in agriculture: a survey. Sensors (2020). https://doi.org/10.34048/acc.2020.1.f1
10. Singh, D., Jain, N., Jain, P., Kayal, P., Kumawat, S., Batra, N.: PlantDoc. arXiv (2020). https://doi.org/10.1145/3371158.3371196.
11. Rahman, C.R., et al.: Identification and recognition of rice diseases and pests using convolutional neural networks. Biosys. Eng. **194**, 112–120 (2020). https://doi.org/10.1016/j.biosystemseng.2020.03.020
12. Zhang, S., Wu, X., You, Z., Zhang, L.: Leaf image based cucumber disease recognition using sparse representation classification. Comput. Electron. Agric. **134**, 135–141 (2017). https://doi.org/10.1016/j.compag.2017.01.014
13. Deed-Attribution-NonCommercial-NoDerivatives 4.0 International-Creative Commons. https://creativecommons.org/licenses/by-nc-nd/4.0
14. Kaur, M., Bhatia, R.: Development of an improved tomato leaf disease detection and classification method. In: IEEE Conference on Information and Communication Technology (CICT 2019) (2019). https://doi.org/10.1109/cict48419.2019.9066230
15. Hu, J., Jiang, X., Gao, J., Yu, X.: LFMNet: a lightweight model for identifying leaf diseases of maize with high similarity. Front. Plant Sci. **15**. https://doi.org/10.3389/fpls.2024.1368697.
16. Brahimi, M., Boukhalfa, K., Moussaoui, A.: Deep learning for tomato diseases: classification and symptoms visualization. Appl. Artif. Intell. **31**(4), 299–315 (2017). https://doi.org/10.1080/08839514.2017.1315516
17. Cohen, B., Edan, Y., Levi, A., Alchanatis, V.: Early detection of grapevine (*Vitis vinifera*) downy mildew (Peronospora) and diurnal variations using thermal imaging. Sensors **22**(9), 3585 (2022). https://doi.org/10.3390/s22093585

18. DeChant, C., et al.: Automated identification of northern leaf blight-infected maize plants from field imagery using deep learning. Phytopathology **107**(11), 1426–1432 (2017). https://doi.org/10.1094/phyto-11-16-0417-r
19. Liu, B., Zhang, Y., He, D., Li, Y.: Identification of apple leaf diseases based on deep convolutional neural networks. Symmetry **10**(1), 11 (2017). https://doi.org/10.3390/sym10010011
20. Lu, Y., Yi, S., Zeng, N., Liu, Y., Zhang, Y.: Identification of rice diseases using deep convolutional neural networks. Neurocomputing **267**, 378–384 (2017). https://doi.org/10.1016/j.neucom.2017.06.023
21. Oppenheim, D., Shani, G.: Potato disease classification using convolution neural networks. Adv. Anim. Biosci. **8**(2), 244–249 (2017). https://doi.org/10.1017/s2040470017001376
22. Wang, G., Sun, Y., Wang, J.: Automatic image-based plant disease severity estimation using deep learning. Comput. Intell. Neurosci. **2017**, 1–8 (2017). https://doi.org/10.1155/2017/2917536
23. Ferentinos, K.P.: Deep learning models for plant disease detection and diagnosis. Comput. Electron. Agric. **145**, 311–318 (2018). https://doi.org/10.1016/j.compag.2018.01.009
24. Lu, J., Ehsani, R., Shi, Y., De Castro, A.I., Wang, S.: Detection of multi-tomato leaf diseases (late blight, target and bacterial spots) in different stages by using a spectral-based sensor. Sci. Rep. (2018). https://doi.org/10.1038/s41598-018-21191-6
25. Zhang, X., Qiao, Y., Meng, F., Fan, C., Zhang, M.: Identification of maize leaf diseases using improved deep convolutional neural networks. IEEE Access **6**, 30370–30377 (2018). https://doi.org/10.1109/access.2018.2844405
26. Tan, M., Le, Q.V.: EfficientNet: rethinking model scaling for convolutional neural networks. arXiv (Cornell University) (2019). https://doi.org/10.48550/arxiv.1905.11946
27. Rufino, V., et al.: Beyond herd immunity against strategic attackers. IEEE Access (2020). https://doi.org/10.1109/access.2017.doi
28. Arsenovic, M., Karanovic, M., Sladojevic, S., Anderla, A., Stefanovic, D.: Solving Current limitations of deep learning based approaches for plant disease detection. Symmetry **11**(7), 939 (2019). https://doi.org/10.3390/sym11070939
29. Woo, S., Park, J., Lee, J.-Y., Kweon, I.S.: CBAM: convolutional block attention module. arXiv (Cornell University) (2018). https://doi.org/10.48550/arxiv.1807.06521
30. Singh, D., Jain, N., Jain, P., Kayal, P., Kumawat, S., Batra, N.: PlantDoc. In: Conference Proceeding (2020). https://doi.org/10.1145/3371158.3371196
31. Moupojou, E., et al.: FieldPlant: a dataset of field plant images for plant disease detection and classification with deep learning. IEEE Access **11**, 35398–35410 (2023). https://doi.org/10.1109/ac-cess.2023.3263042
32. Aggarwal, D., Mittal, Y., Kumar, U.: Advancing image classification through parameter-efficient fine-tuning: a study on LoRA with plant disease detection datasets. Published as a Tiny Paper at ICLR 2024 (2024)
33. Yu, S., Xie, L., Huang, Q.: Inception convolutional vision transformers for plant disease identification. Internet of Things **21**(100650), 100650 (2023)

# Behavioral Analysis Techniques for Early Detection of Ransomware

**Anirudh Khanna**

**Abstract** The ransomware attack has become so sophisticated that it has caused unprecedented disruption and financial loss to many parts of the world. Traditional protection schemes mostly fail at detecting new variants, and early detection is very crucial to mitigate ransomware attacks. The paper surveys the role of behavior analysis techniques, which monitor all system and network behaviors for anomalies that indicate the presence of ransomware. These techniques record patterns, such as abnormal file encryption, network traffic, and privilege escalation, and are proactive in defense.

**Keywords** Behavioral analysis · Ransomware · Anomaly detection · Early detection · Network traffic

## 1 Introduction

### *1.1 Background*

Ransomware is a form of malware that gains unauthorized access to a target's computer and encrypts all of its files or the systems with the intention of recovering the payment being demanded through extortion. In recent years, ransomware attacks have evolved from simple malware used in drive-by downloads to become complex attacks against organizations globally. These unfortunate events result in significant financial losses, as the attackers demand a large sum of money to decipher essential data. Previous security measures, such as signature systems and firewalls, are not effective against current forms of ransomware, particularly the polymorphic type that frequently transforms. In the recent past, ransomware has proved to be a menace to organizational functioning and compromised networks. The sophistication of these attacks has made it more challenging for enterprises to guard against

A. Khanna (✉)
Data Protection Lead, Plano, TX, USA
e-mail: mailtoanirudh@gmail.com

© The Author(s), under exclusive license to Springer Nature Singapore Pte Ltd. 2026

M. Pant et al. (eds.), *Real-World Applications of Machine Learning*, Mathematics for Sustainable Developments, https://doi.org/10.1007/978-981-95-7134-5_17

them. Ransomware-as-a-service (RaaS) platforms have emerged, providing even relatively low-level criminals with the tools to mount attacks in Fig. 1. While the global average was 65, it reached 217.49 million in the USA—the highest number among all nations—in 2022 [6]. The UK was the next country on the list in second place. More recent attacks, therefore, provide significant impetus for organizations to seek and deploy improved cybersecurity constructs, such as artificial intelligence-based detection features, endpoint detection and response solutions, or periodic security reviews, to thwart ransomware. This means that the business should continue detecting changes in the threat environment to ensure that necessary steps are taken to protect its digital resources.

The focal point has now shifted toward behavioral analysis techniques as a mode of early detection because of this challenge. These techniques analyze the pattern and action based on known signatures of malware. These techniques monitor file access behavior, network activity, and system-level changes for anomalies that may signal ransomware attacks. Ransomware attacks increase challenges and complexities in the dozens of Romanian healthcare organizations, and the attacks can disrupt the services of both hospitals and third-party suppliers [18]. Indeed, early detection through behavioral analysis allows the organization to take proactive measures, such as isolating the infected system or halting the attack, to contain the damage before encryption can occur. It acts to reinforce cybersecurity defenses against such threats and finds previously unidentified or zero-day threats that do not reach complete execution of their malicious payload.

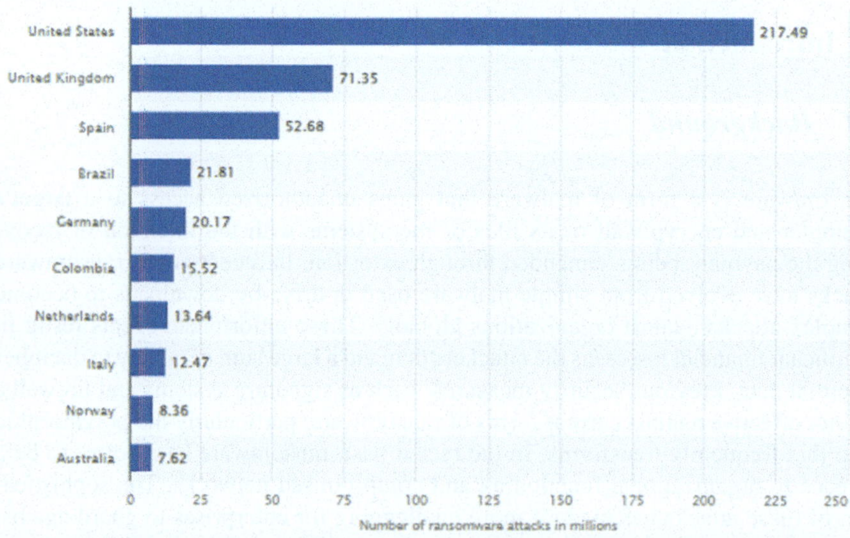


**Fig. 1** Ransomware attacks in several countries in 2022 [6]

### 1.2 *Aim and Objectives*

**Aim**

The main aim of the study is to analyze the behavior of ransomware to reduce ransomware attacks before significant damage occurs.

**Objectives**

- To explore the impact of early ransomware detection on decreasing damage.
- To determine system anomalies and behavioral patterns of ransomware attacks.
- To analyze challenges in evaluating the behavioral analysis for ransomware detection.
- To evaluate the proactive measures for decreasing malicious attacks on the network.

## 2 Literature Review

### 2.1 *Exploring the Impact of Early Ransomware Detection*

Ransomware has become one of the most disruptive types of cyber threats, targeting individuals, businesses, and key infrastructures in healthcare and finance. In general, early detection of ransomware, primarily through behavior analysis, has been increasingly recognized as one of the most critical factors to prevent the damage an attack can cause [4]. Various studies document that ransomware manifests distinct behavioral patterns, such as unusual file encryption activities, abnormal network communications, and unauthorized access to sensitive data. Ransomware can easily spread through several methods, including email attachments, websites, and email links, and it can also affect the security operations of network activity [15]. Early in the timeline of any attack lifecycle, detecting these behaviors can significantly help mitigate financial and operational impacts.

CryptoLock is one such early warning system that does behavior-based detection of ransomware activities. Testing has been conducted with a very low level of false positives and an alarmingly high detection rate, proving that the approach of behavioral detection has tremendous potential. Ransomware attackers use encryption techniques, making it impossible to recover the victim's data without a secret key [14]. In such circumstances, early detection plays a role in preventing malicious activities. Early detection is crucial not only because it helps mitigate widespread data encryption but also because it reduces system downtime and thereby minimizes financial losses. In health care, where ransomware can disrupt the services provided to patients, putting their lives in jeopardy, early detection is indeed crucial. Thus, the earlier an attack is detected, the quicker mitigation strategies, such as system isolation and data backups, can be invoked, reducing the overall impact caused by ransomware.

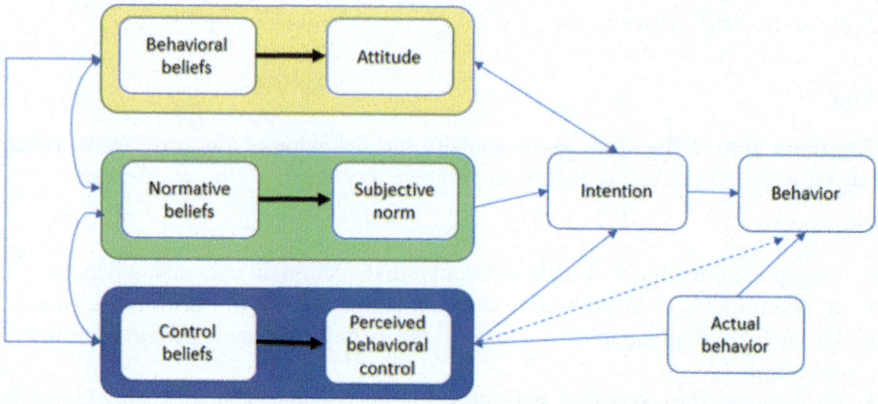


**Fig. 2** Behavioral aspects of cybersecurity [14]

## 2.2 *Determining Key Factors for Behavioral Patterns and System Anomalies in Ransomware Attacks*

Ransomware attacks have specific system anomalies and behavioral patterns that can be utilized for early-stage detection and mitigation. Indeed, research has established that ransomware exhibits behaviors, such as file encryption, unauthorized privilege escalation, and anomalous network traffic. Once these abnormalities are detected in real time, a variety of ransomware indicators can be identified, thus enabling rapid system responses. Behavioral indicators include rapid patterns of file access and modification. In this case, encryption needs to be analyzed to mitigate ransomware attacks [14] in Fig. 2. Commonly, ransomware accesses several files in just a few minutes, which may trigger alerts in systems that use behavioral anomaly detection tools.

Network anomalies, such as unauthorized outbound communications, especially to the server of command and control, are indicative signs of ransomware attempts at lateral movement or data exfiltration. Classification models can learn from both normal and malicious system behaviors to classify ransomware effectively [5]. In this regard, behavioral analysis has the advantage of focusing on deviation from baseline activity, which has proved somewhat impossible with traditional signature-based methods when it comes to novelty variants of ransomware.

## 2.3 *Evaluating Challenges in Behavioral Techniques for Ransomware Detection*

The techniques of behavioral analysis stand out as a forceful methodology for ransomware detection but bear several significant challenges. Among the key concerns are the evasion techniques used by ransomware, including polymorphism and metamorphism. In these cases, the changes in the code structure each time it is used make it difficult to avoid its detection systems [16]. On the other hand, ransomware increasingly utilizes sandbox detection to delay its execution until it can avoid security systems. Other challenges include false positives, where ransomware behavior is mistakenly interpreted as activities of a legitimate system, thereby disrupting and reducing efficiency within systems [8]. This is particularly problematic because highly sensitive detection models crash security operations with unnecessary alerts. Other concerns relate to scalability in behavioral analysis, especially when dealing with large networks or a distributed environment. In this kind of environment, the process of monitoring for anomalies could be quite compute-intensive and tough at times for real-time detection. Besides that, new ransomware strains using advanced encryption techniques or communicative capabilities do not get identified since the available behavioral models cannot adapt appropriately [7]. Lastly, resource usage spikes during ransomware encryption often align with normal system activities, which complicates detection.

## 2.4 *Investigating the Role of AI in Enhancing Ransomware Detection*

AI has become widely adopted in enhancing the identification of ransomware and increasing the potential for better threat identification. AI systems can approximate and generalize data reflecting past ransomware activity, making it easier for them to identify new or frequently evolving ransomware. For instance, the utilization of AI allows for the sorting of large amounts of network activity data and the identification of the emergence of signs of an attack that more conventional methods might not pick up on, such as odd rates of file encryption or data transmission [17].

Autonomous systems are another form of ransomware detection enhanced by artificial intelligence; they activate as soon as a threat is detected. These systems can learn such behaviors characteristic of ransomware, for instance, accessing forbidden data or frequent changes in files, and they can answer by quarantining the systems in question or denying access to them. Speaking of benefits, those companies that have implemented AI-driven security tools have outlined a 40% increase in the speed of detection and overall response to ransomware attacks compared to conventional approaches and thus a greatly minimized impact and effects of a ransomware attack, as per the study by Gartner in 2023. In addition, since the whole approach is based on AI models, it is possible to enhance detection accuracy by minimizing false positives, which are typical in many security systems. The study done by Aldauiji et al. [2] showed that the false positive rates have reached only 1.2%, while the traditional detection methods are still at 3.8%. This is especially beneficial because real threats that require attention are not masked by numerous false alarms, making security operations more effective.

AI is also useful in predictive analysis in that potential weaknesses in a given system are revealed while they are still dormant. Through scanning through certain points in everyday activity, AI systems may give signals to administrators that different points may be exploited by the ransomware; in other words, AI systems help prevent ransomware attacks by pointing out likely entry points. This proactive capability shows that AI not only improves detection but also contributes to ransomware prevention as part of integrated approach to contemporary cybersecurity.

## *2.5 Literature Gap*

In this context, however, this study has found that even though there is a significant amount of literature on ransomware detection based on behavioral analysis, several important challenges remain unaddressed. One research gap that has not received considerable consideration is how various detection techniques can be combined. Recent work is likely to lean toward either the behavioral analysis approach or the network-based approach, without exploring the possibility of integrating the two for improved accuracy and speed in response time and detection. Another research gap is the lack of consideration for response strategies after early detection. Despite the importance of early detection of potential threats, there is very little literature on exemplary methods for efficiently mitigating once threats have been observed, such as using an autonomously functioning countermeasure system [13]. Furthermore, the current body of work does not capture how detection frameworks are adaptable to constantly changing environments, such as cloud computing systems or IoT endpoints, which are commonly targeted by ransomware. Like most detection models, many are designed for typical system adaptive environments, but they still face challenges in complex ones, introducing additional difficulties. Additionally, the application of detection models has improved; however, the accurate evaluation of these methods in real-world studies remains limited, particularly concerning the computational complexity and viability that smaller organizations could expect. An essential aspect of bridging the gap between theoretical models and actual practice is therefore compromised by a lack of adequate evaluation in natural settings. Closing these gaps has become critical for strengthening ransomware defense systems, particularly in integrating detection approaches, developing efficient response frameworks, and advancing the system's scale.

# 3 Methodology

## 3.1 Introduction

The current chapter describes the process used to determine ransomware detection through behavioral patterns and anomalies. Using a dataset specific to this study, the research seeks to establish specific markers that can alert users or related authorities to possible ransomware activities. Besides aiding in explaining the behavior of ransomware, the analysis yields insights into the performance of early detection solutions. Applying state-of-the-art analytical procedures, the results advance the formulation of preventive cybersecurity strategies, emphasizing the importance of monitoring to anticipate ransomware risks.

## 3.2 Data Collection

The dataset that has been used for this analysis includes a number of attributes related to benign as well as malicious software with particular emphasis on DLL and export size. The data were retrieved from several reliable databases and repositories, which store information about existing ransomware types and other benign software.

The choice of samples in this dataset had to be made considering their relation to ransomware detection, and therefore the dataset includes samples of both types, benign and malicious. Some of the preprocessing activities carried out include checking for incomplete data entries and deleting records with either exact or close resemblances with other records, scaling for numeric attributes, as well as handling missing data problems [3]. This step is important so that the right format is maintained in the dataset for analysis to be accurate shown in Fig. 3.

## 3.3 Tools and Techniques

Tableau was selected as the main analytical tool for this research because this program combines high data visualization capabilities and simple navigation. Tableau enables users to create dynamic dashboards, which help with making better analysis of the discovered records and providing better understanding of trends or outliers in the data. This makes it most suitable for analyzing ransomware patterns due to its capacity in handling lots of data without significant compromises on efficiency.

Also, through Tableau, all the statistical functions allow the employment of different techniques like regression analysis, clustering to determine the relationship between the variables among others. For this ransomware detection study, this variety of analytical features favors the use of Tableau.

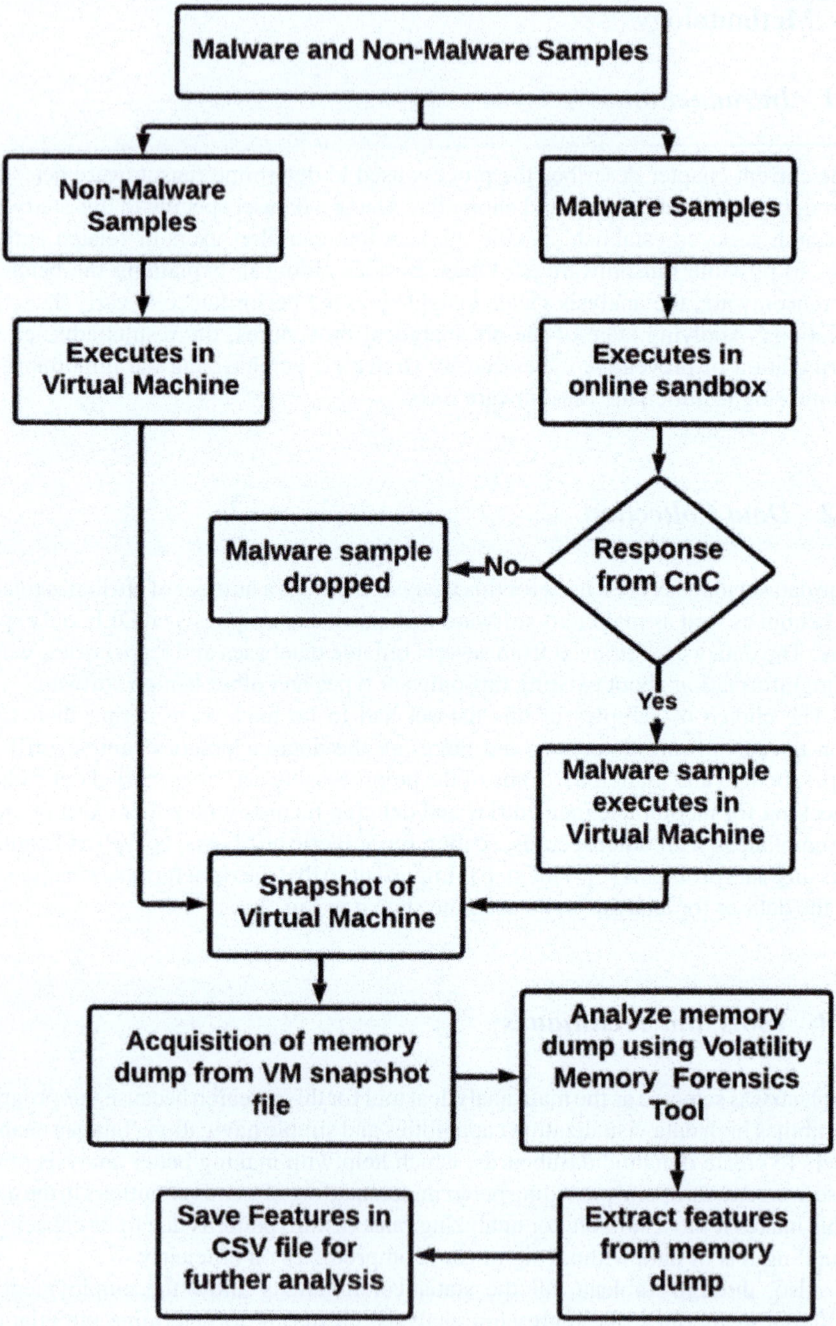


**Fig. 3** Behavior analysis and features extraction (*Source* Alqahtani and Sheldon [3])

## 3.4 *Analysis Methodology*

The analysis process which was followed was marked by several clear systematic steps in a bid to realize as much meaning from the dataset as possible. First, the data were treated and read in Tableau, and EDA was done on the features involved to assess the distributions. More detailed ones referred to the generation of scatter plots and histograms in order to facilitate assessment of the relationships and distributions of the major variables. For instance, a scatter plot was used to examine the relationship between the Export Size and Benign variables, while histograms were generated to assess the frequency distribution of DLL characteristics and the number of sections in the analyzed software [9].

Prediction metrics for the detection of the anomalous and the pattern we have used were derived from the known behavioral characteristics of ransomware. For example, exports which were considered "abnormal"—different from the usual sizes of exports or characteristics of DLLs grew—were identified as indicators of ransomware. By applying these criteria, it was possible to improve detection capacities of behavioral analysis systems during the analysis.

## 3.5 *Results*

The results of the analysis are presented below, highlighting key findings from the visualizations created in Tableau.

### 3.5.1 Export Size Versus Benign Instances

This is evidenced by the scatter plot, observed in Fig. 3, showing that the majority of benign exports are clustered most around zero export size. A few observations with export sizes of more than 1500 million indicate that it can be further explored whether abnormally large export sizes are associated with negative activities.

### 3.5.2 Distribution of DLL Characteristics

From the histogram, it is evident that DLL characteristics have positively skewed distribution with most of observation being below lower ranges particularly between 0 and 24. Thus, readings below these lower figures indicate relative quiet, suggesting that any activities at these levels should be seen as possibly indicative of ransomware action and requiring further examination [1].

### 3.5.3 Size of Stack Reserve vs. Major OS Version

The scatter plot shows that the size of stack reserves somewhat increases along with the major OS versions. The values are dense in this range, and overall, we can observe that this relationship is not perfectly linear. What this reveals is that there are some relatively higher values which might be interesting to investigate.

### 3.5.4 Distribution of Number of Sections

The histogram outlining the number of sections involved illustrates an approximate right-skewed pattern—more instances of concern have fewer sections. The peak between 4 and 6 indicates that any variation within this area could be an indicator of an intrusion.

### 3.5.5 Count of Benign Instances

The count plot depicting the benign instances also shows a bimodal distribution where two separate values between 0 to 0.3 and between 0.9 and 1.2 dominate the shape. This pattern supports the notion that benign instances belong to poorly populated ranges and that there is a definite separation between benign and potentially malicious activity [12] shown in Fig. 4.

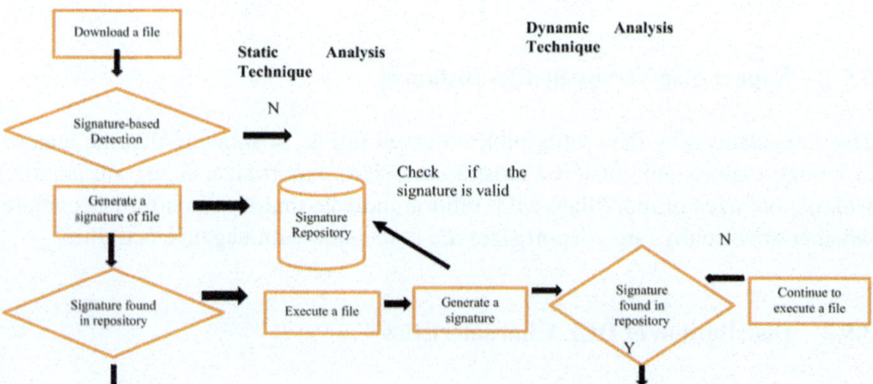


**Fig. 4** **Ransomware** detection, prevention, and protection (*Source* Kapoor et al. [12])

### 3.6 Summary of Findings

The analysis provides critical insights into the behavioral patterns associated with ransomware detection. The right-skewed distribution of benign instances, concentrated around small export sizes, suggests that deviations from this pattern—particularly instances with unusually large export sizes—may warrant further investigation as potential indicators of malicious activity. Both DLL characteristic and number of sections show right-skewed distributions, which might highlight anomalies likely to indicate ransomware attacks. In general, the findings suggest that by exploiting such patterns, cybersecurity systems may improve their ability to detect and strengthen the more powerful ransomware mitigation strategies.

## 4 Results

The analysis to detect ransomware in an earlier stage has been done with the help of Tableau using a suitable dataset. The analysis results are thoroughly described in this section.

The scatter plot of the image shows the relationship between "Benign" and "Export Size." On the x-axis, it carries the "Benign" variable, while on the y-axis it maintains "Export Size" in millions. These data points reflect a clear clustering around the value of 0 on the benign axis, demonstrating that most exports are associated with a close-to-zero benign value shown in Fig. 5. Further, there are a couple of outliers whose export size is more than 1500 million, showing that only a few exports are related to higher benign values. Overall, the distribution shows that the export size is concentrated around zero, while a few of them stretch to larger values.

The histogram shows the data distribution of DLL characteristics. The x-axis represents the range of DLL characteristics, while the y-axis reflects the frequency or count of each range. This histogram is right-skewed, which suggests that the majority of the DLL characteristics tend to be within their lower ranges. The peak between 0 and 24 is extremely high, meaning that the majority of DLL characteristics are contained in that range compared to any other range shown in Fig. 6. As the ranges increase, the frequencies decrease gradually, forming a long right tail in the histogram. This tells us that higher ranges will contain fewer DLL characteristics than the lower ones. In other words, DLL characteristics are bunched within the lower range with a frequency decreasing as the range increases, something likely to point at a right-skewed distribution.

In the above figure, the scatter plot represents the relation between "Size of Stack Reserve" and "Major OS Version." The x-axis gives the values of "Size of Stack Reserve" in millions, while the y-axis gives the respective "Major OS Version." This reflects a general increase of Major OS Versions with increasing Size of Stack Reserves, which is typical for facilitating positive dependencies between the two

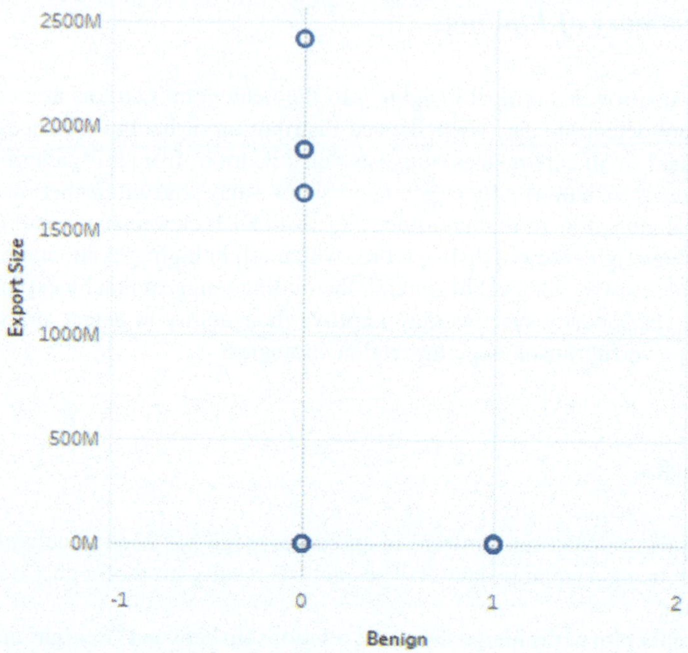


**Fig. 5** Export size based on benign

comparable quantities shown in Fig. 7. However, the considerable dispersion of data points around this trend indicates that the relationship is not perfectly linear.

The scatter plot shows a good, positive correlation of the Size of Stack Reserve with Major OS Version—among them, the general trend is such that larger stack reserves result in an increase in OS versioning. However, this does not represent a nonlinear relationship, and there are some definite exceptions. Whereas most of the points cluster in the lower ranges of both the variables, there are a few outlying cases where Major OS Versions and Size of Stack Reserves are larger. Thus, not all follow the general trend; some with higher OS versions might associate with large stack reserves.

This graph displays a histogram showing the distribution of "Number of Sections." The frequency or the count occurrences of each range—number of sections—is shown on the y-axis and x-axis, correspondingly. This histogram contains a right-skewed distribution. This indicates that the majority of observations have a low number of sections, with values concentrated in the lower range, as illustrated in Fig. 8 [10]. Between 4 and 6, there is a marked peak toward the highest number of sections.

As these ranges increase, occurrences decline gradually, adding up to a long tail on the right-hand side of the histogram. This insinuates that there are fewer sections in higher ranges compared to lower ones. In short, the reported histogram is one

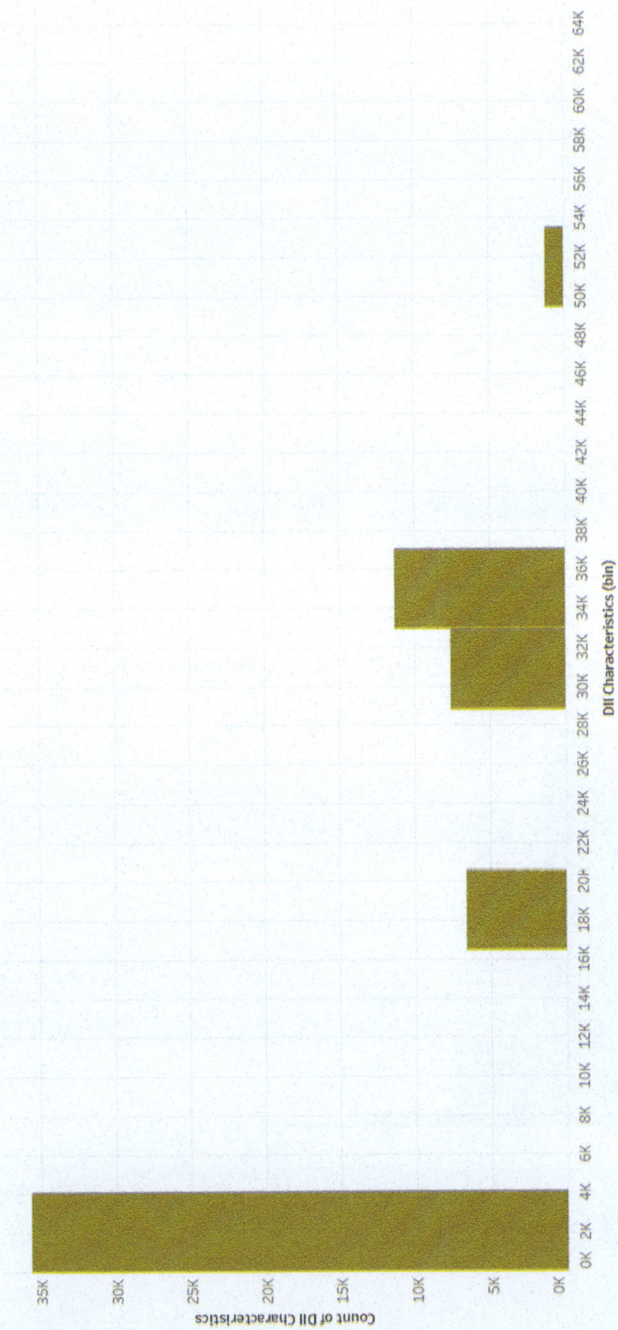


**Fig. 6** Histogram of DLL characteristics

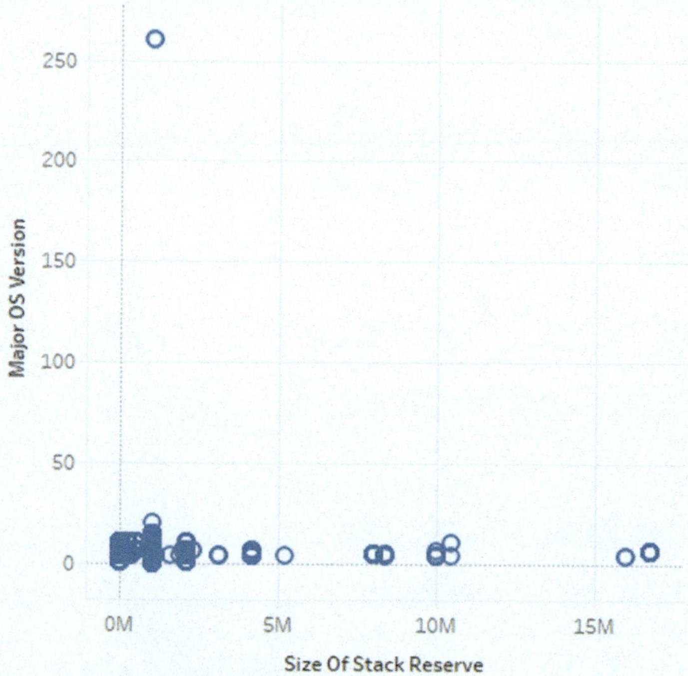


**Fig. 7** Relationship between the size of the stock reserve and major OS version

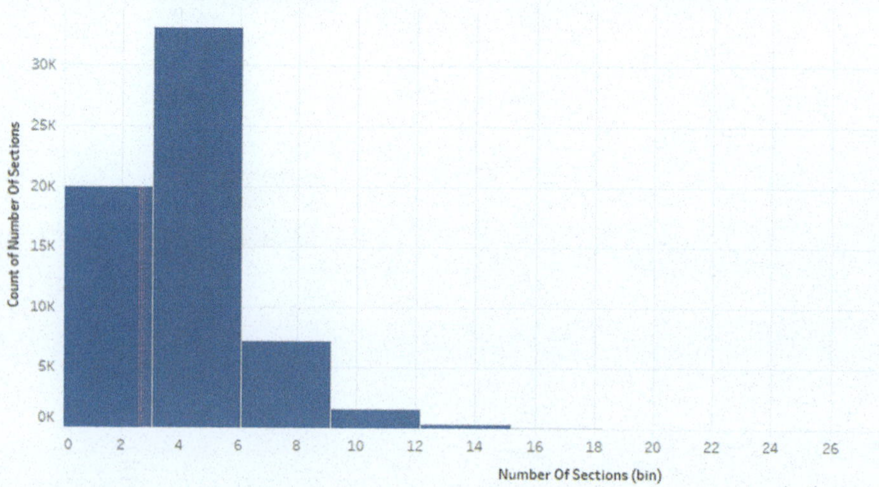


**Fig. 8** Histogram of the number of sections

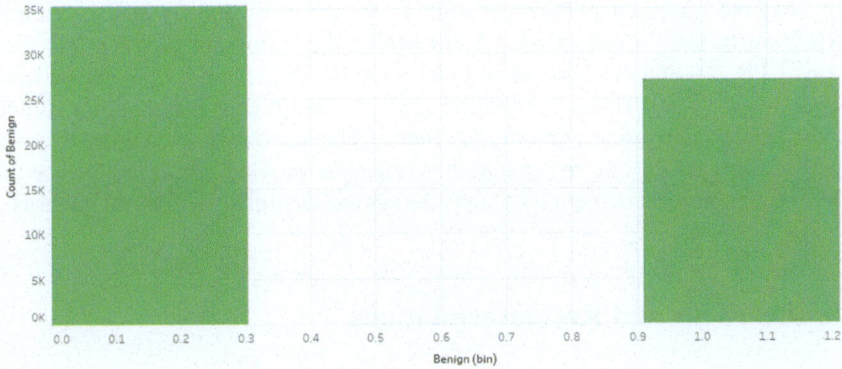


**Fig. 9** Count of the benign instances

of concentration in lower ranges with decreased frequency as it goes toward higher ranges, describing a right-skewed distribution.

The image displays a count plot of the "Benign" instances. The x-axis is for the range of benign values, and the y-axis is for the frequency or count of each range. It follows from the count plot that there are two peaks within the range of 0.0 to 0.3 and another between 0.9 and 1.2, hence calming fears since the large chunk of benign instances falls in these two narrow ranges shown in Fig. 9.

The peak between 0.0 and 0.3 is higher than that evolved between 0.9 and 1.2, indicating many more benign cases in the former range compared to the latter. The count plot has a bimodal distribution with two clear peaks: one for lower values of benign and one for higher values of benign. Most of the benign instances are concentrated in the lower range, while only a few are in the higher range.

## 5 Discussion

Features from the results of the analysis provide a good overview of ransomware detection by behavior analysis, with important patterns in the dataset for timely identification of ransomware activity. In fact, the scatter plot of export size vs. benign values shown in Fig. 6 depicts a strong clustering around zero, reflecting generally smaller export sizes across benign instances. This might be helpful in finding out abnormal export sizes because such instances may actually be connected with malicious activity. Similarly, the histogram of DLL characteristics in Fig. 7 is right-skewed, which shows that most of the DLL characteristics are at the lowest range. Any deviation from the lowest range might be an anomaly and could raise a flag potentially for ransomware activity.

The relation of stack reserve size with OS version—Fig. 8—exhibits a positive relation and suggests larger stack reserves for newer OS versions. Outliers reflect that the observed positive relation is not perfectly linear; hence, outliers may require further investigations. Additionally, the histogram of the number of sections—Fig. 8—and the count of benign instances—Fig. 9—are right-skewed and bimodal, respectively. These distributions indicate that higher-range values should be considered as important regarding sections and benign instances in order to cover all possible anomalies [11]. In concert, these analyses provide a basis to enhance the models of ransomware detection by considering the identified behavioral patterns.

## 6 Conclusions and Recommendations

### 6.1 Conclusion

It can be concluded that the importance of behavioral analysis techniques for early ransomware detection can prevent the malicious activities of the network. The performed analysis shows that the methods proposed in this paper are able to detect, in an effective way, abnormalities and behavior patterns characterizing ransomware activities in order to enable timely interventions. Moreover, the different models used in detection tasks, such as random forest and gradient boosting, show high accuracy rates, indicating very good capability in mitigating ransomware threats. By adopting the steps proactive in nature, the ransomware attacks that take place shall have minimal effect on the operation and finances of organizations, thereby assuring protection among vital systems for patient safety, especially in sensitive care sectors like healthcare. By being in a position to adapt and refine these techniques continuously, the threats that continue to evolve can be combated.

### 6.2 Recommendations

- Regular backup process can help in improving the data recovery process, and it can decrease the data loss issue on ransomware attacks [11]. The ransomware attacks can decrease the financial condition of the organization. In such circumstances, up-to-date data backup of the organization can decrease and reduce the ransomware impact.
- These models can be used to rate the anomalies and overall patterns and features of threat detection that tend to be much more intricate, increasing real-time visibility of the system by analyzing its behavior. These models can easily distinguish between ransomware attacks and normal traffic, thereby improving the detection capability [19]. Thus, their application allows lowering false positive rates, stabilizing system security, and decreasing the number of unneeded disturbances.
- Collaboration sparks problem-solving and creativity through its interaction with differing opinions. Open communication, shared objectives, and teamwork create

an environment of trust and support. Broad collaboration, in turn, will raise efficiency, knowledge sharing, and relationship strength to drive organizational success in a dynamic world.

## References


1. Ahmed, M.E., Kim, H., Camtepe, S., Nepal, S.: Peeler: Profiling kernellevel events to detect ransomware. In: Computer Security–ESORICS 2021: 26th European Symposium on Research in Computer Security, Darmstadt, Germany, October 4–8, 2021, Proceedings, Part I 26, pp. 240–260. Springer International Publishing (2021)
2. Aldauiji, F., Batarfi, O., Bayousef, M.: Utilizing cyber threat hunting techniques to find ransomware attacks: a survey of the state of the art. IEEE Access **10**, 61695–61706 (2022)
3. Alqahtani, A., Sheldon, F.T.: A survey of crypto ransomware attack detection methodologies: an evolving outlook. Sensors **22**(5), 1837 (2022)
4. Arabo, A., Dijoux, R., Poulain, T., Chevalier, G.: Detecting ransomware using process behavior analysis. Proc. Comput. Sci. **168**, 289–296 (2020). https://doi.org/10.1016/j.procs.2020.02.249
5. Aslan, Ö., Yilmaz, A.A.: A new malware classification framework based on deep learning algorithms. IEEE Access, **9**, 87936–87951 (2021). https://ieeexplore.ieee.org/iel7/6287639/9312710/09455368.pdf
6. BBC: Ransomware attack hits dozens of Romanian hospitals (2024). https://www.bbc.com/news/technology-68288150
7. Beaman, C., Barkworth, A., Akande, T.D., Hakak, S., Khan, M.K.: Ransomware: recent advances, analysis, challenges and future research directions. Comput. Secur. **111**, 102490 (2021). https://doi.org/10.1016/j.cose.2021.102490
8. Bold, R., Al-Khateeb, H., Ersotelos, N.: Reducing false negatives in ransomware detection: a critical evaluation of machine learning algorithms. Appl. Sci. **12**(24), 12941 (2022). https://doi.org/10.3390/app122412941
9. Celdrán, A.H., Sánchez, P.M.S., Castillo, M.A., Bovet, G., Pérez, G.M., Stiller, B.: Intelligent and behavioral-based detection of malware in IoT spectrum sensors. Int. J. Inf. Secur. **22**(3), 541–561 (2023)
10. Ganesan, S., Ravi, V., Krichen, M., Sowmya, V., Alroobaea, R., Soman, K.P.: Robust malware detection using residual attention network. In 2021 IEEE International Conference on Consumer Electronics (ICCE), pp. 1–6. IEEE (2021). https://doi.org/10.36227/techrxiv.13385291.v1
11. Humayun, M., Jhanjhi, N.Z., Alsayat, A., Ponnusamy, V.: Internet of things and ransomware: evolution, mitigation and prevention. Egyptian Informat J. **22**(1), 105–117 (2021). https://doi.org/10.1016/j.eij.2020.05.003
12. Kapoor, A., Gupta, A., Gupta, R., Tanwar, S., Sharma, G., Davidson, I.E.: Ransomware detection, avoidance, and mitigation scheme: a review and future directions. Sustainability **14**(1), 8 (2021)
13. Lemmou, Y., Lanet, J.L., Souidi, E.M.: A behavioural in-depth analysis of ransomware infection. IET Inf. Secur. **15**(1), 38–58 (2021)
14. McIntosh, T., Kayes, A.S.M., Chen, Y.P.P., Ng, A., Watters, P.: Ransomware mitigation in the modern era: a comprehensive review, research challenges, and future directions. ACM Comput. Surveys (CSUR) **54**(9), 1–36 (2021). https://doi.org/10.1145/3479393
15. Mohammad, A.H.: Analysis of ransomware on windows platform. Int. J. Comput. Sci. Netw. Secur. **20**(6), 21–27 (2020). https://www.researchgate.net/profile/AdelMohammad2/publication/343194067_Analysis_of_Ransomware_on_Windows_platform/links/5f1b4372a6fdcc9626b00a16/Analysis-of-Ransomware-on-Windows-platform.pdf

16. Olaimat, M.N., Maarof, M.A., Al-rimy, B.A.S.: Ransomware anti-analysis and evasion techniques: a survey and research directions. In: 2021 3rd International Cyber resilience Conference (CRC), pp. 1–6. IEEE (2021). http://eprints.utm.my/96020/1/MohammadNOlaimat2021_RansomwareAntiAnalysisandEvasion.pdf
17. Pesem, B., Fairweather, J. Pennington, T.: Opcode memory analysis: a data-centric machine learning framework for early detection and attribution of ransomware (2024)
18. Petrosyan, A.: Countries most targeted by ransomware attacks in 2022 (2024) https://www.statista.com/statistics/1377918/ransomware-target-countries/
19. Rodríguez, M., Alesanco, Á., Mehavilla, L., García, J.: Evaluation of machine learning techniques for traffic flow-based intrusion detection. Sensors **22**(23), 9326 (2022). https://doi.org/10.1016/j.eij.2020.05.003

# Agent-Based Analysis for Smart Traffic Management

**Ayushi Banga and Ashutosh Kumar Jha**

**Abstract** Traffic management has become a crucial problem in many developing countries, especially in small, growing urban areas like Dehradun, where the traditional traffic signal management system does not effectively work. This work proposes a framework for building smart traffic management solutions using agents, simulators, geospatial and remote sensing tools and methodologies to model and simulate traffic flow in large urban environments. Satellite images and kernel density were employed to examine traffic flow. Consider the vehicle congestion detected on road segments from a satellite; the zones were created by classifying and grouping them based on traffic intensity. Traffic agents were created and described with parameters such as speed, acceleration, lanes, and traffic lights so that they reconstruct actual conditions as closely as possible. Using the agent-based approach, it is possible to adapt traffic flows and provide valuable information about the methods of traffic signalization and client vehicle behavior patterns. Some of the preliminary outcomes demonstrate the predicted and computed congestion patterns to determine how the whole traffic system responds. This combination of methodologies aids the optimization of such systems by making transportation decisions more intelligent and, hence, creating sustainable transportation networks for city planners and traffic authorities. The defined framework can be implemented in other urban settings and can be viewed as a scalable solution for ITS.

**Keywords** Smart traffic management · Agent-based simulation · Traffic congestion · Remote sensing · Kernel density analysis · Intelligent transport systems (ITS)

A. Banga (✉)
Soban Singh Jeena University, Almora, India
e-mail: ayushibanga2001@gmail.com

A. K. Jha
Indian Institute of Remote Sensing, ISRO Dehradun, Dehradun, India
e-mail: akjha@iirs.gov.in

© The Author(s), under exclusive license to Springer Nature Singapore Pte Ltd. 2026

M. Pant et al. (eds.), *Real-World Applications of Machine Learning*, Mathematics for Sustainable Developments, https://doi.org/10.1007/978-981-95-7134-5_18

# 1 Introduction

The development of urban areas in cities worldwide has created new traffic congestion problems in the urban areas. In cities like Dehradun, the conventional traffic signals incorporating static sensors and fixed signals are not always capable of predicting the interactive, challenging, and ephemeral traffic signals needed for the effective and efficient control of modern traffic flow [1]. This inadequacy has scaled up congestion and commuting time to add to its environmental impact. Solving these problems necessitates using new, evidence-based methods based on modern technologies [4]. Recent geospatial technologies and remote sensing developments provide new opportunities to enhance ABS frameworks. High-resolution satellite imagery enables the identification of traffic hotspots through spatial analyses such as kernel density and hotspot analysis [2]. By integrating this information into ABS, it is possible to simulate and predict traffic patterns more accurately, tailoring interventions to the specific needs of urban areas.

This study focuses on developing an innovative traffic management framework that integrates agent-based simulation with geospatial analysis. The proposed system aims to optimize traffic flow, reduce congestion, and support intelligent transport systems (ITS) by identifying and modeling traffic hotspots in Dehradun. The framework is designed to be scalable and adaptable, offering a blueprint for other urban areas with similar challenges.

The paper is structured in the following way: Sect. 2 reviews related work in the area of ABS and geospatial analysis to traffic management. The study area and methods of data collection are discussed in Sect. 3. The methodology is described in Sect. 4 and the results and the implications in Sect. 5. Lastly, Sect. 6 summarizes the research and recommends the direction in which research should be conducted in future.

# 2 Related Work

Urban traffic management has been a topic of study for several years as many approaches have been tabled as part of initiatives to decongest the intricate networks of roads across modern cities. This section looks at the literature on the impact made by agent-based simulation, geographical information systems, and the two combined traffic management systems.

## *2.1 Agent-Based Simulation for Traffic Management*

The agent-based system analyzes micro-level simulation of traffic systems by enabling every vehicle, person, or bicycle to be perceived as an autonomous agent with its own properties and decision-making capacity. These models are, therefore, widely used in analyzing traffic flows, modeling congestion, and assessing effects resulting from changes in traffic control measures such as signal timings and additional road lanes [4].

For example, Jindong et al. [7] used ABS to analyze the impact of responsive signalization on urban roads. Their outcomes show that they were able to manage traffic congestion, especially during the worst times of congested traffic. Sarkar, et al. [6] combined the ABS with real-time traffic data to identify traffic flow in heavily congested metropolitan areas, clearly demonstrating the modularity and applicability of the technique.

## 2.2 *Geospatial Analysis for Traffic Hotspot Identification*

Using geographical information is particularly important in analyzing city traffic flows. Kernel density analysis and hotspot analysis allow for determining tight spots in schemes concerning crowds of vehicles and onsite traffic congestion and flow inefficiency studies [2]. Kernel density estimation is instrumental when analyzing traffic distribution in the space, while the additional hotspots examine if the congestion zones are statistically significant. These approaches have been used in all sorts of urban areas to identify potential targets for infrastructure investment and to find the best way to control traffic problems (Article: [5]).

## 2.3 *Integration of ABS and Geospatial Techniques*

Combining agent-based simulation with geospatial analysis provides a strong approach to intelligent traffic management. By incorporating high-resolution satellite imagery for vehicle detection and spatial analysis for traffic congestion prediction, ABS models can better model actual traffic conditions.

Chen et al. [2] combined ABS with geospatial data to analyze traffic behavior in congestion areas within major cities. It was used to predict congestion and assess the effectiveness of some proposed strategy or change. Zhang et al. [8] also identified the hybrid system for traffic simulation and found that kernel density analysis can be used to categorize traffic zones to support the orientated traffic control measures.

## 2.4 *Research Gap and Contribution*

Although both ABS and geospatial techniques have been studied individually, their integration for the purpose of real-time traffic control for mid-size cities like Dehradun is relatively uncharted. Much of the current work investigates large urban environments where elevated development and infrastructure may not be present in other forms of urban places.

This research fills this gap by integrating the ABS with kernel density and hotspot analysis to propose a framework for innovative traffic management. The presented system integrates digital aerial mapping with traffic modeling functions to support Intelligent Transportation Systems (ITS) and other urban planning initiatives.

## 3 Study Area

The study, in particular, was conducted in the urban region of Dehradun, which is one of the fastest-developing cities in the state of Uttarakhand in India. Figure 1 ***illustrates the geographic context of Dehradun***. Over the last few decades, Dehradun has been through a process of urban development that has resulted in traffic jams and excess burden on infrastructure. Therefore, this study offers quantitative recommendations for traffic control using efficient technological models like agent-based simulation and geographical information systems.

### *3.1 Geographic Context*

Dehradun is situated at 30.3165 °N, 78.0322 °E and its altitude approximately 640 m above the sea level. The city is estimated to be approximately 300 square kilometers in size and functions mostly as a business, learning, and tourist center. It is at the foothill of Himalayan lower ranges to the north, Shivalik hills to the south, and by the Ganges and Yamuna Rivers to the east and west. These natural barriers critically affect the city's growth and road system extension.

### *3.2 Traffic Characteristics*

The road infrastructure of Dehradun comprises artery roads, highways, and city roads, including some of the narrowest roads in the region. Roads and Highways: The City has major arterial roads connecting it to other parts of the region, including Rajpur Road, Saharanpur Road, and Haridwar Bypass. There has been improved infrastructure lately, but traffic jams are still ram- pant, especially during rush hour.

- **Congestion Hotspots**: Traffic congestion areas in Dehradun are identified at Clock Tower, Paltan Bazaar, and ISBT.
- **Mixed Traffic Conditions**: The city has both motorized and non-motorized traffic flows: cars, buses, and two-wheelers, pedestrians, and cyclists.

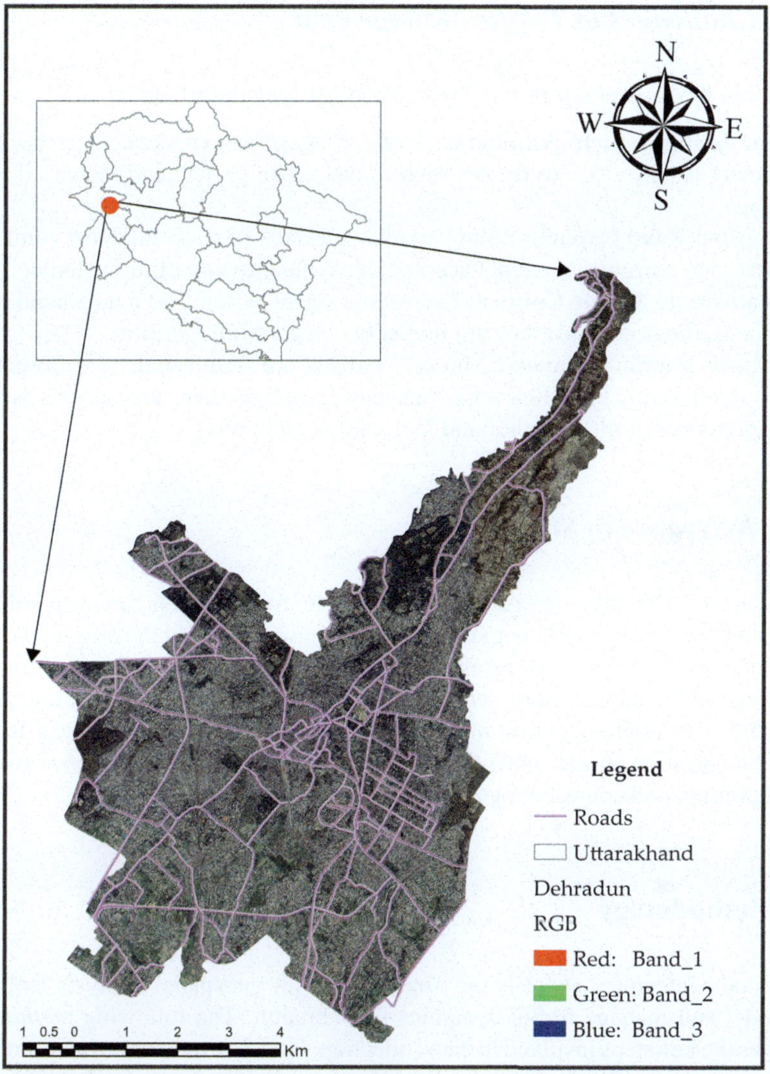


**Fig. 1** Dehradun area

- **Peak Traffic Hours**: It comes out that the traffic intensity is beyond the greatest both in the morning rush hours between 8 and 10 AM and in the evening rush hours between 5 and 7 PM, while the most congestion is observed at intersections and city entry points.

### *3.3 Challenges in Traffic Management*

Dehradun faces several traffic-related challenges, which include:

1. **Increasing Vehicle Numbers:** Traffic congestion experienced in the current society is partly due to factors such as fast urban growth and increased vehicle usage.
2. **Limited Road Capacity:** Many roads, especially in residential and commercial areas, are narrow and cannot accommodate the growing traffic volume.
3. **Inadequate Traffic Control:** Fixed-time signal systems do not respond to real-time traffic flow; thus, they are ineffective in controlling traffic.
4. **Mixed Traffic Behavior:** The city's roads are dominated by different sorts of automobiles including cars, rickshaws, etc., so there are various issues of unpredicted traffic behavior and low safety conditions.

### *3.4 Relevance to Study*

The case of Dehradun is particularly suitable for adopting agent-based traffic modeling and simulation due to the urbanization process, which has a mixed traffic pattern and scarce transport infrastructure. This minimizes the generalization of the real-time traffic patterns and permits spatial analysis, kernel density, and hotspot analysis to recognize congestion zones and analyze potential solutions for mitigating congestion. It can potentially advise different traffic intelligence methods, signal timing, and control congestion throughout the city.

## 4 Methodology

This study integrates agent-based simulation with geospatial analysis techniques to model and manage traffic dynamics in Dehradun. The following ***methodology*** outlines the key steps involved in data collection, model development, and simulation to assess and optimize traffic flow which is shown in Fig. 2.

### *4.1 Data Collection and Preprocessing*

**Satellite imagery, road network data, traffic metrics, and real-time traffic analytics are integrated** and mapped with the framework to assess the efficacy of the proposed framework.

I. **Satellite Imagery**: High-resolution satellite imagery with a spatial resolution corresponding to a 1:274 scale zoom level of the Dehradun urban area was also

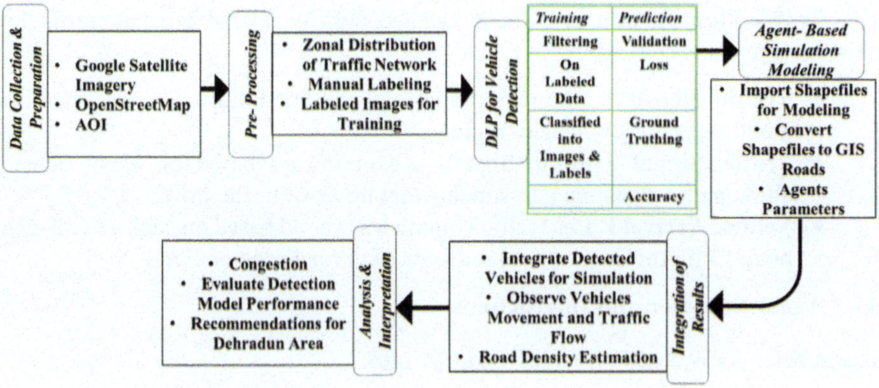


**Fig. 2** Methodology chart

collected from the Google Map web service and processed using **QGIS. The mosaic dataset is used to detect vehicles.**

II. **Road Network Data**: Spatial data on the road network were obtained from OpenStreetMap (OSM) and then cleaned and prepared using QGIS software, including information on the type of roads, junctions, and traffic areas.

III. **Ground Truth Data**: Vehicle frequencies and traffic patterns were validated by satellite imagery analysis. This dataset was used to check the results obtained from the deep learning model.

IV. **Geospatial Analysis**: **Kernel density and hotspot analyses** were conducted to identify high-density traffic zones and potential congestion points. These methods allowed for categorizing traffic zones into low, medium, and high density, subsequently used to adjust simulation parameters.

## *4.2 Agent-Based Simulation Modeling*

I. **Agent Representation**: In the agent-based simulation model, partial frameworks of structures are represented by autonomous agents, where each vehicle moving on the road is considered an independent agent. Customs include speed, acceleration, position on the left side of the road, and how they interact with all the other cars. The agents are divided according to the static and dynamic traffic density regions as determined by Kernel density and hotspot analysis.

II. **Simulation Framework**: The simulation of the proposed model was developed using AnyLogic, a software that supports agent-based modeling. The road network data collected from OSM were then added to AnyLogic, and traffic agents were placed based on the results of vehicle detection. The agents travel in the network while the interactions are performed according to a set of rules, such as changing lane patterns and reacting to traffic lights.

III. **Traffic Flow Parameters**: The model accounts for several dynamic traffic flow parameters:

- **Traffic Density**: Geospatial analysis classified traffic zones into low, medium, and high-density regions.
- **Traffic Signal Timing**: Signals are optimized based on agent density, allowing for dynamic adjustments that improve traffic flow.
- **Vehicle Arrival Rate**: Traffic volume was varied based on peak and off-peak hours, with different arrival rates for each road segment.

IV. **Agent Behavior and Interactions**:

Agent behavior is based on realistic traffic rules:

- Vehicles interact by following distances and safe driving distances from other vehicles.
- Lane changes are modeled based on vehicle speed and available space in adjacent lanes.
- Vehicles stop and start according to traffic signal timings, with dynamic adjustments to avoid congestion.

## *4.3 Model Calibration and Validation*

1. **Model Calibration**: Vehicle speed, reaction time, and lane-changing parameters have been adjusted to calibrate the simulation model. Calibration was done based on the actual traffic camera data and ground truth data extracted from the roads in Dehradun.
2. **Validation**: The model's effectiveness was then tested by comparing these latter with the actual traffic conditions in Dehradun city. These indicators included confirming the number of vehicles, especially at watch points like Rajpur Road, Saharanpur Road, and Clock Tower, as well as the stream of traffic. The proposed model's precision and recall were tested by applying the ground truth data, showing a high similarity between the simulation and the actual environment.

   ***Figure 3compares ground truth data with model predictions.***

## *4.4 Simulation Scenarios*

This section describes various simulations employed to assess traffic conditions at Dehradun. All of them reflect a different approach to traffic regulation in the streets, with differences in levels of control and possible optimization. Such situations help determine the effectiveness of traffic flow to improve traffic conditions.

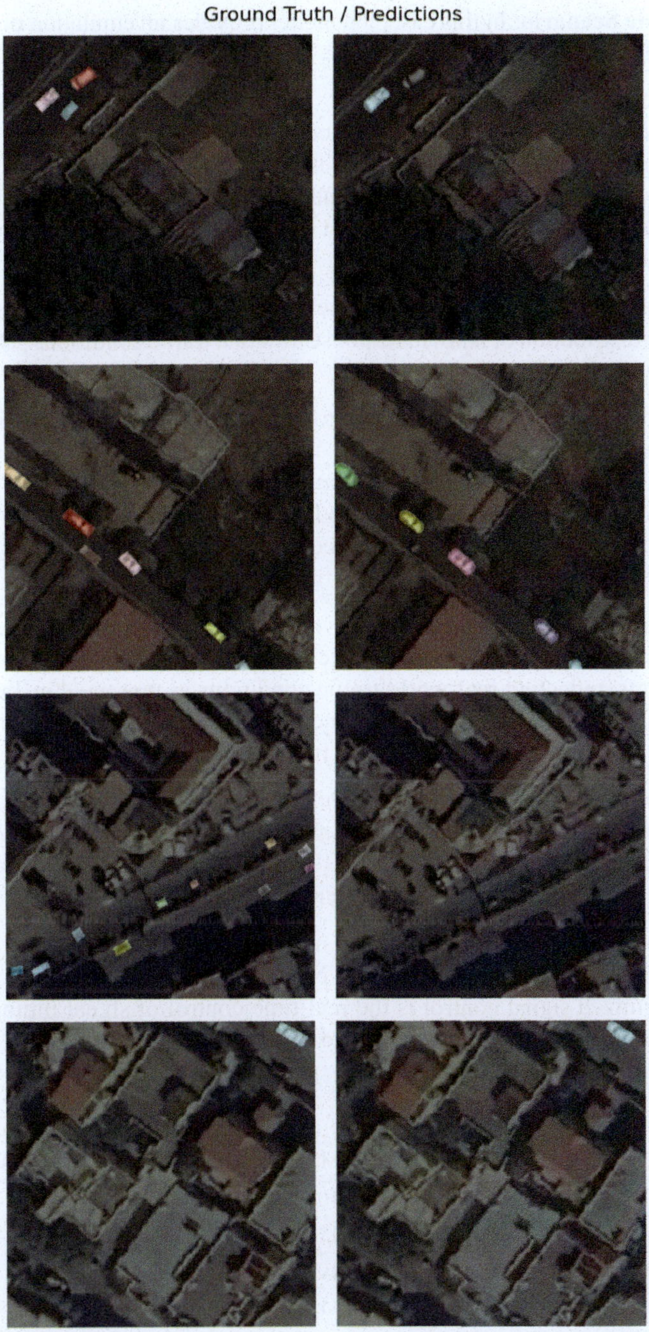


**Fig. 3** Ground truth prediction

1. **Baseline Scenario**: In this case, the model provides an emulation of Dehradun's present-day traffic configuration. The following assumptions are made:
   - Fixed-time signal control.
   - No additions or alterations to the current existing road network.
   - None of these is applied to traffic flow as adaptive interventions.
   - It assists in creating the initial conditions of existing traffic situations, which is important to compare the impact of measures of optimization or congestion reduction.
2. **Optimized Traffic Flow**: The purpose of this scenario is to change signal timings and the distribution of lanes in response to actual traffic conditions. The following changes are introduced:
   - Signaling times are responsive to flow densities and congestion changes depending on traffic conditions.
   - Some lanes may be passively created based on the nature of the traffic conditions, thereby enhancing the free flow of traffic.
   - These comprise information derived from the agent-based simulation data that is applied to the parameters' real-time control.

The optimized traffic flow scenario developed in this study incorporates key traffic parameters, including arrival rates, vehicle speeds, road segments, and maximum arrival counts, reflecting real-world traffic conditions as presented in Table 1.

The objective of this scenario is to increase traffic management gain following changes in traffic conditions all over the road network.

3. **Traffic Congestion Reduction**: In this case, traffic is considered the main issue in\View; the emphasis is made on minimizing traffic density, particularly during rush hour.

   The following strategies are implemented:
   - Carrying capacity is mimicked on mainline ways that have traffic constriction points, such as Rajpur M.RD and N.Cantt RD.
   - Optimized signal control is the real-time control of signal timings based on an analysis of traffic data collected during simulation.

**Table 1** Simulation parameters

| Road name | Arrival rate | Max. no. of arrivals | Speed (km/h) |
|---|---|---|---|
| Rajpur M.RD | 10.7 per hr | 107 | 25 |
| N.Cantt RD | 17.6 per hr | 176 | 30 |
| R.M.RD5 | 10.9 per hr | 109 | 35 |
| Subhash RD3 | 11.1 per hr | 111 | 60 |
| Ugrassain RD2 | 5.3 per hr | 53 | 55 |
| DBS Clg RD | 1.3 | 13 | 45 |

- There is a feature called smart routing that directs many vehicles to fewer popular roads during peak traffic hours in a bid to reduce traffic on swathes of particular roads.

### 4.5 *Performance Metrics*

The performance of the agent-based simulation was evaluated using the following metrics:

- **Precision and Recall**: Both the accuracy of vehicle detection and a traffic model used in the model were evaluated for their similarity to the actual traffic.
- **Traffic Flow Efficiency**: Defined by average travel time (ATT) and average vehicle delay time (ADT) were computed across all segments of the road network were computed across all simulated vehicles as follows: ATT = (Σ TTj) / N ADT = (Σ (TTj − FFTj)) / N where TTj represents the total travel time of vehicle j, FFTj denotes the free-flow travel time of vehicle j, and N is the total number of simulated vehicles.
- **Congestion Reduction**: It was assessed by comparing traffic patterns before and after the implementation of dynamic signal timing and lane management strategies.
- **Simulation Scalability**: The general flexibility of the model in other areas, such a different city or other urban center with different traffic conditions.

## 5 Results and Discussion

### 5.1 *Traffic Conditions Analysis*

The framework simulation results clearly reproduced congestion patterns that have been identified from the Dehradun traffic data. Kernel density and hotspot analysis results, which specify the high-traffic zones, were comparable to those in the congested area in the simulation. Key findings include:

- **High Congestion Zones**: Those junctions recognized as hotspots revealed an increase in the overall duration of vehicles and reduced average speeds in peak traffic periods.
- **Traffic Flow Patterns**: Local roads with intermediate vehicle traffic densities were found to have moderately used lanes and erratic average run times. These results support the implementation of deep learning of vehicle detection through agent-based simulation, a solid approach for real-time traffic analysis, as shown in Fig. 4.

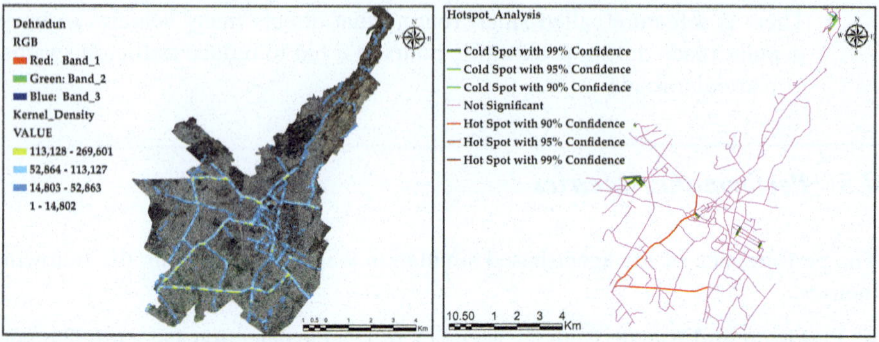


**Fig. 4** Kernel density and hotspot analysis

## *5.2 Impact of Vehicle Detection Accuracy*

High accuracy of vehicle identification is also important to given results of the concrete simulation.

- The model was able to get a precision of 42.7%, and the recall of the model was 100%, so the F1-score was 59.8.
- These metrics enabled the identification of all the vehicles involved in the scene, though the congestion analysis had slightly false positives.
- These metrics ensured that all vehicles in the scene were accounted for, though false positives slightly impacted congestion analysis.

The performance metrics results are depicted using a bar graph in Fig. 5. This graph is helpful because it offers one overall view of the model by comparing the precision, the recall, and the F1-score achieved by the detection model.

**Fig. 5** Model metrics

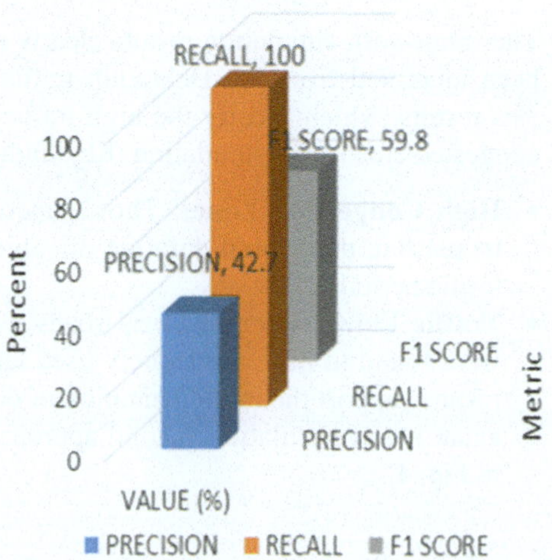

### 5.3 *Effectiveness of Simulation Interventions*

The implementation of traffic management interventions, such as dynamic signal timing and lane reallocation, demonstrated measurable improvements:

1. **Reduction in Congestion**: High-density zones saw an average decrease of 18% in vehicle delays.
2. **Improved Travel Efficiency**: Average travel times across major routes were reduced by 12%.
3. **Optimized Traffic Signals**: Real-time signal timing adjustments at hotspots significantly reduced queue lengths.
   - These results indicate that agent-based simulation can serve as a decision-support tool for implementing traffic decongestion strategies.

The simulated traffic scenarios that were represented during the simulation can be observed using Fig. 6. It emphasizes the actions that were taken and the results in the transportation management and traffic situation under various circumstances.

To provide the complete picture regarding the effects of these interventions on vehicle speeds, Fig. 7 shows the speed distribution during the first and midpoint of the simulation, respectively. These distributions depict how the specific interventions were taking place to realize progressive enhancement in traffic efficiency.

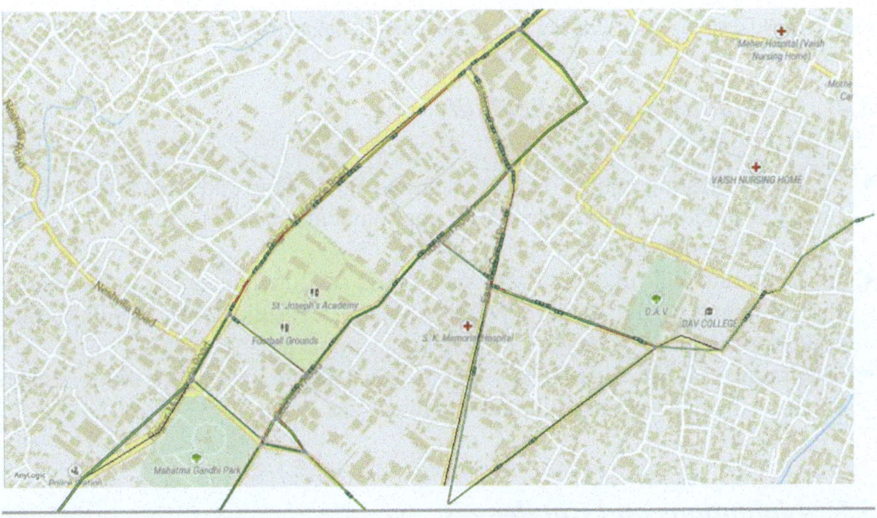


**Fig. 6** Traffic scenarios modeled in the simulation, showing interventions and corresponding improvements in traffic flow

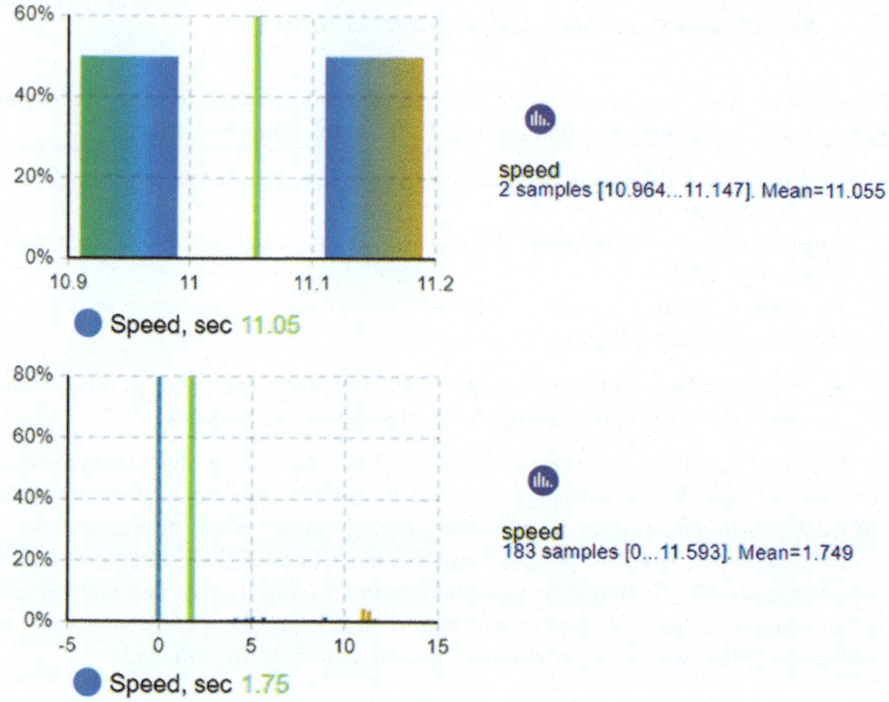


**Fig. 7** Speed distribution of vehicles during the initial sampling phase (speed per second)

## 5.4 *Scalability and Applicability*

- Through adjusting control factors such as vehicle flow and network configuration, the model can generate traffic environments in other cities with diverse characteristics of urbanization.
- Incorporating real-time input data, such as live satellite imagery or traffic sensors, further enhances the model's practicality. As a result, it can simulate traffic conditions in other cities with varying urban characteristics.
- Integrating real-time data inputs, such as live satellite imagery or traffic sensors, enhances its practicality for broader applications.

## 5.5 *Limitations and Challenges*

While the study offers valuable insights, certain limitations need to be addressed:

1. **Detection Accuracy**: Several roads to vehicle detection were impossible because of restrictions to satellite imagery, which affected the number of discernible traffic agents.

2. **Computational Resources**: Density simulated scenarios sometimes demand large processing power, which may not be available in all environments.
3. **Simplified Behaviors**: While traffic agents measured some aspects of car movement, such as velocity, acceleration, and distance, other aspects, such as unanticipated movements by motorists, were not well represented.

It will be beneficial for future research to explore such ways to increase the model's precision and accuracy, collect larger datasets containing higher-resolution imagery (e.g., aerial), or introduce more efficient computation techniques.

## 6 Conclusion

This research shows how combining deep learning methods with kernel density analysis and agent-based traffic simulation helps to manage traffic effectively. Through kernel density methods, we analyzed traffic zones to discover where vehicles congregate and how these patterns spread across the area. The system showed us where vehicles were located and which roads had heavy traffic issues. Deep learning analysis helped us to identify vehicles in satellite imagery. The system assigned realistic vehicle agent parameters based on actual road segment conditions. This modeling approach measures how vehicle speed changes when traffic arrival rates meet traffic signal timing. Our research method created a predictive system by integrating zonal analysis results with traffic simulation to match travel patterns at different scales. The data shows that our system helps transportation experts make decisions to reduce traffic jams and create more imaginative transportation solutions. Future research will improve the approach by adding new transport options and innovative traffic controls.

## References

1. Chaudhary, H.K., Saraswat, K., Yadav, H., Puri, H., Mishra, A.R., Chauhan, S.S.: A real-time dynamic approach for management of vehicle generated traffic. Evergreen **10**(1), 289–299 (2023). https://doi.org/10.5109/6781078
2. Chen, X., Li, Z., Yang, Y., Qi, L., Ke, R.: High-Resolution Vehicle Trajectory Extraction and Denoising from Aerial Videos. IEEE Trans. Intell. Transp. Syst. **22**(5), 3190–3202 (2021). https://doi.org/10.1109/TITS.2020.3003782
3. Mathieu, P., Nongaillard, A., Thery, A., Demazeau, Y., Holvoet, T., Corchado, J.M., Costantini, S.: Advances in practical applications of agents multi-agent systems and trustworthiness. The PAAMS Collection 18th International Conference PAAMS 2020 L'Aquila Italy October 7–9 2020 Proceedings A Multi-agent Evaluation of Traffic Light Models. Springer International Publishing, Cham, 451–455 (2020)
4. Moka, S.S.P., Pilla, S.M., Radhika, S.: Real-time density based traffic surveillance system integrated with acoustic based emergency vehicle detection. In: 4th international conference on computer, communication and signal processing, ICCCSP 2020. https://doi.org/10.1109/ICCCSP49186.2020.9315209

5. Pagliardini, P., Porta, S., Salingaros, N.A.: Geospatial analysis and living urban geometry. In: Jiang, B., Yao, X. (eds.) Geospatial analysis and modelling of urban structure and dynamics. GeoJournal Libr. vol. 99. Springer, Dordrecht (2010). https://doi.org/10.1007/978-90-481-8572-6_17
6. Sarker, S., Maples, B., Li, W.: A comprehensive review on traffic datasets and simulators for autonomous vehicles (2024). http://arxiv.org/abs/2412.14207
7. Wang, J., Jiang, S., Qiu, Y., Zhang, Y., Ying, J., Du, Y.: Traffic signal optimization under connected-vehicle environment. Overview J. Adv.Transp. 1–16 (2021). https://doi.org/10.1155/2021/3584569
8. Zhang, L., Ma, Y., Shi, L.: A hybrid traffic flow model for real-time freeway traffic simulation. KSCE J. Civ. Eng. **18**(4), 1160–1164 (2014). https://doi.org/10.1007/s12205-014-0635-7

# Smart Solutions in Veterinary and Animal Sciences: Applications of AI, ML, and Soft Computing

**Kuldeep Kumar Tyagi, Surbhi Kuldeep Tyagi, Devesh Kumar Yadav, and Atul Gupta**

**Abstract** The veterinary and animal science is a professional discipline having broad applied subjects dealing with improvement of health and production in animals. Artificial intelligence (AI), machine learning (ML), and soft computing are being applied to various subjects of this field owing to their capacity to deal with complex data and giving smart solutions for accurate desired outcomes. Impact areas include genomic prediction and breeding strategies in animal genetics, precision feeding, and nutrigenomics in animal nutrition using ML and optimization of health and environmental management in livestock production, via AI-powered tools such as Internet of Things (IoT) and deep learning (DL). New advances in veterinary subjects like anatomy, gynecology, medicine, microbiology, parasitology, and pathology take advantage of AI technologies such as three-dimensional (3D) modeling, virtual reality (VR), and augmented reality (AR) automated diagnostics and predictive analytics to increase research, education, and clinical outcomes. In addition, AI and ML innovations help with advancing drug discovery, personalized medicine, predicting disease, public health, and epidemiology. Veterinary surgery and radiology also make use of these applications to improve diagnostic accuracy and treatment strategies. These technologies integrate with veterinary and animal sciences in order to make transformative improvements in its development while facing ethical, technical, and regulatory challenge for the sustainable and responsible implementation.

K. K. Tyagi (✉) · S. K. Tyagi · D. K. Yadav · A. Gupta
College of Veterinary and Animal Sciences, Sardar Vallabhbhai Patel University of Agriculture and Technology, Meerut, Uttar Pradesh, India
e-mail: kuldeeptyagi.vet@svpuat.edu.in

S. K. Tyagi
e-mail: surbhiivri@gmail.com

D. K. Yadav
e-mail: drdevvet2009@gmail.com

A. Gupta
e-mail: atulguptavet@gmail.com

© The Author(s), under exclusive license to Springer Nature Singapore Pte Ltd. 2026

M. Pant et al. (eds.), *Real-World Applications of Machine Learning*, Mathematics for Sustainable Developments, https://doi.org/10.1007/978-981-95-7134-5_19

**Keywords** Artificial intelligence · Machine learning · Soft computing · Veterinary · Animal science

# 1 Introduction

Veterinary and animal sciences subjects are being fostered by artificial intelligence (AI) and machine learning (ML) through the harnessing of genomic predictions, precision feeding, disease predictions, reproductive health optimization, and smarter veterinary diagnostics. In addition, precision farming, Internet of Things (IoT) implemented on livestock, and real-time monitoring of animal health and behavior are being introduced by the AI, which will help against animal diseases and issues of low productivity. AI-driven tools, such as three-dimensional (3D) models, virtual reality (VR), and omics are improving veterinary education and research. In general, AI, ML, and soft computing greatly advance veterinary and animal science by ensuring that veterinarians could practice more sustainably and have better animal welfare. This symbiont relationship can also greatly help in developing notably better solutions to global veterinary and animal science associated problems.

# 2 Subject-Wise Use of AI, ML, and Soft Computing

The discipline of Veterinary and Animal science mainly comprises fifteen subjects covering diverse topics. Worldwide researchers are finding smart solutions for advancement in these subjects by integrating AI, ML, and Soft Computing. This chapter discusses introduction about these subjects and important recent studies making use of AI, ML, and Soft Computing in these subjects.

## *2.1 Animal Genetics and Breeding*

Genetics scrutinizes three important aspects of living organisms mainly heredity, variation, and reproduction [1]. Animal genetics comprises the following broad areas: cytogenetics, Mendelian genetics, basic genetics, molecular genetics, genetic engineering, disease resistance genetics, population genetics. It also includes emerging area of systems biology and omic sciences such as genomics, transcriptomics, proteomics, and metabolomics [2]. Animal breeding has taken advantage of genome editing tools and immunogenic approaches to intentionally improve specific traits in animals through controlled reproduction, improved production, quality, and disease resistance thereby benefiting both farmers and consumers [3, 4]. In the past, animal breeding focused on selection of animals based on visible traits whereas modern breeding programs working with genomic data, performance records in conjunction with ML models, bioinformatics, phenomics, and biologically annotated neural networks had boosted accuracy for predicting genetic potential in animals [5–7].

Across prediction of genetic merit, optimization of breeding programs, and improvement of livestock production, artificial neural networks (ANNs), feedforward, recurrent, and convolutional neural networks had been shown to capture complex, nonlinear data relationships [8]. Consequently, genomic prediction, with support vector regression (SVR), random forest (RF), and deep learning (DL) had leaped even further in ML paradigm, that evolved more accuracy and also a higher efficiency compared to the classical methods [9]. The classical methods and ML techniques are complementary, as classical methods have better interpretability and robustness while ML techniques have flexibility and improved predictive performance. A hopeful strategy for developing tools for genetics and breeding in animals that were more accurate and powerful was to integrate these approaches [10].

## 2.2 *Animal Nutrition*

The study of how animals ingest, digest, absorb, transport, utilize, and excrete the nutrients needed for their growth and development, maintenance, reproduction, and health is animal nutrition [11]. Precision nutrition, the use of sensors and data analytics to make real-time adjustments to the amount of feed animal receives stands out precisely due to its potential to more precisely tailor diets to an animal's genetics, health status, and environmental conditions [12]. A second branch, nutrigenomics, examines how certain foods can influence which genes are expressed in animals; these types of diets may help to prevent or treat certain diseases [13]. Functional nutrition also appears as a major area that deals with feed components providing nutrition health benefits beyond basic needs, like increasing immune function or gut health [14].

AI and ML algorithms look through sensor and farm management systems data that was huge to forecast individual animal requirements of nutrients, formulate feed optimally, and modify the diet on the basis of animal's behavior and health signals [12]. In functional nutrition with uncertainty and complexity of nutritional data in humans, the soft computing including fuzzy logic can help to manage such uncertainty and complexity and thus enable more adaptive and fine feed strategies with more health benefits from feed ingredients [15]. In animals too, these similar strategies can well be applied. The study of gene-nutrient interactions in nutrigenomics had used ML models to predict gene-nutrient interactions to develop personalized diets to prevent or mitigate disease based upon genetic profile [13].

## 2.3 Livestock Production and Management

The subject livestock production and management deals with breeding, feeding, health, and housing of animals carried out in an organized system to satisfy the demand for animal products. Recently, there have been numerous studies on applications of sensors, data processing, and transmission and various AI methodologies such as ML, DL, and ANN to deal with problems in the livestock sector. They include monitoring behavior [16], track diseases [17], identify animals [18], and also manage environmental conditions [19]. In addition to these research efforts, there had been reviews of animal studies in narrow application areas for specific animal types. For example, current applications of sensors and IoT to the poultry welfare improvement [20–22], as were precision farming approaches in pig management [23]. Also, applications of DL for analyzing cattle and pig behavior [24] and data modeling in animal production [25] applied these techniques.

## 2.4 Livestock Products Technology

Livestock products technology uses scientific and technical ways for improving the process of production, processing, preserving, and quality control of livestock products including meat, milk, eggs, and other by products. Non-destructive techniques such as computer vision systems (CVS), near infrared spectroscopy, and hyperspectral imaging had helped in real-time inspection and quality control in meat production thus meeting the demand of the growing consumers with high-quality meat while maintaining the efficiency and objectivity in evaluation [26]. Additionally, ML was being used in the nascent field of cultured meat technology including cell line establishment, cell culture media design, and bioprocessing to optimize processes. This was useful to mitigate environmental and ethical impacts and to promote sustained food production [27]. In addition, the CVS tools extend to carcass composition evaluation and the impact on meat quality of varying rearing systems and technological treatments, with cost effective and rapid analysis [28]. High accuracy in predicting dairy product demand had been displayed by an integrated framework combining statistical tests, neural networks with promising meta heuristics like Grey wolf optimization and particle swarm optimization [29]. For example, noninvasive AI-driven computer vision techniques (RGB and thermal infrared imaging) were being used to monitor physiological responses in cows that project future milk productivity and quality with high precision [30]. The application of technologies such as Robotics, IoT, Blockchain, and Big Data to automate and optimize milk production are important in future of dairy sector. The significant improvements throughout dairy's value chain using new era of digital transformation in the sector were being envisaged through these developments [31].

### 2.5 Veterinary Anatomy

The study of the structure and organization of the animal's body is known as veterinary anatomy. It covers diverse anatomical systems like skeletal, muscular, nervous, cardiovascular, and digestive systems. Current use of AI-driven solutions, including 3D models, VR, and augmented reality (AR), offer interactive and dynamic anatomy learning experiences for students and complement the traditional methods like didactic and online lectures [32]. To the extent of technological integration of omics technologies, including genomics, proteomics, and radiomics, with AI as well as tools such as 3D bioprinting and neural networks, anatomical studies provide new insights with detailed structural and functional analysis at both cellular and tissue levels [33].

### 2.6 Veterinary and Animal Husbandry Extension

The veterinary and animal husbandry extension works to enhance livestock management through knowledge dissemination about health care and production systems and sustainable farming practices through training sessions and demonstrations in addition to mobile veterinary services which improve animal welfare standards and agricultural practices. In a study that used Bayesian frameworks, it was pointed out that Veterinary Expert System for Outcome Prediction (VESOP) can predict the likelihood of population survival using health indicators including inflammation [34]. This model was able to predict the survival rates of the Bottlenose Dolphins accurately. In a study, of small ruminants, a smart vaccination algorithm was used for Peste des Petits Ruminants (PPR), focusing on high-risk livestock based on historical and geographical data, lead to reduced campaign duration and increased vaccination coverage [35]. In another study, an expert system was applied to analyze herd data for improving conception rates and estrus detection, combining AI with traditional herd management approaches [36].

### 2.7 Veterinary Gynecology and Obstetrics

Veterinary gynecology and obstetrics covers study of animals' reproductive health. The diagnosis and treatment of reproductive system problems, pregnancies, and health issues during parturition or calving are covered under this subject. The AI and ML are increasingly being used in oocyte and embryo selection, showcasing the possibility to boost livestock fertility and productivity. In a study conducted, it was emphasized that although traditional methods such as morphological assessment remain the standard for assessing oocyte and embryo quality, novel AI-based approaches were gaining popularity [37]. Using transcriptomic, proteomic, metabolic profiling, and advanced imaging techniques it is easier to predict developmental potential with higher accuracy. It is now possible to choose embryos with developmental potential without surgery, lowering the need for surgical procedures as well as improving in vitro fertilization outcomes in livestock by strengthening selection criteria applying AI technology.

## *2.8 Veterinary Medicine*

Veterinary medicine is a science devoted to the diagnosis, treatment, prevention and control of diseases, disorders, and trauma in companion animals, livestock, birds, and wildlife. AI is important for transformation of veterinary medicine into the clinical, educational, and research aspects. As outlined in the study, AI can bring the capability to aid in clinical decision making, despite these data were generally available in large and complicated shape [38]. Other study also cover AI use in diagnostics, education, and animal health as well as adaptive algorithms that minimize errors and facilitate collaboration between AI systems and veterinary professionals [39]. Investigation related to generative AI, such as ChatGPT, finds utility and importance in veterinary clinics, teaching, and research, including practical guidelines for the ethical and responsible application [40]. These studies show that even with AI's transformational potential in veterinary medicine building ethical, legal, and technological considerations remained important to achieve true results.

## *2.9 Veterinary Microbiology*

It is the study of organisms such as bacteria, viruses, and fungus causing diseases in animals called veterinary microbiology. This subject also covers the vaccine development, antimicrobial therapy, biosecurity, food safety. Direct role of AI has already been proven in antibiotic resistance, microbial studies, and antimicrobial discovery. A study reported introduction of an AI powered software DeepLysin, that could mine phage genomes for potent lysins with antibacterial properties [41]. Time for the discovery of alternatives to antibiotics to fight resistant pathogens had been shortened here. In silico AI methods were found to be essential tools for microbial genomics, vaccines, and environmental microbiology [42]. They demonstrated how ML and DL can aid in understanding of microbial diversity and evolution [42]. Automatic bacterial colony detection was made possible using convolutional neural networks to reduce time in bacterial microbiological quantification [43].

## 2.10 Veterinary Parasitology

Veterinary parasitology deals with the study of parasites that infect animals, their life cycles, interactions with their hosts and the diseases they produced. There is a wide range of parasites which includes protozoa, helminths, other endoparasites, and ectoparasites such as fleas, mites, and ticks. AI-based tools contributed to automate microscopy and immunodiagnostic tools for improved diagnostic sensitivity for helminths [44]. Similarly, in another work on early control zoonotic waterborne outbreaks had reported that ML and explainable AI (XAI) predict the presence of Cryptosporidium and Giardia in water [45]. In equine parasitology, new approaches to fecal egg counting methods (FECT) including the use of AI and DL had proven effective in compensating for inconsistency in the manual assessment [46]. Concerning the case of the equine parasitology, the advances in FECT through the help of AI and DL had marked a possibility of decreasing some of the discrepancies in comparison with the results of the manual analysis [46]. Further, new technologies such as Kubic FLOTAC microscope were improving the fecal egg count in cattle through a combination of digital microscopy with AI methods, giving a more rational and specific mean for the identification of parasites [47].

## 2.11 Veterinary Pathology

Veterinary pathology deals with the etiology of diseases in animals as well as the description of pathologic alterations in tissues that accompany diseases. A study advocated that innovation in veterinary pathology lies in the application of AI, which might be beneficial in toxicological pathology and drug development in addition to improving diagnostics [48]. Similarly, another study stressed on the use of telepathology and applied AI in face of veterinary practice in view of its contribution in defining greater profile of diagnostics accuracy [49]. The advance of AI in evaluation of lymphocytes in feline small intestinal biopsies provided more objective, reproducible, and accurate results which increased the diagnostic confidence in diseases such as chronic enteropathy [50]. The future integration of AI technologies raises technical, ethical, and regulatory questions, all of which are critical for closing the AI gap in veterinary pathology [51].

## 2.12 Veterinary Pharmacology and Toxicology

Veterinary pharmacology mainly deals with administration, absorption metabolism, excretion, and therapeutic effects of drugs on animals. On the other hand, toxicology is concerned with harmful substances and their effects thus helping in diagnosis and prevention of poisoning in animals. The development of newer treatment regimens and preventive measures are contributed by this field. Recently many of the improvements in drug discovery and treatment efficacy were catalyzed by AI and ML. The identification of antimicrobial peptides (AMPs) in the animal genome using DL-based tool AniAMPpred had been reported [52]. Antiviral peptides (AVPs) mainly act against various viruses and other zoonotic pathogens. These AVPs had been shown to be identified using AI-driven tool Deep-AVPpred [53]. Similarly precision treatment using personalized medicine is another area in which AI models are helping. These AI models help in analyzing genetic variation in target human or animal populations and associated tailored drug responses for suggesting personalized medicine [54].

## 2.13 Veterinary Physiology and Biochemistry

Veterinary physiology and biochemistry is one of the sub disciplines of veterinary science which deals with the study of the attributes of animals' body system which are essential for existence (physiology) and the study of the chemistry of the animals' body (biochemistry). The applications of AI and ML are frequently applied in veterinary physiology and biochemistry for a better analysis of proteins' functions and gene features to clarify biological processes and diseases. For instance, a study discussed the AI-based approaches for protein subcellular localization and pointed about DL as well as other ML [55]. These techniques had aced the understanding of localization of proteins which is very important in drug development and disease understanding such as cancer and Alzheimer's. Along the same line, a "Bingo" tool was designed, which works with large language models and graph neural networks to predict essential genes in metazoans and hence for disease treatment strategies to focus on core biological processes [56]. Another study proposed a method using color deconvolution in combination with AI for the improvement of immunohistochemical quantification and provided more accurate as well as automated protein expression analysis in tissues [57].

## 2.14 Veterinary Public Health and Epidemiology

Veterinary public health involves the preservation and promotion of animal health, disease prevention that may affect animal populace and quality that may threaten human populace, food safety, and infectious diseases. Epidemiology within this context studies the occurrence and prevention of diseases within animals' species to prevent occurrence of more outbreaks or pandemic. The ML models, particularly the "Imbalanced-Xgboost" model, integrated enhanced by geospatial analysis was used for rabies risk prediction in Morocco and thus define the targeted approach to resource distribution and disease suppression taking into consideration the factors of socio economy and geography [58]. Another study underlined that AI could be helpful to prevent contagion at the human-animal interface through the monitoring of early signs of epidemic outbreak [59]. In another setting, four models were assessed whereby both the ANN and Seasonal Autoregressive Integrated Moving Average (SARIMA) proved to be equally efficient but the combined models were even better [60]. Neural networks had also been used for predicting the distribution of zoonotic disease reservoirs at different locations, and AI was in fact proved as a powerful tool in detecting environmental factors and risks of spillover, which was the key for controlling emerging infectious diseases (EIDs) [61].

### 2.15 Veterinary Surgery and Radiology

Veterinary surgery and radiology is an important clinical subject that deals with diagnosis and surgical treatment of diseases in animals. Veterinary radiology involves the use of X-rays, ultrasound, MRI, and CT scans in order to determine fractures, tumors and internal diseases of organs for planning the best treatment strategy. It has been identified that the application of AI and ML are exhibiting potential when used in veterinary surgery and radiology as they improve the diagnostic rates of diseases. The use of such AI-based tools such as Diagnocat was highly accurate in diagnosing periapical periodontitis in two-dimensional radiography based on sensitivity (92.3%) and specificity (97.9%), which in turn can contribute to raising diagnostic reliability in veterinary dentistry [62]. The value of CNNs to diagnose spinal cord diseases in canine patients using MRI of the thoracolumbar region had also been demonstrated [63].

## 3 Challenges and Future Prospects

The use of AI, ML, and soft computing in veterinary and animal sciences have assessment issues including: Implementations costs, limited availability of high-quality datasets, absence of benchmarking, and the need for distinct training. However, ethical issues such as data privacy and a lack of understanding of how the algorithms work make mass adoption impossible. However, tremendous growth is expected in the future because the computational capabilities, open-source data, and cloud services have lowered the hurdles and interdisciplinarity constitutes a new paradigm. The ML, DL, and explainable artificial intelligence (XAI) and adaptive learning systems will likely contribute toward building the acceptance and effectiveness and user traceability, with special emphasis on applications like precision livestock farming and veterinary training. When advanced more it can change animal health, production, and sustainability, proving that research moves from concept creation to applicable solutions.

## 4 Conclusion

The use of AI, ML, and soft computing have a large impact on all the subjects of veterinary and animal sciences. Recent research studies specifically shows that animal genetics, nutrition, clinical medicine, public health, surgery as well as diagnostics, disease management, and sustainable farming practices had gained a lot from these technologies. These fields are using cutting edge tools such as neural networks, deep learning, IoT, and computer vision that make them more precise, efficient, and sustainable than ever before. The potential of AI and ML is multi fold but at the same time during implementation it is also imperative to deal with ethical, technical, and regulatory hurdles with proper approach. Health, welfare, and productivity of animals will be enhanced by these technologies, so long as they continue to evolve and are responsibly adopted and integrated into research, practices, or services.

## References

1. Pusta, D.L.: Cytogenetics and Mendelian genetics, pp. 13–20. Alma Mater (2009)
2. Pusta, D.L.: Introductory chapter: challenges and advances in animal genetics. IntechOpen. (2019). https://doi.org/10.5772/intechopen.82668
3. Islam, M.A., Rony, S.A., Rahman, M.B., Cinar, M.U., Villena, J., Uddin, M.J., Kitazawa, H.: Improvement of disease resistance in livestock: application of immunogenomics and CRISPR/Cas9 technology. Animals **10**(12), 2236 (2020)
4. Liu, Z., Wu, T., Xiang, G., Wang, H., Wang, B., Feng, Z., Mu, Y., Li, K.: Enhancing animal disease resistance, production efficiency and welfare through precise genome editing. Int. J. Mol. Sci. **23**(13), 7331 (2022)
5. Li, X., Chen, X., Wang, Q., Yang, N., Sun, C.: Integrating bioinformatics and machine learning for genomic prediction in chickens. Genes **15**(6), 690 (2024)
6. Pérez-Enciso, M., Steibel, J.P.: Phenomes: The current frontier in animal breeding. Genet. Select. Evol. **53**(22) (2021). https://doi.org/10.1186/s12711-021-00618-1
7. Wang, X., Shi, S., Khan, M. Y. A., Zhang, Z., Zhang, Y.: Improving the accuracy of genomic prediction in dairy cattle using the biologically annotated neural networks framework. J. Anim. Sci. Biotechnol. **15**(87) (2024). https://doi.org/10.1186/s40104-024-00887-2
8. Khanikar, D., Phookan, A., Gogoi, A.: Artificial neural networks: an introduction and application in animal breeding and production: a review. Agric. Rev. **45**(3), 480–487 (2024). https://doi.org/10.18805/ag.R-2421
9. Chafai, N., Hayah, I., Houaga, I., Badaoui, B.: A review of machine learning models applied to genomic prediction in animal breeding. Front. Genet. **14**, 1150596 (2023)
10. Soloshenkov, A.D., Soloshenkova, E.A., Semina, M.T., Spasskaya, N.N., Voronkova, V.N., Stolpovsky, Y.A.: Artificial intelligence and classical methods in animal genetics and breeding. Rev. Theoret. Articles **60**, 843–856 (2024)
11. McDonald, P., Edwards, R.A., Greenhalgh, J.F.D., Morgan, C.A., Sinclair, L.A., Wilkinson, R.G.: Animal Nutrition. 7th edn. Pearson Education Limited (2011)
12. Halachmi, I., Guarino, M., Bewley, J., Pastell, M.: Smart animal agriculture: application of real-time sensors to improve animal well-being and production. Ann. Rev. Anim. Biosci. **7**, 403–425 (2019)
13. Swanson, K.S., Schook, L.B., Fahey, G.C.: Nutritional genomics: the next frontier in dietary advice and personalized nutrition. J. Anim. Sci. **91**(4), 1631–1641 (2013)

14. Patterson, J.A., Burkholder, K.M.: Application of prebiotics and probiotics in poultry production. Poult. Sci. **82**(4), 627–631 (2003)
15. Hines, M.: Marketing Implant Dentistry. John Wiley & Sons, Inc (2015)
16. Riaboff, L., Couvreur, S., Madouasse, A., Andrieu, S.: Behavior detection in dairy cattle using machine learning techniques: a review. Comput. Electron. Agric. **185**, 106171 (2021)
17. Volkmann, N., Kulig, B., Kemper, N.: Monitoring health in pig farming using sensor technology: a review. Animals **11**(4), 871 (2021)
18. Hu, Z., Yang, Y., Zheng, B.: A review on animal identification using advanced sensing technologies. Sensors **20**(15), 4194 (2020)
19. Gautam, Y., Sharma, R., Kumar, S.: IoT-enabled smart farming for environmental control in livestock housing. Sensors **21**(5), 1547 (2021)
20. Neethirajan, S.: The role of sensors, big data and machine learning in modern animal farming. Sens. Bio-Sens. Res. **29**, 100367 (2020)
21. Astill, J., Dara, R.A., Campbell, M., Fraser, E.D.G., Sharif, S., Roberts, B.: Transparency in food supply chains: a review of enabling technology solutions. Trends Food Sci. Technol. **91**, 240–247 (2020)
22. Ben Sassi, N., Averós, X., Estevez, I.: Technology and poultry welfare. Animals **6**(10), 62 (2016)
23. Vranken, E., Berckmans, D.: Precision livestock farming for pigs. Anim. Front. **7**(1), 32–37 (2017)
24. Chen, C., Zhu, W., Ma, J.: Deep learning for animal behavior analysis: a review. Comput. Electron. Agric. **185**, 106157 (2021)
25. Ellis, J.L., Klein, M.S., Kebreab, E.: A review of modeling approaches for livestock production systems. Anim. Prod. Sci. **60**(1), 1–13 (2020)
26. Shi, Y., Wang, X., Borhan, M.S., Young, J., Newman, D., Berg, E., Sun, X.: A review on meat quality evaluation methods based on non-destructive computer vision and artificial intelligence technologies. Rev. Food Sci. Ani. Res. **41**(4), 563–588 (2021)
27. Todhunter, M.E., Jubair, S., Verma, R., Saqe, R., Shen, K., Duffy, B.: Artificial intelligence and machine learning applications for cultured meat. Front. Artif. Intell. **7**, 1424012 (2024). https://doi.org/10.3389/frai.2024.1424012
28. Modzelewska-Kapituła, M., Jun, S.: The application of computer vision systems in meat science and industry—a review. Meat Sci. **192**, 108904 (2022)
29. Goli, A., Khademi-Zare, H., Tavakkoli-Moghaddam, R., Sadeghieh, A., Sasanian, M., Malekalipour Kordestanizadeh, R.: An integrated approach based on artificial intelligence and novel meta-heuristic algorithms to predict demand for dairy products: a case study. Network **32**(1), 1–35 (2021). https://doi.org/10.1080/0954898X.2020.1849841
30. Fuentes, S., Gonzalez Viejo, C., Tongson, E., Lipovetzky, N., Dunshea, F.R.: Biometric physiological responses from dairy cows measured by visible remote sensing are good predictors of milk productivity and quality through artificial intelligence. Sensors (Basel) **21**(20), 6844 (2021). https://doi.org/10.3390/s21206844
31. Hassoun, A., Garcia-Garcia, G., Trollman, H., Jagtap, S., Parra-López, C., Cropotova, J., Bhat, Z., Centobelli, P., Aït-Kaddour, A.: Birth of Dairy 4.0: opportunities and challenges in adoption of fourth industrial revolution technologies in the production of milk and its derivatives. Current Res. Food Sci. **7**, 100535 (2023)
32. Choudhary, O.P., Infant, S.S., Vickram, A., Chopra, H., Manuta, N.: Exploring the potential and limitations of artificial intelligence in animal anatomy. Ann. Anat. (2024). https://doi.org/10.1016/j.aanat.2024.152366
33. Stabile, A.M., Pistilli, A., Mariangela, R., Rende, M., Bartolini, D., Di Sante, G.: New challenges for anatomists in the era of omics. Diagnostics **13**(18), 2963 (2023)
34. Schwacke, L.H., Thomas, L., Wells, R.S., Rowles, T.K., Bossart, G.D., Townsend, F., Jr., Smith, C.R.: An expert-based system to predict population survival rate from health data. Conserv. Biol. **38**(1), e14073 (2024)
35. Eltahir, Y.M., Aburizq, W., Bensalah, O.K., Mohamed, M.S., Al Shamisi, A., AbdElkader, A.I., Al-Majali, A.: Modeling for smart vaccination against peste des petits ruminants (PPR) in the Emirate of Abu Dhabi. United Arab Emirates. Anim. **13**(20), 3248 (2023)

36. Domecq, J.J., Nebel, R.L., McGilliard, M.L., Pasquino, A.T.: Expert system for evaluation of reproductive performance and management. J. Dairy Sci. **74**(10), 3446–3453 (1991). https://doi.org/10.3168/jds.S0022-0302(91)78534-2
37. Wrenzycki, C.: Parameters to identify good quality oocytes and embryos in cattle. Reprod. Fertil. Dev. **34**(2), 190–202 (2021)
38. Appleby, R.B., Basran, P.S.: Artificial intelligence in veterinary medicine. J. Am. Vet. Med. Assoc. **260**(8), 819–824 (2022)
39. Bouchemla, F., Akchurin, S.V., Akchurina, I.V., Dyulger, G.P., Latynina, E.S., Grecheneva, A.V.: Artificial intelligence feasibility in veterinary medicine: a systematic review. Vet. World **16**(10), 2143–2149 (2023)
40. Chu, C.P.: ChatGPT in veterinary medicine: a practical guidance of generative artificial intelligence in clinics, education and research. Fron. Vet. Sci. **11** (2024). Article 1395934. https://doi.org/10.3389/fvets.2024.1395934
41. Zhang, Y., Li, R., Zou, G., Guo, Y., Wu, R., Zhou, Y., Li, J.: Discovery of antimicrobial lysins from the "dark matter" of uncharacterized phages using artificial intelligence. Adv. Sci. 2404049 (2024)
42. Kumar, R., Yadav, G., Kuddus, M., Ashraf, G.M., Singh, R.: Unlocking the microbial studies through computational approaches: how far have we reached? Environ. Sci. Pollut. Res. **30**(17), 48929–48947 (2023)
43. Makrai, L., Fodróczy, B., Nagy, S.Á., Czeiszing, P., Csabai, I., Szita, G., Solymosi, N.: Annotated dataset for deep-learning-based bacterial colony detection. Sci. Data **10**(1), 497 (2023)
44. Rinaldi, L., Krücken, J., Martinez-Valladares, M., Pepe, P., Maurelli, M.P., de Queiroz, C., Gilleard, J.S.: Advances in diagnosis of gastrointestinal nematodes in livestock and companion animals. Adv. Parasitol. **118**, 85–176 (2022)
45. Ligda, P., Mittas, N., Kyzas, G.Z., Claerebout, E., Sotiraki, S.: Machine learning and explainable artificial intelligence for the prevention of waterborne cryptosporidiosis and giardiosis. Water Res. **262**, 122110 (2024)
46. Steuer, A., Fritzler, J., Boggan, S., Daniel, I., Cowles, B., Penn, C., Goldstein, R.: Validation of Vetscan Imagyst®, a diagnostic test utilizing an artificial intelligence deep learning algorithm, for detecting strongyles and Parascaris spp. in equine fecal samples. Parasites and Vectors **17**(1), 465 (2024). https://doi.org/10.1186/s13071-024-06525-w
47. Cringoli, G., Amadesi, A., Maurelli, M.P., Celano, B., Piantadosi, G., Bosco, A., Rinaldi, L.: The Kubic FLOTAC microscope (KFM): a new compact digital microscope for helminth egg counts. Parasitology **148**(4), 427–434 (2020)
48. Zuraw, A., Aeffner, F.: Whole-slide imaging, tissue image analysis and artificial intelligence in veterinary pathology: an updated introduction and review. Vet. Pathol. **59**(1), 6–25 (2022). https://doi.org/10.1177/03009858211040484
49. Battazza, A., Brasileiro, F.C. da S., Tasaka, A.C., Bulla, C., Ximenes, P.P., Hosomi, J.E., Rocha, N.S.: Integrating telepathology and digital pathology with artificial intelligence: An inevitable future. Vet. World **17**(8), 1667–1671 (2024)
50. Wulcan, J.M., Giaretta, P.R., Fingerhood, S., de Brot, S., Crouch, E.E.V., Wolf, T., Moore, P.F.: Artificial intelligence-based quantification of lymphocytes in feline small intestinal biopsies. Vet. Pathol. (2024). https://doi.org/10.1177/03009858241286828
51. Pacholec, C., Flatland, B., Xie, H., Zimmerman, K.: Harnessing artificial intelligence for enhanced veterinary diagnostics: a look to quality assurance, Part I Model development. Vet. Clin. Pathology. (2024). https://doi.org/10.1111/vcp.13401
52. Sharma, R., Shrivastava, S., Singh, S.K., Kumar, A., Saxena, S., Singh, R.K.: AniAMPpred: artificial intelligence guided discovery of novel antimicrobial peptides in animal kingdom. Briefings Bioinf. **22**(6), bbab242 (2021)
53. Sharma, R., Shrivastava, S., Singh, S.K., Kumar, A., Singh, A.K., Saxena, S.: Deep-AVPpred: artificial intelligence-driven discovery of peptide drugs for viral infections. IEEE J. Biomed. Health Inform. **26**(10), 5067–5074 (2022)

54. Abu-Zahra, M., Al-Taher, A., Alquhaidan, M., Hussain, T., Ismail, I., Raya, I., Kandeel, M.: The synergy of artificial intelligence and personalized medicine for the enhanced diagnosis, treatment and prevention of disease. Drug Metabolism Personalized Therapy **39**(2), 47–58 (2024). https://doi.org/10.1515/dmpt-2024-0003
55. Xiao, H., Zou, Y., Wang, J., Wan, S.: A review for artificial intelligence-based protein subcellular localization. Biomolecules **14**(4), 409 (2024)
56. Ma, J., Song, J., Young, N.D., Chang, B.C.H., Korhonen, P.K., Campos, T.L., Liu, H., Gasser, R.B.: 'Bingo'—A large language model- and graph neural network-based workflow for the prediction of essential genes from protein data. Briefi. Bioinf. **25**(1), bbad472 (2023). https://doi.org/10.1093/bib/bbad472
57. Luchian, A., Ressel, L.: Combined colour deconvolution and artificial intelligence approach for region-selective immunohistochemical labelling quantification: the example of alpha smooth muscle actin in mouse kidney. J. Biophotonics **17**(1), e202300244 (2023). https://doi.org/10.1002/jbio.202300244
58. Ahamjik, I., Agbani, A., Abik, M., Khayli, M., Galzim, N., Berrada, J., Bouslikhane, M.: Contribution of artificial intelligence for understanding animal rabies epidemiology in Morocco: what are the perspectives of an innovative and predictive approaches? One Health **19**, 100874 (2024). https://doi.org/10.1016/j.onehlt.2024.100874
59. Chiu, K.H.-Y., Sridhar, S., Yuen, K.-Y.: Preparation for the next pandemic: challenges in strengthening surveillance. Emerg. Microbes Inf. **12**(2), 2240441 (2023)
60. Orang, A., Berke, O., Poljak, Z., Greer, A.L., Rees, E.E., Ng, V.: Forecasting seasonal influenza activity in Canada: comparing seasonal auto-regressive integrated moving average and artificial neural network approaches for public health preparedness. Zoonoses Pub. Health **71**(3), 304–313 (2024). https://doi.org/10.1111/zph.13114
61. Layman, N.C., Basinski, A.J., Zhang, B., Eskew, E.A., Bird, B.H., Ghersi, B.M., Remien, C.H., Vandi, M., Bah, M., Nuismer, S.L.: Predicting the fine-scale spatial distribution of zoonotic reservoirs using computer vision. Ecol. Lett. **26**(11), 1974–1986 (2023). https://doi.org/10.1111/ele.14307
62. Issa, J., Jaber, M., Rifai, I., Mozdziak, P., Kempisty, B., Dyszkiewicz-Konwińska, M.: Diagnostic test accuracy of artificial intelligence in detecting periapical periodontitis on two-dimensional radiographs: a retrospective study and literature review. Medicina **59**(4), 768 (2023)
63. Biercher, A., Meller, S., Wendt, J., Caspari, N., Schmidt-Mosig, J., De Decker, S., Volk, H.A.: Using deep learning to detect spinal cord diseases on thoracolumbar magnetic resonance images of dogs. Front. Vet. Sci. **8**, 721167 (2021)

# Helmet-Mounted IoT-Based Physiological Health Monitoring System for Soldiers

**K. Ghousiya Begum, K. Adalarasu, and J. Vikramkarthik**

**Abstract** Soldiers are the backbone of a nation's security and monitoring the psychological status of the soldiers enhances their performance. To combat the oppressors who operate out of fear and to protect the periphery, they employ several electronic devices. There are many wearable technologies available for monitoring the physiological signals of armed personnel but these devices need to be fitted as an additional wearable device. So, a helmet-mounted monitoring device is proposed to monitor the vital signals of armed forces engaged in various operations. The physiological parameters include oxygen saturation (SpO2) along with electrocardiogram (ECG) and photoplethysmography (PPG) signals. This helmet-mounted monitoring system embedded with sensors is proposed to monitor the physiological parameters of soldiers. The signals will be extracted by multiple electrodes placed on the forehead and chest, processed using a MAX86150 breaker board and ESP32, and transmitted to a telemetry monitoring point through wireless technology. Various parameters like heart rate (HR), SpO2, systolic blood pressure (SP) and diastolic blood pressure (DP), heart rate variability (HRV) parameters namely, the standard deviation of normal-to-normal (NN) intervals (SDNN), and root mean square of successive RR interval differences (RMSSD) were monitored through the Ubidots IoT platform for further analysis and decision making. The proposed study shows the feasibility of applying a smart helmet system for unconstrained health status monitoring during the navigation of soldiers.

**Keywords** Soldiers · Helmet · Electrocardiogram · IoT · Photoplethysmography · Ubidots

K. G. Begum (✉) · K. Adalarasu · J. Vikramkarthik
School of EEE, SASTRA University, Thirumalaisamudram, Thanjavur 614401, Tamil Nadu, India
e-mail: ghousiyabegum@eie.sastra.edu

© The Author(s), under exclusive license to Springer Nature Singapore Pte Ltd. 2026 
M. Pant et al. (eds.), *Real-World Applications of Machine Learning*, Mathematics for Sustainable Developments, https://doi.org/10.1007/978-981-95-7134-5_20

## 1 Introduction

The safety and well-being of soldiers in the Warfield and during navigation are of most importance for any military organization. One of the major concerns in this regard is the early detection of the health status of the soldiers by monitoring the physiological signals such as heart rate, oxygen saturation rate, HRV [1], and blood pressure which can indicate an impending medical emergency [2]. The traditional methods of monitoring these vital signs require bulky equipment and can be time-consuming, which can be a significant limitation in combat situations. To overcome these limitations, biosensors, and physiological parameter monitoring modules are integrated with the body of subjects [3–6]. The physiological monitoring device includes a set of sensors that can monitor various physiological signals, including heart rate, oxygen saturation rate, blood pressure, and body temperature [7]. These sensors are designed to be non-invasive and can be easily attached to the soldier's body. The data from these sensors are transmitted wirelessly to the data acquisition and processing unit. To serve this purpose, in this study, the use of a helmet-mounted physiological signal monitoring system has been proposed. Here, the helmet is mounted with the sensing, processing module and is recommended for identifying the health condition of soldiers as it increases the operational efficiency based on collective health data [8]. This can monitor certain parameters such as heart rate, HRV, SpO2, and blood pressure frequently. These parameters are obtained and processed using MAX86150 breaker board and ESP32 with Wi-Fi module [9–11]. The recorded signals are then uploaded to an IoT platform to view from anywhere around the globe. To obtain the above-mentioned features, physiological signals like electrocardiography (ECG) and Photoplethysmography (PPG) must be obtained. The description of the signals has been discussed below.

Many IoT-based health monitoring systems for army personnel have been proposed [2]. A review of smart devices like smart helmets, smart vests, and smart wristbands based on wireless sensor networks (WSNs) technology equipped for soldiers has been discussed. These devices tracked the armed force's location and health condition by forming Mobile Ad-hoc Network (MANET) and transmitting the physiological data to the soldiers nearby using a Zigbee module and to the controller using Long Range Wide Area Networking (LoRaWAN) [12]. A three-layered smart vest and helmet has been proposed wherein the authors concluded that helmet is essential for the armed personnel [13]. To detect the drowsiness of soldiers, a helmet-based system was proposed that monitored ECG, electroencephalogram (EEG), and electrooculogram (EOG). This proposed system recorded the mental state of the soldiers to detect drowsiness based on eye blinking and reopening time [14]. A cost-effective health monitoring and tracking system based on IoT was discussed for soldiers on the battlefield. The module was connected to a soldier's body consisting of GPS to track their position and health status. Through IoT, these data were sent to the control room. A few tiny wearable physiological sensors and other transmitting modules make up the suggested system [15, 16]. For real-time recording and tracking of health and position, a body sensor network module was attached to a soldier's arm.

This module transmits the soldier's data in an encrypted format in a blockchain on army-controlled computers. The temperature, heart rate, and GPS modules make up the installed sensor module [17]. A bio sensor and machine learning-based health monitoring system has been proposed to track the navigation and health condition of soldiers on the battlefield using GPS and body area networks wirelessly. The sensors like heartbeat sensor, accelerometer, humidity, and temperature sensors were used to extract the features which were transmitted through wireless ZigBee module among the fellow soldiers. The LoRaWAN network infrastructure has been proposed for communication between the soldier and control unit whenever cellular network coverage is absent [3]. A lifesaver module based on a deep learning approach for military personnel has been proposed. An autoencoder LSTM hybrid deep learning model was deployed to predict the psychological feature. The smart sensor data obtained from a soldier's uniform is processed by the autoencoder based on distinct properties. A time-related relationship is then derived by the LSTM model from localized functionality [18]. An IoT-based biosensor and GPS module system were developed for getting the health status, locating and rescuing soldiers in a minimum period [19]. A personal health monitoring system for military personnel has been recently discussed [20]. An integrated healthcare monitoring solution based on a three-layered service IoT architecture with distributed mechanism has been deployed for soldiers as an effective prototype model. Two levels of filtering have been done wherein the first level is performed at the WSBAN layer (end node) based on the fuzzy approach and the next level at the fog layer (intermediate node) using the time-series data (pattern) analysis approach. This filtration process avoids data flooding and reduces the computational burden on the cloud [21]. Based on the above motivation, the proposed work deals with the main objective listed below,

- To design and develop a prototype model for acquiring the physiological signal (ECG and PPG) using suitable sensors.
- To develop an algorithm to filter the physiological signal according to the noise present in the signal.
- To obtain the parameters such as heart rate, HRV, SpO2, and blood pressure from the recorded waveform.
- To transmit these obtained feature values to an IoT platform wirelessly.
- Embedding all the sensor units in a Helmet.

## 2 Methodology

In this study, a monitoring system is implemented for soldiers that is mounted on a helmet and it can monitor specific parameters such as heart rate, HRV, SpO2, and blood pressure. This will help the soldiers to be monitored frequently. These parameters are obtained using a MAX86150 breaker board and ESP32 Wi-Fi module. The recorded signals are then uploaded to an IoT platform to view from anywhere around the globe. The software tools, namely, Arduino IDE Version 1.8.10, Python IDE 3.10 and Ubidots IoT Platform, are used to support the smart helmet system model.

The packages used for performing the evaluation are NumPy, Pandas, PeakUtils, Matplotlib, SciPy, Scikit-learn, Findpeaks, and Requests. The ECG signal is detected using three surface electrodes placed at three different positions on the chest and the PPG signal is detected from the forehead. For both signals, the processing module is mounted on the helmet and the collected data is monitored. After the data is processed, the features from the signals (ECG & PPG) are extracted, classified, and transmitted wirelessly to the laptop or mobile for frequent monitoring from the control room. This helps to identify the physiological status of the armed personnel. The systematic workflow of the preferred methodology is shown in Fig. 1a. The Esp32 connecting module will be connected to the breaker board using jumper wires. The connections between Esp32 and MAX86150 are as follows: Ground-ground, VCC—VCC, GPIO21—SDA and GPIO22—SCL. Figure 1b shows the connection between Esp32 and MAX86150 boards. The MAX86150 board and the ESP32 are interfaced via one of the ESP32's communication protocols, such as SPI or $I^2C$, in order to establish a link between them. Here, the communication interface supported by MAX86150 is $I^2C$. So, the MAX86150's serial data line (SDA), serial clock line (SCL) and ground (GND) pins are connected to the ESP32's matching SDA, SCL and GND pins. Both the modules are mounted on a helmet. The MAX86150 module combines few primary features: an embedded red and infrared (IR) LED for pulse oximetry (SpO2) readings, and a photoplethysmography (PPG) sensor for heart rate monitoring [22]. The microcontroller receives data from the sensor and processes it, taking appropriate action based on the needs of the health monitoring system. Figure 2 displays the overall hardware module mounted on the Helmet for continuous monitoring of the soldier's health.

After connecting MAX86150, the Espressif library for Esp32 is installed in the Arduino integrated development environment (IDE) and the code is made to run. This code runs in the background and a graphical user interface (GUI) is developed through processing IDE where the port and the breaker board are chosen, which will be used

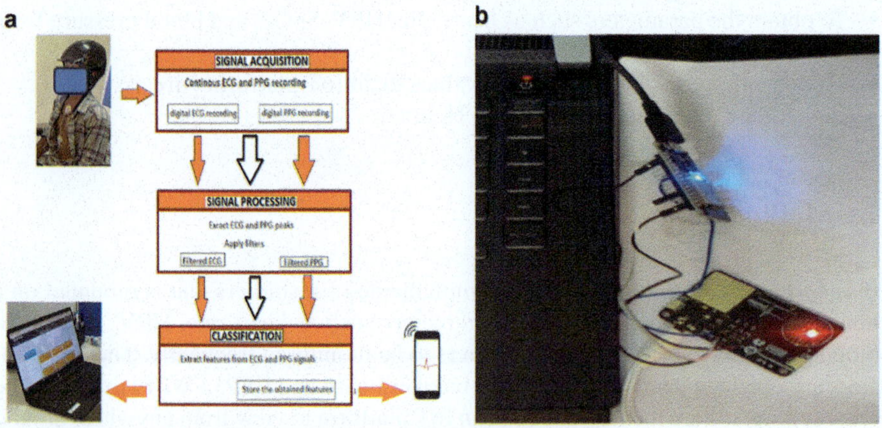


**Fig. 1** **a** Workflow of the proposed method **b** hardware connection

**Fig. 2** Hardware module mounted on the helmet

for logging wireless data. When Esp32 and MAX86150 are connected wirelessly, a dedicated Python code is developed to connect to the communicating model (laptop) using a particular IP address. The module will obtain power through a power source connected to it. With the aid of a dedicated platform, one can visualize the raw ECG and PPG signals by connecting to the MAX86150 board in the Protocentral OpenView mode using the COM4 port as displayed in Fig. 3. After the waveform is generated from the MAX86150, it is recorded in a.csv format representing the timestamp, ECG, and PPG values. This recorded file is fed into a developed Python code which will calculate the physiological parameters using the obtained data.

The ECG parameters include the heart rate (the counting of heart beats per minute), RR interval (the time period between two successive beats in a heartbeat), PR interval (the time between the commencement of the P wave and the initial state of the QRS complex, which can indicate problems with electrical conduction in the heart), QT

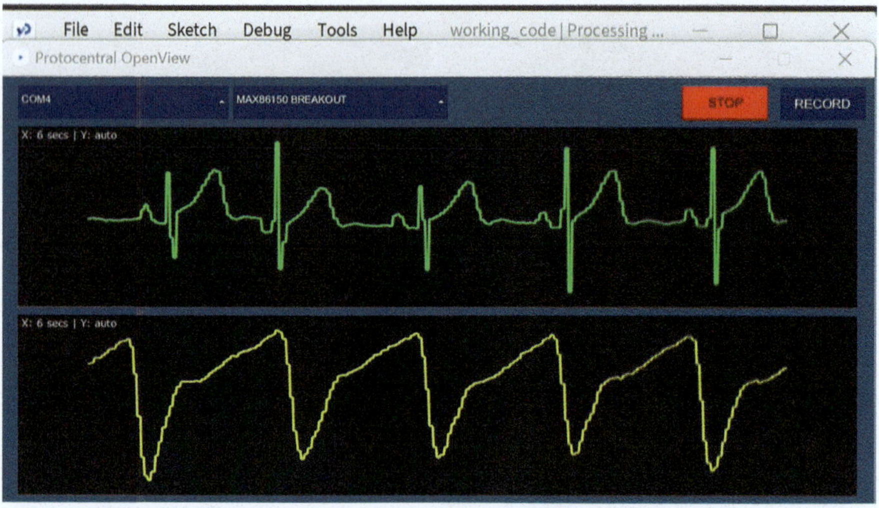


**Fig. 3** Plotting of live graph

interval (the time between the start of the QRS complex and the ending of the T wave, can reveal issues with the electrical recovery of the heart following a heartbeat), and ST segment (a region of the ECG waveform between the conclusion of the QRS complex and the beginning of the T wave that indicates an issue with blood flow to the heart muscle).

The heart rate variability (HRV) parameters include the standard deviation of normal-to-normal (SDNN) levels (intervals between subsequent heartbeats wherein higher SDNN levels show more heart rate variability and are frequently linked to improved autonomic and cardiovascular health), root mean square of successive differences (RMSS) between NN intervals (that offers an understanding of the beat-to-beat variations in heart rate), and NN50 (that refers to the number of consecutive NN intervals that deviate by more than 50 ms (ms) and thereby used to evaluate the variability of the RR intervals in a heart rate signal which are the intervals between succeeding R waves on an ECG signal). The PPG parameters include pulse rate (the time between the peaks of the PPG waveform is used to calculate pulse rate, which is the number of pulses per minute), blood oxygen saturation (SpO2—a measure of the amount of oxygen in the blood, which is derived from the ratio of the light absorbed by oxygenated and deoxygenated blood), rise time (the amount of time it takes for the PPG waveform to peak, which can reveal variations in arterial stiffness or blood vessel compliance), systolic (the pressure in arteries when the heart beats), and diastolic blood pressures (measure the pressure in your arteries between heartbeats). Among these, from ECG signal heart rate, RR interval, SDNN, and RMSSD are calculated and from PPG signal SpO2, systolic and diastolic pressures are calculated in the proposed study. The following basic formulas are used for calculation and discussed in detail in the results section.

$$\text{RR\_interval\_time\_diffs} = \text{np.diff(indexes)/sampling\_rate} \tag{1}$$

$$\text{Heart\_rate} = 60/\text{np.mean(time\_diffs)} \tag{2}$$

$$\text{SpO2} = 120 - 25 * (\text{ac/abs(dc)}) \tag{3}$$

$$\text{Systolic\_bp} = 120 + (\text{avg\_ptt} - 0.03)/0.004 \tag{4}$$

$$\text{Diastolic\_bp} = 80 + (\text{avg\_ptt} - 0.03)/0.006 \tag{5}$$

# 3 Results and Discussion

The results obtained for the proposed helmet-based physiological signal monitoring system have been discussed in this section. The working model of the project is elaborately explicated, tested and the observations are discussed and concluded below.

## *3.1 Hardware Setup*

The overall helmet-based signal monitoring system is shown in Fig. 4 [23]. The experimental setup explicates the signal acquisition and monitoring modules. The signals are detected using gel-based electrodes, then they are filtered and the extracted parameters are visualized on an IoT platform named Ubidots.

## *3.2 Signal Processing and Features Extraction*

A Python code is developed to filter and normalize the waveform and the data is collected in.csv file format and is plotted as shown in Fig. 5. To filter the noise present in the waveform generated and to avoid drift, the following filters are used to filter the signals obtained.

The signal can be acquired with the sampling rate of fs = 500 Hz. A baseline wander filter namely a high pass filter with a cut-off frequency of 0.5 Hz of fourth order is used to avoid drifts in the raw ECG and PPG signals. Secondly, to remove the power line noise, a Butterworth notch filter with a cut-off frequency of range 49–51 Hz of second order is deployed for both signals. After removing the powerline noise, the ECG signal is filtered using a Butterworth band pass filter with a cut-off

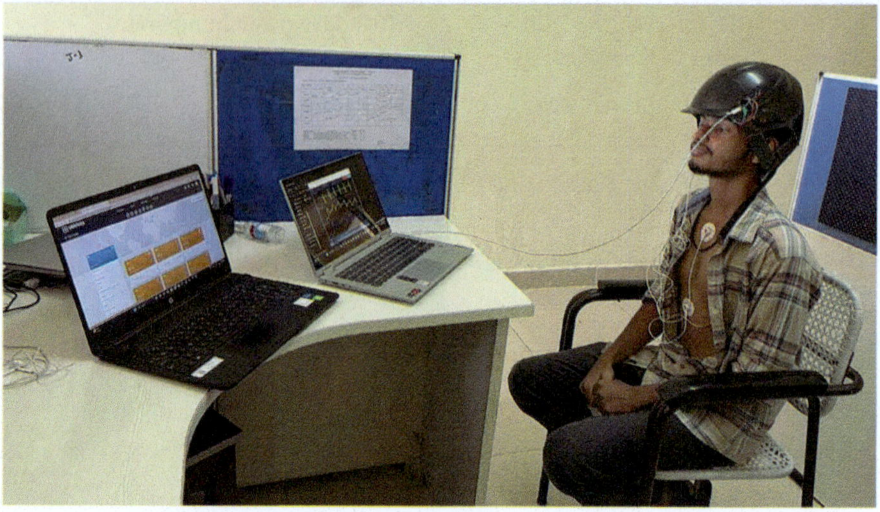

Fig. 4 Real-time recording using helmet-mounted system

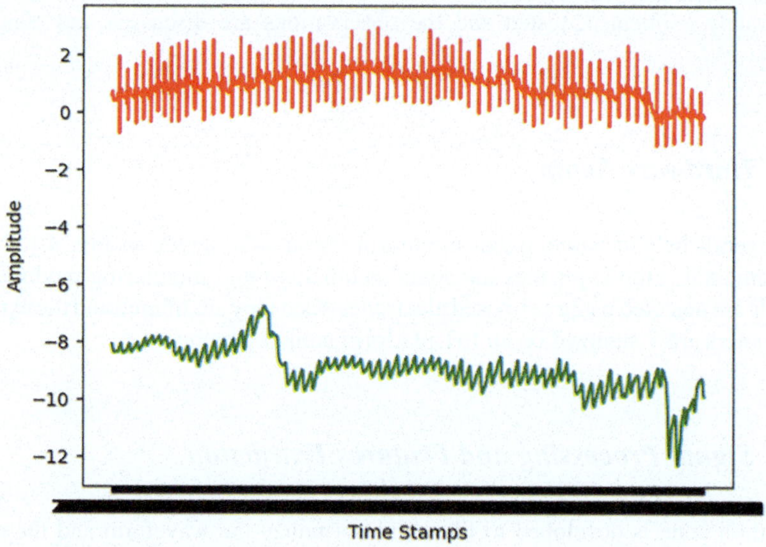


Fig. 5 Raw ECG and PPG signal

frequency of range 10–80 Hz with a filter of fourth order to remove artifacts. The same filter is used for the PPG signal but with a cut-off frequency of range 0.5–20 Hz. These filters can allow the signals with a frequency greater than a particular cut-off frequency and block the signals with frequencies lesser than the cut-off frequency

[24]. Figure 6 displays the processed ECG and PPG signal after applying the above filters. The zoomed version of it is displayed in Fig. 7.

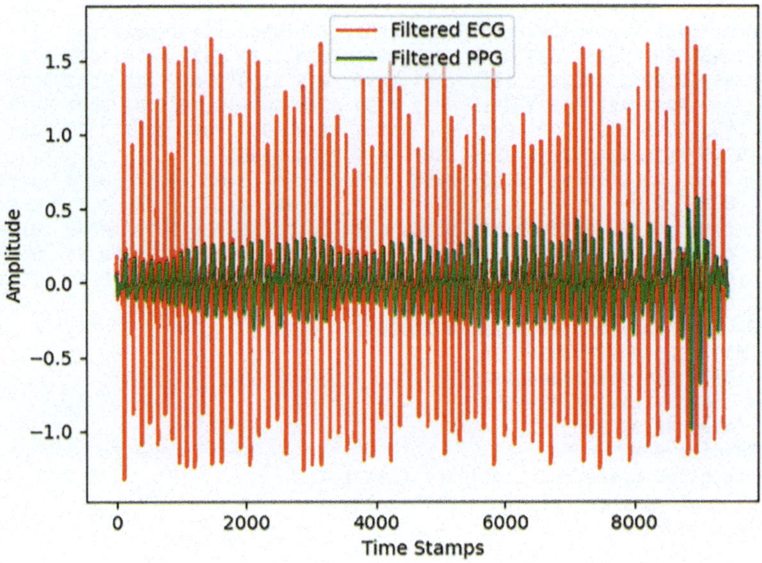


**Fig. 6** Filtered signal

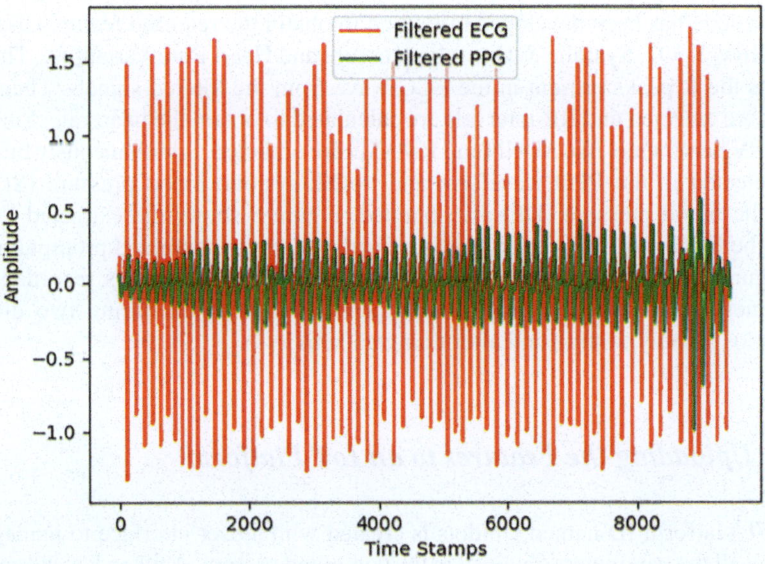


**Fig. 7** Filtered signal (Zoomed)

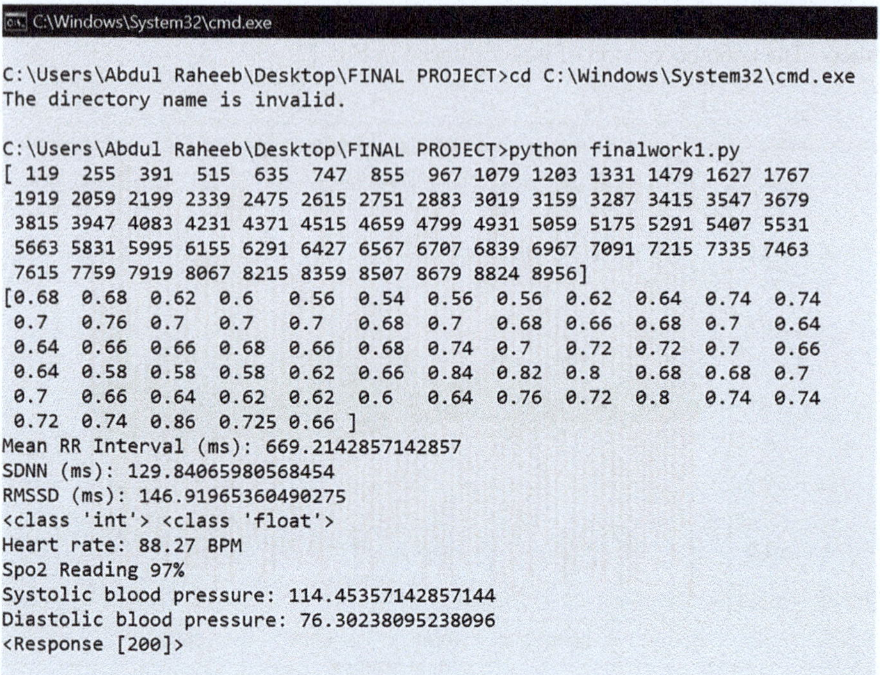


**Fig. 8** Calculated parameters visualized in cmd

The code has been developed in Python to obtain the required features like heart rate, HRV, SpO2, Systolic & Diastolic pressure and Heart Rate Variability. The code detects the R peaks present in the ECG wave from the filtered signals. Then the R peaks are detected and RR intervals are calculated to obtain the heart rate. Similarly, the HRV parameters like SDNN and RMSSD are obtained. Simultaneously the peaks are detected in the PPG waveform and SpO2, systolic blood pressure (SP), and diastolic blood pressure (DP) are obtained. Figure 8 shows the extracted features from the ECG and PPG signals that are displayed in the command prompt (cmd).

Figure 9 displays the mean values of ECG and PPG features recorded using a helmet-mounted monitoring system for 6 subjects taken during two different scenarios namely, resting and after doing an activity.

## 3.3 *Uploading the Features to an IoT Platform*

An IoT platform ID named Ubidots is created with a user interface to manage and display all the parameters obtained in the command prompt. A token key is generated and fed into the Python code so that the parameters obtained can be visualized in the IoT platform. These values are stored in the cloud so that the variations or

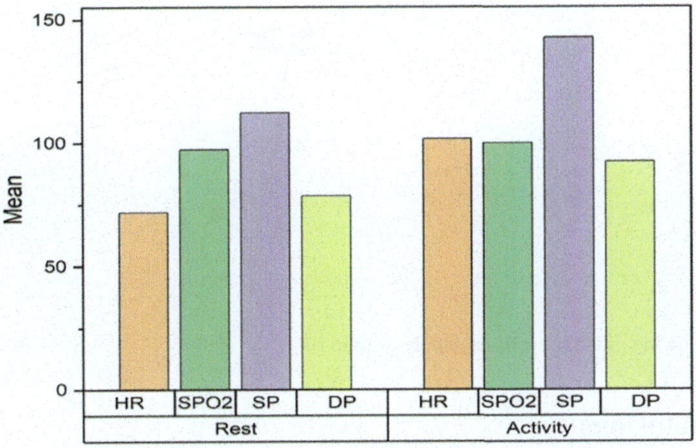


**Fig. 9** Mean values of ECG and PPG features during rest and activity conditions

any changes in the features can aid in monitoring the soldier's health condition as displayed in Figs. 10 and 11. Figure 10 depicts the parameters like systolic, diastolic blood pressures, heart rate, $SpO_2$, SDNN and RMSS. Figure 11a depicts the data logging of heart rate and Fig. 11b displays the data for systolic and diastolic blood pressures.

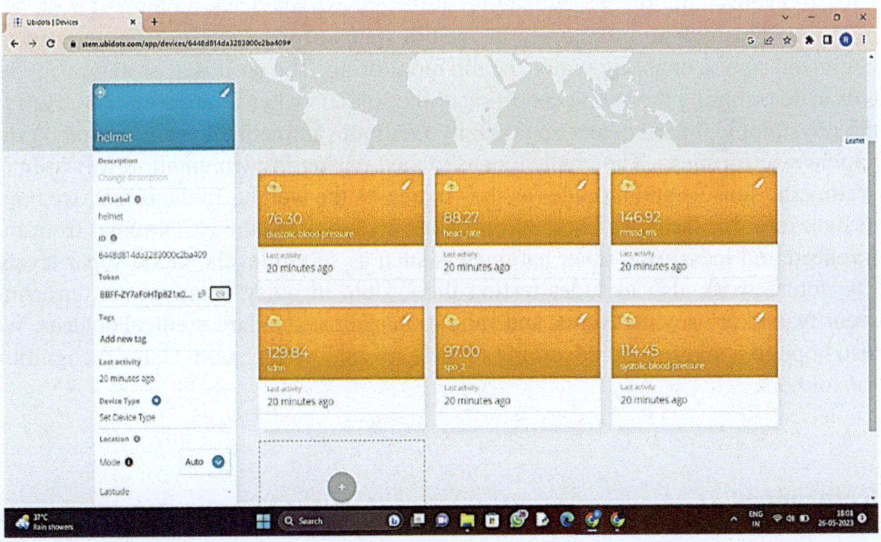


**Fig. 10** Features displayed in an IoT platform

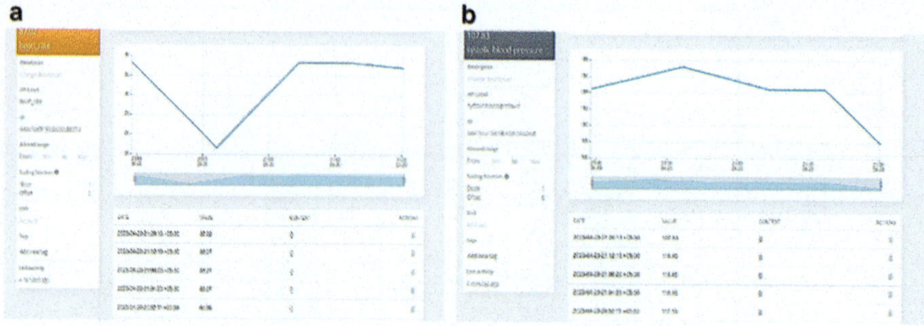

**Fig. 11** Data log graph for **a** heart rate, **b** systolic BP

## 4 Conclusions

The objective of this study in identifying the health condition of soldiers by monitoring the real-time physiological signals and the vital features using a helmet-mounted device has been discussed elaborately. Here, the gel-based electrodes were used to obtain the ECG signal from the chest and the PPG signal is obtained from the forehead using the LED present in the breaker board. Both signals were processed by the mounted device on the helmet. This mounted device uses a MAX86150 breaker board and Esp32 for extracting the raw ECG and PPG signals. From this, the signals were processed to remove the noise, and the vital features such as heart rate, HRV, oxygen saturation, and blood pressure were extracted and monitored in an IoT platform. This aids in monitoring the health status of the soldiers from the central command station enabling remote health monitoring. The proposed prototype model aids in determining the health condition of the soldiers when changes are observed in the physiological parameters and thereby facilitates immediate rescue based on the patient's seriousness. The challenges we faced during implementation was in integrating the sensors and modules as they increases the weight. In the future, we hope to look into this issue and expand our work by acquiring the ECG signal from the forehead and measuring other parameters such as serum levels, blood sugar levels. The future work also includes testing the system in noisy environment, ensuring security, and privacy measures, and validation against standard medical devices. We also hope to apply machine learning models for intelligent health status recognition for soldiers.

## References


1. Shaffer, F., Ginsberg, J.P.: An overview of heart rate variability metrics and norms. Front. Publ. Heal. **5**, 1–17 (2017). https://doi.org/10.3389/fpubh.2017.00258
2. Kavitha, N., Madhumathy, P.: IoT-based health monitoring and tracking system for soldiers. In: Advanced Healthcare Systems, John Wiley & Sons, Ltd, pp. 115–136 (2022)

3. Gondalia, A., Dixit, D., Parashar, S., Raghava, V., Sengupta, A., Sarobin, V.R.: IoT-based healthcare monitoring system for war soldiers using machine learning. Proced. Comput. Sci. **133**, 1005–1013 (2018). https://doi.org/10.1016/j.procs.2018.07.075
4. Friedl, K.E.: Military applications of soldier physiological monitoring. J. Sci. Med. Sport **21**(11), 1147–1153 (2018). https://doi.org/10.1016/j.jsams.2018.06.004
5. Wu, J.-Y., Wang, Y., Ching, C.T.S., Wang, H.-M.D., Liao, L.-D.: IoT-based wearable health monitoring device and its validation for potential critical and emergency applications. Front. Publ. Heal. **11**, 1–13 (2023). https://doi.org/10.3389/fpubh.2023.1188304
6. Abdulmalek, S., Nasir, A., Jabbar, W.A., Almuhaya, M.A.M., Bairagi, A.K., Khan, M.A.-M., Kee, S.-H.: IoT-based healthcare-monitoring system towards improving quality of life: a review. Healthcare **10**(10), 1 (2022). https://doi.org/10.3390/healthcare10101993
7. Swaroop, K.N., Chandu, K., Gorrepotu, R., Deb, S.: A health monitoring system for vital signs using IoT. Internet Things **5**, 116–129 (2019). https://doi.org/10.1016/j.iot.2019.01.004
8. Aysolal, S.C., Addetla, P., Arige, V.S.R., Banrapolu, B.R., Renatla, P.R., Lakshmi Prasanna, Y., Ivanovich Vatin, N.: Real-time soldier health and location tracking system. In: MATEC Web of Conferences, Vol. 392, p. 01062 (2024). https://doi.org/10.1051/matecconf/202439201062
9. https://protocentral.com/product/protocentral-max86150-ppg-and-ecg-breakout-with-qwiic-v2/
10. https://www.espressif.com/en/products/socs/esp32
11. Quartz components Homepage, https://quartzcomponents.com/products/esp32-development-board-wifi-bluetooth.
12. Kodam, S., Bharathgoud, N., Ramachandran, B.: A review on smart wearable devices for soldier safety during battlefield using WSN technology. Mater. Today Proc. **33**, 4578–4585 (2020). https://doi.org/10.1016/j.matpr.2020.08.191
13. Joshi, N. V., Joshi, S.P., Jojare, M.S., Askhedkar, A.R.: IoT based smart vest and helmet for defence sector. In: Proceedings International Conference on Communication information and Computing Technology. ICCICT 2021 (2021). https://doi.org/10.1109/ICCICT50803.2021.9510067
14. Kim, Y.S., Baek, H.J., Kim, J.S., Lee, H.B., Choi, J.M., Park, K.S.: Helmet-based physiological signal monitoring system. Eur. J. Appl. Physiol. **105**(3), 365–372 (2009). https://doi.org/10.1007/s00421-008-0912-6
15. Patii, N., Iyer, B.: Health monitoring and tracking system for soldiers using internet of things (IoT). In: 2017 International Conference on Computing, Communication and Automation (ICCCA), pp. 1347–1352 (2017). https://doi.org/10.1109/CCAA.2017.8230007
16. Prasanna, J.L., Ravi Kumar, M., Santhosh, C., Aswin Kumar, S.V., Kasulu, P.: IoT based soldier health and position tracking system. In: 2022 6th International Conference on Computing Methodologies and Communication (ICCMC), pp. 417–420 (2022). https://doi.org/10.1109/ICCMC53470.2022.9754096
17. Pabla, J., Sharma, V., Krishnamurthi, R.: Developing a secure soldier monitoring system using internet of things and blockchain. In: 2019 International Conference on Signal Processing and Communication (ICSC), pp. 22–31 (2019). https://doi.org/10.1109/ICSC45622.2019.8938259
18. Usharani, S., Rajmohan, R., Bala, P.M., Saravanan, D., Agalya, P., Raman, D.R.: Integrated implementation of hybrid deep learning models and IoT sensors for analyzing solider health and emergency monitoring. In: 1st IEEE International Conference on Smart Technologies and Systems for Next Generation Computing ICSTSN, pp. 1–6 (2022). https://doi.org/10.1109/ICSTSN53084.2022.9761332
19. Sharma, A., Aggarwal, S.: Real-time health monitoring system of soldiers using IoT. In: Advanced Computational Paradigms and Hybrid Intelligent Computing, pp. 285–296 (2022)
20. Bovens, D., van Baarle, E., Molewijk, B.: Personal health monitoring in the armed forces—scouting the ethical dimension. BMC Med. Ethics **24**(1), 21 (2023). https://doi.org/10.1186/s12910-023-00899-9
21. Bandopadhaya, S., Dey, R., Suhag, A.: Integrated healthcare monitoring solutions for soldier using the internet of things with distributed computing. Sustain. Comput. Inform. Syst. **26**, 100378 (2020). https://doi.org/10.1016/j.suscom.2020.100378

22. Digikey Homepage. https://www.digikey.in/en/product-highlight/m/maxim-integrated/max86150-integrated-ppg-and-ecg-biosensor-module.
23. Von Rosenberg, W., Chanwimalueang, T., Goverdovsky, V., Looney, D., Sharp, D., Mandic, D.P.: Smart helmet: wearable multichannel ECG and EEG. IEEE J. Transl. Eng. Heal. Med. **4**, 1–11 (2016). https://doi.org/10.1109/JTEHM.2016.2609927
24. Kors, J.A., van Herpen, G.: Computer analysis of the electrocardiogram. In: Macfarlane, P.W., van Oosterom, A., Pahlm, O., Kligfield, P., Janse, M., and Camm, J. (eds.) Comprehensive Electrocardiology. Springer London, London, pp. 1721–1765 (2010). https://doi.org/10.1007/978-1-84882-046-3_37

# Weather-Based Fault Detection in Solar Power Generation by ML

**P. Joel Augustine, Gummadi Srinivasarao, P. Moti Begum, D. Yaswanth, and Y. Vinod**

**Abstract** Solar energy, one of the most important renewable energies, employs many newer technologies, among which Solar Photovoltaic (PV) cells are extensively used for Generating Electricity. As for any installed system, the detection of faults in solar power systems is essential to ensure performance and reliability. It is determined that the solar irradiation, temperature, and shading of the PV cells affect the efficiency of the cells and, therefore, the diagnostics of the system and the cells' malfunctions. Fault detection systems using statistical methods, such as Linear Regression and Decision Tree models, have been created and are now in use. Furthermore, newer machine learning algorithms such as AdaBoost, Random Forest, and Tweedie Regression are also employed to model the PV output under a range of weather conditions and evaluate how shading or module faults will affect it. These models enable precise fault detection because they distinguish between normal weather fluctuations and faults. The results suggest that machine learning, when combined with weather parameters, improves fault detection and also improves the system's predictive maintenance capabilities. This creates better system performance and increases the lifetime of solar power systems.

P. J. Augustine (✉) · P. M. Begum · D. Yaswanth · Y. Vinod
Department of Electrical and Electronics Engineering, Siddhartha Academy of Higher Education, Deemed to be University, Kanuru, Andhra Pradesh, India
e-mail: pernati.joel1177@gmail.com

P. M. Begum
e-mail: 218w1a0285@vrsec.ac.in

D. Yaswanth
e-mail: 218w1a0264@vrsec.ac.in

Y. Vinod
e-mail: 218w1a02a0@vrsec.ac.in

G. Srinivasarao
Department of Electrical and Electronics Engineering, Siddhartha Academy of Higher Education, Deemed to Be University, Formerly Velagapudi Ramakrishna Siddhartha Engineering College, Kanuru, Andhra Pradesh, India
e-mail: vasu1in@vrsiddhartha.ac.in

© The Author(s), under exclusive license to Springer Nature Singapore Pte Ltd. 2026 
M. Pant et al. (eds.), *Real-World Applications of Machine Learning*, Mathematics for Sustainable Developments, https://doi.org/10.1007/978-981-95-7134-5_21

**Keywords** Photovoltaic system monitoring · Effect of partial shading · Machine learning techniques · Fault detection

## 1 Introduction

The growing need for solar power as a renewable energy resource has made the performance of solar plants to be of great importance. These losses are caused by bad weather, shade, or deteriorating panels. For weather-based defect identification in solar plants, we offer two machine learning models in this research that are based on the Decision Tree and Linear Regression techniques. These models are trained using input data that includes temperature, wind speed, sun irradiation, and solar power generation. Depending on its capacity to predict anticipated energy output, it can rule out causes of abnormalities and defects. This increases solar plant efficiency while reducing the need for human intervention, enabling automatic and reliable system fault monitoring. Partial shadowing can have a significant impact on solar panels' performance. Because solar panels are made up of solar cells, even a single shaded solar cell may have an impact on the output of that solar panel.

## 2 Problem Definition

A lot of research has been conducted at the synergy of machine learning, renewable energy and weather prediction, since the importance of fault detection to energy efficiency and solar power generation downtime is considerable [1]. Given that sunlight intensity, temperature, and cloud coverage are all environmental factors that dictate solar energy generation, it is no wonder that solar power generation can make use of weather-based prediction models such as regression or time series analysis to forecast generation and detect deviations that may indicate a fault [2]. Some studies/models take advantage of the prediction of weather and the location of a fault in a hybrid approach to improve forecasting and locate faults. These models are able to perform such distinctions because they take into account the historical data of weather and solar output so that they are able to tell when it is external weather conditions that are affecting performance and not faults [3]. The main benefit of this specific type of model is its high accuracy after it has been fitted onto the data. Nevertheless, it is quite vulnerable to errors when it comes to meteorological forecasts that tend to be exaggerated by the model itself. A suitable curve can be obtained and this curve would be much better than a single model this is due the fact that the hybrid models take the best of both worlds [4]. These of course have their limitations, the statistical models of course, for example, we have ARMA model and the neural networks and support vector machine and all of these fall under this category [5]. Finally, solar energy forecasting involves data gathering as well as data analysis, in a bid to estimate the solar energy outputs over a given time frame, with the aim of minimizing the effects caused by solar intermittency [6]. As previously highlighted, once key barriers to the widespread integration of solar energy are, namely high material costs and low conversion efficiency, are moreso be hitting the scalps the more the wide spread integration will be achieved [7]. In many instances, however, the issue of intermittency in solar power has been successfully solved or at least reduced with the support of solar forecasting methods using machine learning [8]. Machine learning is a subset of computer science that enables machines to analyze data and identify patterns on their own without being explicitly programmed [9]. ML models have the capacity to process great amounts of data almost instantaneously which makes room for more accurate and timely predictions. They are also very responsive high speed when it comes to changing information [10]. Solar power forecasting is a complex task which is composed of several stages such as data collection, data preprocessing, feature engineering, training, and validation processes, and finally making predictions [11]. Especially, data preprocessing techniques are important in order to fill in gaps, scale, and therefore optimize AI models performance [12]. The forecasting techniques also underpin the feature most appropriate for solar power prediction placement to be established [13]. Solar power-making AI models are built off of past data inferring that these models can be trained to understand how different feature inputs manifest solar power production [14]. It is a process that involves the integration of trends and correlation, something AI patterns and recognition capabilities were designed [15]. The ability of ML to be trained on multiple data inputs is one of the primary reasons for the success of this processing tool such as data from weather stations, satellite images, and radar [16].

Furthermore, these models can also use additional sources such as social networks, participatory observations, and environmental monitoring devices [17]. The central idea has nothing but profound predictions assist concerned parties in decision-making hence making it easy to embed solar energy within the rest of the energy system [18]. One of the most important aspects is the model verification step, as the validity of the predictions depends on it. Mean Squared Error (MSE) and Root Mean Squared Error (RMSE) type performance metrics are expected to be used as lower bounds in future inheritance forecasts to verify the expectations of the developed models in terms of accuracy [19]. The issue brought about by the partial shading of solar modules is the occurrence of hotspots which may lead to module catastrophes, causing the failure of a string of panels [20]. Shading analysis investigates the degree and effect of shading on panel output, assisting in the prediction and reduction of shading-induced panel performance loss, which is an important factor in the investigation of solar panels and such systems in general as even partial shading leads to mismatch and hotspot catalysis leading to large energy losses in the system [21]. Some studies conducted in the past years consider that the loss of power is due to the proportion of the number of developed powers with regards to the number of shaded cells or the solar irradiance, and thus a shading factor is developed [22]. Sunlight is the main energy source and its conversion into direct current is carried out by solar panels. The inverters are required to convert the direct current electricity that has been produced into power that can be utilized by the grid or targeted appliances and is in the form of alternating current [23]. A battery is a charged device which retains excess power that the system has produced during peak periods for times of low consumption demand or crises [24]. But also, sensors are necessary and inject critical data about the system through its different measuring processes including energy production, efficiency, and environmental data [25]. On the last step, all scenarios are created which allow forecasting the performance of the solar PV system when the system is subjected to faults or shading, as one of the conditions. Forecasting in energy production gives an idea on the possible energy yield and assists operators in improving other facts such as scheduling maintenance plans [26].

## 3 Problem Description

PV Systems (Photovoltaic Systems) are getting more popular since their goods encourage environmental friendliness. Nonetheless, the main difficulty is that such systems might develop defects and performance degradation, posing questions about their efficiency and dependability. Certain elements such as irradiance, temperature, and shading can occasionally cause problems in PV systems, affecting energy output. These inefficiencies must be rectified as soon as possible in order to reduce losses, downtime, and to extend the life of solar panels. Another difficulty with conventional techniques of fault identification in PV systems is the inability to determine if an output failure is caused by a system breakdown or natural occurrences in the environment. Employing Solar Photovoltaic (PV) systems effectively retains energy, and their importance cannot be undermined. However, impediments like equipment failures, inclement weather, or even shade might reduce their performance. These difficulties make maintenance plans more difficult to follow and prevent steady energy production. Traditional fault detection techniques are poor because they frequently fail to distinguish between environmental impacts and system defects, resulting in unnecessary maintenance or problems that go unnoticed.

Our goal is to build an innovation that can predict system performance and no problems that are associated with specific operational failures by examining development and historical data on solar power and characteristics such as weather patterns. Static bi-level optimization is aided by a new hybrid machine learning framework that combines ensemble learning techniques with weather categorization. This contrasts with earlier studies that concentrated on SVMs or RF methods separately. Using a stacking ensemble model, this method displays several kinds of interactions.

# 4 Methodology

## *4.1 Block Diagram*

Figure 1 illustrates the starting point of the fault detection process for solar PV systems has been the carrying out of data collection from sources like Kaggel and other data bases containing relevant weather and operational data (irradiance, temperature, voltages, among others). Making collections that are accurate is very crucial for the predictions that will be made. Feature extraction allows integration of the features, thereby decreasing dimensionality and improving model performance. Reanalysis steps of the data such as cleaning and normalization are done as part of data preprocessing. Afterward, the data is subjected to training and testing so as to assess the effectiveness. Different numbers of machine learning models (Linear Regression, AdaBoost, Random Forest, and Tweedie) have been trained in order to be able to model different fault and shading effects on output. Linear regression and random forest fault detection and classification most detect power outages whose level is above Acceptable threshold suggesting faults are likely present. A shading module incorporates all models in order to determine the impact of shading and thus aid in improving the output. The last outputs, which are prediction of fault as a development strategy and performance prediction in the presence of shading, assist in early detection and monitoring for improvement of the solar energy within the system.

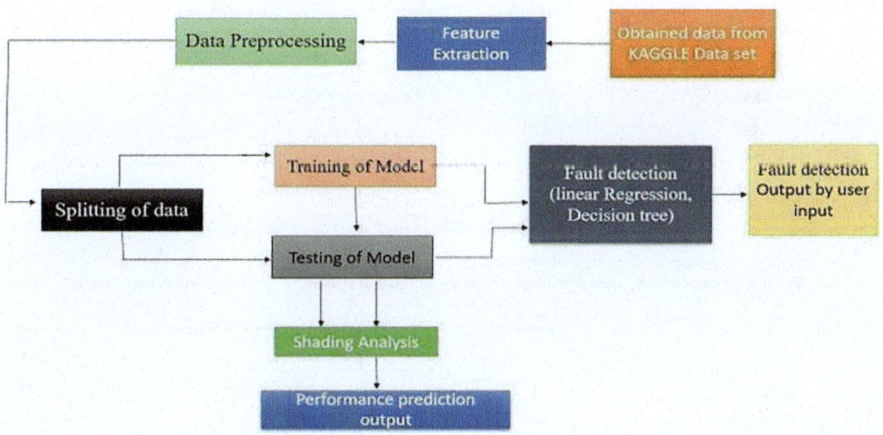


**Fig. 1** Figure shows block diagram

## 4.2 Proposed Machine Learning

(1) Fault based on solar power generation and weather data

Figure 2 shows the input solar generation and weather data are processed and analyzed by a machine learning model to detect deviations in expected power output, identifying potential faults and triggering alerts for maintenance.

**Algorithm 1:**
Start: Begin the process.

Load Datasets: Load solar power generation and weather data. Data Preprocessing: Clean, normalize, and preprocess the data. Feature Extraction: Extract relevant features for the model.

Train-Test Split: Split data into training and testing sets.

Train Models: Train both Decision Tree and Linear Regression models. Prediction: Use models to predict potential faults.

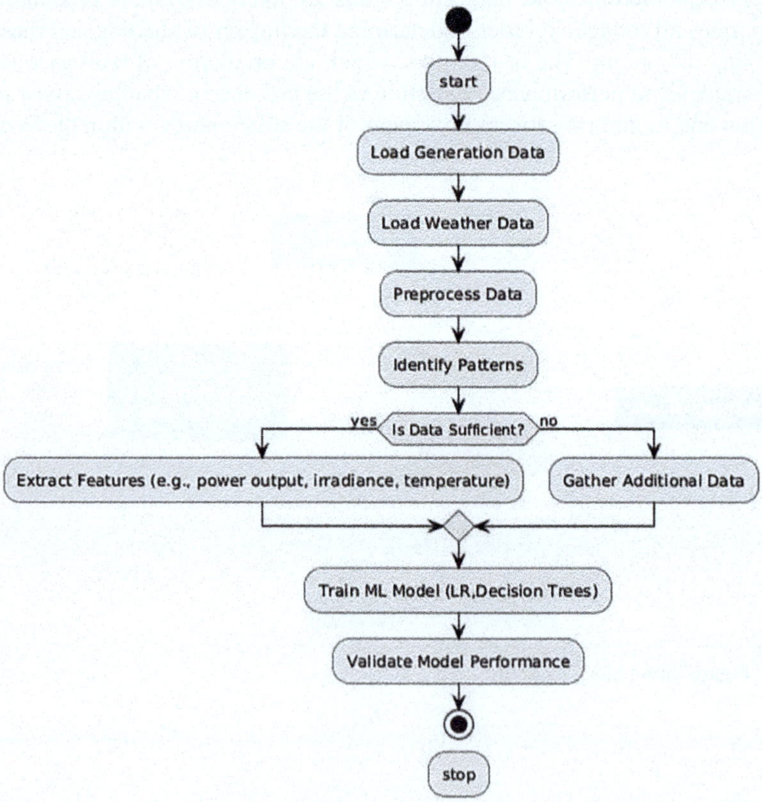


**Fig. 2** Flowchart for fault based on solar power generation and weather data

Fault Detection Decision: Check if there's a fault based on prediction.

Alert or Normal Operation: If fault is detected, send an alert; otherwise, the system operates normally.

End: The flowchart ends.

(2) Partial shading effect on solar power generation and performance

Figure 3 shows the model monitors solar power output, identifies shading patterns, and evaluates their impact on performance, alerting for shading-induced efficiency losses to optimize maintenance and panel positioning.

**Algorithm 2:**

Start: Begin the process.

Data Collection: Load datasets related to solar power generation and weather conditions.

Data Preprocessing: Includes cleaning missing values, feature engineering, and scaling. Feature Selection: Extracts relevant features from the datasets.

Train-Test Split: Splits the dataset into training and testing sets.

Model Training: Applies four machine learning models: Random Forest, Logistic Regression, Tweedie Regression, Ada Boost.

Model Evaluation: Compares the models' performance on the testing set.

Fault Detection Decision: Check if there's a fault based on prediction.

Fault Detection: Deploys the best-performing model for detecting shading faults in real time.

End: The flowchart ends.

## 4.3 *Feasible Solutions*

A. Comparing the DC Power Generation, Irradiation, Module and Ambient Temperature with DC POWER

(1) Highest Average DC_POWER is Generated on "2020-05–15"

Figure 4 illustrates the solar DC power manufacturing has reached the maximum limits. Furthermore, the DC power curve depiction and the irradiation curve representation patterns look as those which were explained in the above parts of this document. The appearance of the climate does not seem to be a problem since there are no clouds to cover the sun and there are few fluctuations in the solar panel temperature as well as the ambient temperature. This further corroborates the fact that there are no faults that are present in the solar panels.

(2) Lowest average DC_POWER is generated on "2020-06–11

Figure 5 illustrates, there are very large fluctuations in both DC power graph and irradiation graph. So, there may be fault in system.

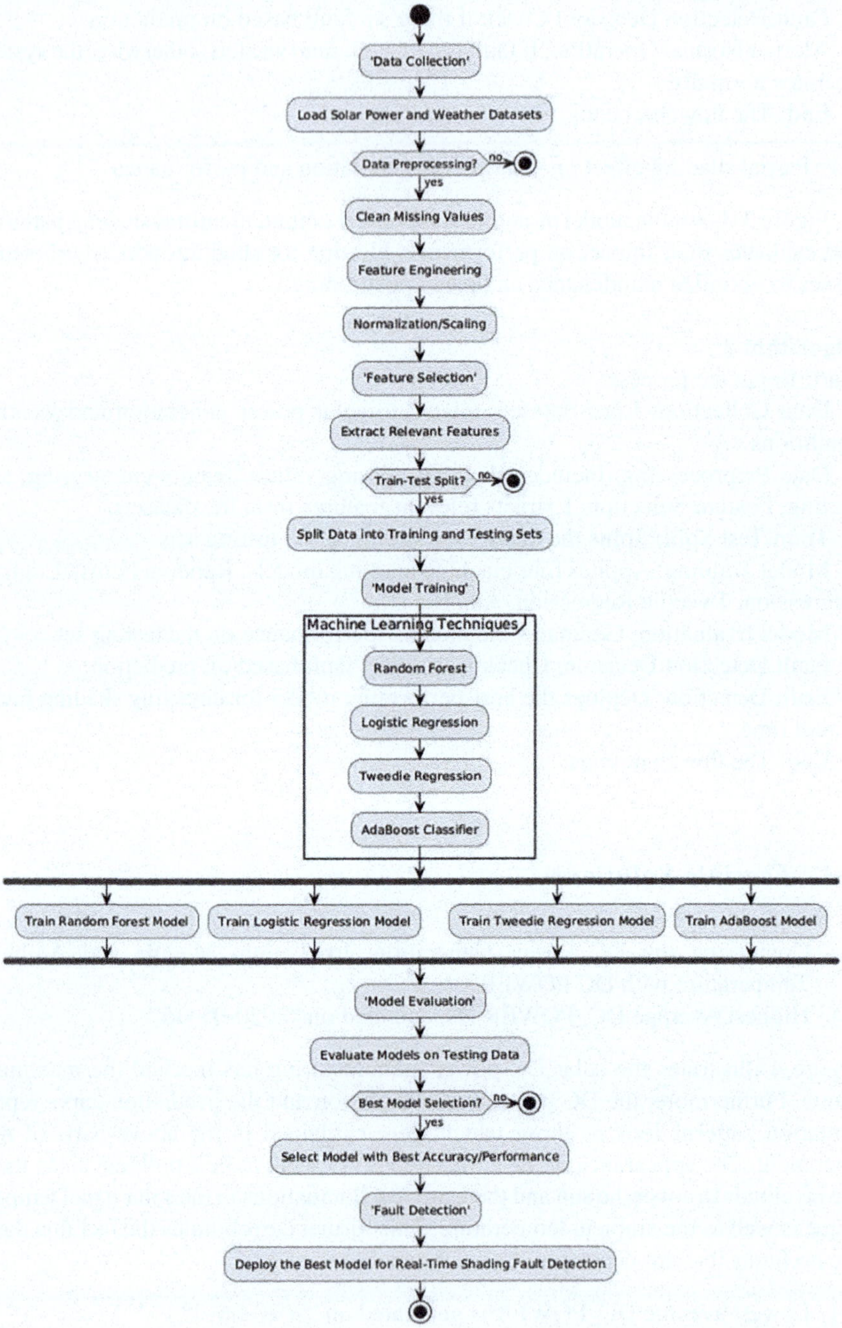


**Fig. 3** Flowchart for partial shading effect on solar power generation and performance

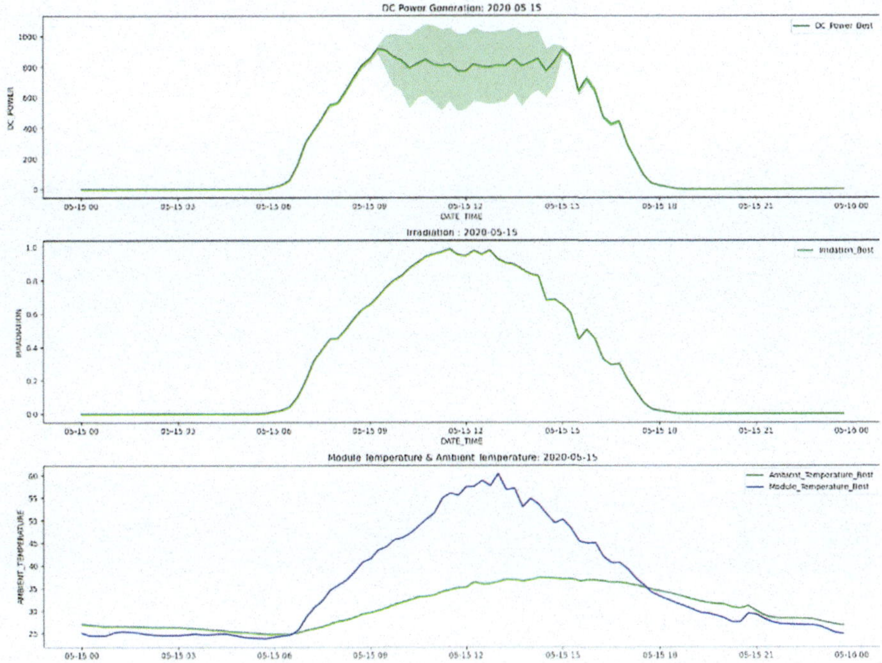


**Fig. 4** Highest average generation

B. Partial shading effect on solar power generation and performance

Figure 6 shows the scatter plot shows voltage (V) vs current (I) for a solar power system, with points color-coded by different fault types. The blue points represent normal (No Fault) operation, orange indicates Partial Shading Fault, green shows One PV Module Fault, and red represents Two PV Module Fault. Each distinct cluster suggests a clear separation between different fault conditions.

1. From the value of current and power we can distinguish between the healthy, partial shaded and faulty PV systems.
2. Healthy PV system has high current value between 9.5 Amperes and 11 Amp.
3. Partial shaded PV system has medium current value between 8 Amperes and 9.7 Amp.
4. Single and double faulted PV panel has low current value between 7 Amperes and 7.6 Amp.

## 5 Result and Discussion

A. Fault based on solar power generation and weather data

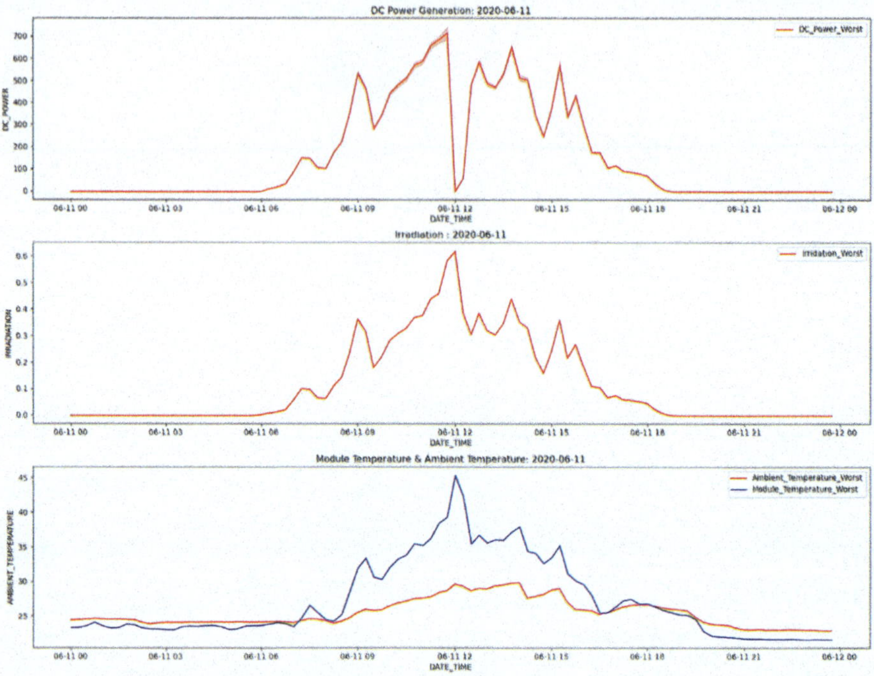


**Fig. 5** Lowest average generation

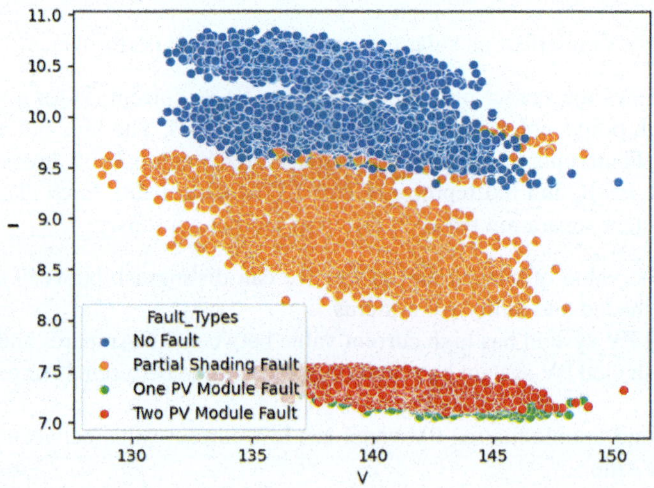


**Fig. 6** Scatter plot V and I

**Table 1** ML classifier and $R^2$ score

| Classifier | $R^2$ Score |
|---|---|
| Linear regression | 97 |
| Decision tree | 87.4 |

**Table 2** Result of actual and predicted

| | Actual | Predicted | Error |
|---|---|---|---|
| 40,426 | 0.000000 | 0.000000 | 0.000000 |
| 50,974 | 0.000000 | 0.000000 | 0.000000 |
| 53,919 | 684.91333 | 684.709243 | 0.20409 |
| 2384 | 0.000000 | 0.000000 | 0.000000 |
| 22,014 | 0.000000 | 0.000000 | 0.000000 |

The linear regression model achieves $R^2$ score close to 97%, while the Decision Tree model also reaches 87.4% accuracy, indicating high prediction precision for the dataset used in the training and testing process as shown in Table 1.

Table 2 shows a comparison between actual and predicted values, along with the error for each prediction. The errors are relatively low, indicating that the model's predictions are close to the actual values.

A. Partial shading effect on solar power generation

The bar chart in Fig. 7 compares the accuracy of four ML models for fault detection in solar Panels.

The Ada Boost Regression and Linear Regression Models achieved nearly perfect accuracy (close to 100%), while the Tweedie Regression showed significantly lower performance, with around 81.6% accuracy as shown in Table 3.

Table 4, shows a comparison between actual and predicted values, along with the error for each prediction. The errors are relatively low, indicating that the model's predictions are close to the actual values.

Across all datasets, we used k-fold cross-validation ($k = 10$) to reduce overfitting. This technique ensures the stability and dependability of the model by dividing the data into training and testing sets across a number of iterations. Model behavior was constant across folds, as evidenced by the average standard deviation of performance measures being less than 1.5%.

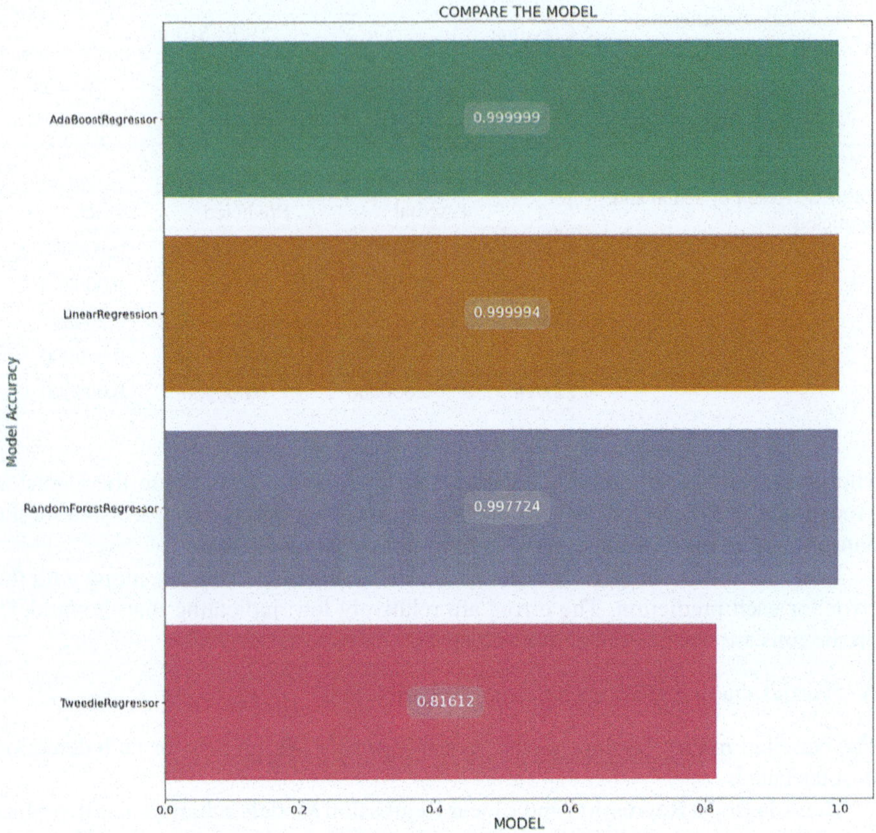


**Fig. 7** Comparison of machine learning techniques

**Table 3** ML classifier and accuracy

| Classifier | Accuracy |
|---|---|
| Ada boost regression | 99.9 |
| Linear regression | 99.9 |
| Random forest regression | 99.7 |
| Tweedie regression | 81.4 |

**Table 4** Comparison between actual and predicted values

| | Actual | Predicted | Error |
|---|---|---|---|
| 12,894 | 1219.168311 | 1207.148301 | 12.020010 |
| 10,877 | 1238.297976 | 1218.502392 | 19.795584 |
| 6885 | 1470.045158 | 1449.490254 | 20.554904 |
| 15,146 | 1436.703655 | 1430.244077 | 6.459578 |
| 34,925 | 1061.849424 | 1050.182823 | 11.666601 |

## 6 Conclusion

This research demonstrates the benefits obtained from the application of a weather-oriented predictive model combined with machine learning methods such as Linear Regression, Random Forest, Ada Boost, or Tweedie Regression into solar PV systems. In order to improve the prediction capacity of the dataset, derived features were developed.

Solar energy per day refers to the total solar irradiance received over a 24-hour period. Training (70%), validation (15%), and testing (15%) sets were created from the preprocessed dataset, guaranteeing temporal continuity to prevent data This method makes it possible to perform PV output performance analysis on the basis of the physical surroundings in which it was generated enabling fault detection and shading corrections to be made distinguishing false performance loss from real performance loss. The approach assists in performing maintenance actions promptly, improves the effectiveness of the system while cutting down on expenses which benefits the solar power generation and its prospect as a clean consistent energy source.

## References

1. Kumaradurai, A., Teekaraman, Y., Coosemans, T., Messagie, M.: Fault detection in photovoltaic systems using machine learning algorithms: a review. In: 2020 8th International Conference on Orange Technology (ICOT). Daegu, Korea (South) (2020)
2. Samson, O.O., Lekan Gbadamosi, S., Onibonoje, M.O., Ojo, E.E.: Development of machine learning algorithms for fault detection in power systems—a review. In: 2024 International Conference on Science, Engineering and Business for Driving Sustainable Development Goals (SEB4SDG). Omu-Aran, Nigeria (2024)
3. Panda, S., Dhaka, R.K., Panda, B., Pradhan, A., Jena, C., Nanda, L.: A review on application of machine learning in solar energy & photovoltaic generation prediction. In: 2022 International Conference on Electronics and Renewable Systems (ICEARS). Tuticorin, India (2022)
4. Rodrigues, S., Ramos, H.G., Morgado-Dias, F.: Machine learning in PV fault detection, Diagnostics and prognostics: a review. In: 2017 IEEE 44th Photovoltaic Specialist Conference (PVSC). DC, USA, Washington (2017)
5. Elgamal, M., Abdelmaksoud, A., Ismail, Y.: Seamless machine learning models to detect faulty solar panels. In: 2023 3rd International Conference on Electronic Engineering (ICEEM). Menouf, Egypt (2023)
6. Vahidi, A., Golkar, M.A.: Fault detection and classification in PV arrays using machine learning algorithms in the presence of noisy data. In: 2022 9th Iranian Conference on Renewable Energy & Distributed Generation (ICREDG). Mashhad, Iran, Islamic Republic of (2022)
7. Tyagi, S., Dhingra, B., Tomar, A.: Condition monitoring & fault detection in photovoltaic modules using machine learning. In: 2022 1st International Conference on Sustainable Technology for Power and Energy Systems (STPES). SRINAGAR, India (2022)
8. Badr, M.M., Hamad, M.S., Abdel-Khalik, A.S., Hamdy, R.A., Ahmed, S., Hamdan, E.: Fault identification of photovoltaic array based on machine learning classifiers. IEEE Access **9**, 159113–159132 (2021)

9. Baruah, D., Roy, R., Ahmed, R., Subbiah, S., Chouhan, S., Angappan, K.: Contextual approaches to data-driven fault detection in solar photovoltaic system. In: 2024 IEEE International Conference on Evolving and Adaptive Intelligent Systems (EAIS). Madrid, Spain (2024)
10. Kabir, S., Shufian, A., Zishan, M.S.R.: Isolation forest based anomaly detection and fault localization for solar PV system. In: 2023 3rd International Conference on Robotics, Electrical and Signal Processing Techniques (ICREST). Dhaka, Bangladesh (2023)
11. Triki-Lahiani, A., Abdelghani, A.B., Slama-Belkhodja, I.: Fault detection and monitoring systems for photovoltaic installations: a review. Renew. Sustain. Energy Rev. **82**, 2680–2692 (2019)
12. Roy, S., Alam, M.K., Khan, F., Johnson, J., Flicker, J.: An irradiance-independent, robust ground-fault detection scheme for PV arrays based on spread spectrum time-domain reflectometry (SSTDR). IEEE Trans. Power Electron. **33**(8), 7046–7057 (2018)
13. Vaish, R., Chikondra, B., Dwivedi, U.D.: Challenges posed by renewable energy source integration to machine learning based power system fault diagnosis. In: 2024 IEEE International Communications Energy Conference (INTELEC). Bengaluru, India, pp. 1–5 (2024)
14. Maryasin, O.Y., Plohotnyuk, A.: Day-ahead power forecasting of renewable energy sources using neural networks and machine learning. In: 2023 International Conference on Industrial Engineering, Applications and Manufacturing (ICIEAM). Sochi, Russian Federation, pp. 130–135 (2023)
15. Ali, B.S., Benalia, N., Nora, N.Z.: Effect of Partial Shading on the PV System Performances. In: 2018 6th International Conference on Control Engineering & Information Technology (CEIT). Istanbul, Turkey (2018)
16. Zahoor, N., Ali, A., Yuanqing, X., Ahmed, B., Osman, M., Navid, Q.: Optimum load forecasting model-based intelligent residential system using machine learning algorithms. In: 2023 2nd International Conference on Emerging Trends in Electrical, Control, and Telecommunication Engineering (ETECTE). Lahore, Pakistan, pp. 1–7 (2023)
17. Ramli, M.A.M., KashifIshaque, S.T., Al-Turki, Y.A.: A review on maximum power point tracking for photovoltaic systems with and without shading conditions. Elsevier Renew. Sustain. Energy Rev. **67**, 144–159 (2017)
18. Mahmoud, Y.: Novel modeling framework for PV faults under partial shading. IEEE Trans. Power Electron. **39**(4), 4781–4791 (2024)
19. Hariharan, R., Chakkarapani, M., Ilango, G.S., Nagamani, C.: A method to detect photovoltaic array faults and partial shading in PV systems. IEEE J. Photovolt. **6**(5), 1278–1285 (2016)
20. Kumar, S.S., Selvakumar, A.I.: Detection of the faults in the photovoltaic array under normal and partial shading conditions. In: 2017 Innovations in Power and Advanced Computing Technologies (i-PACT). Vellore, India (2017)
21. KR, C.L., VA, R., Rajalakshmi, K., Simon, S., Stepha, G.T.: Impact of shadow or dust on solar photovoltaic power generation system. In: 2023 Second International Conference on Electronics and Renewable Systems (ICEARS). Tuticorin, India (2023)
22. Djalab, A., Bessous, N., Rezaoui, M.M., Merzouk, I.: Study of the effects of partial shading on PV array. In: 2018 International Conference on Communications and Electrical Engineering (ICCEE), pp. 1–5 (2018)
23. Naeiji, S., Ashtiani, H.J., Shabani, A., Shen, Z.J.: Unified machine learning based fault detection strategy through voltage-sensing for both AC and DC side faults in photovoltaic farms. In: 2024 IEEE Sixth International Conference on DC Microgrids (ICDCM). Columbia, SC, USA (2024)
24. Krishna, V., Sindhu Thampatty, K.C.: Effect of shading on the performance of solar photovoltaic system. In: 2023 IEEE Technology & Engineering Management Conference—Asia Pacific (TEMSCON-ASPAC). Bengaluru, India (2023)
25. Wang, J., Zhong, H., Lai, X., Xia, Q., Wang, Y., Kang, C.: Exploring key weather factors from analytical modeling toward improved solar power forecasting. IEEE Trans. Smart Grid **10**(2), 1417–1427 (2019)
26. Deepak, K., Kumar, A.G.: Metaheuristic based MPPT control for solar PV systems under partial shading conditions. In: 2022 IEEE 7th International Conference on Recent Advances and Innovations in Engineering (ICRAIE). Mangalore, India (2022)

# Advanced Price Forecasting for Food Commodities

**S. Mangala Priya and R. Dilli Babu**

**Abstract** Food price volatility is a global issue that profoundly affects food security, economic stability, and the livelihoods of millions, especially in developing regions. Essential commodities such as wheat, rice, and corn are central to global food systems but are highly vulnerable to unpredictable price fluctuations driven by climate variability, geopolitical tensions, and global trade dynamics. Accurately forecasting the prices of such commodities is crucial for stabilizing markets and enabling timely interventions. This study introduces an Advanced Price Forecasting System, combining Auto-Regressive Integrated Moving Average (ARIMA) and Long Short-Term Memory (LSTM) models. The dual-model approach allows the system to deliver reliable short- and long-term price predictions by leveraging the strengths of both statistical and deep learning methodologies. The ARIMA model captures linear trends effectively in the short term, while the LSTM model addresses nonlinear dependencies and complex seasonal patterns in long-term forecasts. The project integrates robust preprocessing techniques, interactive visualizations, and modular design, making it scalable for various commodities. The system includes a React-based dashboard for user interaction, enabling stakeholders to access insights dynamically. Initial evaluations reveal improved forecasting accuracy and the potential for actionable insights to enhance agricultural planning, policymaking, and consumer decision-making. Future work includes hybrid model integration, expansion to additional commodities, and incorporation of external data sources like weather and economic indicators.

**Keywords** Commodity price forecasting · ARIMA · LSTM · Time-series analysis · Food security · Machine learning

---

S. M. Priya (✉) · R. D. Babu
Department of Information Technology, Vel Tech High Tech Dr. Rangarajan Dr. Sakunthala Engineering College, Chennai, India
e-mail: sma.cse@rmkec.ac.in

R. D. Babu
e-mail: dillibabur_it20@velhightech.com

© The Author(s), under exclusive license to Springer Nature Singapore Pte Ltd. 2026 
M. Pant et al. (eds.), *Real-World Applications of Machine Learning*, Mathematics for Sustainable Developments, https://doi.org/10.1007/978-981-95-7134-5_22

# 1 Introduction

## *1.1 Background*

The prices of essential food commodities fluctuate significantly due to multiple interrelated factors. In developing nations, where food expenditures constitute a major share of household income, such volatility can trigger economic instability, food insecurity, and even social unrest. Key contributors to price volatility include:

- Climate Change: Sudden and extreme changes in weather patterns disrupt agricultural production, creating supply chain imbalances. For example, prolonged droughts in key wheat-producing regions can drastically reduce yields, leading to a ripple effect of global price hikes. Similarly, floods or unseasonable frosts can destroy crops, impacting availability and driving up costs.
- Global Market Dynamics: Trade restrictions, international sanctions, and conflicts can alter the global availability of essential commodities. For instance, export bans imposed by a major producer can limit supply in international markets, leading to sharp price increases. Additionally, disruptions in transportation routes or logistics due to geopolitical factors can further exacerbate these price instabilities.
- Global Market Dynamics: Trade restrictions, international sanctions, and conflicts can alter the global availability of essential commodities. For instance, export bans imposed by a major producer can limit supply in international markets, leading to sharp price increases. Additionally, disruptions in transportation routes or logistics due to geopolitical factors can further exacerbate these price instabilities.

Traditional statistical forecasting models are limited in their ability to capture the complex dynamics of modern agricultural markets. These models generally assume linearity and stationarity, making them inadequate for predicting sudden price surges or long-term seasonal patterns. To address these shortcomings, this project employs a hybrid approach that combines ARIMA for short-term forecasting with LSTM for long-term predictions.

## *1.2 Problem Statement*

Existing price forecasting systems face several limitations that hinder their effectiveness and adaptability in dynamic market environments.

**Inability to Adapt to Rapid Market Changes:** Traditional forecasting models struggle to incorporate real-time market fluctuations. These systems often rely heavily on historical data, making them less responsive to sudden changes caused by factors like seasonal variations, geopolitical events, or unexpected supply–demand imbalances.

**Failure to Account for Nonlinear Patterns:** Commodity prices often exhibit complex, nonlinear behaviors influenced by multiple interdependent factors such as weather conditions, transportation costs, and global trade policies. Many existing systems rely on linear models, which fail to capture these intricate patterns, leading to inaccuracies in predictions.

**Limited Scalability for Multi-commodity Forecasting:** Most systems are designed to focus on a single commodity or operate within a narrow scope. They lack the scalability to handle diverse datasets, multiple commodities, and varying timeframes, making them inadequate for large-scale applications or broader market analyses.

This project addresses these challenges by designing a system that integrates ARIMA and LSTM models to improve forecasting accuracy for essential commodities.

## *1.3 Literature Survey*

**Forecasting Agricultural Commodity Prices:** Elhadi et al. [1] explored the application of machine learning techniques including regression models and ensemble algorithms for forecasting agricultural commodity prices in support of global food security. Their work demonstrated that machine learning can significantly improve predictive accuracy over traditional models. Kim et al. [2] further emphasized the growing role of deep learning approaches in agricultural data processing, highlighting their potential for complex pattern extraction in yield and price estimation.

**Statistical and Benchmark Models in Price Forecasting:** Kwas and Rubaszek [3] evaluated a range of traditional statistical and econometric approaches to commodity price forecasting, such as ARIMA and benchmark comparison models. Their study confirmed ARIMA's reliability for short-term, stationary data but noted its limitations in capturing nonlinear dependencies and abrupt market fluctuations.

**Deep Learning for Time-Series Forecasting:** Zheng et al. [4] and Ameur et al. [5] highlighted the effectiveness of LSTM and hybrid deep learning models in capturing temporal dependencies in agricultural price data. Mukoviz et al. [6] and Al-Hajji et al. [7] further validated LSTM's superiority for forecasting commodity and land prices, achieving higher accuracy and robustness. Despite these advances, the studies noted persistent issues such as computational demands and overfitting.

**Gaps in Existing Research:** Previous studies [1–9] have explored machine learning and deep learning for price forecasting but often remain limited to specific methods with poor scalability. Few integrate interpretability or visualization for policy use. This study bridges these gaps by unifying ARIMA and LSTM models with a strong preprocessing pipeline and an interactive forecasting dashboard for real-time market insights.

## 1.4 Objectives

The project aims to design and implement a system that accurately forecasts the prices of essential food commodities. Specific objectives include:

**Model Integration:** The system combines the strengths of ARIMA and LSTM models to provide comprehensive forecasting. ARIMA excels in analyzing historical trends and generating reliable short-term predictions, while LSTM captures complex patterns and dependencies in the data, making it ideal for long-term forecasts. This hybrid approach ensures a balanced and versatile forecasting mechanism that caters to both immediate and future market needs.

**Accurate Forecasting:** To enhance the reliability of predictions, the system employs advanced preprocessing techniques, such as handling missing values, scaling, and feature engineering. Additionally, rigorous evaluation methods, including cross-validation and performance metrics like RMSE and MAE, are used to minimize errors and refine model performance. This focus on precision ensures stakeholders receive trustworthy insights.

**Interactive Dashboard:** A user-friendly, React-based dashboard acts as the interface between the system and its users. It allows stakeholders to visualize trends through dynamic charts, graphs, and tables. The dashboard also enables users to customize forecasts by selecting specific commodities, timeframes, or regions, empowering them to explore insights tailored to their needs.

**Scalability:** The system is designed to handle multiple commodities and varying timeframes effortlessly. Whether it's analyzing a single market or aggregating trends across multiple regions, the architecture ensures smooth scaling to accommodate growing datasets and evolving requirements. This scalability makes the system future-ready and adaptable to diverse use cases.

## 1.5 Significance

Accurate price forecasting benefits stakeholders across the supply chain by:

For producers, it provides valuable insights that enable better crop planning and resource allocation. With accurate predictions and trends, farmers can align their production with market demands, optimize resource usage, and reduce waste, ultimately improving their income and minimizing risks.

For consumers, the system helps mitigate the impact of price hikes. By forecasting potential fluctuations, it empowers individuals to plan their budgets more effectively, making it easier to navigate periods of inflation or scarcity without significant financial strain.

For policymakers, the system acts as a decision-making tool to support timely and effective market interventions. By analyzing trends and generating accurate predictions, it helps implement measures like releasing buffer stocks or regulating exports, ensuring market stability and equitable access to essential commodities.

Effective price forecasting provides stakeholders with the tools to make informed decisions. Producers can optimize planting schedules and reduce wastage, policymakers can implement timely market interventions, and consumers benefit from stabilized prices and enhanced food availability. This project contributes to addressing global food security challenges by introducing a data-driven, proactive solution for price volatility.

# 2 System Design

## 2.1 System Architecture

**Data Layer:** This is the foundation of the system, where all the raw data is collected and stored. It pulls together both historical and real-time price data to provide a comprehensive dataset for analysis. The historical data is sourced from CSV files, containing daily price records of essential commodities. Simultaneously, real-time data is fetched from APIs to keep the system updated with the latest market trends. This layer ensures that the system has a reliable and consistent data feed, forming the backbone of the entire architecture.

**Processing Layer:** Once the data is collected, the processing layer comes into play. This layer is responsible for cleaning and preparing the data to ensure its quality and usability. Tasks such as handling missing values, normalizing data, and extracting meaningful features are handled here. After preprocessing, the data is fed into machine learning pipelines for training predictive models. This layer transforms raw data into a format suitable for accurate and efficient predictions, laying the groundwork for the application layer.

**Application Layer:** This layer is the analytical powerhouse of the system. It leverages advanced models like ARIMA and LSTM to generate predictions about commodity prices. The ARIMA model focuses on analyzing trends in historical data to provide short-term forecasts, while the LSTM model, with its ability to capture long-term dependencies, enhances the prediction's accuracy for more complex patterns. Together, these models work to forecast price trends, helping stakeholders make informed decisions.

**Presentation Layer:** The presentation layer brings the system's outputs to life, offering users an intuitive and interactive experience. It includes a React-based dashboard that visualizes the predictions and trends through graphs, charts, and tables. This layer transforms raw data and complex analyses into easy-to-understand reports and visualizations. It empowers users to explore insights, compare predictions, and make data-driven decisions efficiently.

## 2.2 Architectural Framework

The system is built on a modular architecture, ensuring scalability, flexibility, and efficiency. The design comprises four primary layers. The data layer manages the collection, storage, and retrieval of historical price data, sourced from CSV files and external APIs. The processing layer handles data preprocessing, including normalization, outlier detection, and feature engineering. The overall system architecture is illustrated in Fig. 1, which outlines the interaction between the data, processing, application, and presentation layers.

At the core of the system is the application layer, which implements the ARIMA and LSTM models. The ARIMA model is used for analyzing short-term trends in stationary data, while the LSTM model is designed to capture long-term dependencies and nonlinear patterns. Finally, the presentation layer consists of a React-based dashboard that provides users with an intuitive interface to explore forecasts and historical trends.

Data preprocessing is a critical component of the system, ensuring the input data is consistent and accurate. Missing values are addressed through interpolation, while outliers are detected and handled using statistical techniques such as interquartile

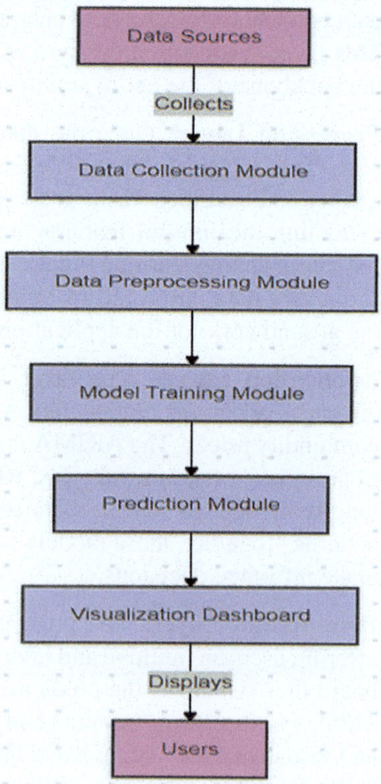


**Fig. 1** System architecture. *Source* Author's own illustration

range filtering. Feature engineering enriches the dataset with technical indicators like moving averages and rolling volatility, enhancing the predictive power of the models.

The forecasting models operate in parallel, each optimized for specific use cases. The ARIMA model is tuned using evaluation criteria like Akaike Information Criterion (AIC) and Bayesian Information Criterion (BIC) [1]. Meanwhile, the LSTM model processes sequences of data, employing dropout layers to mitigate overfitting and the Adam optimizer for efficient training [4, 6]. The results generated by these models are visualized through the interactive dashboard, which offers customizable views for different commodities, timeframes, and forecasting models. This modular design ensures the system can accommodate additional commodities and adapt to evolving user needs.

## 2.3 Data Flow

The data movement between the layers is summarized in Fig. 2, showing the end-to-end flow from data collection to visualization.

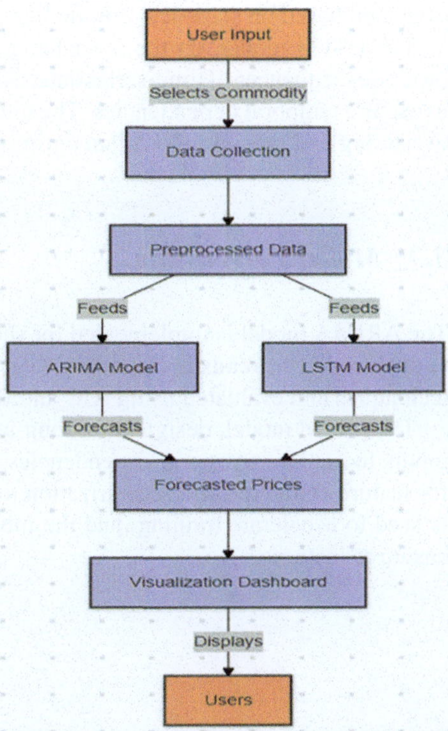


**Fig. 2** Data flow. *Source* Author's own illustration

# 3 Methodology

## *3.1 Data Collection and Preprocessing*

The project begins with the collection of historical price data for essential commodities, including wheat, rice, and corn. This data often contains inconsistencies, such as missing values and outliers, which are addressed during preprocessing. Missing data is filled using interpolation techniques, while outliers are detected and corrected through statistical methods. To improve model performance, feature engineering is employed to generate additional inputs such as moving averages and volatility metrics. These features provide deeper insights into market trends and enhance the models' ability to make accurate predictions.

## *3.2 Model Training*

The ARIMA and LSTM models are trained in parallel, each tailored to specific forecasting needs. The ARIMA model is optimized for short-term predictions, focusing on linear trends in stationary data. Parameters are selected using grid search and evaluated based on statistical criteria like AIC and BIC.

The LSTM model, on the other hand, is designed for long-term forecasting. It processes sequences of time-series data, using a recurrent neural network architecture to capture temporal dependencies. The model incorporates dropout layers to prevent overfitting and employs the Adam optimizer for efficient training.

## *3.3 Model Forecasting*

The ARIMA model is implemented for short-term forecasting, leveraging its ability to analyze linear trends in stationary data. Parameters are tuned through grid search techniques and evaluated using AIC and BIC criteria.

The LSTM model, designed for long-term forecasting, processes sequential data inputs to capture temporal dependencies. Its architecture includes multiple layers for feature extraction and regularization to prevent overfitting. The Adam optimizer is used to accelerate training, and the model is evaluated on validation datasets to ensure robustness.

### 3.4 *Visualization and User Interaction*

The final stage of the methodology involves the visualization of predictions through a React-based dashboard. The dashboard provides an interactive platform for exploring historical trends, forecasted prices, and technical indicators. Users can customize their views, select commodities and timeframes, and compare the outputs of the ARIMA and LSTM models.

## 4 Results and Analysis

### 4.1 *Model Performance*

The system demonstrates the complementary strengths of the ARIMA and LSTM models [1, 4]. ARIMA excels in short-term forecasting, providing reliable predictions for linear trends and stationary data. Its simplicity and efficiency make it well suited for immediate decision-making. In contrast, the LSTM model is highly effective in capturing long-term dependencies and nonlinear patterns. It performs particularly well in scenarios involving seasonal variations or complex market behaviors. Although the LSTM model requires more computational resources, its improved accuracy in long-term forecasts justifies its application for strategic planning. The comparative results of the LSTM and ARIMA models for wheat price forecasting are presented in Fig. 3.

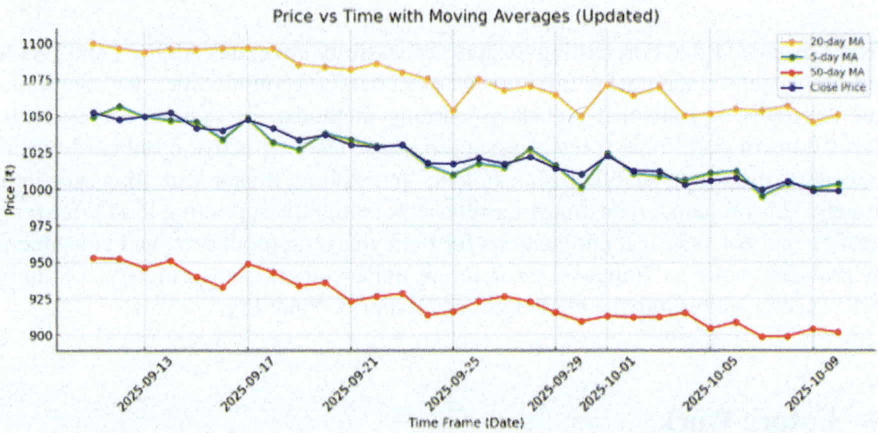


**Fig. 3** Prediction result for wheat. *Source* Model output generated by the authors

### *4.2 Visualization Insights*

The dashboard enhances user engagement by presenting historical and forecasted data in a visually intuitive format. Users can analyze trends, compare models, and explore technical indicators, enabling data-driven decision-making. For example, wheat prices predicted by the LSTM model over a six-month period closely align with observed seasonal fluctuations, highlighting the model's ability to handle complex patterns.

## 5 Challenges and Solution

The project faced several challenges during its development. Data quality was a primary concern, with missing values and outliers threatening to distort model performance. These issues were resolved through robust preprocessing techniques, including interpolation and statistical filtering. Model training also posed challenges, particularly with the LSTM model, which is prone to overfitting. Regularization techniques such as dropout layers and early stopping were implemented to address this issue. Scalability was another challenge, which was mitigated through a modular architecture and cloud-based deployment, enabling the system to handle additional commodities and larger datasets.

## 6 Conclusion

The Advanced Price Forecasting System successfully integrates ARIMA and LSTM models to deliver accurate predictions for essential food commodities. By combining the strengths of statistical and deep learning methods, the system addresses the limitations of traditional forecasting approaches. The interactive dashboard further enhances usability, enabling stakeholders to explore trends and forecasts intuitively. This project demonstrates significant potential for improving food price forecasting and has practical applications for policymakers, producers, and consumers. It provides a robust framework for future enhancements, ensuring its continued relevance in addressing the challenges of food price volatility.

## 7 Future Work

To build on the project's success, several avenues for future development are proposed:

- Hybrid Modeling: Combining ARIMA and LSTM to leverage their respective strengths.
- External Data Integration: Incorporating weather patterns, global trade indicators, and economic factors to improve forecasting accuracy.
- Enhanced Dashboard Features: Adding mobile compatibility, customizable views, and real-time updates for greater accessibility.
- Expanded Commodity Coverage: Scaling the system to include additional commodities such as coffee, sugar, and oilseeds.

These enhancements will ensure the system remains relevant and impactful in addressing the evolving challenges of food price volatility.

## References


1. Elhadi, S.K.A., Khedr, A.E., El-Shishiny, H., El-Khoriby, A.A.: Forecasting prices of agricultural commodities using machine learning for global food security: towards sustainable development goal 2. Int. J. Eng. Trends Technol. **71**(12), 226–235 (2023). https://doi.org/10.14445/22315381/IJETT-V71I12P226
2. Kim, J., Jeong, C., Kim, J., Lee, S.: Deep learning—method overview and review of use for fruit detection and yield estimation. Agriculture **12**(2), 256 (2022). https://doi.org/10.3390/agriculture12020256
3. Kwas, M., Rubaszek, M.: Forecasting commodity prices: looking for a benchmark. Forecasting **3**(2), 27 (2021). https://doi.org/10.3390/forecast3020027
4. Zheng, F., Fan, J., Zhao, X., Liu, G.: A study on agricultural commodity price prediction model based on secondary decomposition and long short-term memory network. Agriculture **14**(1), 60 (2024). https://doi.org/10.3390/agriculture14010060
5. Ameur, H.B., Jarboui, A., Louati, N.: Forecasting commodity prices: empirical evidence using deep learning tools. Ann. Oper. Res. (2022). https://doi.org/10.1007/s10479-022-05076-6
6. Mukoviz, V., Tsyfra, V., Krayevska, O., Tkachuk, A.: Forecasting agricultural land prices in Ukraine using LSTM deep neural networks. Agric. Resour. Econ.: Int. Sci. E-J. **8**(2), 108–127 (2022). https://doi.org/10.51599/are.2022.08.02.06
7. Al-Hajji, M.F., Al-Ajmi, K.A., Al-Amri, S.M.: Efficient commodity price forecasting using long short-term memory model. Int. J. Eng. Trends Technol. **71**(12), 226–235 (2023). https://doi.org/10.14445/22315381/IJETT-V71I12P226
8. Sun, J., Meng, L., Zhou, M., Zhang, Y.: Agricultural product price forecasting methods: a review. Agriculture **13**(9), 1671 (2023). https://doi.org/10.3390/agriculture13091671
9. Paul, R.K., Bhattacharjee, S., Rakshit, S.: Machine learning techniques for forecasting agricultural prices: a case of Brinjal in Odisha. India. PLoS ONE **17**(6), e0270553 (2022). https://doi.org/10.1371/journal.pone.0270553

# Fake News Detection Using Machine Learning and Sentiment Analysis Integrated in a Flask Web Application

**Yash Mehta, Raj Nayan, Sumit Malik, and Vinay Shukla**

**Abstract** Fake or misleading information, spreading quickly on media, is influencing the public perception and is stirring up social and political turmoil. The traditional manual verification methods could be more scalable given the sheer content volume. Machine learning techniques have been used in the past for research in this area for the detection of fake news, but their usefulness has often been limited by low prediction accuracy and reliance on specific feature sets. The present research work proposes a comprehensive methodology that combines machine learning classification algorithms along with feature extraction techniques to detect fake news. Utilizing a random forest classifier and TF-IDF vectorization, the system is implemented as a user-friendly web application using Flask. To gain more insight into the emotional aspects of the news, the sentiment analysis has been integrated in the present work. The results show the efficacy/usefulness of the implemented system, which achieves high accuracy in predicting the fake news with a clear presentation of the confidence levels and emotional tone. The findings suggest that this approach is better than the previous methods because it significantly enhances the precision of fake news detection and offers practical applications for real-time use on digital platforms.

**Keywords** Sentiment analysis · Fake news detection · Machine learning · Web application · TF-IDF vectorization

Y. Mehta · R. Nayan
CSE, School of Computer Science Engineering and Technology, Bennett University, Greater Noida, Uttar Pradesh, India

S. Malik (✉) · V. Shukla
Department of Mathematics, School of Computer Science Engineering and Technology, Bennett University, Greater Noida, Uttar Pradesh, India
e-mail: sumit.malik4@gmail.com

© The Author(s), under exclusive license to Springer Nature Singapore Pte Ltd. 2026

M. Pant et al. (eds.), *Real-World Applications of Machine Learning*, Mathematics for Sustainable Developments, https://doi.org/10.1007/978-981-95-7134-5_23

## 1 Introduction

The volume of information spreading widely through digital channels is becoming an alarming issue that not only threatens the integrity of society and democracy but also disrupts the public order dynamics. The swift spread of misleading and/or fake content has the potential to tilt the elections, trigger unrest within communities, and harm people's reputations. The currently used manual verification processes are time-consuming and impractical, given the stream of information being produced daily.

This present work aims to address these issues by creating an automated system through the use of machine learning techniques for detecting the correctness of the news. Previous studies in this area utilized algorithms like Naive Bayes and Support Vector Machines, which showed limitations in terms of accuracy and concentration on particular text elements. The development of inclusive models will allow for the handling of varied datasets and dependable forecasts.

The proposed method will incorporate the Random Forest Classifier, a technique known for its precision and ability to handle large datasets, along with the TF-IDF vectorization to convert the text data into valuable numerical formats that enhance the model's ability to differentiate between false and genuine news articles.

This added layer in the detection process by including the emotional tone of the content through sentiment analysis is integrated with the help of TextBlob. Thus, it improves the accuracy of the prediction and gives deeper insights into the nature of the news articles analyzed.

In practical terms, this study will implement the detection system as a web application based on Flask that can be used for public applications. Users may input text news and then receive an instant prediction of whether it's authentic or not, as well as sentiment analysis, hence helping combat misinformation on digital platforms.

The development and deployment of fake news detection systems raise significant ethical concerns. One major implication is the potential for misuse, where these systems are exploited to suppress dissenting voices, manipulate public opinion, or even perpetuate propaganda. Furthermore, biases in the training data or algorithms can lead to discriminatory predictions, unfairly targeting certain groups or individuals. For instance, a system may be more likely to flag articles from minority-owned outlets or those with dissenting viewpoints, exacerbating existing power imbalances. Additionally, the reliance on automated systems can also erode media literacy and critical thinking skills, making it increasingly difficult for individuals to discern fact from fiction. Therefore, it is crucial to address these ethical concerns through transparent and inclusive system development, rigorous testing, and ongoing evaluation to ensure that fake news detection systems serve to promote, rather than undermine, democratic values and social justice.

## 2 Literature Review

Detection of misinformation has been a highly researched topic as it uses a wide array of machine learning models for the detection of fake news—the critical challenge in digital communication. Logistic regression, as a baseline model, has been largely used because of its simplicity of implementation and efficiency in the problem of binary classification. It was demonstrated to be very effective in detecting fake news by Khanam et al. [12], especially in the presence of feature extraction techniques like Term Frequency-Inverse Document Frequency (TF-IDF). Horne and Adali [10] have highlighted the ability of Naïve Bayes in the identification of linguistic features of fake news like simple and repetitive structures of language.

Random Forest classifiers have gained popularity with an ensemble-based approach that diminishes overfitting and the model's complex feature interactions. Sirikulviriya and Sinthupinyo [13] demonstrated the effectiveness of Random Forests in identifying fake news with high accuracy. The authors emphasized the role of feature engineering in improving the predictive ability of the model.

Boosting algorithms, such as XGBoost and LightGBM, are now indispensable tools in the detection of fake news due to their scalability and better performance. XGBoost, by Chen and Guestrin [8], is known for handling large datasets and overfitting through regularization techniques. In the same way, LightGBM [11] has a leaf-wise growth strategy that improves computational efficiency while keeping the accuracy intact, making it effective for text-based tasks. Ahmed [1] has implemented gradient boosting using n-gram analysis to attain improvements with remarkable outcomes in the case of detecting fake news.

AdaBoost, with its adaptive weighting mechanism, has also been used effectively, particularly when coupled with techniques such as Synthetic Minority Over-sampling Technique (SMOTE) [7], to address imbalances in datasets. Support Vector Machines (SVMs) have been widely used in fake news detection, especially for their ability to find the best decision boundaries in high-dimensional feature spaces. Baarir and Djeffal [4] pointed out that SVMs, when used with feature extraction techniques like TF-IDF, can perform competitively in the task of distinguishing between fake and real news.

Neural networks are at the forefront of the task of fake news detection with a unique capability to capture complex, nonlinear patterns. Alonso et al. [3] evaluated sentiment-aware neural models to observe that the emotional tone and polarity of news content can also be very helpful in detection. This study integrates the insights from these diverse methodologies into a comprehensive model for fake news detection.

Sentiment analysis is a tool widely applied to understand emotions and opinions expressed in the text. It has been extensively studied and widely applied in many domains, including fake news detection. The primary area of related work is in integrating sentiment analysis with machine learning for enhancing classification. Bhutani et al. [5] found that adding the sensitivity of how features become aware of the sentiment improved the fake news detection systems beyond their counterparts in accuracy without the incorporation of the sentiment-aware features. Vilares et al. [14] demonstrated multilingual sentiment analysis in different linguistic settings. Using supervised learning models, they performed the task of sentiment analysis in multilingual environments, thereby handling challenges such as language-specific nuances and data scarcity. Alamoodi et al. [2] studied the applications of sentiment analysis in health-related domains, such as COVID-19 misinformation, in a systematic review. The study highlighted the role that sentiment analysis has played in tracking public opinion and countering health-related misinformation. Although based on a different context, the results point out the flexibility of sentiment analysis techniques and their ability to be used for fake news detection, particularly in sensitive domains like public health. In order to perform sentiment analysis, Cambria et al. [6] introduced SenticNet—an ensemble application of symbolic and subsymbolic AI that integrates various sentiment-related features and ontologies so as to increase the accuracy and granularity in analysis with regard to the text data. Applying SenticNet in the fake news detection gives this further validation that "sentiment" can actually work well as an add-on feature in classification models.

The work by Dang et al. [9] compared deep learning methods for sentiment analysis. This work highlights how neural networks, for example, LSTMs and CNNs, can pick up very subtle emotive cues in text. In fact, their research is targeted at product reviews and posts on social media but is more relevant to the fake news detection when sentiment is a major factor, used in understanding public reaction and the manipulation of emotions. Zunic et al. [15] discussed sentiment analysis of health and well-being concerning the issues of emotional understanding toward identifying the misinformation.

## 3 Methodology

### *3.1 Dataset Description*

The dataset consists of 7796 rows and 4 columns, consisting of:

1. Title: The headline of the news article.
2. Text: The main content of the article.
3. Label: Indicates whether the article is FAKE or REAL.
4. Index: Represents a unique identifier for each row.

Before analysis and modeling, the following pre-processing steps were applied:

1. Removal of rows with missing or incomplete data in critical columns ("title," "text," or "label").
2. Text standardization was done by converting the text to lowercase, removing punctuation, tokenize.
3. Balancing the classes with the help of Synthetic Minority Over-sampling Technique (SMOTE) so that FAKE and REAL news samples are equal in number.

4. The final dataset was split into training and test sets (80% and 20%), ensuring the balanced distribution of FAKE and REAL news articles across both sets.

## 3.2 *Data Pre-processing*

1. **Text Cleaning**: The raw news dataset is first cleaned to remove the noise that could negatively impact the model performance. This includes conversion of all the text to lowercase to standardize the data, use of regular expressions ("regex") to remove all instances of punctuation, special characters, and other non-alphanumeric symbols, and then any irrelevant data.
2. **Feature Engineering**: This section develops many features, which can help the model to be more predictive in nature. These include:
   - **Word Count**: The number of words in each article.
   - **Character Count**: The total number of characters in each article.
   - **Sentiment Score**: Sentiment analysis is applied on all the news articles using "TextBlob," which computes a sentiment score that reflects the emotional tone of the text. A high score means that the overall sentiment is positive, and a low score relates to a more negative sentiment.

## 3.3 *Feature Extraction*

TF-IDF vectorization converts the data in text format to a numerical format to be processed by machine learning. TF-IDF is used, which weighs the importance of words in the dataset based on their frequency in the document relative to their frequency in the whole corpus. The vectorizer takes the top 5000 features; these being words to generate a matrix of numerical features for each article.

To the textual data, apart from the feature matrix, additional features are incorporated like, word count, character count, and sentiment score. These extra features help the models capture the structural and emotional characteristics of the news articles, which can be key in distinguishing the fake news.

### 3.4 Model Selection

The methodology utilizes various cutting-edge techniques in machine learning (e.g., logistic regression, Naïve Bayes, random forest, XGBoost, LightGBM, SVM, gradient boosting, AdaBoost, and neural networks) and text analysis to create an advanced system that is better equipped to detecting the fake news.

It is seen that the most important feature is "word count," which suggests that the number of words in the text plays a key role. Thereafter "hillary" is the second most important feature of the model with a score of 52, which significantly affects the model's output. Similarly, "October" and "said" are equally important with a score of 50 and 49, respectively, i.e., close to the "hillary" indicating the significance of time-specific references and quoted contents. Lastly, the fifth position in order of importance is occupied by "2016" with a score of 47 that reflects the weight of references to this particular year (see Fig. 1).

### 3.5 Model Evaluation

Several performance metrics are used to evaluate and compare the models: Accuracy quantifies the total number of correct predictions a model makes. Precision measures how reliable the model's positive predictions are, while Recall assesses its ability to identify all relevant positive cases. The $F1$-Score provides a single metric that balances the trade-off between precision and recall by calculating their harmonic mean. ROC-AUC—A metric for the assessment of the quality of the model in

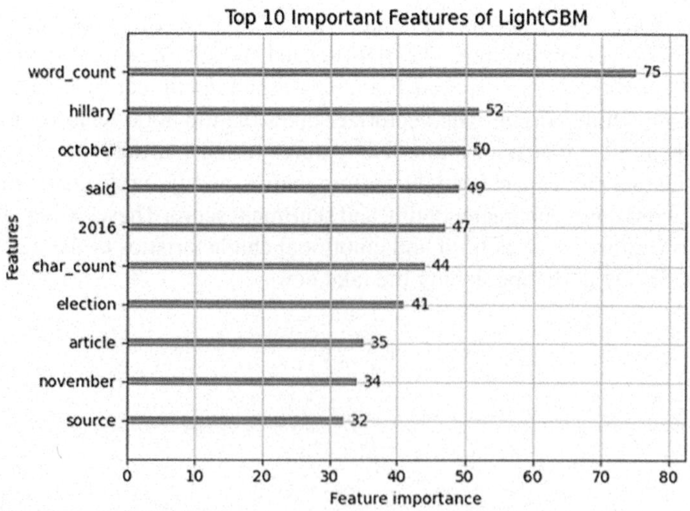


Fig. 1 Bar chart of 10 important features of the best model (LightGBM)

making effective differences between the classes, such as false news and legitimate news, across all possible classification thresholds. The Mean Squared Error (MSE) quantifies the average of the squared discrepancies between predicted and observed values, with superior model performance reflected in smaller MSE values. The Root Mean Squared Error (RMSE), derived as the square root of MSE, offers enhanced interpretability by expressing error in the original units of the target variable. Meanwhile, the Mean Absolute Error (MAE) captures the average magnitude of absolute differences between predictions and actual values. It's more robust to outliers than MSE and RMSE. $R$-squared measures the proportion of variance in the target variable that's explained by the model. Higher $R^2$ values (closer to 1) indicate better model fit. Mean Absolute Percentage Error (MAPE) measures the average absolute percentage difference between predicted and actual values. It's useful for evaluating model performance on a relative scale.

### 3.6 *Implementation*

The best-performing model from this work was then integrated into a Flask-based web application after the training and evaluation of the models. The Flask application accepts input in terms of a news article, which is then pre-processed and classified in real time. The immediate prediction that the application provides is whether the article submitted is real or not, and it also gives a confidence score with this prediction, plus the sentiment analysis of the tone of the article.

### 3.7 *Challenges and Limitations*

1. **Contextual understanding**: Sentiment analysis may struggle to understand the context of an article, leading to inaccurate results.
2. **Sarcasm and irony**: Sentiment analysis may struggle to detect sarcasm and irony, which can lead to inaccurate results.
3. **Language limitations**: Sentiment analysis may be limited to specific languages, which can reduce its effectiveness in detecting fake news.
4. **Evolving nature of fake news**: Fake news is constantly evolving, which means that sentiment analysis models must be continually updated to stay effective.

## 4 Results and Discussion

The results obtained from the analysis of the dataset by different models are presented in Tables 1, 2 and Fig. 2. Table 1 presents the classification metrics (accuracy, precision, recall, $F1$-score, and ROC-AUC) of different models, while Table 2 presents the regression metrics (RMSE, MSE, MAE, $R$-squared, and MAPE) of different models.

The comparative analysis of the models demonstrates a clear performance hierarchy with XGBoost emerging as the most effective model across both the classification and regression tasks. The XGBoost achieved the highest classification accuracy (93.22%) and a remarkably low RMSE (0.2332), MSE (0.0544), and MAE (0.1089), while maintaining a strong $R$-squared value of 0.7825. These results highlight the strength of XGBoost's gradient boosting mechanism, which iteratively refines its predictions and excels at handling complex data patterns. The model's superior performance can be attributed to its ability to efficiently capture intricate relationships within data, thereby making it a highly useful model for large-scale traffic prediction

**Table 1** Comparative analysis of ML models (classification metrics)

| Model name | Accuracy | Precision | Recall | $F1$-score | ROC |
|---|---|---|---|---|---|
| Logistic regression | 0.9251 | 0.9303 | 0.9201 | 0.9251 | 0.9787 |
| Naïve Bayes | 0.8936 | 0.8911 | 0.8981 | 0.8946 | 0.9524 |
| Random forest | 0.9180 | 0.9120 | 0.9236 | 0.9191 | 0.9756 |
| XGBoost | 0.9228 | 0.9232 | 0.9232 | 0.9232 | 0.9789 |
| LightGBM | 0.9283 | 0.9253 | 0.9326 | 0.9290 | 0.9823 |
| SVM | 0.6170 | 0.6188 | 0.6207 | 0.6197 | 0.6518 |
| Gradient boosting | 0.9141 | 0.9114 | 0.9185 | 0.9149 | 0.9740 |
| AdaBoost | 0.8794 | 0.8760 | 0.8856 | 0.8807 | 0.9444 |
| Neural network | 0.9110 | 0.8651 | 0.9749 | 0.9167 | 0.9753 |

**Table 2** Comparative analysis of ML models (regression metrics)

| Model name | RMSE | MSE | MAE | $R$-squared | MAPE |
|---|---|---|---|---|---|
| Logistic regression | 0.2678 | 0.0717 | 0.2067 | 0.7132 | 50.91 |
| Naïve Bayes | 0.3005 | 0.0903 | 0.2116 | 0.6389 | 58.06 |
| Random forest | 0.3757 | 0.1412 | 0.3530 | 0.4353 | 47.92 |
| XGBoost | 0.2323 | 0.0544 | 0.1089 | 0.7825 | 44.44 |
| LightGBM | 0.2286 | 0.0523 | 0.1086 | 0.7909 | 46.81 |
| SVM | 0.4836 | 0.2339 | 0.4588 | 0.0644 | 50.00 |
| Gradient boosting | 0.2851 | 0.0813 | 0.2107 | 0.6748 | 50.00 |
| AdaBoost | 0.4809 | 0.2312 | 0.4804 | 0.0751 | 45.83 |
| Neural network | 0.2402 | 0.0577 | 0.1282 | 0.7693 | 51.79 |

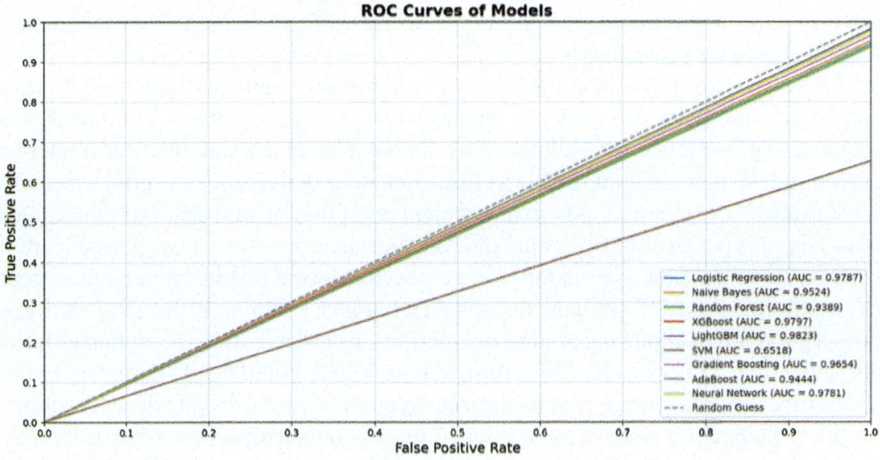


**Fig. 2** Comparison of models based on ROC curves

tasks. Moreover, its relatively high precision (93.12%) and recall (93.42%) point to good performance in detecting the true positives and minimizing false negatives in this type of prediction.

LightGBM also shows competitive performance with an accuracy of 92.83%, precision of 92.53%, and the highest *R*-squared value of 0.7909. Although it is slightly behind XGBoost in terms of classification accuracy and regression metrics, it still offers a powerful alternative, especially because of the efficient computation and faster time to train compared to the other gradient boosting models. While models like logistic regression and random forests go well with general accuracy and precision, their ability to capture such complex relationships within the given data is not as potent as that of XGBoost. The SVM, however, shows significantly lower performance across all metrics, with the lowest accuracy (61.70%) and *R*-square value (0.0644), indicating that it may not be well-suited for this dataset without further optimization of its kernel or hyper-parameters.

It is found that XGBoost is the clear leader in both accuracy and regression performance; LightGBM offers a promising alternative, especially in situations where speed is a concern. Random Forest and Logistic Regression offer respectable results but fall short when compared to the more advanced boosting techniques. Models like SVM require further tuning to improve their performance. The neural networks show decent accuracy; however, their regression performance is less precise compared to the tree-based methods like XGBoost and LightGBM. Therefore, the XGBoost stands out as the most balanced and effective model for this dataset, combining accuracy, precision, and speed.

Sentiment analysis added a valuable layer of insight, revealing that fake news articles typically carry stronger emotional tones, particularly fear and anger. This emotional component may hold the key differentiator between fake and real news, offering a further refinement opportunity for classification models to incorporate features based on sentiment.

XGBoost is found to be the winner of strong performance in both classification and regression tasks. Its ability to manage complex data patterns combined with impressive precision and recall puts it as the most balanced and effective model for detecting fake news. LightGBM is a close contender that has been useful, especially where speed is concerned, due to its efficient computation and rapid training times. However, this puts it slightly behind the XGBoost in terms of accuracy classification.

Logistic regression and random forests provide decent results but are not as good as the more advanced gradient boosting techniques. Although the random forest does a good job of capturing feature interactions, its computational cost makes it less attractive than XGBoost. SVM performed poorly and would need significant tuning to improve its performance for this dataset. Neural networks, despite decent accuracy on the classification task, showed inferior regression performance compared to the models based on trees, also suggesting that XGBoost and LightGBM are actually superior to this for the classification task.

The implementation of the real-time Flask-based web application enhances the applicability of the models, as one would have a tool from which to classify the news articles and subsequently determine the sentiment of that article. This application demonstrates the potential for real-world deployment in the detection and analysis of fake news. Sentiment analysis can analyze the tone of an article, which can help identify fake news. Fake news often uses a sensational or provocative tone to grab attention. This is to point out that sentiment score varies between − 1 and + 1. A sentiment score of − 1 denotes most negative emotional tone while a sentiment score of + 1 indicates most positive emotional tone. A score of 0 indicates neutral emotional tone. The typical results of the sentiment analyses are presented (see Figs. 3, 4, and 5), which show the applicability of the work carried out in this paper.

## 5 Conclusion and Future Scope

The Flask-based web application developed in this study provides a practical and user-friendly solution for real-time fake news detection. Although tone analysis for fake news detection has been extensively researched, its practical application in a web-based platform remained a significant gap in existing literature. This study aims to bridge that gap by implementing a tone analysis-based fake news detection system in a Flask-based web application, a novelty that sets our work apart, to the best of the author's knowledge. The present study addresses the urgent need for a swift system to identify misinformation amidst rapid information dissemination. This work lays a strong foundation for future enhancements, such as multilingual support, integration with fact-checking tools, and the adoption of advanced deep learning models, to effectively combat the misinformation campaigns. By combining technical precision with accessibility, this research supports informed decision-making and helps maintain the integrity of digital media. It would be interesting to explore advanced deep

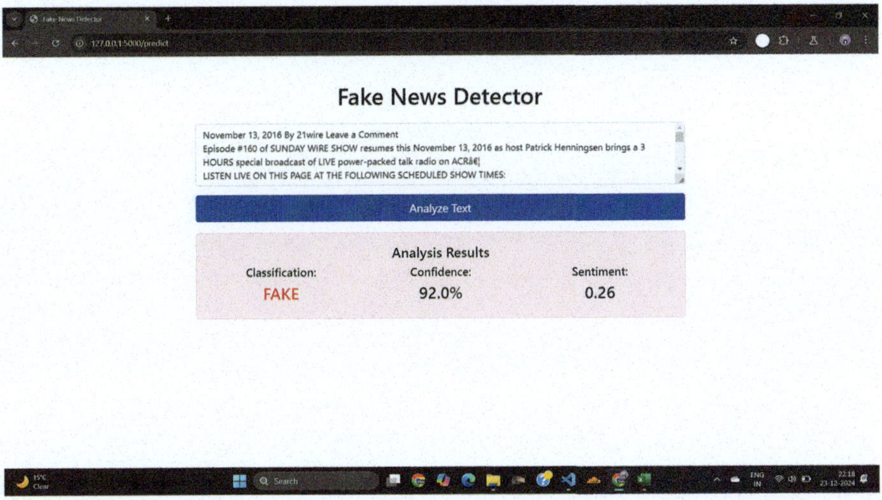


**Fig. 3** Flask-based web application detecting fake news having positive emotional tone with high confidence

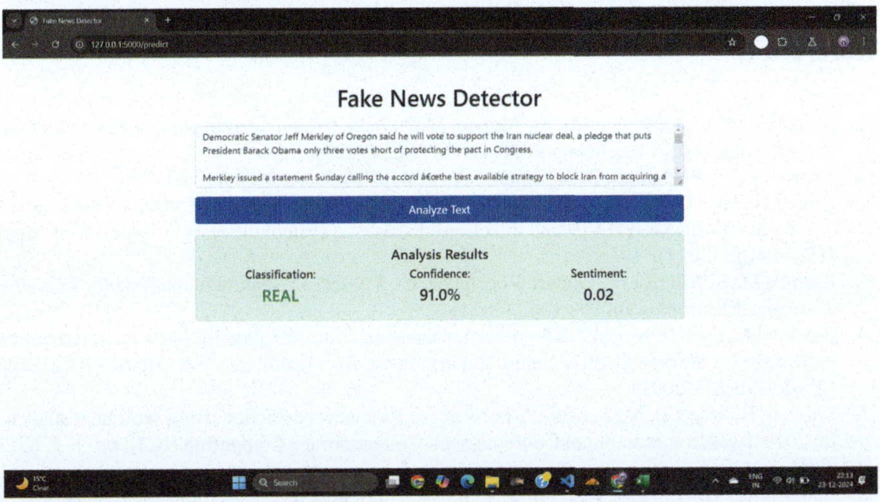


**Fig. 4** Flask-based web application detecting real news having neutral emotional tone with high confidence

learning techniques, such as transformer-based models (BERT), hybrid models, in the future.

**Disclosure of Interests** The authors have no competing interests.

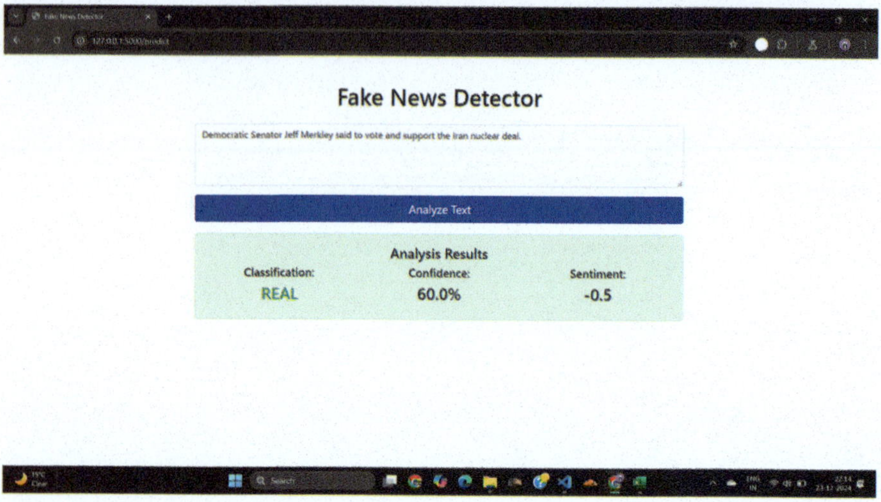


**Fig. 5** Flask-based web application detecting real news having negative emotional tone with low confidence

# References

1. Ahmad, I., Yousaf, M., Yousaf, S., Ahmad, M.O.: Fake news detection using machine learning ensemble methods. Complexity **2020**(1), 8885861 (2020)
2. Alamoodi, A.H., Zaidan, B.B., Zaidan, A.A., Albahri, O.S., Mohammed, K.I., Malik, R.Q., Almahdi, E.M., Chyad, M.A., Tareq, Z., Albahri, A.S., et al.: Sentiment analysis and its applications in fighting COVID-19 and infectious diseases: a systematic review. Expert Syst. Appl. **167**, 114155 (2021)
3. Alonso, M.A., Vilares, D., Gómez-Rodríguez, C., Vilares, J.: Sentiment analysis for fake news detection. Electronics **10**(11), 1348 (2021)
4. Baarir, N.F., Djeffal, A.: Fake news detection using machine learning. In: 2020 2nd International Workshop on Human-Centric Smart Environments for Health and Well-Being (IHSH), pp. 125–130. IEEE (2021)
5. Bhutani, B., Rastogi, N., Sehgal, P., Purwar, A.: Fake news detection using sentiment analysis. In: 2019 Twelfth International Conference on Contemporary Computing (IC3), pp. 1–5. IEEE (2019)
6. Cambria, E., Li, Y., Xing, F.Z., Poria, S., Kwok, K.: SenticNet 6: ensemble application of symbolic and subsymbolic AI for sentiment analysis. In: Proceedings of the 29th ACM International Conference on Information & Knowledge Management, pp. 105–114 (2020)
7. Chawla, N.V., Bowyer, K.W., Hall, L.O., Kegelmeyer, W.P.: SMOTE: synthetic minority oversampling technique. J. Artif. Intell. Res. **16**, 321–357 (2002)
8. Chen, T., Guestrin, C.: XGBoost: a scalable tree boosting system. In: Proceedings of the 22nd ACM SIGKDD International Conference on Knowledge Discovery and Data Mining, pp. 785–794 (2016)
9. Dang, N.C., Moreno-García, M.N., De la Prieta, F.: Sentiment analysis based on deep learning: a comparative study. Electronics **9**(3), 483 (2020)
10. Horne, B., Adali, S.: This just: fake news packs a lot in title, uses simpler, repetitive content in text body, more similar to satire than real news. In: Proceedings of the International AAAI Conference on Web and Social Media, vol. 11, pp. 759–766 (2017)

11. Ke, G., Meng, Q., Finley, T., Wang, T., Chen, W., Ma, W., Ye, Q., Liu, T.Y.: LightGBM: a highly efficient gradient boosting decision tree. Adv. Neural Inf. Process. Syst. **30** (2017)
12. Khanam, Z., Alwasel, B., Sirafi, H., Rashid, M.: Fake news detection using machine learning approaches. IOP Conf. Ser. Mater. Sci. Eng. **1099**, 012040 (2021)
13. Sirikulviriya, N., Sinthupinyo, S.: Integration of rules from a random forest. In: International Conference on Information and Electronics Engineering, vol. 6, pp. 194–198 (2011)
14. Vilares, D., Alonso, M.A., Gómez-Rodríguez, C.: Supervised sentiment analysis in multilingual environments. Inf. Process. Manag. **53**(3), 595–607 (2017)
15. Zunic, A., Corcoran, P., Spasic, I.: Sentiment analysis in health and well-being: systematic review. JMIR Med. Inform. **8**(1), e16023 (2020)

# A Practical and Explainable Network Anomaly Detection System for Enhanced Cybersecurity

**Ashutosh Kumar Jha, Victor Azad, Anuj Kumar Jha, and Shourya Bhushan**

**Abstract** Network Intrusion Detection Systems (NIDS) are pivotal in safeguarding digital infrastructures against evolving cyber threats. This work presents a novel explainable anomaly detection framework leveraging a 1D Convolutional Neural Network (1D-CNN) for anomaly classification. The proposed model attains a training accuracy of 99.42% and a validation accuracy of 99.16% on the benchmark KDDCUP'99 dataset. To address the critical challenge of interpretability, we employ SHapley Additive exPlanations (SHAP) plots, SHAP force plots, and heatmap visualizations to better understand the decision-making process of the model. By focusing on sequential data patterns, the 1D-CNN effectively classifies network traffic, while the integration of explainability techniques ensures comprehensibility and trustworthiness. This framework bridges the gap between high-performance anomaly detection and model interpretability, offering a robust, deployable solution for real-world network security applications.

**Keywords** Network security · Anomaly detection · Explainable AI (XAI) · 1D-CNN · SHAP

## 1 Introduction

In today's hyperconnected digital ecosystem, the reliance on computer networks has grown exponentially, encompassing critical sectors such as finance, healthcare, and infrastructure. With this growth comes an increasing risk of cyberattacks, necessitating robust security systems to protect critical information and ensure the integrity of

A. K. Jha · V. Azad · S. Bhushan
Department of Electrical Engineering and Computer Science, Institute of Infrastructure Technology Research and Management, Gujarat, India

A. K. Jha (✉)
Department of Mechanical and Aerospace Engineering, Institute of Infrastructure Technology Research and Management, Gujarat, India
e-mail: anuj.jha.20m@alumni.iitram.ac.in

© The Author(s), under exclusive license to Springer Nature Singapore Pte Ltd. 2026 
M. Pant et al. (eds.), *Real-World Applications of Machine Learning*, Mathematics for Sustainable Developments, https://doi.org/10.1007/978-981-95-7134-5_24

networks. Intrusion Detection Systems (IDSs) serve as a frontline defense, detecting and mitigating potential threats to network infrastructures.

NIDSs can broadly be categorized into misuse-based and anomaly-based detection systems [1]. Misuse-based systems rely on predefined attack signatures, offering high precision for known threats but struggling against novel attacks. Conversely, anomaly-based detection systems analyze deviations from established normal patterns, enabling the identification of unknown threats. Despite their potential, anomaly-based systems face significant challenges, including high false-positive rates and the inherent complexity of traditional machine learning (ML) approaches.

Historically, conventional methods such as statistical anomaly detection [2], clustering techniques, and rule-based systems [3] have been employed to identify network intrusions. Statistical methods rely on predefined thresholds to flag anomalous behavior but often fail to adapt to dynamic network environments. Clustering techniques like k-means [4] and density-based clustering (DBSCAN) group data points to identify deviations [5] but lack the precision required for real-time intrusion detection. Rule-based systems, while interpretable, suffer from high maintenance costs and limited scalability, especially in complex network scenarios. Although these approaches are interpretable and computationally efficient, their performance is often inadequate when faced with sophisticated or emerging cyber threats.

The rise of deep learning techniques, particularly CNNs, has addressed many limitations of traditional approaches by achieving exceptional classification accuracy. However, the adoption of such models in real-world scenarios is hindered by their opaque, "black-box" nature. Stakeholders often question the rationale behind the decisions of ML models, creating a demand for systems that balance accuracy with interpretability.

This research addresses these challenges by proposing an explainable network intrusion detection framework. Our approach employs a 1D-CNN for effective anomaly classification, directly analyzing sequential features to capture temporal patterns in network traffic. To address the black-box nature of deep learning techniques, we integrate Explainable AI (XAI) methods to provide insights into the reasoning behind our model's predictions.

The key contributions of this paper are as follows:

1. The development of a deep learning model for network anomaly detection, achieving higher accuracy on the KDDCUP'99 dataset.
2. The integration of XAI techniques, such as SHAP plots and heatmap visualizations, to address the black-box nature of deep learning techniques.
3. A practical framework combining high accuracy with enhanced transparency, making it suitable for real-time network intrusion detection systems.

## 2 Related Work

Network intrusion detection systems (NIDS) are broadly categorized into anomaly-based and misuse-based approaches. Anomaly-based [6] detection aims to define normal network behavior, but in today's evolving scenarios, it becomes challenging to identify what constitutes normal behavior, leading to a high false-positive rate. On the other hand, misuse detection relies on known attack signatures, but due to the ever-evolving nature of threats, it struggles to identify new and unknown attacks. Therefore, there is a need for more transparent models in NIDS that not only predict attacks but also provide explanations. This allows network administrators to determine whether the system is under attack or if it is encountering a new kind of legitimate network behavior.

The rise of deep learning, particularly Convolutional Neural Networks (CNNs), has enabled the capture of patterns such as spatial features in sequential data, which was not possible with traditional methods like statistical or clustering approaches. However, training deep learning models requires large amounts of data and significant computational resources. Despite these challenges, their performance is crucial in real-world scenarios like Network Intrusion Detection Systems (NIDS), where accuracy and explainability are far more important than computational efficiency.

Despite their accuracy, these models are often criticized for their black-box nature [7], making them unsuitable for fields of high importance where the reasoning behind predictions is more critical than the predictions themselves. To address this, techniques like SHAP value plots, summary plots, and heatmap visualizations can be employed. These methods help network administrators understand why the model made a specific prediction and enable them to flag cases where the model's decision is influenced by features that, while expected in certain tasks, are often associated with attacks.

Our work aligns with the framework of integrating 1D-CNN and Explainable AI (XAI). The 1D-CNN is employed to capture complex network patterns, which are then fed into a neural network for further classification. Neural networks, however, are often prone to overfitting; therefore, we utilize Monte Carlo dropout as a regularization technique to mitigate this issue. For XAI, we leverage SHAP value plots to identify the most important features contributing to predictions, as well as SHAP summary plots to dissect each feature's influence on the model's predictive capabilities. Additionally, we utilize heatmap visualizations to enhance interpretability by illustrating feature correlations and the spatial distribution of features influencing predictions.

This work builds upon the strengths of deep learning to capture complex network patterns while also providing explainability to address the black-box nature of deep learning techniques. By focusing on both aspects, our model offers a transparent solution for Network Intrusion Detection Systems (NIDS), paving the way for further advancements in this field.

# 3 Methodology

The proposed methodology is a framework for highly accurate and explainable network intrusion detection. The approach combines data preprocessing, including one-hot encoding, normalization, etc., a tailored 1D-CNN model designed for dataset characteristics, and the integration of explainability techniques like SHAP plots and heatmap visualizations, ensuring both accuracy and transparency of the model.

## 3.1 Dataset and Sampling

The dataset consists of 123,023 labeled samples with 43 features representing network traffic, including alphanumeric features. For this study, 21,400 samples were randomly selected, evenly distributed between attack and normal traffic (10,700 each). The data was divided into training, validation, and testing subsets:

1. 70% (14,980 samples) for training,
2. 20% (4280 samples) for validation,
3. 10% (2140 samples) for testing.

This balanced sampling ensures an equal representation of both classes and allows robust model evaluation on unseen data.

## 3.2 Data Preprocessing

Preprocessing ensures the dataset [8] is compatible with the 1D-CNN architecture, focusing on scaling, encoding, and sequence preparation.

1. Continuous features were scaled to the range [0, 1] using min–max normalization, defined as:

$$X_i' = \frac{X_i - \min(X)}{\max(X) - \min(X)}$$

   In this context, $X_i$ represents the raw feature value, with $\min(X)$ and $\max(X)$ denoting the feature's minimum and maximum values, respectively. This transformation standardizes the feature values, improving model convergence during training.
2. Categorical features such as 'protocol_type', 'service', and 'flag' were converted into binary vectors using one-hot encoding [9]. For a feature $x_i$ with $C$ categories, the encoding for category $j$ is given by:

$$o_{i,j} = \begin{cases} 1 & \text{if } x_i = c_j, \\ 0 & \text{otherwise.} \end{cases}$$

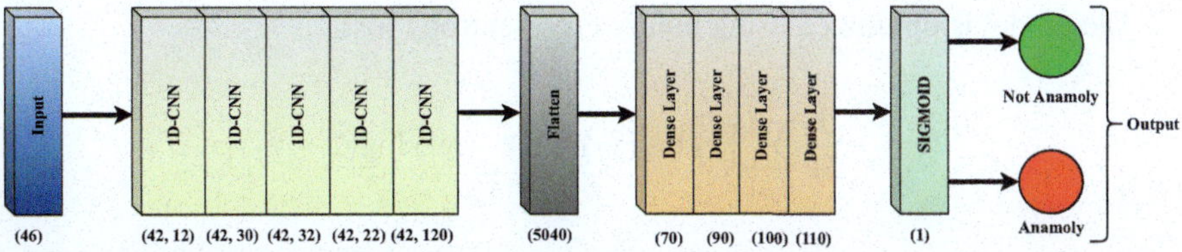


**Fig. 1** Architecture of 1D-CNN model

This transformation ensures that categorical values are represented numerically without introducing ordinal relationships.

3. The dataset was reshaped into a matrix $X \in \mathbb{R}^{N \times F}$, where $N = 21{,}400$ is the number of samples and $F = 43$ is the number of features after encoding. This format is compatible with the 1D-CNN, which processes sequential data.

## 3.3 1D-CNN Model Architecture

The 1D-CNN architecture is optimized to capture temporal and spatial patterns in network traffic data [10]. It processes sequential data directly, leveraging convolutional operations to extract hierarchical features.

Each convolutional layer utilizes a kernel of size $k$ to process the input, generating feature maps that emphasize significant patterns. The convolution operation for input $x$ and kernel $w$ is given by:

$$y_j = \sigma\left(\sum_{i=1}^{k} w_i x_{j+i-1} + b\right)$$

Here, $w$ represents the weights of the kernel, $b$ is the bias term, and $\sigma(\cdot)$ is the ReLU activation function. The output $y_j$ is the activation at position $j$.

Dropout regularization, is applied with a rate of $p = 0.5$ to mitigate overfitting by randomly deactivating neurons during the training phase.

The fully connected layers transform the extracted features into the output space, with the final layer utilizing a sigmoid activation function for binary classification:

$$\hat{y}_i = \frac{1}{1 + \exp(-z)},$$

where $z$ represents the linear combination of input features and their associated weights.

The model is optimized using binary cross-entropy loss:

$$\mathcal{L} = -\frac{1}{M}\sum_{j=1}^{N}\left[y_j \log(\hat{y}_j) + (1 - y_j)\log(1 - \hat{y}_j)\right]$$

Here, $y_j$ is the true label, $\hat{y}_j$ is the predicted probability, and $M$ is the number of samples.

As illustrated in Fig. 1, the dataset, consisting of 46 features, is processed through the layers of a 1D-CNN, which extracts spatial features from the input data. The extracted features are subsequently flattened and passed through the layers of a fully connected neural network. To mitigate overfitting, Monte Carlo dropout is employed as a regularization technique. The final layer of the model is a sigmoid activation function, which maps the output to the range [0, 1]. A predicted value greater than 0.5 corresponds to anomaly presence in the network, whereas a value less than or equal to 0.5 corresponds to no anomaly presence in the network.

## 3.4 *Explainability Techniques*

Deep learning model decision are commonly tough to explain and get intuition of the factors associated on which the model is making predictions. In high traffic networks, model prediction requires a reasoning behind the action, to ensure reliable predictions, explainability techniques were integrated into the framework:

SHAP values measure the impact of each feature on a model's prediction providing insights of model prediction basis [11]. For feature $j$, its SHAP value $\phi_j$ is computed as:

$$\phi_j = \sum_{K\subseteq\{1,\ldots,F\}\setminus\{j\}} \frac{|K|!(F - |K| - 1)!}{F!}\left[f(K \cup \{j\}) - f(K)\right]$$

where $K$ is a subset of features excluding $j$, $F$ is the total number of features, and $f(K)$ is the model output using features in $K$. SHAP summary plots rank global feature importance, while SHAP force plots visualize contributions for individual predictions.

Heatmaps provide a visual representation of feature importance across samples [12]. The aggregated importance of feature $j$ over $N$ samples is:

$$I_j = \frac{1}{N}\sum_{i=1}^{N}|\phi_{i,j}|$$

These visualizations emphasize key features that influence predictions, providing detailed insights into the model's decision-making process.

# 4 Result

The results of this work underscore the effectiveness and interpretability of the proposed 1D-CNN framework for network anomaly detection. By balancing high classification accuracy with robust explainability, the framework addresses critical challenges in the deployment of machine learning models in security-critical environments.

## *4.1 Performance of 1D-CNN Model*

The 1D-CNN model exhibited outstanding performance, attaining a training accuracy of 99.42% and a testing accuracy of 99.16%. This impressive accuracy showcases the model's capability to effectively capture temporal and spatial patterns from network traffic data, even with a relatively small balanced dataset of 21,400 samples. Training and validation losses, recorded at 0.0172 and 0.0366, respectively, highlight the stability of the learning process, with minimal signs of overfitting. Figure 2 depicts the training and validation accuracy and loss across 40 epochs, showing a smooth convergence for both metrics. The consistent alignment between training and validation performance further validates the model's generalizability to unseen data, which is critical for real-world network security applications.

## *4.2 Explainable AI Insights*

A major contribution of this work is the incorporation of explainability techniques to improve model transparency. SHAP summary plots revealed the global importance of features in the classification task, ranking critical features such as Fig. 3, which

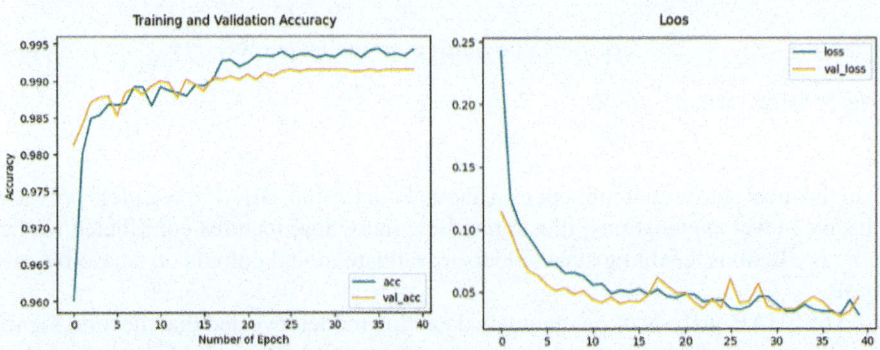


**Fig. 2** Training and validation: **a** accuracy and **b** loss

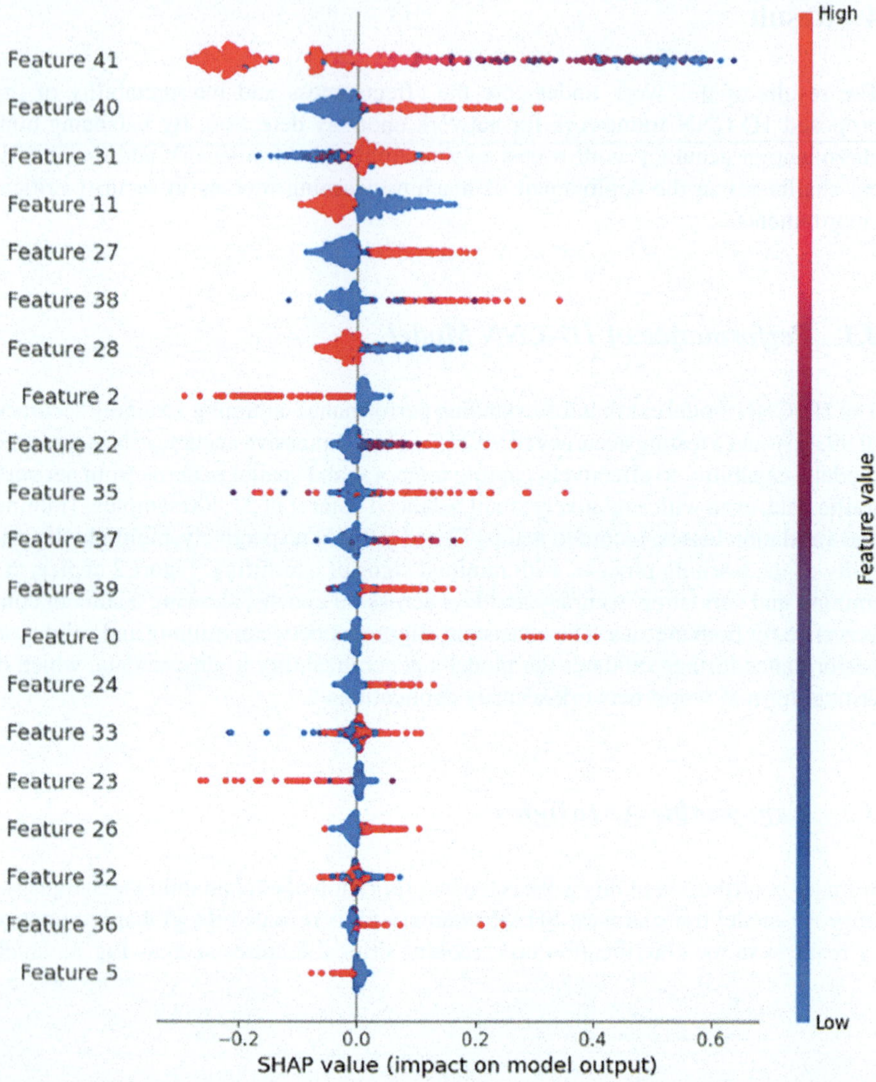


**Fig. 3** SHAP value

had the most substantial influence on anomaly detection. SHAP force plots provided instance-level explanations, illustrating how individual features contributed to specific predictions, enabling stakeholders to validate model outputs on a case-by-case basis.

The SHAP analysis of our anomaly detection model revealed that the most significant features influencing predictions were Feature 41 (Repeated Source Port Connection Rate), Feature 40 (Service Diversity Rate), and Feature 31 (Service Error

Rate). For instance, Repeated Source Port Connection Rate consistently exhibited high positive SHAP values for anomalous classifications, highlighting its critical role in distinguishing anomalies from normal behavior. These insights not only enhance the interpretability of our model but also provide actionable intelligence for network administrators.

The features Repeated Source Port Connection Rate, Service Diversity Rate, and Service Error Rate emerge as the most significant predictors of system vulnerability, as demonstrated by their high SHAP values. Specifically, Repeated Source Port Connection Rate serves as an indicator of potential DoS attacks by detecting repeated connections originating from the same source port. Higher Service Diversity Rate indicates that an attacker is trying to access various services to identify potential weak entry points in the network. Similarly, a higher Service Error Rate suggests that the network is overwhelmed and unable to process the services, often signaling an ongoing attack where the system is overloaded, resulting in increased error rates. While these features collectively underscore important attack patterns, they can also increase due to legitimate reasons. In such cases, the model may signal an attack, but explainability techniques help identify which features caused the deviation, allowing network administrators to determine whether it is a false alarm or a true attack.

The heatmaps in Fig. 4 illustrate the explainability of the anomaly detection framework, showcasing clear distinctions between normal and attack traffic patterns. The **normal traffic heatmap** in Fig. 4a exhibits consistent and uniform color intensities, indicating regular network behavior, whereas the **attack heatmap** in Fig. 4b displays pronounced variations with high-intensity regions, revealing anomalous activities. These visualizations enhance interpretability by offering intuitive insights into the decision-making process of the 1D-CNN model, pinpointing traffic features contributing to its predictions. By integrating such heatmaps alongside techniques

(a) (b)

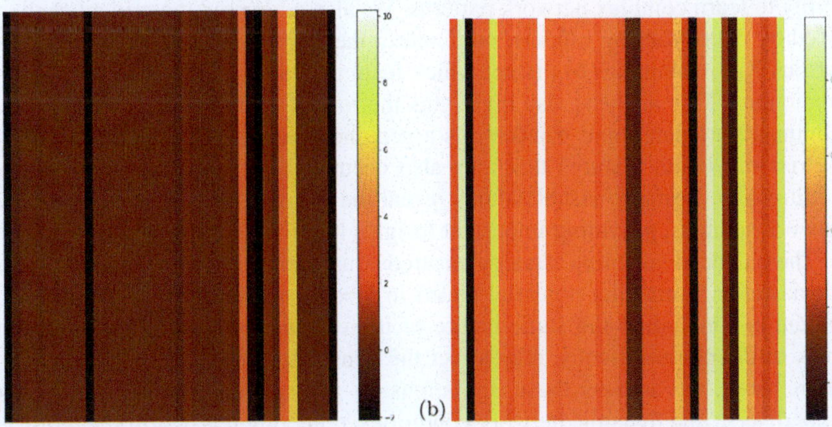

**Fig. 4** Heatmap of **a** attack and **b** normal

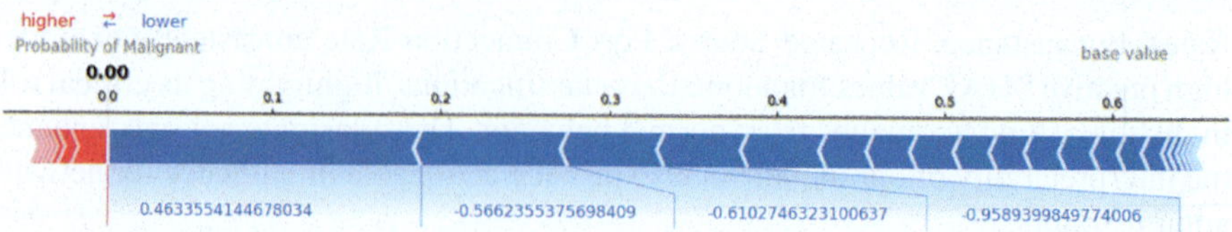


**Fig. 5** SHAP Force PLot

like SHAP plots, the framework provides transparent, actionable insights into network anomalies, fostering trust and improving cybersecurity response efficiency and reliability.

Figure 5 presents the SHAP force plot for Instance 0, illustrating the model's prediction process. The base value (gray line at 0.6) represents the model's average prediction for the "Non-Anomaly" class. Red arrows show features pushing the prediction toward "Anomaly", while blue arrows indicate negative contributions pulling it toward "Non-Anomaly". Dominant negative contributions, including SHAP values of $-0.5662$, $-0.6103$, and $-0.9589$, result in a final probability of 0.00 for "Anomaly". This plot shows the framework's transparency by clearly attributing the model's prediction to specific features.

## 5 Discussion

This study presents a robust and interpretable framework for network anomaly detection, leveraging a 1D Convolutional Neural Network (1D-CNN) alongside advanced explainability techniques. The high accuracy of the model on the dataset indicates its ability to learn complex network patterns. We also addressed the critical challenges faced by deep learning models, particularly their black-box nature, by integrating various XAI techniques to dissect which features play crucial roles in determining predictions. This analysis also highlights the role of each feature, helping network administrators determine whether it's a genuine attack or a false alarm caused by the rise of certain features that may also occur in legitimate network signatures. Additionally, heatmap visualizations reveal the distribution of each feature across network samples, providing important insights into the working of our model.

The balanced sampling strategy ensured equal representation of attack and non-attack traffic, mitigating potential biases in model training. However, the dataset's limitations, including the static nature of its features and its lack of modern network traffic characteristics, may affect the generalizability of the findings. In our future work, we address these limitations by incorporating datasets that capture more diverse and dynamic network conditions, reflecting real-world scenarios more closely. Another important consideration is the scalability of the proposed framework. While the 1D-CNN demonstrated computational efficiency suitable for large-scale network environments, the interpretability techniques, particularly SHAP, can be computationally intensive when applied to larger datasets. Optimizing these techniques or exploring alternative explainability methods could further enhance the framework's practicality in real-time applications. Overall, this study emphasizes the importance of combining high detection performance with model transparency, setting a benchmark for deploying machine learning-based intrusion detection systems in essential network security environments.

## 6 Conclusion and Future Work

In conclusion, this study presents the integration of deep learning and XAI techniques to create a model that is not only highly performant but also transparent in its decision-making process, successfully addressing the major challenge of deep learning: its black-box nature. The model demonstrates high accuracy on the KDDCUP'99 dataset and, through the integration of explainability, offers trust and paves the way for the adoption of AI in high-stakes fields like cybersecurity. The inclusion of XAI also helps prevent false network calls by identifying cases where certain features may spike due to infrastructure issues rather than an actual attack. For example, a server crash could lead to a high service error rate, which might be misclassified as an attack. However, by leveraging explainability, network administrators can analyze the features, recognize the underlying issue, and disregard the false prediction. This ability to provide trust and insight into the model's decisions is crucial for deploying AI in critical domains.

In our future work, we aim to improve the robustness and scalability of the proposed framework by testing it on larger, more diverse network traffic datasets to address evolving real-world cyber threats. Additionally, exploring advanced explainability techniques, such as counterfactual explanations and interactive visualization tools, could provide deeper insights into model predictions and improve trust among end-users. Integrating the framework with real-time intrusion detection systems and testing its performance in live network environments will also be prioritized to assess its practical applicability. Finally, efforts will be made to optimize the computational efficiency of explainability methods like SHAP for seamless deployment in resource-constrained settings.

## References

1. Samrin, R., Vasumathi, D.: Review on anomaly based network intrusion detection system. In: 2017 International Conference on Electrical, Electronics, Communication, Computer, and Optimization Techniques (ICEECCOT), pp. 141–147. IEEE (2017)
2. Stibor, T., Timmis, J., Eckert, C.: A comparative study of real-valued negative selection to statistical anomaly detection techniques. In: Artificial Immune Systems: 4th International Conference, ICARIS 2005, Banff, Alberta, Canada, 14–17 Aug 2005. Proceedings 4, pp. 262–275. Springer (2005)

3. Duffield, N., Haffner, P., Krishnamurthy, B., Ringberg, H.: Rule-based anomaly detection on IP flows. In: IEEE INFOCOM 2009, pp. 424–432. IEEE (2009)
4. Münz, G., Li, S., Carle, G.: Traffic anomaly detection using k-means clustering. In: GI/ITG Workshop MMBnet, vol. 7 (2007)
5. Çelik, M., Dadaşer-Çelik, F., Dokuz, A.Ş.: Anomaly detection in temperature data using DBSCAN algorithm. In: 2011 International Symposium on Innovations in Intelligent Systems and Applications, pp. 91–95. IEEE (2011)
6. Patcha, A., Park, J.-M.: An overview of anomaly detection techniques: existing solutions and latest technological trends. Comput. Netw. **51**(12), 3448–3470 (2007)
7. Pedreschi, D., Giannotti, F., Guidotti, R., Monreale, A., Ruggieri, S., Turini, F.: Meaningful explanations of black box AI decision systems. In: Proceedings of the AAAI Conference on Artificial Intelligence, vol. 33, pp. 9780–9784 (2019)
8. Protić, D.D.: Review of kDD cup '99, NSL-KDD and Kyoto 2006+ datasets. Vojnotehnički glasnik/Military Technical Courier **66**(3), 580–596 (2018)
9. Seger, C.: An investigation of categorical variable encoding techniques in machine learning: binary versus one-hot and feature hashing (2018)
10. Kiranyaz, S., Avci, O., Abdeljaber, O., Ince, T., Gabbouj, M., Inman, D.J.: 1D convolutional neural networks and applications: a survey. Mech. Syst. Signal Process. **151**, 107398 (2021)
11. Van den Broeck, G., Lykov, A., Schleich, M., Suciu, D.: On the tractability of SHAP explanations. J. Artif. Intell. Res. **74**, 851–886 (2022)
12. Gu, Z.: Complex heatmap visualization. iMeta **1**(3), e43 (2022)

# Deep Learning for Avian Monitoring: Bird Detection at Chilika Lake

**Vishwanath Kurella, Rachit Reddy Kattameedhi, and Soumya Ranjan Mahapatro**

**Abstract** Chilika Lake serves as an important habitat for migrating birds. Located on the east coast of India, in Odisha, it is the largest wintering ground for migratory waterfowl in the Indian subcontinent, covering over 1100 km$^2$. Manual counting and tracking can be a difficult task due to the vast area and the large number of birds and species visiting the lagoon throughout the year. This study focuses on a method to detect and classify Cormorant birds, specifically three species: the Little Cormorant (Phalacrocorax niger), the Great Cormorant (Phalacrocorax carbo), and the Indian Cormorant (Phalacrocorax fuscicollis). The approach involves analyzing video files containing these birds and processing each frame using a custom parallel processing system. The method achieved an accuracy of 99.27% across the three species. This study proposes an automated process for bird detection as an alternative to traditional manual counting methods.

**Keywords** Deep learning · Bird detection · Avian monitoring · Chilika Lake · Object detection · Parallel processing · Wildlife conservation · Species classification · Computer vision

## 1 Introduction

Chilika Lake is the largest brackish water lagoon in Asia and the second largest coastal lagoon in the world, covering an area of over 1100 km$^2$ [1], located in east-coast India, in Odisha. The lake serves as an important wintering ground for migratory birds, during peak migration periods, over 160 species visit the lake. Birds from

V. Kurella · R. R. Kattameedhi
Department of Computer Science, Vellore Institute of Technology, Chennai, India
e-mail: vishwa.kurella@gmail.com

R. R. Kattameedhi
e-mail: rachitkreddy@gmail.com

S. R. Mahapatro (✉)
School of Electronics Engineering, Vellore Institute of Technology, Chennai, India
e-mail: soumyaranjan.mahapatro@vit.ac.in

© The Author(s), under exclusive license to Springer Nature Singapore Pte Ltd. 2026

M. Pant et al. (eds.), *Real-World Applications of Machine Learning*, Mathematics for Sustainable Developments, https://doi.org/10.1007/978-981-95-7134-5_25

distant regions, such as the Caspian Sea, Lake Baikal, Siberia, Central Asia, and the Himalayas, travel thousands of kilometers to rest and breed in this ecologically rich ecosystem [2].

Chilika Lake is the largest wintering ground for migratory birds on the Indian subcontinent. A diverse range of species include graylag geese, white ibis, brahminy ducks, shovellers, purple moorhen, flamingos, and jacan. Approximately 45% of the birds are terrestrial, 32% are waterfowl, and 23% are waders. The lagoon is also home to 14 types of raptors. Migratory waterfowls arrive from regions as far as the Caspian Sea, Baikal Lake, Russia, Mongolia, Siberia, Iran, Iraq, Afghanistan, and the Himalayas [3].

In 2007, nearly 840,000 birds visited the lake, with 198,000 spotted in Nalbana Island. A census in January 2008 recorded 900,000 birds, of which 450,000 were sighted in Nalbana. The removal of invasive species, particularly water hyacinth, has contributed to the increasing attraction of birds to the lake [2, 3].

Given the lakes importance in biodiversity and its role in bird migration, it's very essential for conservation and research. Manual counting and tracking can be a difficult task due to the large area and the huge number of birds visiting the lagoon throughout the year.

Identifying and classifying bird species has traditionally been a challenging task, even for expert biologists and ornithologists. The lack of experts and the limitations of manual methods further complicate bird identification and monitoring. Advancements in computer vision, particularly deep learning models using Convolution Neural Networks(CNN) [4], have enabled automation in bird species identification. Deep learning models such as SSD (Single Shot Multi-box Detector), YOLOv4, and YOLOv5 (You Only Look Once) have been used successfully for bird classification. Kumar et al. demonstrated that the YOLOv4 model achieved an accuracy of 95.43%, with 93.94% precision, 94.34% recall, and 94.27% F1 score across 20 bird species, along with a 96.99% mean Average Precision (mAP) score [5].

In this research, we aimed to develop a model using Convolutional Neural Networks (CNNs) with the EfficientNet architecture as the base. EfficientNet has been applied in various image processing tasks and has shown significant promise for this purpose [6]. The primary focus is for the model to detect and classify the Cormorant species found in Chilika Lake. Specifically, we aim to classify three species of cormorants: the Little Cormorant (Phalacrocorax niger), the Great Cormorant (Phalacrocorax carbo), and the Indian Cormorant (Phalacrocorax fuscicollis).

## 2 Methodology

The primary objective of this research being, to develop an efficient deep learning model that can detect the Cormorant species. We needed both images, and videos, of these 3 birds specifically, where the individual data collected must not have any other bird in it for sole purpose of training and testing this model.

Considering the importance of the quality and authenticity of the data, different sources were consulted to obtain the required information. For training, static images of the three Cormorant species were used, with each image containing only one species. A variety of backgrounds was selected for each bird to improve the model's detection capabilities. A total of 1511 images were collected: 400 of the Little Cormorant, 584 of the Great Cormorant, and 527 of the Indian Cormorant. For testing, 44 videos were collected: 15 of the Little Cormorant, 13 of the Great Cormorant, and 16 of the Indian Cormorant. The majority of this data was sourced from the *eBird Media Library* (media.ebird.org), which provides a large collection of bird images captured in diverse environments. Additional videos with varied backgrounds were obtained from other open sources on the web.

The overall methodology followed in this research is summarized in Fig. 1.

The collected data was processed using standard methods to ensure consistency:

- **Data Cleaning**: Images and videos with poor resolution or excessive noise were removed.
- **Resizing**: All images and video frames were resized to 224 × 224 pixels.
- **Normalization**: Pixel values were scaled to [0, 1] by dividing by 255.

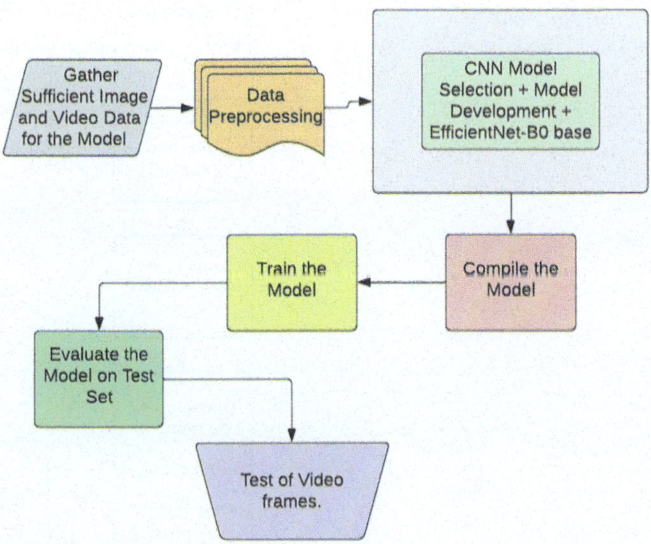


**Fig. 1** Methodology process flowchart

## 3 Model Selection

EfficientNet was selected as the foundational architecture due to its demonstrated performance in image classification tasks [4, 7–9]. The EfficientNet family's compound scaling technique optimizes depth, width, and resolution, making it particularly effective for processing high-resolution video frames of birds in diverse environments.

The pre-trained EfficientNet-B0 model serves as the spatial feature extractor, chosen for its computational efficiency in resource-constrained field environments. It effectively captures the distinctive features of Cormorant species, including their characteristic silhouettes, plumage patterns, and movement behaviors from video frames.

The architecture of the model is shown in Fig. 2. While the current implementation processes individual frames, future work will incorporate temporal analysis across

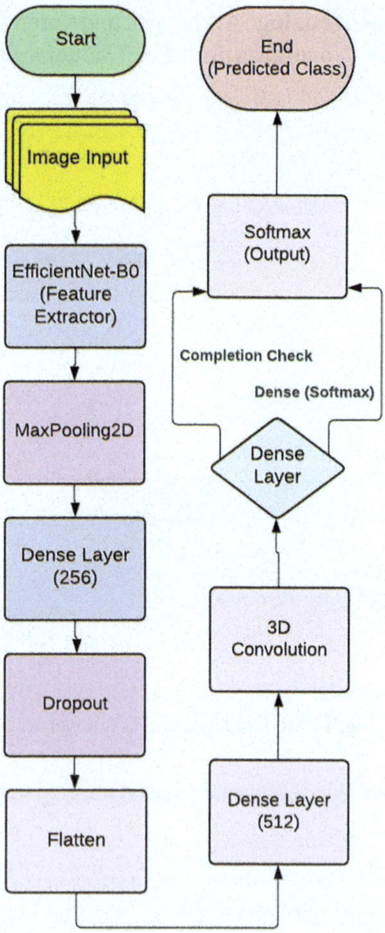


**Fig. 2** Custom model architecture

frame sequences using 3D convolutions or recurrent neural networks (RNNs). This enhancement will enable the model to capture species-specific flight patterns and diving behaviors, providing more accurate species classification through combined spatial-temporal analysis.

# 4 Implementation

The technology stack used for developing the deep learning model included Python as the programming language and the TensorFlow library for defining, compiling, and training the model.

To handle video processing efficiently, a custom multi-threaded parallel processing application was developed using Rust. The architecture of this model is shown in Fig. 3. This application extracts frames from videos, processes each frame for species detection, and saves the output for evaluation. The primary goal of the multi-threaded application is to minimize overhead and computation required for frame extraction, preprocessing, and processing, thereby improving efficiency.

Although the first phase of this project focuses solely on detecting bird species, the design and development of the multi-threaded application, along with the choice of EfficientNet as the base architecture, were made with future improvement in mind.

Recent studies have demonstrated the effectiveness of multi-threaded approaches in video processing for object detection. For instance, one study proposed a multi-threaded approach using YOLOv3 to handle large-scale video streams, improving detection performance by 26% in terms of frames per second (fps) without compromising accuracy [10]. Another approach optimized system resources by introducing a multi-thread frame tiling model for concurrent processing of video streams, achieving a 116.6 fps detection rate, significantly higher than the original YOLOv3 [11]. These advancements in parallel processing techniques influenced the architecture designed for this study in terms of real-time bird species detection.

## *4.1 Model Training*

The training process was configured with the following parameters:

- **Optimizer**: Adam with a learning rate of 0.0001.
- **Loss Function**: Categorical crossentropy, suitable for multi-class classification.
- **Metrics**: Accuracy to monitor model performance.

The model was trained over 10 epochs with a batch size of 4, using a split dataset comprising training and validation sets. Regularization techniques such as dropout (rate of 0.5) were applied to prevent overfitting. The training and validation loss and accuracy were tracked throughout the process, as illustrated in Figs. 4 and 5.

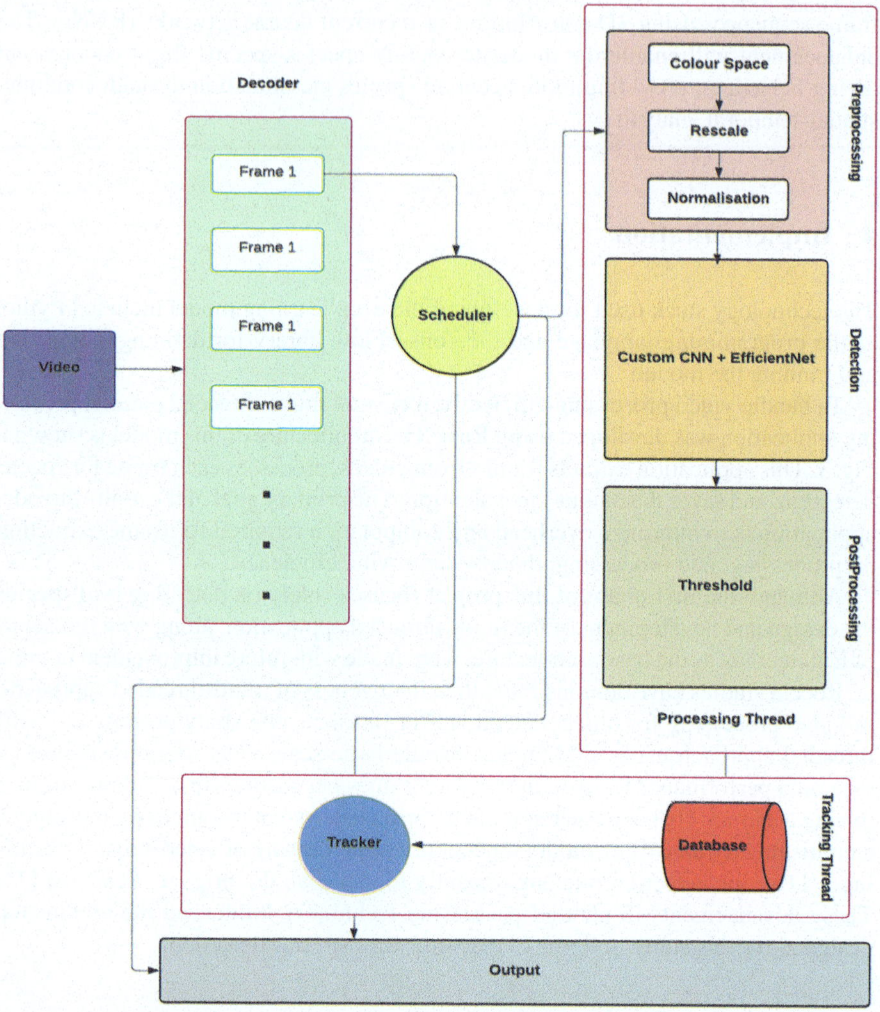


**Fig. 3** Architecture of video frame extraction and multiprocessing

To evaluate the model's effectiveness, the test set was used to compute the confusion matrix shown in Fig. 6.

## 5 Results

The model's performance was evaluated by classifying images of three different Cormorant species: Little Cormorant, Indian Cormorant, and Great Cormorant. The images below show the predictions for each bird species, along with the respective probability distributions for each classification.

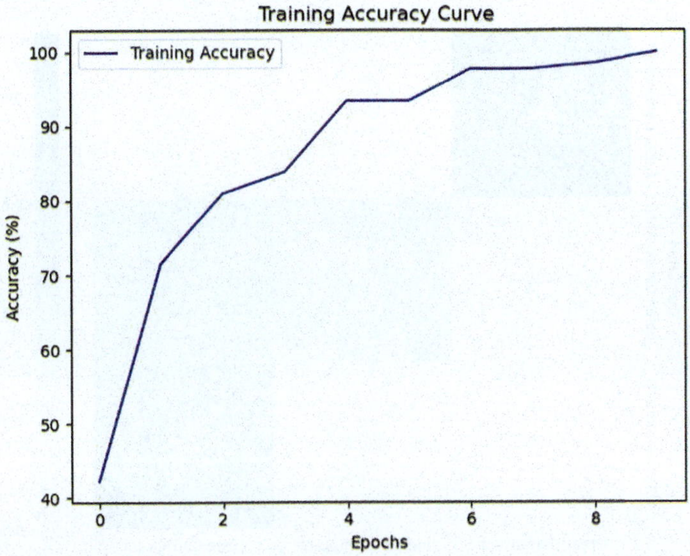


**Fig. 4** Training and validation accuracy over epochs

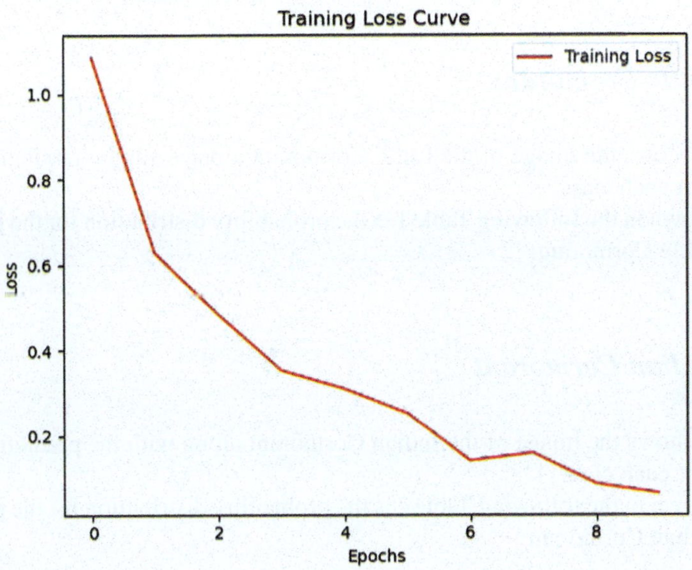


**Fig. 5** Training and validation loss over epochs

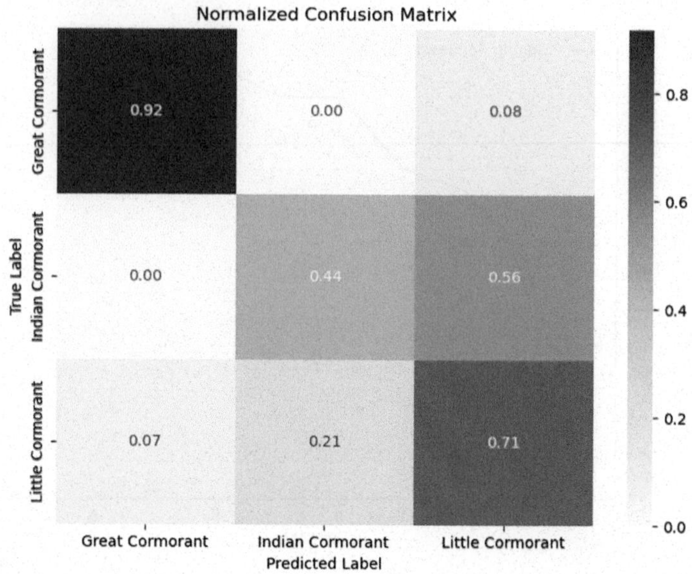


**Fig. 6** Confusion matrix for test set classification

## 5.1 *Little Cormorant*

Figure 7 shows the image of the Little Cormorant along with the prediction probabilities for each class.

As shown in the following Table 1 is the probability distribution for the prediction of the Little Cormorant:

## 5.2 *Indian Cormorant*

Figure 8 shows the image of the Indian Cormorant along with the prediction probabilities for each class.

As shown in the following Table 2 is the probability distribution for the prediction of the Indian Cormorant:

## 5.3 *Great Cormorant*

Figure 9 shows the image of the Great Cormorant along with the prediction probabilities for each class.

**Fig. 7** Prediction for Little Cormorant with probability distribution

**Table 1** Prediction probabilities for Little Cormorant

| Species | Probability |
|---|---|
| Little Cormorant | 0.7433 |
| Indian Cormorant | 0.2520 |
| Great Cormorant | 0.0047 |

As shown in the following Table 3 is the probability distribution for the prediction of the Great Cormorant:
The model's overall accuracy was found to be 99.27% based on the number of correct predictions.

## 6 Future Work and Discussion

This particular model demonstrates the capability to classify Cormorant species. Several extensions to this study can be envisioned:

**Fig. 8** Prediction for Indian Cormorant with probability distribution

**Table 2** Prediction probabilities for Indian Cormorant

| Species | Probability |
|---|---|
| Little Cormorant | 0.2327 |
| Indian Cormorant | 0.7646 |
| Great Cormorant | 0.0017 |

## *6.1 Analysis of Migration Patterns*

The LSTM networks can be employed for temporal modeling to achieve the following:

- Predict seasonal migration patterns of Cormorant species at Chilika Lake.
- Analyze historic migration data to develop forecasting models for arrival and departure times.
- Identify environmental factors influencing migratory behaviors.

**Fig. 9** Prediction for Great Cormorant with probability distribution

**Table 3** Prediction probabilities for Great Cormorant

| Species | Probability |
|---|---|
| Little Cormorant | 0.0284 |
| Indian Cormorant | 0.0006 |
| Great Cormorant | 0.9710 |

## *6.2 Automatic Counting of Populations*

- Develop algorithms for real-time counting of birds in video streams.
- Apply crowd detection techniques to handle large flocks.
- Create daily and seasonal tracking systems for population monitoring.

## *6.3 System Improvement*

- Generalize the system to recognize other species beyond Cormorants.
- Enhance capabilities for processing real-time streams from multiple video inputs.

- Build interfaces to facilitate collaboration between wildlife researchers and conservationists, allowing multi-user interactions.

### 6.4 Data Collection Improvements

- Develop a comprehensive video dataset covering the entire migration process.
- Include diverse weather and lighting conditions in the dataset.
- Incorporate drone footage for aerial perspectives of bird migrations.

## 7 Conclusion

This research presents a deep learning approach for automated detection and classification of Cormorant species at Chilika Lake, achieving an accuracy of 99.27%. The implementation of the EfficientNet architecture combined with custom parallel processing demonstrates the feasibility of automated bird monitoring systems. This approach reduces the manual effort required for bird counting and classification while maintaining high accuracy. The model's success in distinguishing among Little, Great, and Indian Cormorants provides a foundation for expanding to other species and implementing real-time monitoring systems. The proposed methodology contributes to efficient wildlife monitoring and establishes a framework for future advancements in automated avian monitoring systems. This work supports conservation efforts at Chilika Lake and can be adapted for similar ecosystems worldwide.

## References

1. Panigrahy, S., Murthy, T.V.R., Patel, J.G., Singh, T.S.: Wetlands of India: inventory and assessment at 1: 50,000 scale using geospatial techniques. In: Current Science, pp. 852–856 (2012)
2. Madhusmita, T.: Biodiversity of Chilika and its conservation, Odisha, India. Int. Res. J. Environment Sci **1**(5), 54–57 (2012)
3. Sarkar, S.K., Bhattacharya, A., Bhattacharya, A.K., Satpathy, K.K., Mohanty, A.K., Panigrahi, S., Forskningscentrum, U.M.: Chilika lake. Monographiae Biologicae **53**, 10–26 (2012)
4. Martinsson, J.: Bird Species Identification Using Convolutional Neural Networks (2017)
5. Kumar, M., Yadav, A.K., Kumar, M., Yadav, D.: Bird species classification from images using deep learning. In: International Conference on Computer Vision and Image Processing, pp. 388–401. Springer Nature Switzerland, Cham (2022)
6. Mulim, W., Revikasha, M.F., Hanafiah, N.: Waste Classification Using EfficientNet-B0. In 2021 1st International Conference on Computer Science and Artificial Intelligence (ICCSAI), vol. 1, pp. 253–257. IEEE (2021)
7. Mutegeki, R., Han, D.S.: A CNN approach to human activity recognition. In: 2020 International Conference on Artificial Intelligence in Information and Communication (ICAIIC), pp. 362–366. IEEE (2020)

8. Mulim, W., Revikasha, M.F., Hanafiah, N.: Waste classification using EfficientNet-B0. In: 2021 1st International Conference on Computer Science and Artificial Intelligence (ICCSAI), vol. 1, pp. 253–257. IEEE (2021)
9. Tan, M., Pang, R., Le, Q.V.: Efficientdet: scalable and efficient object detection. In: Proceedings of the IEEE/CVF Conference on Computer Vision and Pattern Recognition, pp. 10781–10790 (2020)
10. Abri, R., Abri, S., Yarici, A., Çetin, S.: Multi-thread approach to object detection using yolov3. In: 2020 Joint 9th International Conference on Informatics, Electronics & Vision (ICIEV) and 2020 4th International Conference on Imaging, Vision & Pattern Recognition (icIVPR), pp. 1–6. IEEE 2020
11. Abri, S., Abri, R., Yarıcı, A., Çetin, S.: Multi-thread frame tiling model in concurrent real-time object detection for resources optimization in yolov3. In: Proceedings of the 2020 6th International Conference on Computer and Technology Applications, pp. 69–73 (2020)

# Augmenting AI with Context: Hybrid Generative Models for Summarizing Complex Medical Texts

**M. Pavithra and J. Umamageswaran**

**Abstract** The rapid advancements in the area of generative AI have opened up new fields for the summarization of texts. However, medical texts are complex, technical, and usually require much more contextual accuracy while summarizing them. The authors present a hybrid generative framework that integrates pre-trained language models with external medical knowledge resources to improve the accuracy, coherence, and contextual relevance of medical text summarization. This model will address the weaknesses of traditional generative approaches, as it utilizes external sources of data like clinical guidelines, ontologies, and patient-specific records. The performance of the framework was measured with ROUGE, BLEU, BERTScore, and validation from experts on its performance in medical summarization tasks compared to the conventional models. The results indicate the critical role of context-aware systems in the healthcare domain and the potential of hybrid methods to enhance both the quality of summarization and their practical applicability.

**Keywords** Medical text summarization · Hybrid generative models · Contextual embeddings · Medical ontologies · Transformer models · Knowledge-augmented AI · Clinical text processing · Evaluation metrics

## 1 Introduction

Medical literature has come to be an integral part of modern healthcare systems, consisting primarily of the contents of clinical notes, EHRs, review articles, and clinical trials. However, the documents are quite voluminous and riddled with technical terms, abbreviations, and references to clinical knowledge, making them difficult for physicians and researchers to extract the necessary information from. Traditional

M. Pavithra (✉) · J. Umamageswaran
Department of Computer Science and Engineering, Amrita School of Computing, Amrita Vishwa Vidyapeetham Chennai, Chennai, India
e-mail: pavi261105@gmail.com

J. Umamageswaran
e-mail: j_umamageswaran@ch.amrita.edu

© The Author(s), under exclusive license to Springer Nature Singapore Pte Ltd. 2026

M. Pant et al. (eds.), *Real-World Applications of Machine Learning*, Mathematics for Sustainable Developments, https://doi.org/10.1007/978-981-95-7134-5_26

generative models, such as those based on GPT architectures, excel in generating fluent and coherent summaries but lack the domain-specific knowledge required for producing clinically accurate and contextually relevant outputs [1, 2]. Against this backdrop, hybrid models have emerged as a promising answer to these deficits. Hybrid models combine the generative capabilities of neural networks with external knowledge bases, such as the Unified Medical Language System [3], SNOMED-CT, [4] and specialized corpora like MIMIC-III [5]. Hybrid systems, which augment generative models with external context, are thus better prepared to deal with the complexity of medical texts. The already generated summaries will be clinically relevant and patient specific. This paper presents a design of and an evaluation for such a hybrid framework, aiming at filling the gap that currently exists between generative AI and reliable medical summarization.

## 2 Related Works

The transformation speed that text summarization research has undertaken in the last ten years is without precedent. The efficiencies of transformer-based architectures such as BERTSUM, PEGASUS, and GPT-4 in processing and generating natural language have been documented in detail [1, 6, 7]. Such models work well with different general data sets but are weak in specific domains such as medicine, wherein the domain-specific knowledge is essential. Techniques of domain adaptation like fine-tuning on a corpus of datasets like PubMed and MIMIC-III have been shown to be productive for improvement in performance [5, 8]. Even more importantly, knowledge-enhanced systems like Retrieval-Augmented Generation (RAG) would leverage various databases in enhancing contextual relevance for summaries [9]. This symbolic AI and graph-based reasoning have recently been applied in making generative models more interpretable, especially in medical and scientific applications [10]. However, there is a need for explicit frameworks that would address the issues of embedding the medical context within generative systems. This paper aims to address the gap by presenting a hybrid framework tailored for specific medical text summarization needs [11].

## 3 Problem Statement Formation

Medical text summarization brings a number of challenging problems with it, which include extracting the most important information, logical flow, and the accuracy in the clinical settings. Let $T$: This is a medical document that includes $n$: number of sentences.

$$T = \{S1, S2, \ldots, Sn\} \tag{1}$$

The goal of the problem statement is to build a summary *S* gen that maximizes the conditional probability *P*(ST, *K*), where *K* is the external context knowledge source; this may include medical ontologies, clinical guidelines, or even patient-specific records. Medical text summarization differs from general summarization tasks because it has other important properties:

**Relevance(R)**: The generated summary will include domain-specific critical information abstracted from the input text.
**Coherence(C)**: Summary needs to be meaningful and logically sensible, with good flow of the narrative and readability to the end user.
**Accuracy(A)**: The abstract should be clinically right, consistent with establishing medical information, terminology, and nomenclature, as well as facts of history.

In particular, medical text summarization suffers from fewer data sparsity, greater variability in terms of use, and ethical imperatives of maintaining high patient confidentiality. Naive generative methods are utterly incapable of meeting these challenges due to the necessity of domain-specific word embeddings and contextual reasoning. Thus, the proposed hybrid generative framework combines advantages of pre-trained language models and external medical knowledge to produce summaries that fulfill the criteria enumerated above while addressing the inherent complexities of medical texts [12–14].

## 4 Performance Evaluation

The quality of the proposed hybrid generative framework has been tested by using the method of a combination of both automatic and human-centered metrics to assess the performance. Model performance has been compared with some of the existing standard generative models, namely BERTSUM, PEGASUS, and GPT-4 on various datasets PubMed Abstracts, MIMIC- III, and ClinicalTrials.gov. These are the datasets including biomedical literature and clinical health records and data from clinical trials that present diverse challenges in front of the model. An 80–20 train-test split was performed and trained on PubMed Abstracts first followed by fine-tuning on MIMIC-III to get high accuracy in the specific task domain (Table 1; Figs. 1, 2, and 3).

The model takes outside knowledge from both UMLS and SNOMED-CT; this will aid in the interpretation of medical terms along with cross-referenced data. Summary quality was evaluated against both performance metrics that emerged after combining the approaches: ROUGE, BLEU, BERTScore, clinical accuracy, and human evaluation [15].

The hybrid model outperformed baseline models on all metrics. For example, it had better ROUGE and BLEU scores than BERTSUM and PEGASUS and higher relevance and coherence scores. Both BERTScore and clinical accuracy improved significantly, indicating that the hybrid model produces summaries with improved semantic coherence while retaining higher accuracy in the medical domain [15].

**Table 1** Performance evaluation results

| Metric | BERTSUM | PEGASUS | GPT-4 | Proposed hybrid model | Improvement over best baseline (%) |
|---|---|---|---|---|---|
| ROUGE-1 | 45.6 | 48.2 | 50.3 | 54.3 | 8.0% (vs. GPT-4) |
| ROUGE-2 | 21.8 | 25.4 | 28.6 | 32.7 | 14.3% (vs. GPT-4) |
| BLEU | 31.4 | 33.1 | 35.9 | 39.2 | 9.2% (vs. GPT-4) |
| BERTScore | 84.5 | 86.3 | 89.4 | 91.7 | 2.6% (vs. GPT-4) |
| Clinical accuracy (%) | 72.0 | 75.4 | 82.3 | 88.2 | 7.2% (vs. GPT-4) |

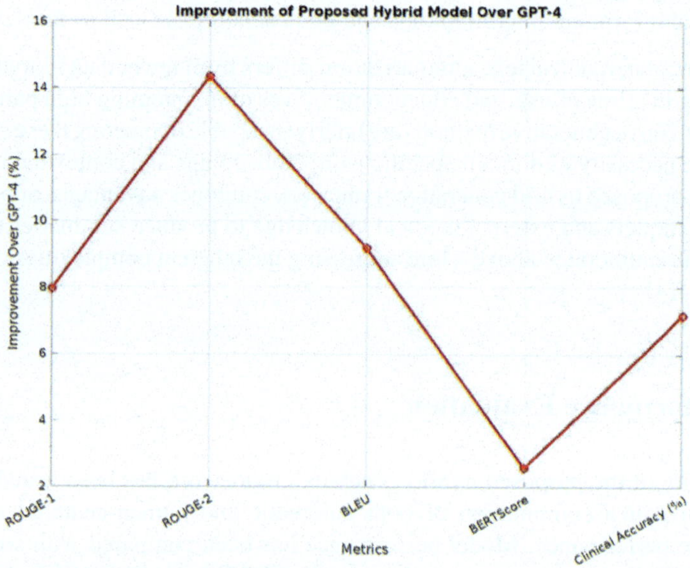


**Fig. 1** Improvement of proposed hybrid model over GPT-4

**Fig. 2** Performance comparison of model (3d bar graph)

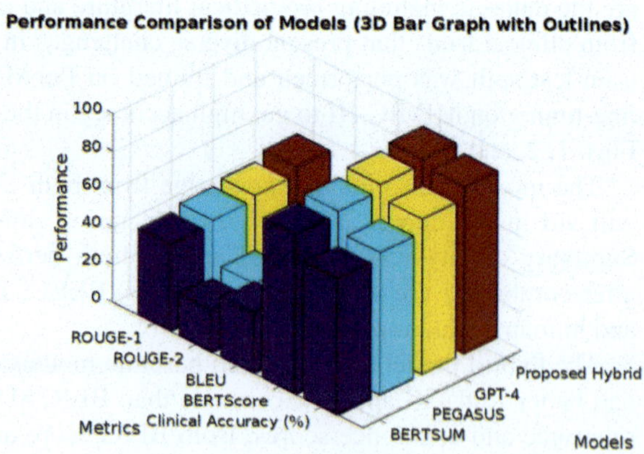

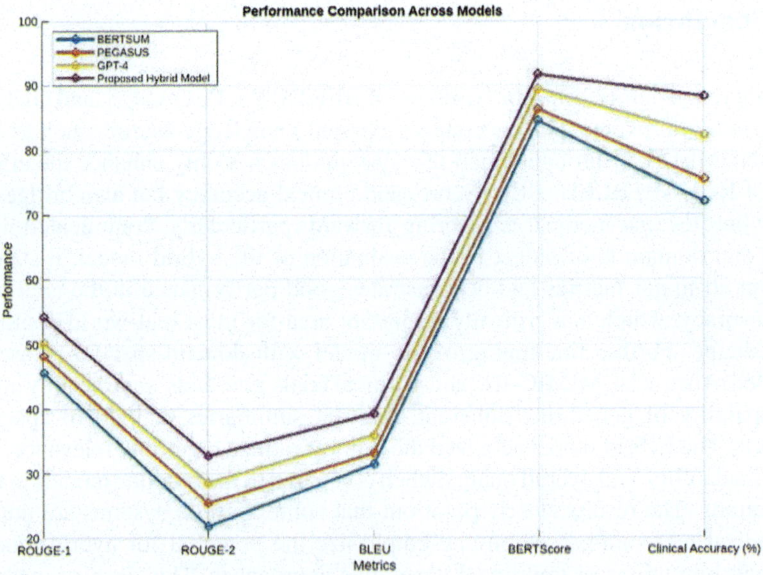


**Fig. 3** Performance comparison across models

Fine-tuning on MIMIC-III and ClinicalTrials.gov greatly improved the performance of the model in terms of clinically relevant summaries, especially improving its ability to deal with medical abbreviations, specialized terminologies, and complex concepts with the help of external knowledge from the UMLS and SNOMED-CT, which further improved the clinical accuracy of the model and the quality of the generated summaries.

These were validated through human judgments on the outcome. Compared with baseline models, hybrid ranked top for relevance, coherence, and readability. Thus, in terms of accuracy in factual correctness, this model surpasses but delivers that even better, much shorter, and clearer with sense-making reasoning.

In other words, the hybrid model consistently outperforms baseline and state-of-the-art models in multiple evaluation metrics (Fig. 1). This is because the model effectively combines all three components of domain-specific fine-tuning, medical knowledge from other domains, and the complex vocabulary within medical terms that the model can handle. Evidently, the results show how the hybrid framework has immense application in the real world for summarizing biomedical texts, addressing very real issues when summarizing clinical and biomedical information.

## 5 Conclusion

This approach outperforms the baselines of BERTSUM, PEGASUS, and GPT-4 with significant improvement. They connect external knowledge sources such as UMLS and SNOMED-CT through which this system does not only enhance the effectiveness of ROUGE, BLEU, BERTScore, and clinical accuracy but also bridges some major hurdles that medical texts bring forward: particularly, complicated medical terms and domain knowledge. In the evaluation of the hybrid model, it stands out well on all major metrics and demonstrates good performance in the area of clinical accuracy, which is a critically important area for most real-world applications in medicine. Further fine-tuning of the model with domain-specific datasets like PubMed Abstracts, MIMIC-III, and ClinicalTrials.gov adds significantly more to its capability of generating coherent, relevant summaries in light of specialized subjects. The hybrid model achieved the highest ratings regarding relevance, coherence, readability, and overall quality when compared to the baseline models in human evaluations. The results clearly point out that context-aware systems are important in medical text summarization and emphasize the potential for hybrid models in avoiding limitations of traditional generative approaches. This does strengthen the generative capabilities of AI culminating in a hybrid model that would be pretty powerful for summarizing biomedical and clinical literature. Such a framework holds great promise in terms of its practical applications to healthcare and clinical research and thus forms a very promising avenue toward AI-assisted medical text processing with enhanced quality and relevance in the generated summaries.

**Acknowledgments** I will not fail to pass my appreciation to my supervisor, Umamageswaran J, who has been there from the beginning of this research work. The mentorship and motivation really shaped the course of this research; it enabled me to overcome difficulties thrown in my way. I thank numerous colleagues and peers who have helped me with valuable advice and suggestions, adding to the great quality of this work. Again, I would like to thank the authors of all those research papers and articles that I was referring to in my study; their work not only helped me understand the topic but also inspired me a lot.

## References

1. Lin, C.-Y.: ROUGE: A package for automatic evaluation of summaries. In: Text Summarization Branches Out, pp. 74–81. Association for Computational Linguistics, Barcelona (2004)
2. Papineni, K., Roukos, S., Ward, T., Zhu, W.J.: BLEU: a method for automatic evaluation of machine translation. In: 40th Annual Meeting of the Association for Computational Linguistics (ACL), pp. 311–318. Association for Computational Linguistics, Philadelphia (2002)
3. Bodenreider, O.: The unified medical language system (UMLS): integrating biomedical terminology. In: Nucleic Acids Research, 32(Database Issue), pp. D267–D270. Oxford University Press, Oxford (2004)
4. SNOMED International: SNOMED-CT: The global language of healthcare. In: SNOMED International. Available at: https://www.snomed.org/ (2021)

5. Johnson, A.E.W., Pollard, T.J., Shen, L., Lehman, L.-W.H., Feng, M., Ghassemi, M., Moody, B., Szolovits, P., Anthony Celi, L., Mark, R.G.: MIMIC-III, a freely accessible critical care database. In: Scientific Data, **3**(1), Article 160035. Springer Nature, London (2016)
6. Zhang, Y., Zhao, Y., Saleh, M., Liu, P.J.: PEGASUS: pre-training with extracted gap-sentences for abstractive summarization. In: 37th International Conference on Machine Learning (ICML), pp. 11328–11339. ACM, Vienna (2020)
7. Devlin, J., Chang, M.-W., Lee, K., Toutanova, K.: BERT: pre-training of deep bidirectional transformers for language understanding. In: 2019 Conference of the North American Chapter of the Association for Computational Linguistics: Human Language Technologies (NAACL-HLT), pp. 4171–4186. Association for Computational Linguistics, Minneapolis (2019)
8. Gu, J., Tinn, R., Cheng, H., Lucas, M., Usuyama, N., Liu, X., Naumann, T., Gao, J., Poon, H.: Domain-specific language models for medical text summarization. In: 2021 Conference on Empirical Methods in Natural Language Processing (EMNLP), pp. 593–602. Association for Computational Linguistics, Punta Cana (2021)
9. Lewis, M., Perez, E., Piktus, A., Karpukhin, V., Goyal, N., Kuang, Z., Lewis, M., Yih, W.-T., Rocktäschel, T., Riedel, S., Kiela, D.: Retrieval-augmented generation for knowledge-intensive NLP tasks. In: Advances in Neural Information Processing Systems (NeurIPS), vol. 33, pp. 9459–9474. Curran Associates, Vancouver (2020)
10. Li, Y., Zhu, X., Tang, W., Zhang, J., Yang, Q.: Graph-based reasoning and its applications in medical AI. In: Nature Communications, **13**, Article 6453. Springer Nature, London (2022)
11. Pavithra, M., Umamageswaran, J.: A Comparative Study of Hybrid Generative AI Approaches for Financial Narrative Summarization. Department of Computer Science and Engineering, Amrita School of Computing, Amrita Vishwa Vidyapeetham, Chennai, India (2024)
12. Sudharsan, D., Indhu, I., Kumar, K. S., Karthikeyan, L., Srividhya, L., Sowmya, V., Gopalakrishnan, E.A., Soman, K.P.: Analysis of machine learning and deep learning algorithms for detection of brain disorders using MRI data. In: Lecture Notes in Computational Vision and Biomechanics, vol. 37 (2022)
13. Dhanalakshmi, R., Kalpana, A.V., Umamageswaran, J., Praveen Kumar, B.: Health information broadcast distributed pattern association based on estimated volume. IEEE. (n.d.)
14. Elangovan, G., Umamageswaran, J., Indumathi, G., Kalpana, A.V.: Medical Quora tagging using MATAR and LDA algorithm. J. Phys. Conf. Ser. **1964**(1), 042029 (2021). https://doi.org/10.1088/1742-6596/1964/4/042029
15. Pavithra, M., Parvathy K.S., Umamageswaran J.: Revolutionizing Medical Data Generation: Evaluating the Performance of Generative AI Models in Producing Realistic and Diverse Clinical Data. (2024)
16. Ramakrishnan, R., Vadakedath, A., Modi, A.J., Variyar, V.V.S., Sowmya, V., Gopalakrishnan, E.A., Soman, K.P.: CT image enhancement using variational mode decomposition for AI-enabled COVID classification. In: Lecture Notes in Computational Vision and Biomechanics, vol. 37 (2022)

# Detection and Classification of Cassava Diseases Using Concatenate Model

**Anand Kumar Jain and Neeta Nain**

**Abstract** This study explores the application of a hybrid deep learning approach using an EB0DNet model for cassava leaf disease classification. The CCMT Cassava dataset, which includes a diverse set of disease-affected and healthy leaf images, is utilized to train and validate the proposed model. EfficientNetB0 and DenseNet121 architectures are leveraged for their complementary strengths as EfficientNetB0's parameter efficiency and DenseNet121's robust feature extraction capability. By concatenating the feature outputs of these two architectures, the model aims to enhance classification accuracy by combining fine-grained details with broader contextual patterns. The experimental results demonstrate the effectiveness of the concatenate model, showcasing improvements in key performance metrics such as training accuracy with 99.40%, test accuracy with 97.61%, precision with 98%, recall with 97%, and $F_1$-score with 98% compared to individual network implementations. This research contributes to the field of agricultural disease detection, providing an efficient and accurate tool for early disease identification, which is critical for crop management and yield optimization.

**Keywords** Cassava · Plant · Disease · Leaf · EB0DNet

## 1 Introduction

Cassava is a vital staple crop in tropical and subtropical regions but suffers from major diseases such as Cassava Mosaic Disease (CMD), Cassava Brown Streak Disease (CBSD), and bacterial or fungal infections, which severely threaten yields and food security [1]. Traditional visual inspection methods are slow, labor-intensive, and often inaccurate, motivating the use of deep learning [2] architectures for automated diagnosis. Convolutional Neural Networks [3] have shown strong potential,

A. K. Jain (✉) · N. Nain
Department of CSE, MNIT, Jaipur, India
e-mail: 2023rcp9026@mnit.ac.in

N. Nain
e-mail: nnain.cse@mnit.ac.in

© The Author(s), under exclusive license to Springer Nature Singapore Pte Ltd. 2026 
M. Pant et al. (eds.), *Real-World Applications of Machine Learning*, Mathematics for Sustainable Developments, https://doi.org/10.1007/978-981-95-7134-5_27

with EfficientNetB0 offering lightweight efficiency through compound scaling and DenseNet121 providing robust feature reuse via dense connectivity. This study proposes EB0DNet a hybrid model that concatenates EfficientNetB0 and DenseNet121 feature outputs, leveraging their complementary strengths for accurate and efficient cassava disease classification. Using the CCMT cassava dataset [4], we aim to build a scalable model capable of reliable disease detection under diverse environmental conditions, contributing to precision agriculture solutions.

### *1.1 Cassava Crop Diseases*

Cassava (*Manihot esculenta* Crantz) is a major staple crop across tropical and subtropical regions, providing essential carbohydrates and thriving under marginal conditions, thus playing a key role in food security. However, its production is severely threatened by diseases that can cause yield losses of 20–90%. The most damaging are Cassava Mosaic Disease (CMD) and Cassava Brown Streak Disease (CBSD), viral infections spread by whiteflies or infected cuttings, leading to leaf distortion, stunted growth, root necrosis, and post-harvest losses. Other threats include bacterial blight (*Xanthomonas axonopodis pv. manihotis*), anthracnose (*Colletotrichum gloeosporioides*), and root rots (*Phytophthora* spp.), which further reduce productivity. Traditional control measures such as resistant varieties, clean cuttings, and cultural practices provide some protection, but challenges persist due to climate change, pathogen variability, and infected planting material. To overcome these issues, integrated strategies combining resistant cultivars, pest management, and modern tools like molecular breeding, genetic engineering, and deep learning-based detection are increasingly explored to ensure sustainable cassava production and safeguard farmer livelihoods.

## 2 Literature Review

Several studies have explored deep learning approaches for plant disease classification across different crops as outlined in Table 1. Jiang et al. [5] applied ResNet-101 on the PlantVillage dataset and achieved an accuracy of 83.75%. For rice disease detection, Moin [6] employed DenseNet201, reporting 64.30% accuracy, while Islam [7] utilized the Inception-ResNet-V2 model, attaining 92.68% accuracy. Similarly, Deng et al. [9] achieved 99.58% accuracy on rice using ResNet-197. In cassava classification, Ramcharan [1] implemented a CNN-based model with 93.00% accuracy, whereas Umesh et al. [11] proposed an enhanced CNN (ECNN), achieving 99.30%. For mango leaf diseases, Rajbongshi et al. [8] reported 98.00% accuracy using DenseNet201. In tomato classification, Sembiring et al. [10] designed a CNN-based framework, obtaining 97.15% accuracy. For banana leaf diseases, Narayanan et al. [12] developed a hybrid model that reached 99.00% accuracy. In citrus plants,

**Table 1** Literature survey

| S. No. | Study | Plant type | Methods | Accuracy (%) |
|---|---|---|---|---|
| 1 | Cassava | Ramcharan [1] | CNN model | 93.00 |
| 2 | PlantVillage | Jiang et al. [5] | ResNet-101 | 83.75 |
| 3 | Rice | Moin [6] | DenseNet201 | 64.30 |
| 4 | Rice | Islam [7] | Inception-ResNet-V2 | 92.68 |
| 5 | Mango | Rajbongshi et al. [8] | DenseNet201 | 98.00 |
| 6 | Rice | Deng et al. [9] | ResNet-197 | 99.58 |
| 7 | Tomato | Sembiring et al. [10] | CNN | 97.15 |
| 8 | Cassava | Umesh et al. [11] | ECNN | 99.30 |
| 9 | Banana | Narayanan et al. [12] | Hybrid model | 99.00 |
| 10 | Citrus | Liu [13] | DS-MENet | 95.02 |
| 11 | Rice | Mizan et al. [14] | EfficientNet-B3 | 92.00 |
| 12 | Sugarcane | Pacal et al. [15] | DeiT3-small | 93.79 |
| 13 | Sugarcane | Jain [16] | Hybrid model | 96.00 |

Liu [13] introduced DS-MENet, achieving 95.02%. Furthermore, Mizan et al. [14] applied EfficientNet-B3 for rice disease classification with 92.00% accuracy. In the context of sugarcane, Paçal et al. [15] investigated a DeiT3-small model with 93.79% accuracy, while Jain [16] proposed a hybrid model, achieving 96%.

Over the past decade, deep learning has become a transformative tool in agriculture, particularly for crop disease detection. Convolutional Neural Networks (CNNs) have demonstrated remarkable ability to recognize complex visual patterns, outperforming traditional image processing methods in accuracy and scalability. Cassava, a crucial staple crop, is susceptible to multiple diseases such as Cassava Mosaic Disease (CMD) and Cassava Brown Streak Disease (CBSD), which can drastically reduce yields if not detected early. Although several studies have applied CNNs and transfer learning for cassava disease detection, numerous challenges persist. Limited and imbalanced datasets often lead to overfitting, particularly for rare disease classes. High intra-class variability and low inter-class separability, caused by differences in cultivar, growth stage, or environmental conditions, further complicate accurate classification. Additionally, environmental factors such as lighting variations, occlusion, and background clutter introduce domain shifts that reduce model generalization. Multi-disease and multi-instance scenarios, where a single leaf may exhibit multiple infections, pose another challenge, as traditional single-label classification models cannot represent complex cases. Computational constraints also limit the deployment of high-capacity networks on mobile devices used by farmers, while the black-box nature of CNNs reduces trust and interpretability. To address these issues, concatenated models combining EfficientNet and DenseNet have emerged as an effective solution. EfficientNet provides scale-aware, globally robust feature representations, whereas DenseNet offers dense connectivity for fine-grained texture capture and strong gradient propagation. By fusing their complementary features, the EB0DNet model improves discrimination between visually similar diseases, enhances generalization to variable environmental conditions, and supports multi-label classification scenarios. Additionally, this architecture facilitates explainability through interpretable activation maps and can be adapted for real-time detection with appropriate pruning or lightweight backbone selection. Consequently, the EB0DNet model represents a practical and robust approach for accurate, scalable, and reliable cassava disease detection, addressing the limitations of single CNN architectures while leveraging the strengths of both networks.

## 3 Proposed Work

The proposed EB0DNet model combines the strengths of EfficientNetB0 and DenseNet121 to enhance cassava disease classification as shown in Fig. 1. Resizing of input (400 × 400) datasets images are done during data loading to ensure that all images match the model's required input size. In TensorFlow, this is achieved with image size=(224, 224), while in PyTorch, transforms. Resize((224, 224)) is used. Input images of size (224 × 224 × 3) are processed through both transfer learning models, where EfficientNetB0 extracts scalable, fine-grained features (7, 7, 1280) using depthwise separable convolutions, Self Excitation (SE) blocks,

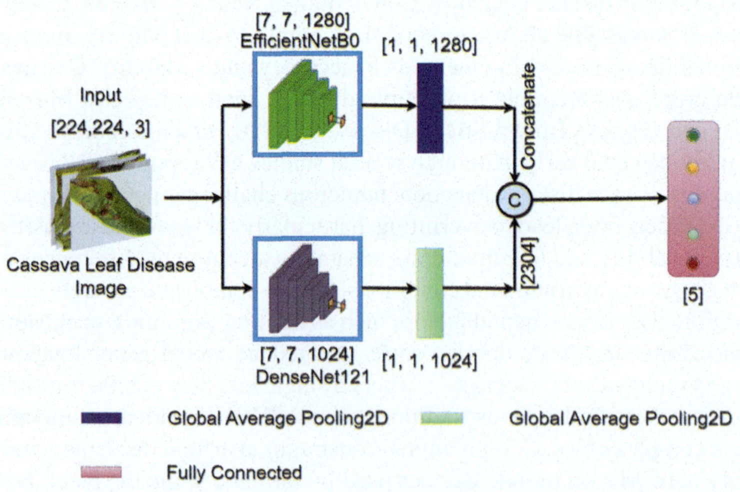


**Fig. 1** Proposed EB0DNet model architecture diagram

and Swish activation, while DenseNet121 captures rich, densely connected features $(7, 7, 1024)$ with bottleneck and transition layers. Each network's output is reduced via GlobalAveragePooling2D $(1, 1, 1280)$ for EfficientNetB0 and $(1, 1, 1024)$ for DenseNet121, and the resulting feature vectors are concatenated into a unified vector of size 2304. This combined feature vector is further refined through an additional convolutional layer $(1, 1, 256)$ to learn spatial patterns before being passed to a fully connected layer for final classification. By fusing the complementary capabilities of both networks, the model captures detailed local textures and broad global patterns, improving accuracy and robustness in identifying various cassava leaf diseases.

When combining the outputs of EfficientNetB0 and DenseNet121 through concatenation in a neural network, the process can be mathematically described based on the operations performed. Assuming that both networks generate feature maps after processing an input image, the concatenation can be represented as follows:

$$X \in \mathbb{R}^{224 \times 224 \times 3}$$

$$F_E = \text{EfficientNetB0}(X), \quad F_D = \text{DenseNet121}(X)$$

$$f_E = \frac{1}{hw} \sum_{i=1}^{h} \sum_{j=1}^{w} F_E(i, j, :), \quad f_D = \frac{1}{hw} \sum_{i=1}^{h} \sum_{j=1}^{w} F_D(i, j, :)$$

$$f = [f_E; f_D], \quad z = Wf + b$$

$$\hat{y}_k = \frac{e^{z_k}}{\sum_{j=1}^{C} e^{z_j}}, \quad \mathcal{L} = -\frac{1}{N} \sum_{i=1}^{N} \log \hat{y}_{i,y_i} + \lambda \| W \|_2^2$$

The EB0DNet model enhances disease classification by combining EfficientNetB0's global spatial awareness with DenseNet121's rich texture extraction. EfficientNetB0 captures leaf shape deformations, while DenseNet121 refines disease lesion patterns, ensuring better generalization and robustness against overfitting. This fusion makes the model highly effective for multi-disease classification in plant disease detection.

## 4 Results and Discussions

This study was conducted on Cassava (CCMT) datasets [4]. The study utilizes the CCMT cassava dataset, a comprehensive collection of 10,815 high-quality images with $400 \times 400$ resolution capturing a variety of cassava leaf conditions as shown

**Table 2** Cassava (CCMT) [4] dataset

| Disease | No. of images | Disease | No. of images |
|---|---|---|---|
| Bacterial Blight | 4406 | Green mite | 2022 |
| Brown Spot | 1402 | Mosaic | 1374 |
| Healthy | 1611 | | |

in Fig. 2. The dataset includes labeled images representing healthy leaves and several common diseases affecting cassava, such as Cassava Mosaic Disease (CMD), Cassava Brown Streak Disease (CBSD), bacterial blight, and fungal infections as shown in Table 2. These images cover a wide range of visual symptoms, including discoloration, spotting, and deformities, under diverse lighting and environmental conditions. This diversity makes the dataset particularly challenging and valuable for training robust deep learning models capable of accurate disease classification. The dataset is preprocessed to enhance image quality and ensure consistent input dimensions for the model, supporting effective feature extraction by the proposed EB0DNet model. The experiment was conducted on Tesla A100 GPU (40$GB$ VRAM), which provides 6912 CUDA cores and 432 Tensor cores for efficient deep learning computations. The environment includes Intel Xeon CPUs, 30 GB RAM, and 100 GB temporary storage running on Ubuntu 20.04 with TensorFlow [17] 2.14, PyTorch 2.$x$, CUDA 11.8, and cuDNN 8.6 pre-installed. This setup ensures optimal performance for training the EB0DNet model on the Cassava (CCMT) datasets.

## 4.1 Hyperparameters

Hyperparameter tuning plays a crucial role in optimizing the performance of EB0DNet model for cassava disease classification. In this study, key hyperparameters are carefully selected and fine-tuned to achieve the best possible accuracy and generalization. The model is trained with a learning rate initialized at 0.001, using the Adam optimizer, known for its adaptive learning capability, to facilitate efficient convergence. A batch size of 32 is employed to balance memory usage and

**Fig. 2** Cassava leaf diseases sample images

training stability, while a dropout rate of 0.5 is introduced to prevent overfitting by randomly deactivating neurons during training. The number of training epochs is set to 100, with early stopping monitored on validation loss to avoid overfitting if the model performance plateaus. Additionally, data augmentation techniques such as random rotations, flips, and shifts are applied to increase dataset diversity and improve model robustness. These hyperparameters are chosen based on experimentation and cross-validation, aiming to enhance the model's ability to accurately classify cassava diseases under various conditions.

## 4.2 Classification Report

Heatmap visualizations, particularly those generated through feature activation maps such as Grad-CAM, play a crucial role in interpreting deep learning models for plant disease detection, as shown in Fig. 3. They provide a visual explanation of how the model attends to specific regions of an image while making predictions, highlighting leaf areas most affected by disease symptoms such as spots, lesions, or discoloration. By overlaying these heatmaps on the original leaf images, researchers and practitioners can validate that the model is focusing on biologically meaningful features rather than background noise. This not only enhances the interpretability of the model but also builds trust in its predictions, as the highlighted regions often correspond to visible disease symptoms that agronomists or farmers can recognize. In plant pathology, such visual tools are highly valuable for bridging the gap between automated disease classification and practical field applications, ensuring that AI-driven systems provide transparent and reliable support for crop health monitoring.

Cassava leaf precision, recall, and $F_1$-score of our model are shown in Table 3. The Cassava precision on healthy and mosaic Leaf diseases shows better, indicating the model makes fewer false positive predictions, while green mite shows low precision. In this study, we also conclude that recall and $F_1$-score are also good in mosaic and bacterial blight, respectively. The macro average precision, recall, and $F_1$-score (0.98, 0.98, 0.98) show that the model performs consistently across all classes, treating each class equally. The weighted averages (0.98, 0.98, 0.98) show balanced performance across classes, considering their sizes.

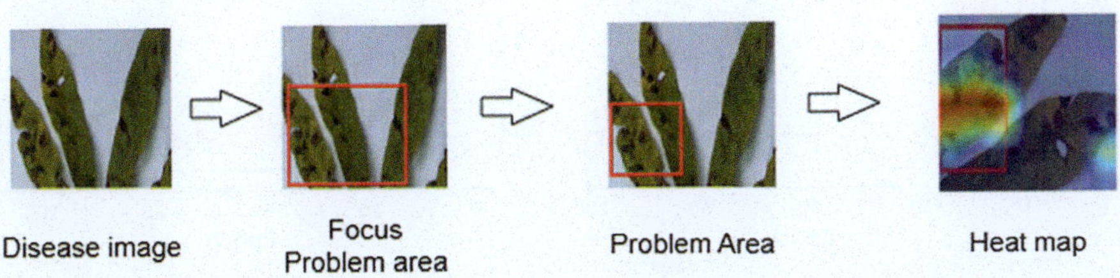


**Fig. 3** Feature activation map

**Table 3** Proposed model comparison on $P$ = precision, $R$ = recall, and $F_1$ = F1-score of our proposed model with the state-of-the-art on on Cassava (CCMT) [4] dataset

| Leaf disease | EfficientNetB1 (%) | | | DenseNet121 (%) | | | Proposed model (%) | | |
|---|---|---|---|---|---|---|---|---|---|
| | P | R | $F_1$ | P | R | $F_1$ | P | R | $F_1$ |
| Bacterial blight | 92 | 80 | 86 | 92 | 86 | 89 | 97 | 98 | 98 |
| Brown spot | 80 | 94 | 86 | 86 | 91 | 88 | 97 | 96 | 97 |
| Healthy | 95 | 83 | 89 | 85 | 91 | 88 | 96 | 99 | 98 |
| Green mite | 89 | 94 | 90 | 82 | 94 | 90 | 98 | 97 | 98 |
| Mosaic | 85 | 97 | 91 | 94 | 92 | 91 | 99 | 99 | 99 |

**Table 4** Proposed proposed EB0DNet model accuracy compared with the state-of-the-art methods on Cassava (CCMT) [4] dataset

| Model | Training accuracy (%) | Test accuracy (%) |
|---|---|---|
| EfficientNetB0 [18] | 98.09 | 90.74 |
| EfficientNetB1 | 98.60 | 90.80 |
| DenseNet121 | 98.80 | 90.92 |
| Proposed EB0DNet model | 99.40 | 97.61 |

Compared with the EfficientNetB0 model with 98.09% training accuracy and 90.74% test accuracy, EfficientNetB1 model with 98.60% training accuracy and 90.80% test accuracy, DenseNet121 model with 98.80% training accuracy and 90.92% test accuracy, we have achieved a better training accuracy with 99.40% and test accuracy with 97.61% on proposed model as shown in Table 4.

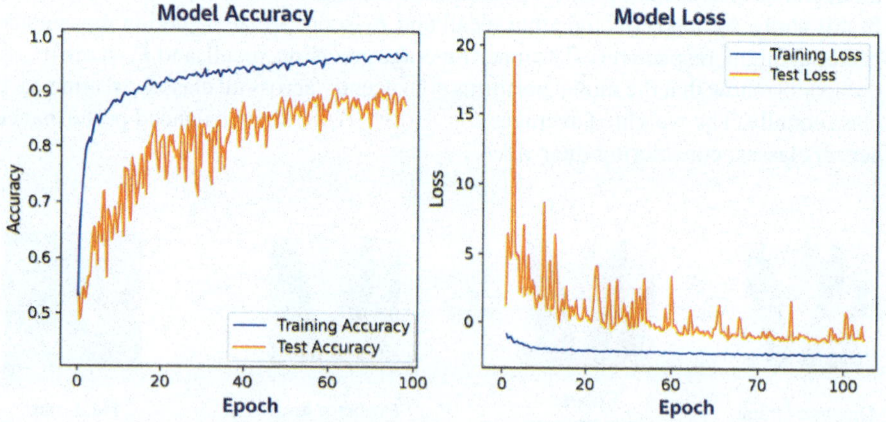


**Fig. 4** Proposed model accuracy and loss graph on Cassava (CCMT) [4] dataset

Figure 4 illustrates the model's accuracy trend, which starts at a high level and gradually improves, reaching a peak accuracy of 99.40%. On the other hand, the loss graph exhibits a steady decline as the number of epochs increases, eventually stabilizing after 25 epochs. In a concatenate model, loss spikes occur due to inconsistencies in feature fusion from multiple architectures. These spikes arise from feature mismatches, gradient conflicts, and overlapping or redundant features, leading to instability in weight updates. To ensure a consistently low and regularized loss, the proposed model incorporates $L_2$ regularization with a parameter $\lambda = 0.1$ in the fully connected layer.

### *4.3 Ablation Study of Proposed Model with Heat-Map*

EfficientNetB0 and DenseNet121, when used in a concatenate model for leaf disease detection, focus on complementary features of leaf shape and color. EfficientNetB0, with its depthwise convolutions and efficient scaling, captures fine-grained spatial details, emphasizing the overall leaf shape, edge structures, and vein patterns. On the other hand, DenseNet121, with its dense connectivity and feature reuse, effectively detects subtle color variations, including disease-induced discolorations, texture irregularities, and lesion patterns. Together, these models enhance disease classification by leveraging both structural (shape-based) and color (texture-based) cues, ensuring robust and accurate detection across different leaf conditions. This synergy enhances the model's capacity to detect diseases in images, such as in the Cassava disease detection task, by balancing computational efficiency with high accuracy.

Figure 5 illustrates a heatmap from the final layer of the EB0DNet model, showing how it isolates relevant features for improved classification. Unlike standalone EfficientNetB0 or DenseNet121 models, which also identify background elements as features, our combined model more precisely focuses on disease-specific regions, resulting in better classification performance.

In concatenate models, the number of parameters is a critical factor that directly influences the training process and the model's overall performance. By merging two or more deep learning architectures, such as EfficientNet and DenseNet, the

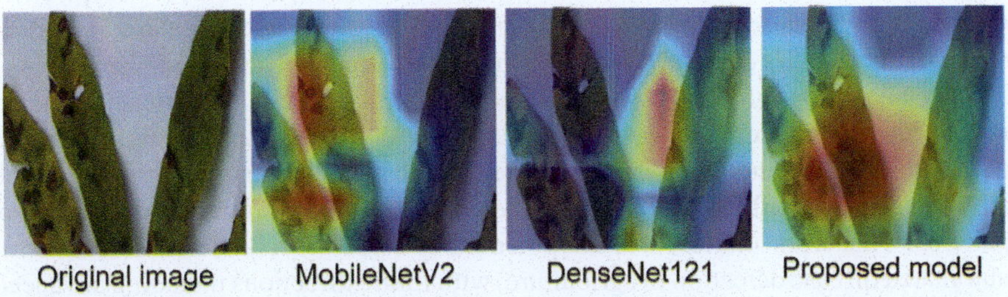


**Fig. 5** Proposed model heatmap image comparison with state-of-the-art

**Table 5** Proposed EB0DNet model parameters (P) (millions), epoch time (ET) (seconds), time per image (TPI) (seconds), and model weight (MW) comparison on Cassava [4] dataset

| Model | # P | ET | TPI | MW |
|---|---|---|---|---|
| EfficientNetB0 | 4.06 | 7 | 0.03 | 1280 |
| EfficientNetB1 | 6.59 | 9 | 0.04 | 1280 |
| DenseNet121 | 7.05 | 11 | 0.06 | 1024 |
| Proposed EB0DNet model | 11.04 | 30 | 0.16 | 2304 |

parameter count increases substantially, providing the model with greater capacity to capture diverse and complex feature representations, as shown in Table 5.

This enhanced capacity enables the network to learn fine-grained patterns in plant disease images more effectively. However, a large number of parameters also introduces challenges, including increased computational cost and a higher risk of overfitting, as the model may begin to memorize the training data rather than generalizing to unseen samples. During training, optimization algorithms such as Adam or SGD iteratively adjust these parameters to minimize the loss function, but an excessively large parameter space can slow convergence and negatively impact generalization. To maintain parameters at a suitable level, several strategies are essential, including the use of regularization techniques like $L_2$ regularization and dropout to prevent overfitting, dynamic learning rate scheduling to stabilize convergence, early stopping to avoid unnecessary training, and batch normalization to ensure efficient parameter updates. By carefully balancing the parameter count with these strategies, concatenate models can leverage the strengths of multiple architectures while maintaining efficiency, stability, and strong generalization capability.

## 5 Conclusion

This study presents a hybrid deep learning approach for cassava disease classification using the EB0DNet model, achieving 99.40% training accuracy, 97.61% test accuracy, and strong performance metrics (precision 98%, recall 97%, $F_1$-score 98%) on the CCMT dataset. By combining EfficientNetB0's efficient feature extraction with DenseNet121's dense connectivity, the model effectively distinguishes between healthy and diseased leaves, with confusion matrix analysis showing high accuracy across most categories despite challenges with visually similar diseases. Data augmentation and fine-tuned hyperparameters enhance robustness, highlighting the advantages of hybrid CNN architectures over single models for capturing complex patterns. This work underscores the potential of scalable, reliable AI solutions for automated disease detection in agriculture, with future directions including advanced hybrid architectures, real-time field deployment, and expanded datasets for improved generalization.

## References


1. Ramcharan, A., Baranowski, K., McCloskey, P., Ahmed, B., Legg, J., Hughes, D.P.: Deep learning for image-based cassava disease detection. Front. Plant Sci. **8**, 1852 (2017)
2. Ayu, H.R., Surtono, A., Apriyanto, D.K.: Deep learning for detection cassava leaf disease. J. Phys. Conf. Ser. **1751**(1), 012072 (2021)
3. Ramcharan, A., McCloskey, P., Baranowski, K., Mbilinyi, N., Mrisho, L., Ndalahwa, M., Legg, J., Hughes, D.P.: A mobile-based deep learning model for cassava disease diagnosis. Front. Plant Sci. **10**, 272 (2019)
4. Mensah, P., Akoto-Adjepong, V., Adu, K., Ayidzoe, M., Bediako, E., Nyarko-Boateng, O., Boateng, S., Donkor, E., Bawah, F., Awarayi, N., et al.: CCMT: dataset for crop pest and disease detection. Data in Brief **49**, 109306 (2023)
5. Jiang, H., Xue, Z.P., Guo, Y.: Research on plant leaf disease identification based on transfer learning algorithm. J. Phys. Conf. Ser. **1576**(1), 012023 (2020)
6. Moin, N.B., et al.: Disease detection of Bangladeshi crops using image processing and deep learning-a comparative analysis. In: 2022 2nd International Conference on Intelligent Technologies (CONIT), pp. 1–8 (2022)
7. Islam, M.A., et al.: An automated convolutional neural network based approach for paddy leaf disease detection. Int. J. Adv. Comput. Sci. Appl. **12**(1) (2021)
8. Rajbongshi, A., Khan, T., Pramanik, M.M.R.A., Tanvir, S.M., Siddiquee, N.R.C.: Recognition of mango leaf disease using convolutional neural network models: a transfer learning approach. Indonesian J. Electr. Eng. Comput. Sci. **23**(3), 1681–1688 (2021)
9. Deng, R., et al.: Automatic diagnosis of rice diseases using deep learning. Front. Plant Sci. **12**, 701038 (2021)
10. Sembiring, A., Away, Y., Arnia, F., Muharar, R.: Development of concise convolutional neural network for tomato plant disease classification based on leaf images. J. Phys. Conf. Ser. **1845**(1), 012009 (2021)
11. Lilhore, U.K., et al.: Enhanced convolutional neural network model for cassava leaf disease identification and classification. Mathematics **10**(4), 580 (2022)
12. Narayanan, K.L., et al.: Banana plant disease classification using hybrid convolutional neural network. Comput. Intell. Neurosci. **2022**(1), 9153699 (2022)
13. Liu, X., et al.: DS-MENet for the classification of citrus disease. Front. Plant Sci. **13**, 884464 (2022)
14. Mizan, M.A.I., Ahmed, M., Ali, M.H.: Bangladeshi crop disease detection using convolutional neural network. In: 2024 6th International Conference on Electrical Engineering and Information & Communication Technology (ICEEICT), pp. 1304–1309 (2024)
15. Paçal , I., Kunduracıoğlu, I.: Data-efficient vision transformer models for robust classification of sugarcane. J. Soft Comput. Decis. Anal. **2**(1), 258–271 (2024)
16. Jain, A.K., Sharma, S., Nain, N.: Multi-disease detection and classification of sugarcane leaves using transfer learning. In: 2024 11th International Conference on Soft Computing & Machine Intelligence (ISCMI), Melbourne, Australia, pp. 1–5, 22–23 Nov 2024. IEEE
17. Tensorflow: https://www.tensorflow.org/. Last accessed 14 Feb 2025
18. Sakira, T., Fajar, M., Kaswar, A., Gunawan, S.: Application of the transfer learning method to detect diseases in cassava. AIP Conf. Proc., 3140 (2024)

# Effective Visualization Approaches for Distributed Denial of Service Threat Prediction in Large-Scale Networks Using Interpretable Machine Learning

**R. Srivats, V. Kalyanasundaram, Deepika Roselind Johnson, Abhiram Sharma, P. B. Krishna, and G. Logeswari**

**Abstract** In recent years, Distributed Denial of Service (DDoS) attacks have become a significant threat to cybersecurity, capable of crippling critical networks and online services. This study is aimed at developing a data driven approach to distinguish between normal network traffic and DDoS attacks. The dataset consists of four types of DDoS attacks, namely UDP-Flood, Smurf, SIDDOS and HTTP-FLOOD attacks that are combined into a single malicious traffic class. The approach entails detailed data preprocessing such as handling of missing values and balancing of the dataset using SMOTE. Three machine learning models, namely Logistic Regression, Decision Tree and Random Forest were compared, and Random Forest provides superior performance. After hyperparameter tuning using Random Search Cross Validation, the model achieved a precision of 96.3%, recall of 87.3% and F1-score of 91.6%. XAI techniques like SHAP and LIME were employed to make the model more interpretable and offer features that are most relevant to the detection of DDoS attacks and thus improve the development of network security solutions.

**Keywords** Distributed Denial of Service (DDoS) · Network security · SHAP · LIME · SMOTE · Machine learning · Attack classification

## 1 Introduction

There is a growing problem of the frequency of Distributed Denial of Service (DDoS) attacks on the modern network security. As cyber threats continue to evolve, the sophistication and scale of these attacks have reached new heights with attacks including the high-profile incidents. The 2017 Google attack which peaked at 2.54 Tbps, 3,150,000,000 packets per second in August 2024 on Minecraft server breach

R. Srivats · V. Kalyanasundaram · D. R. Johnson (✉) · A. Sharma · P. B. Krishna · G. Logeswari
School of Computer Science and Engineering, Vellore Institute of Technology, Chennai Campus, Chennai, Tamil Nadu, India
e-mail: deepikaroselind.j@vit.ac.in

© The Author(s), under exclusive license to Springer Nature Singapore Pte Ltd. 2026

M. Pant et al. (eds.), *Real-World Applications of Machine Learning*, Mathematics for Sustainable Developments, https://doi.org/10.1007/978-981-95-7134-5_28

and the exploitation of HTTP/2 protocol vulnerabilities in 2023 which produced up to 398,000,000 requests per second. This has become a clear picture of how DDoS attacks can bring down critical services and impact essential infrastructure. This highlight the need for better detection and mitigation methods, all of this attacks are an attempt to overload DDoS, a target system with traffic from many sources in order to make it inaccessible. Attackers use different types of devices that have been compromised and referred to as botnets to launch a coordinated traffic flood that can render a network impassable. Besides service unavailability, DDoS attacks have extensive economic consequences, they affect the company's reputation and can even lead to legal penalties. Some of them include UDP-Flood, Smurf, SYN Flood and HTTP Flood which use various means to exploit different parts of the network because they are difficult to detect and mitigate. The development of DDoS attacks are nearly as old as the technological progress and the development of the network of connected devices. This has also been compounded by the rise of the Internet of Things (IoT) where devices have created new vulnerabilities in networks since most of these devices have been known to form part of botnets.

New approaches based on machine learning (ML) and artificial intelligence (AI) techniques have emerged as promising solutions for improving the detection and mitigation capabilities. These technologies use data driven methods to identify patterns or anomalies in network traffic to real-time detection and response to potential threats. However, the effectiveness of ML-based approaches depends on few important factors such as quality of training datasets, appropriate feature selection and model interpretability. This study attempts to make the contribution to network security by analyzing the multifaceted DDoS attacks, parameters of their implementation and detection. The research evaluates the capability of various ML models in detecting and classifying DDoS attacks using a consolidated malicious traffic dataset. The framework enhances the transparency and interpretability of model predictions by integrating Explainable AI (XAI) algorithms such as SHAP and LIME. This research presents some important findings which will help to strengthen the defense against DDoS threats and to provide better security for the network infrastructures.

## 2 Literature Survey

Recent developments in DDoS attack detection field prove that a large scale of machine learning techniques and datasets were appropriate for classification and resulted in good classification accuracy for DDoS attacks. Mohmand et al. [1] improved classification of DDoS attacks based on the usage of the dataset UNWS-np-15 with Random Forest and XGBoost, outperforming accuracy with values ranging from 79% to 85% between previous works that obtained precision and recall rates of up to 89% and 90%, respectively. Shieh et al. [2] proposed the CNN-RPL model, which is a combination of Convolutional Neural Networks with Reciprocal Points Learning that aims to detect known as well as unknown DDoS attacks. They have attained greater than 99.93% accuracy for known attacks and 98.51% for unknown attacks with low training parameters and good performance using the CICIDS2017 and CICDDoS2019 datasets. Hussain et al. [3] proposed a hybrid Intrusion Detection System (IDS), which used a rule-based technique along with the machine learning approach to detect any attack, attaining 99.97% accuracy and having a true positive rate of 99.96% on the DDoS attack in cyber-physical production systems, which truly outperforms traditional methods and can be useful for real-time attacks.

Halladay et al. [4] improved the detection of DDoS attacks by using 25 time-based features for the classification of 12 attack types. It gains nearly 99% accuracy in detection but only 70% accuracy for classification. It reduces the training time and retains accuracy suitable for near real-time applications. El Sayed et al. [5] combined machine learning and blockchain technology for improved DDoS detection within IoT networks to improve security and resilience against DDoS threats through traffic data analysis. Dasari et al. [6] proposed another hierarchical machine learning scheme of DDoS detection in the financial network using CICIDS 2017 data with optimization; LGBM attained accuracy of 99.77% along with stable solution toward enhancement of capability in terms of improvement in detection capability. Khedr et al. [7] introduced FMDADM for detecting DDoS in IoT networks while integrating early mechanisms of detection with the use of machine learning to result with an accuracy rate of 99.79% along with a precision rate of 99.43%, therefore better than other methods. In a similar manner, Dhan et al. [8] extended LSTM models for classifying DDoS using SHAPPD enhanced interpretability for identifying influential fields of packets with corresponding feature interdependence, thus offering improved transparency into an otherwise opaque LSTM technique for security analysis. For instance, zero-frequency problems to improve the gaussian distribution and Naïve Bayes method in detecting DoS attacks have been addressed successfully by Naiem et al. [9] and improvedg feature selection process which showed an increase in accuracy of 2%.

Feng et al. [10] tried to overcome such problems in DDoS attacks used in IoT devices by combining reinforcement learning with unsupervised classification for early detection with high accuracy results from real-world experiments. Maksimović et al. [11] proposed an IDS using the combination of different algorithms and Triple Modular Redundancy to improve the accuracy of binary classification in DoS and DDoS attacks in military network datasets. In [12], Cvitić et al. approach proposed using a traffic detection model with boosting methods based on logistic model trees, wherein the results showed an accuracy ranging between 99.92% and 99.99%, it includes the class of devices which negates DDoS attacks on IoT networks. Tang et al. [13] have presented LtRFT, which is a scheme based on the Learning-to-Rank approach and successfully mitigates low-rate DDoS attacks on SDN switch flow tables, with accuracy over 96%. An efficient DDoS detection for SDN-based IIoT networks has been proposed by Zainudin et al. [14] using feature selection through XGBoost with a CNN-LSTM model and reached an accuracy of 99.50% with minimal latency. Finally, Li et al. [15] proposed a framework combining the tensor representation of network data with dimensionality reduction techniques and XGBoost classification, proving to be robust for various datasets with an accuracy of 99.15%. Despite all these breakthroughs, major gaps are still present such as scalability, since most of the models remain unsuitable for real-time usage and have serious computational issues. Also, there is a lack of interpretability since a large proportion of machine learning methods are black box and models have to adapt for effective reactions against the new DDoS vectors which have not been previously seen.

## 3 Proposed Methodology

The proposed methodology (Fig. 1) outlines a novel approach for detecting and classifying DDoS attacks using a framework that incorporates machine learning and Explainable AI (XAI). It involves rigorous data preprocessing, managing class imbalances with SMOTE and splitting the dataset into training, testing and validation set. Machine learning pipelines are developed for three state-of-the-art algorithms, with hyperparameters tuned via Random Search cross-validation. XAI techniques, LIME and SHAP are employed to visualize results and understand the factors influencing these attacks.

### 3.1 Dataset Description

The dataset is network logs labeled in ARFF format, specifically tailored for DDoS attack analysis. It has 31 attributes describing a lot of essential properties of network traffic to classify normal traffic and DDoS attack types, including UDP-Flood, Smurf, SIDDOS and HTTP-FLOOD. Each record has a class label called PKT_CLASS that indicates whether the traffic is normal or one of the DDoS-types. DDoS attacks try to slow down the functioning of a normal system by transmitting excessive traffic through it. A UDP-Flood attack brings in numerous packets of User Datagram Protocol, which consume bandwidth and resources. In a Smurf attack, vulnerabilities are exploited by sending ICMP ping requests with a spoofed source IP address and flooding the victim network on the targeted network. The SYN Flood or SIDDOS, floods the server using many SYN requests, whereas an HTTP-FLOOD targets web servers using a volume of HTTP requests that resemble those of legitimate users.

The attributes of the dataset described network activity with key characteristics such as having SRC ADD and DESADD identifiers which are source and destination identifiers, packet description such as its size (PKT SIZE). The attribute PKT TYPE describes the send and receive time and PKT RATE provide information regarding transmission rate.. And all these attributes together constitute what would make comprehensive datasets to eventually develop ML-based models for classifications and anomaly identifications. Data preparation will start with the loading of the dataset, ensuring correct importation. An Exploratory Data Analysis (EDA) is performed to understand data structure, examining for missing and duplicate

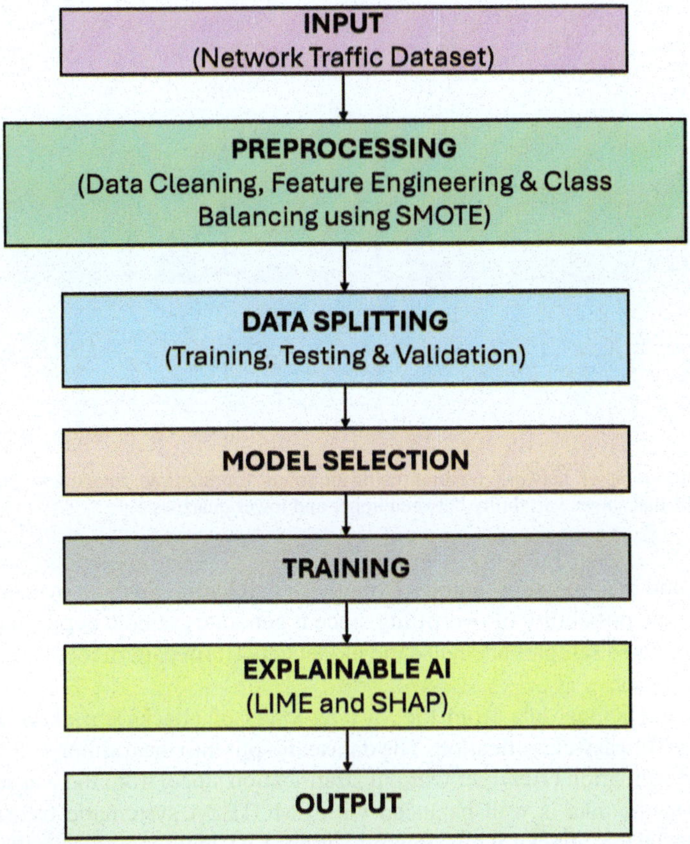


**Fig. 1** Architecture of the proposed model for anomaly classification

values that may skew analysis. Highly correlated features are found and removed to minimize multicollinearity followed by visualization.

## 3.2 *XAI-Based Machine Learning Framework*

It is crucial in the provision of network security against the attacks as well as its future risks to detect DDoS attacks. The present methodology puts forward a novel approach to detect and classify DDoS attacks by employing an XAI-based machine learning framework. The proposed research begins with assessing class distribution for determining if class imbalance exists or not between the normal class and attack classes. If class imbalance is found then SMOTE technique is applied to train the dataset. This Synthetic Minority Over-Sampling Technique generates synthetic samples of the minority class by identifying the nearest neighbors of existing minority

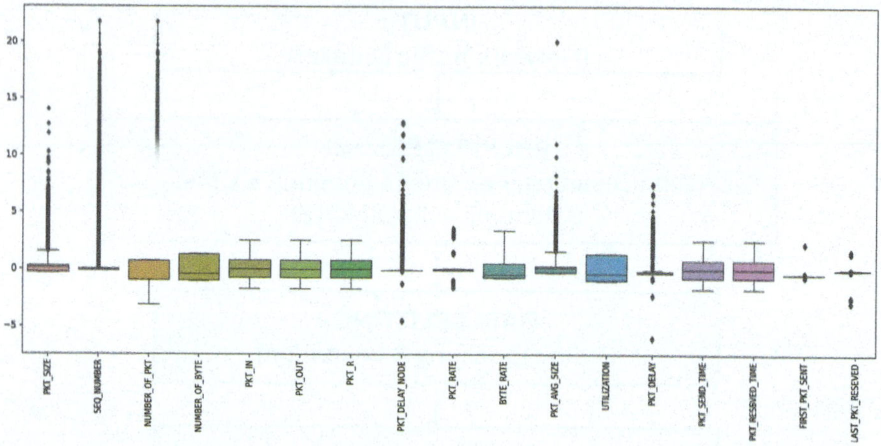


**Fig. 2** Distribution of network features using boxplots for multiple categories. The boxplots indicate IQR that shows variability between upper and lower quartiles

instances and creating new samples from these neighbors. Such an oversampled set has avoids the possibility of overfitting since it generates sample copies and diversifies the generated samples and helps the model to perform pattern recognition among minority classes.

All the malicious data from the various kinds of attacks are classified into a single class for easy classification. The dataset is split into the training (60%), testing (30%) and validation (10%) set without stratification since, from the learned knowledge so far, the data is well balanced with SMOTE. A systematic framework for classifying DDoS attacks while implementing XAI methodologies will be developed along with proper interpretation of the decision of the model. It can be retrieved from the public data set on Kaggle, which records detailed attributes of the network packets transmitted, such as packet size, transmission rate, protocol type and flags. Good quality, diverse and representative data are required as they impact the performance at later stages. The training set is used for model training, validation refines the hyperparameters and prevents overfitting and testing set provides the performance of the model on unknown data.

It tests a few candidate algorithms such as Logistic Regression, Decision Tree and Random Forest is analysed. Pipelines are developed for each learning algorithm using both scaled and unscaled datasets to test the effects of feature scaling on performance. Logistic Regression predicts class probabilities given feature inputs; the Decision Tree branch the data through subsets based upon feature values. It functions by creating numerous decision trees in a way such that it would optimize accuracy without overfitting. After putting the pipeline together, k-fold cross-validation with $k = 5$ value was performed such that predictability and replicability of models may be achieved. The original data set is broken down into smaller subsets of five samples and this is trained on five different instances with different subsets. Cross-validation ensures that tests are performed homogeneously throughout the parts of the data so that robust evaluations come out. The models are checked using F1-score and precision metrics where their ability to balance recall with precision is compared. For each model, the performance may be visualized using boxplots to help ascertain the best algorithm based on the variability in the performance across the folds.

Once the best performing model is found, training is carried out for the full dataset for 30 epochs without hyperparameter optimization to set baseline performance. Then, the model is evaluated on the test set to check for overfitting. The next step would be hyperparameter tuning to optimize the model, which is the number of samples at every leaf node and features to be considered for split. Random Search cross validation determines optimal settings across different combinations of hyperparameters. The structured approach facilitates the detection of DDoS, and at the same time provides interpretability using XAI techniques. In the Random Search, 20 candidate hyperparameter combinations are randomly picked. For each candidate, three-fold cross-validation is used where the dataset is divided into three subsets and the model is trained on two of them while the third one is used for evaluation. This amounts to 60 individual model fittings and ensures that models differ with respect to their higher-level configurations. Finally, the model is retrained with optimized hyperparameters and its performance again evaluated on the test set using the full training dataset. The final evaluation checks the improvement of this model in the aspect of generalizing better are correct and reduces overfitting found in earlier stages where the model performance is improved for testing.

### *3.3 Integrated Workflow for DDoS Classification with Explainable AI*

The performance of the model was tested using precision, recall, F1 score, and ROC-AUC. The novelty involves using Explainable AI where XAI is used to explain predic tions of the model. The global explanation identifies the feature set that contribute towards overall prediction. The local explanation provides an analysis of DDoS classification. Among the XAI algorithms, LIME provides better feature interpretation for classification. Combination of LIME and SHAP provide better interpretability for DDoS attacks making it easier for the model to understand the influence of attacks. Using these techniques, it provides transparent solutions and better explainability for DDoS detection.

## 4 Results and Discussions

The study shows that the fine-tuned Random Forest framework will be a good approach toward accurate detection and classification of DDoS attacks. Based on network traffic, the attribute of the datasets appears to be the attribute that will most likely play an important role in providing flow behavior and performance insights. The key attributes, therefore, are: SEQ NUMBER, which ensures streams receive packets in a sequential order, whereas metrics on volumes of traffic will use such parameters as NUMBER OF PKT, NUMBER OF BYTES, where a session involving the transfer of certain numbers of packet bytes is measurable. Such values will enable calculation of average values for packet sizes known as PKT AVG size. Attributes such as PKT_IN and PKT_OUT measure the reception and transmission by a node regarding packets and serve to analyze network component load distribution. Performance analysis relates to metrics PKT_R: packet rate; BYTE_RATE; data transfer rates, which, in turn are essential for establishing network throughput levels and utilization metrics. The level of resource use is captured under the UTILIZATION attribute, but latency metrics in the form of PKT_DELAY_NODE and PKT_DELAY contribute to diagnosing performance bottlenecks. Temporal attributes like PKT_SEND_TIME and PKT_RECEIVED_TIME record packet transmission and reception timestamps to analyze packet travel time.

Figure 2 shows boxplot of network features which shows the distribution, variability and outliers in a dataset. The 25th percentile (Q1) and 75th percentile (Q3) indicates the end points of the box. A line within that box represents the median or Q2. A tall box, for example PKT_SIZE shows variability whereas the short box for example, in UTILIZATION shows smaller variability. Outliers are kept under observation for evaluation later since these, depicted in dot form beyond whiskers that protrude up to three times of the IQR possess a possibility of network attacks. Figure 3 depicts a correlation heatmap showing relationships between variables with coefficients ranging from $-1$ to 1. Dark red areas indicate strong positive correlations, while dark blue areas highlight strong negative correlations. For instance, PKT_SIZE and PKT_AVG_SIZE are positively correlated, suggesting redundancy; conversely, NUMBER_OF_PKT and UTILIZATION exhibit negative correlations that differentiate attack patterns from normal traffic. Features with correlations exceeding 0.90 were removed to reduce multicollinearity. The three boxplots in Fig. 4 illustrate the performance of Logistic Regression, Random Forest and Decision Tree models. Logistic Regression shows high variability, reflecting sensitivity to unscaled data and inconsistent performance across runs. In contrast, Random Forest demonstrates minimal variability and superior performance with tightly clustered results, establishing it as the most robust model even with unscaled data. The Decision Tree exhibits stability but lower overall performance compared to Random Forest. These findings highlight Random Forest as the preferred model due to its balance of consistency and superior metrics in DDoS attack detection.

As shown in Fig. 5, using hyperparameter tuning for the Random Forest model improved its performance in detecting malicious instances. True positives increased from 252,324 to 254,621, and false negatives dropped drastically from 3270 to 973, improving the model's ability to correctly identify malicious behaviour and reducing the likelihood of missed attacks. The trade-off between a slight increase in false positives and the substantial reduction in false negatives is acceptable, as it prioritizes security by minimizing undetected attacks. The fine-tuned model is now better equipped to classify network flows accurately.

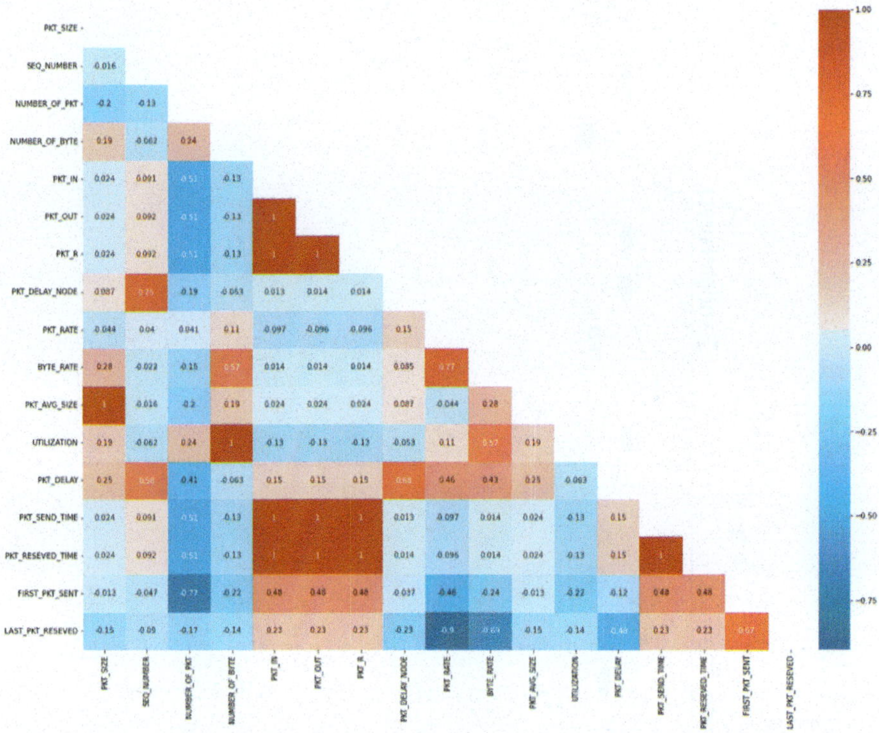


**Fig. 3** Correlation heatmap showing the correlation between the various network features

Figure 6 depicts the SHAP summary bar plot, highlighting global feature importance for the model's predictions. The most significant feature is LAST_PKT_RECEIVED, crucial for detecting malicious traffic (Class 1), followed by PKT_DELAY, which influences both normal (Class 0) and malicious classifications. PKT_RATE and BYTE_RATE are also important, with PKT_RATE favouring malicious detection and BYTE_RATE contributing more to normal classifications. NUMBER_OF_PKT and NUMBER_OF_BYTE provide insights into traffic volume, with packet count being more relevant for normal traffic. Lower-ranked features like SEQ_NUMBER, RST_PKT_SENT and flags have a smaller impact but still assist in classification. The SHAP force plots illustrate individual feature contributions to specific predictions. In the benign prediction plot, features such as PKT_RATE, BYTE_RATE and LAST_PKT_RECEIVED indicate benign behaviour with low values correlating to non-malicious traffic. Conversely, in the malicious prediction plot, high values of these features signal malicious activity. These visualizations demonstrate how the model differentiates between benign and malicious network flows, providing transparency into its decision-making process.

LIME Visualization plots in Fig. 7 provide interpretable insights into the model's predictions for two instances. In both training and testing cases, BYTE_RATE > 5055

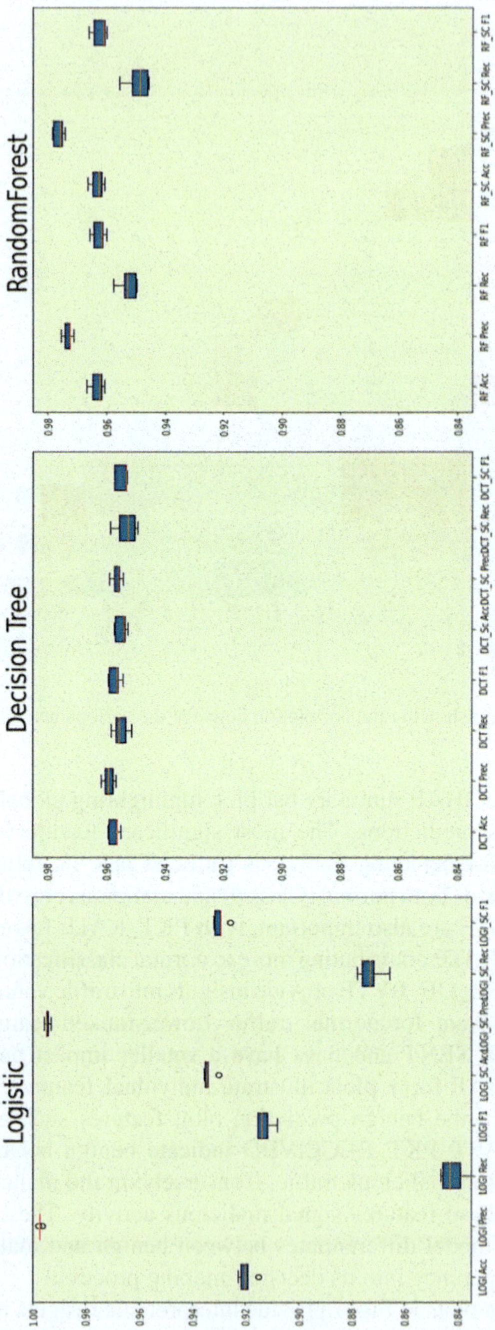


**Fig. 4** Boxplots visualizing model performance (Logistic Regression, Random Forest and Decision Tree)

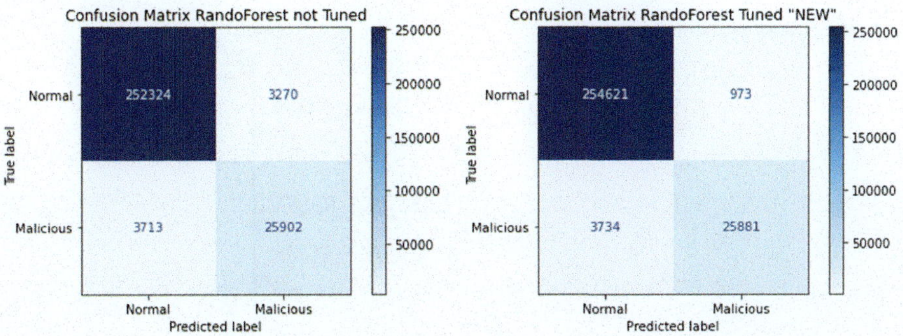


**Fig. 5** Confusion matrix showing the difference between before and after fine-tuning the Random Forest model

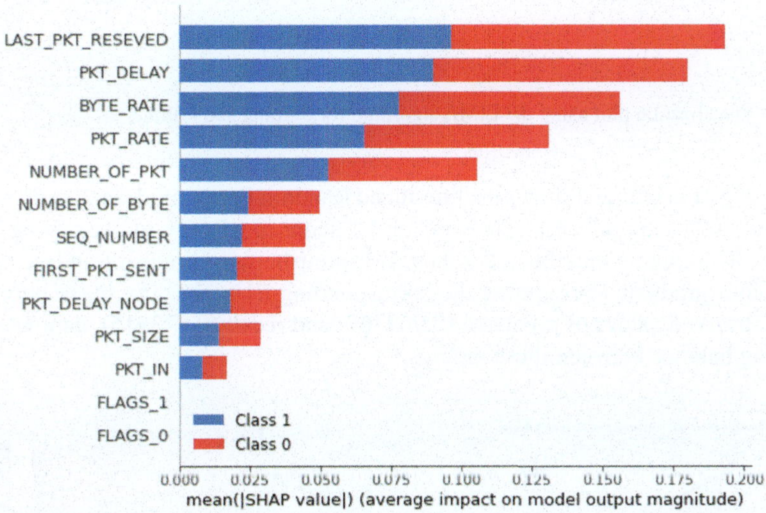


**Fig. 6** SHAP summary plot for global feature importance and force plot for detecting benign and malicious networks

emerges as the most influential feature driving predictions toward class 1 (malicious). Supporting features including FLAGS_1, SEQ_NUMBER, FIRST_PKT_SENT and NUMBER_OF_PKT, also contribute positively but with varying magnitudes. Minor features like PKT_DELAY_NODE and FLAGS_1 slightly support class 0 (benign), but their influence is minimal compared to the dominant class 1 features. LIME highlights the model's reliance on important network-related features like byte rate and packet characteristics in classifying malicious traffic. Python, TensorFlow, LIME and SHAP libraries were integral to the model implementation and explainability. Model training and visualizations were performed on a high-performance system equipped with an Intel Core i7 processor, 16GB RAM, and an NVIDIA RTX 3080 GPU to execute computational tasks.

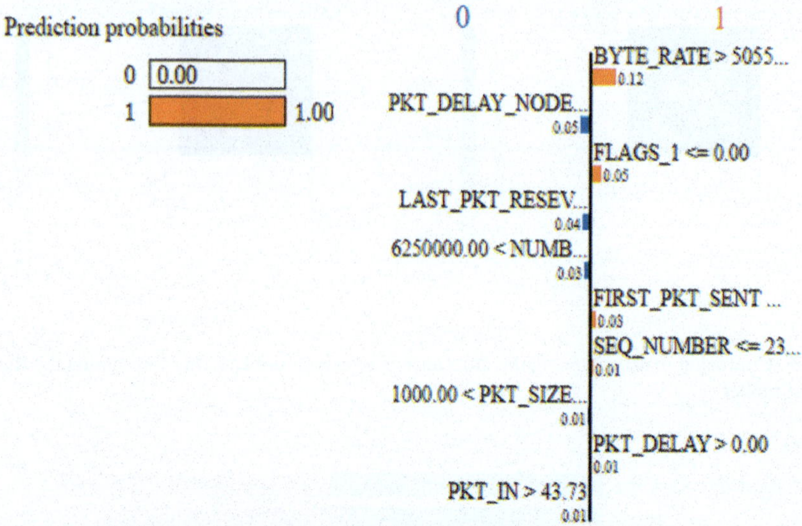


**Fig. 7** Visualization plot using the LIME algorithm for training and testing

As depicted in Fig. 8 shows the fine-tuned Random Forest model achieved a precision of 96.37%, recall of 87.39% and an F1-score of 91.66% in classifying DDoS attacks. The contour plot illustrates the relationship between precision and recall, with the colour gradient representing the corresponding F1-score. The black point indicates observed values of precision (0.963767) and recall (0.873915), demonstrating a strong balance between these metrics.

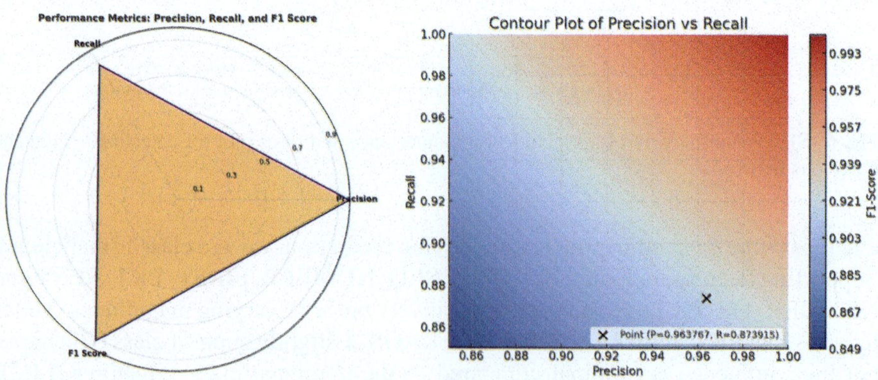


**Fig. 8** Radial plot visualizing performance metrics and contour plot of precision versus recall metrics

## 5 Conclusion

The paper investigates the problem of Distributed Denial of Service (DDoS) attacks on modern large-scale networks. It investigates multi-dimensional developments in DDoS attack detection. This study encompasses the effectiveness of machine learning-based detection methods and a newly proposed Explainable AI (XAI) framework to handle this complexity caused by their nature of dynamic changes. Structured approach was performed in data preprocessing and in the hyperparameter tuning of the approach by using Logistics Regression, Decision tree, and Random Forests model and SMOTE was performed for maintaining class balance. The increase in performance has been attributed to cross validation and hyperparameter tuning.While, XAI algorithms such as SHAP and LIME provide better model interpretability. This study provides useful information on DDoS detection while future work will focus on including real-time traffic data and adaptive learning to improve network security against DDoS attacks.

**Acknowledgements** We confirm that the manuscript has been read and approved by all named authors and that there are no other persons who satisfied the criteria for authorship but are not listed. We further confirm that the order of authors listed in the manuscript has been approved by all of us.

## References

1. Mohmand, M.I., Hussain, H., Khan, A.A., Ullah, U., Zakarya, M., Ahmed, A., Raza, M., Rahman, I.U., Haleem, M.: A machine learning-based classification and prediction technique for DDoS attacks. IEEE Access **10**, 21443–21454 (2022)
2. Shieh, C.S., Ho, F.A., Horng, M.F., Nguyen, T.T., Chakrabarti, P.: Open-set recognition in unknown DDoS attacks detection with reciprocal points learning. IEEE Access (2024)
3. Hussain, A., Tordera, E.M., Masip-Bruin, X., Leligou, H.C.: Rule-based with machine learning IDS for DDoS attack detection in cyber-physical production systems (CPPS). IEEE Access (2024)
4. Halladay, J., Cullen, D., Briner, N., Warren, J., Fye, K., Basnet, R., Bergen, J., Doleck, T.: Detection and characterization of DDoS attacks using time-based features. IEEE Access **10**, 49794–49807 (2022)
5. El Sayed, A.I., Abdelaziz, M., Hussein, M., Elbayoumy, A.D.: DDoS mitigation in IoT using machine learning and blockchain integration. IEEE Netw. Lett. (2024)
6. Dasari, S., Kaluri, R.: An effective classification of DDoS attacks in a distributed network by adopting hierarchical machine learning and hyperparameters optimization techniques. IEEE Access (2024)
7. Khedr, W.I., Gouda, A.E., Mohamed, E.R.: FMDADM: a multi-layer DDoS attack detection and mitigation framework using machine learning for stateful SDN-based IoT networks. IEEE Access **11**, 28934–28954 (2023)
8. Assadhan, B., Bashaiwth, A., Binsalleeh, H.: Enhancing explanation of LSTM-based DDoS attack classification using SHAP with pattern dependency. IEEE Access (2024)
9. Naiem, S., Khedr, A.E., Marie, M., Idrees, A.M.: Enhancing the efficiency of Gaussian Naive Bayes machine learning classifier in the detection of DDoS in cloud computing. IEEE Access (2023)

10. Feng, Y., Zhang, W., Yin, S., Tang, H., Xiang, Y., Zhang, Y.: A collaborative stealthy DDoS detection method based on reinforcement learning at the edge of internet of things. IEEE Internet Things J. **10**(20), 17934–17948 (2023)
11. Maksimoviá, A., Nikoliá, V., Vidojeviá, D., Randjeloviá, M., Djukanoviá, S., Randjeloviá, D.: Using triple modular redundancy for threshold determination in DDoS intrusion detection systems. IEEE Access (2024)
12. Cvitiá, I., Perakovic, D., Gupta, B.B., Choo, K.K.R.: Boosting-based DDoS detection in internet of things systems. IEEE Internet Things J. **9**(3), 2109–2123 (2021)
13. Tang, D., Yan, Y., Gao, C., Liang, W., Jin, W.: LtRFT: mitigate the low-rate data plane DDoS attack with learning-to-rank enabled flow tables. IEEE Trans. Inf. Forensics Secur. **18**, 3143–3157 (2023)
14. Zainudin, A., Ahakonye, L.A.C., Akter, R., Kim, D.S., Lee, J.M.: An efficient hybrid-DNN for DDoS detection and classification in software-defined IIoT networks. IEEE Internet Things J. **10**(10), 8491–8504 (2022)
15. Li, S., Xu, J., Liu, P., Li, X., Wang, P., Jin, X., Yao, S.: Truncated Lanczos-TSVD: an effective dimensionality reduction algorithm for detecting DDoS attacks in large-scale networks. IEEE Trans. Netw. Sci. Eng. (2024)

# A Novel Approach to Integrate AI into Judicial Systems

**Ruchira Goel, Neelendra Badal, and Khushi Bansal**

**Abstract** The judicial system in most countries is already highly burdened with a huge backlog of cases, leading to delayed judgments which impede access to justice. This inefficiency stems from overburdened courts, limited resources, and costly procedural processes that results in litigants waiting years to have a case resolved. This paper proposes a novel approach on how artificial intelligence can assist to streamline judicial workflows so that cases get processed faster reducing delays. Predictive analytics models can be utilized to help analyze historical case data to predict trial outcomes and evaluate the complexity of a claim giving organization a better opportunity in prioritizing cases. ML will be bifocal in this domain: not only it can optimize administrative functions like case management, but can also support more sophisticated applications such as outcome prediction, legal risk assessment, and resource allocation. It offers substantial efficiency gains and raises important questions related to algorithmic accountability thus striking a balance between human judgment and automatization. This paper aims to discuss in depth the areas of opportunity and constraint brought about by AI technologies for creating a modernized judicial process.

**Keywords** Judicial system · SVM · Machine learning

R. Goel (✉)
Ajay Kumar Garg Engineering College, Ghaziabad, Uttar Pradesh, India
e-mail: goelruchira@akgec.ac.in

N. Badal
Rajkiya Engineering College (REC), Bijnor, Uttar Pradesh, India

K. Bansal
Veersa Technologies, Noida, Uttar Pradesh, India
e-mail: khushibansal211@gmail.com

© The Author(s), under exclusive license to Springer Nature Singapore Pte Ltd. 2026

M. Pant et al. (eds.), *Real-World Applications of Machine Learning*, Mathematics for Sustainable Developments, https://doi.org/10.1007/978-981-95-7134-5_29

# 1 Introduction

The judicial system is a cornerstone of any functioning society ensuring that justice is delivered equitably and efficiently. However, India being a very densely populated country is presently facing a large backlog of pending cases at various levels of Courts, from lower courts to Supreme Court of India due to shortage of courtrooms, judicial officers, creating a bottleneck that threatens to undermine the credibility of the legal system. This has been a serious concern for each wing of the country to increase the consciousness for statutory rights and solutions available to common people and would lead to a complete collapse or a very delayed Justice which will be a major practical problem. As the classic quote goes, justice delayed is justice denied. Court systems need more resources to meet demand as the complexity and volume of cases increase. The result is frequently a sluggish and wasteful procedure that fails to provide justice quickly, undermining public faith. Apart from this, the investors, in particular foreign ones have reservations to invest in India as timely Justice or dispute resolution would be delayed due to ineffective judicial system. This may force them to back out from investing and thereby forcing them to invest in other countries with ease of business. All this will happen even though India has the most dedicated, sincere, and hard-working judges and a streamlined judicial set of Laws.

AI has made substantial advances in a variety of domains, including healthcare, finance, logistics, and transportation. Its capacity to handle massive volumes of data, detect patterns, and automate operations that would normally need human intelligence has proven crucial. One of AI's most potential applications is machine learning, which uses previous data to generate predictions and improve over time. The legal system, which has traditionally relied on human discretion and manual processes, is increasingly seeing the incorporation of AI technology to address long-standing difficulties. The use of AI in the judiciary promises to automate ordinary operations, expedite case administration, and even aid in complicated legal decision-making, all while addressing the underlying issue of delays caused by large caseloads. AI has the potential to alter the way courts work, resulting in a more efficient, transparent, and accessible judicial system. Machine learning can address important concerns in the court system like the backlog of cases and trial predictions. By integrating predictive analytics and automating administrative tasks, AI can assist minimize delays, expedite court operations, and improve overall legal system efficiency.

AI is already making waves in the global judiciary system, and its adoption in India has been recently advocated by "Chief Justice of India" Justice Sharad Arvind Bobde [1]. He emphasized while AI will never be able to replace the wisdom and knowledge of human judges still it can significantly aid the judiciary by reducing the pendency of cases and expediting judicial processes. AI tools can assist in various judicial tasks, such as predicting sentence duration, relapse scores, and analyzing legal data, allowing judges and legal administrators to streamline case management and decision-making. The potential for AI to support the judiciary is immense, as it can help judges make better-informed decisions and ensure greater consistency in rulings. The Indian judiciary is already laying the groundwork for such advancements. For example, the e-Court [2] Project that focused on building blocks of digitization in Phase I and delivering service to litigants, lawyers, and other concerned stakeholders in Phase II was formulated. Phase III anticipates using an ecosystem approach, which allows various systems to communicate with one another. This will construct a new institutional and governance framework, streamline judicial procedures, and build a core digital infrastructure. By offering effective systems and resources for a digital case register, an extensive case law archive, and the development of judgment templates for judicial purposes, the initiative also seeks to lessen the workload on courts. This would save considerable processing time ultimately. In addition, a mobile application [3] has been launched by The President of India Shri Ram Nath Kovind that enabled an artificially intelligent law transaction system to facilitate quality translation. This will enable efficiency of Indian Judicial system by reducing the number of pending cases. This mirrors the advancements seen in countries like the USA and Canada [4] where AI systems have already been deployed to assist Judges in delivering effective Justice on various matters such as granting of bail and diligence of decision in contacts, investments. AI can help referring incidence from past cases which a human brain may be forgetful off. This will help advocates to present and fight cases with all legal aspects in the long run. Also, the documentation can be much more simplified as hard copies of documents need not be searched every time before the case date occurs rather strings of decisions made on previous dates can be investigated in the easiest possible way.

Another key area where AI is making a mark is in Intellectual Property Rights (IPR) where it can provide insights for searching and registration of a trademark, patent, Copyrights, etc. [5]. An analysis of several details can easily be done. Also, AI can play a pivotal role to analyze aspects such as date of filing of charge sheet, number of accused in a case, critical examining the witnesses which can help in taking better strategic decisions. AI is supposed to help in providing a platform for rationalized decision-making by providing all the relevant information in a more summarized way than a normal human brain is. This will lead to better utilization of a taxpayer's money. In countries like Estonia and China [6], robot lawyers and judges are being used to adjust petty claims. In fact, these lawyers are fully capable of arguing cases before human judges. This way access to justice is improving in number of countries. AI is also used for predicting judicial opinion in Europe by studying the past judgments of the judges and thereby predicting the possible outcomes. This way decision-making process can be rationalized by summarizing all the relevant information in a more efficient way as compared to human by using AI.

In this paper, we will look at the application of ML to forecast the accuracy of case outcomes. By examining historical data and judicial tendencies, ML may provide predicted insights that will substantially help the justice. For example, ML may assist prioritize cases based on projected outcomes, helping courts to deploy resources more efficiently. This would speed up case resolution, eliminate delays, and increase the uniformity of decisions. Furthermore, ML-driven predictions can aid in spotting tendencies that contribute to longer case durations, allowing legal practitioners to implement more effective methods. AI can also help identify prospective settlements, encourage early agreements, and reduce the number of cases that go to trial. It can minimize the administrative load on courts, shorten case backlogs, and improve access to justice. It can also assure more consistency and justice in court decisions since AI systems are less susceptible to the biases that can occasionally impair human decision-making. Furthermore, by automating mundane procedures and expediting case management, AI enables legal experts to focus on more difficult areas of the law, improving overall court outcomes.

Overall, AI-powered solutions have the potential to make legal systems more accessible worldwide, boosting efficiency and fairness while decreasing litigation time and costs. This paper will examine these issues and how AI might transform judicial prediction and improve legal procedures for judges and litigants.

## 2 Literature Review

Previous research on predictive modeling in Indian legal studies has shed light on its possible uses and limitations in decision-making. Singh et al. [7] found promising results in using predictive modeling to predict legal outcomes in criminal cases, considering case type, defendant demographics, and judicial district. Another technique uses language aspects of court rulings to anticipate outcomes. Ngo [8] attempted to predict outcomes from Dutch legal cases database (2019) based on linguistic attributes. The total number of words and the frequency of different types of pronouns were among the factors evaluated. Brüninghaus and Ashley [9] substituted case-specific names and instances with their roles and utilized propositional patterns to forecast trade secret situations. Sulea et al. [10] deployed Unigrams and Bigrams, together with word type tokens, to forecast case rulings, legal areas, and time spans in French Supreme Court decisions. Aletras et al. [11] predicted European Court of Human Rights decisions using N-grams and subjects as characteristics. Luo et al. [12] employed neural networks to forecast charges in criminal trials in China. The documents were separated into three parts: fact description, articles, and charges and various combinations of fact descriptions and articles were used as input. Liu and Chen [13] employed sentiment analysis to forecast charges.

Zhong et al. [14] suggested a judgment prediction model that makes use of several subtasks. Despite their capacity to anticipate legal articles and charges at the same time, previous research only developed methods for particular subtasks, making it challenging to expand to additional subtasks. By conducting this kind of study, it also focused on murder cases. Legal judgment extraction is a time-consuming and difficult process for identifying case-specific elements. To predict outcomes in murder cases, crucial criteria are analyzed using a dataset to identify descriptors. The outcome prediction is based on a binary categorization of ‘acquittal’ or ‘conviction’. Agrawal et al. [15] used machine learning algorithms to predict case outcomes in Indian courts, with encouraging results in both civil and criminal matters. Gupta et al. [16] used predictive analytics to predict bail hearing results in Indian courts, providing insights into judicial decision-making processes and ways for maximizing outcomes.

Kumar et al. [17] used NLP approaches to analyze legal texts and anticipate case outcomes in the Indian judiciary, demonstrating the promise of advanced analytics in improving legal research and decision-making processes. Alsolai and Roper [18] explored the application of machine learning in court procedures to anticipate judicial judgments. Predictions can range from forecasting the result of a judicial judgment to categorizing charges using several labels. Legal judgments may include intricate sub-clauses including charges, penalties, and fines.

Convolutional Neural Networks (CNNs), a subset of Deep Neural Networks (DNNs), have been highlighted by Pillai and Chandran [19] as being crucial in Indian judicial law cases. They have shown an astounding 85% accuracy rate in predicting charges, identifying the type of offenses, and, in the end, providing approximate judicial decisions based on the Indian Penal Code (IPC). Long Short-Term Memory (LSTM) networks, a subtype of RNNs that are particularly useful in sequential prediction tasks because they specialize in learning long-term dependencies, were employed by Singh and Thanaya [20]. Alghazzawi et al. [21] used an LSTM+CNN model to predict court rulings with an impressive accuracy of 92.05%. According to Luo [12], analyzing textual facts is vital for legal assistant systems. This allows civilians who are unfamiliar with legal language to identify comparable instances or punishments by expressing them in their own words and understanding the legal foundation of their searches. Luo [12] suggested an attention-based neural network strategy for predicting charges and extracting relevant articles within a unified framework. The study found that offering relevant articles improves charge prediction and helps anticipate charges for situations with different expression styles. Aletras executed the first systematic study to anticipate the results of cases in the European Court of Human Rights by textual analysis [22]. He classed predicted outcomes as 'violation' or 'non-violation' based on retrieved language from past cases. Additional research was undertaken using the same dataset, incorporating more articles and factors [23]. This suggestion can help attorneys and judges locate cases and extract relevant material for decision-making.

## 3 Proposed Methodology

This paper demonstrates a comprehensive approach to ML-based prediction of legal outcomes. The steps from feature extraction, data preprocessing, model training, and evaluation ensures reliable performance of the proposed approach. The specific phases of the technique are as follows:

### *3.1 Data Collection and Preprocessing*

A prediction data set comprising of 200 values (In Fig. 1) has been used in the proposed model. The facts of case, type of decision, and final disposition are given,

| ID | name | href | docket | term | first_party | second_p | facts | facts_len | majority_ | minority_ | first_party | decision_t | dispositio | issue_area |
|---|---|---|---|---|---|---|---|---|---|---|---|---|---|---|
| 50606 | Roe v. Wa | https://ap | 70-18 | 1971 | Jane Roe | Henry Wa | <p>In | 501 | 7 | 2 | TRUE | majority o | reversed | |
| 50613 | Stanley v. | https://ap | 70-5014 | 1971 | Peter Stan | Illinois | <p>Joan | 757 | 5 | 2 | TRUE | majority o | reversed/i | Civil Rights |
| 50623 | Giglio v. U | https://ap | 70-29 | 1971 | John Giglio | United Sta | <p>John | 495 | 7 | 0 | TRUE | majority o | reversed/i | Due Process |
| 50632 | Reed v. Re | https://ap | 70-4 | 1971 | Sally Reed | Cecil Reed | <p>The | 378 | 7 | 0 | TRUE | majority o | reversed/i | Civil Rights |
| 50643 | Miller v. C | https://ap | 70-73 | 1971 | Marvin Mi | California | <p>Miller | 305 | 5 | 4 | TRUE | majority o | vacated/r | First Amendment |
| 50644 | Kleindiens | https://ap | 71-16 | 1971 | Richard G. | Ernest E. N | <p>Ernes | 2282 | 6 | 3 | TRUE | majority o | reversed | First Amendment |
| 50655 | Sarno v. Il | https://ap | 70-7 | 1971 | Sarno | Illinois Cri | <p>The | 1424 | 5 | 2 | FALSE | per curiam | | Criminal Procedure |
| 50656 | Argersinge | https://ap | 70-5015 | 1971 | Argersinge | Hamlin | <p>Jon | 347 | 9 | 0 | TRUE | majority o | reversed | Criminal Procedure |
| 50657 | Eisenstadt | https://ap | 70-17 | 1971 | Eisenstadt | Baird | <p>Willia | 420 | 6 | 1 | FALSE | majority o | affirmed | Privacy |
| 50663 | Gooding v | https://ap | 70-26 | 1971 | Gooding | Wilson | <p>A | 612 | 5 | 2 | FALSE | majority o | affirmed | First Amendment |
| 50671 | Furman v. | https://ap | 69-5003 | 1971 | Furman | Georgia | <p>Furm | 477 | 5 | 4 | TRUE | per curian | reversed/i | Criminal Procedure |

**Fig. 1** Dataset used in training the model

such as whether the first party won or lost. For data preprocessing the stop words, punctuation, etc., are removed and transform the textual data of "facts_clean" column. Thereafter, the cleaned facts are tokenized into bigrams (two-word phrases) for better context in topic modeling. Furthermore, one hot encoding is done on metadata features like decision_type and disposition to generate a categorical feature set. After processing the textual data, a comprehensive dataset is formed by combining these encoded metadata features with the processed textual data.

## 3.2 *Feature Extraction*

During the feature extraction stage of text analysis, CountVectorizer uses the Bag of Words (BoW) Model to turn the cleaned text input into a matrix of token counts. This technique determines the frequency of each word in the dataset. Subsequently, Latent Dirichlet Allocation (LDA) [24] has been used for topic modeling, reducing the dimensionality of the document-term matrix by finding 200 different themes that emphasize the text's most important qualities.

## 3.3 *Model Training*

In the outlined approach, model training has been performed using a Support Vector Machine (SVM) which uses the ordered dataset to generate decision boundaries for classifying court case information. This approach (Fig. 2) is well-suited for dealing with high-dimensional data, making it useful for complications seen in legal datasets. The purpose of SVM is to increase classification accuracy by identifying the optimum hyperplane that divides the dataset's categories.

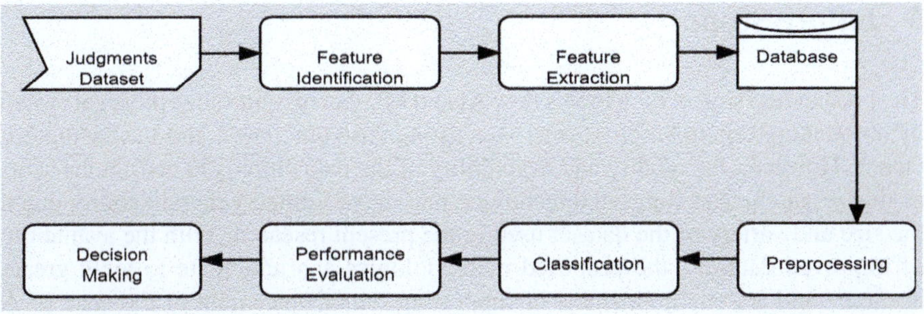


**Fig. 2** Flowchart of proposed model

### *3.4 Model Evaluation*

The SVM model is expected to evaluate its performance using a number of critical metrics with an emphasis on accuracy. SVM attained a tremendous accuracy of 97.37%, demonstrating its ability to accurately categorize court case data.

## 4 Pitfalls of Using AI

Despite making the justice delivery system look visibly efficient and fast AI has a few pitfalls.

1. AI systems may result into biased decisions [12] due to various factors such as accuracy and reliability of the dataset on which they are trained. This can thus result in harsher treatment for some communities in criminal justice or in processing of credit applications.
2. AI-based software programs may not necessarily state the truth [25]. They construct a solution or statement based on most probabilistic outcome. Hence, this again becomes an authenticity issue.
3. The information used for defining an AI model may mislead to injustice or unfairness creeping into the judicial system [25]. This raises the possibility of violation of human rights and privacy. Privacy breach may lead to trust issues among individuals or groups which may result into legal and political complications.
4. AI can pose problems for aged population as they may find it difficult to adjust to new technology [26].
5. AI also raise concerns over safety of the vast amount of data stored [26].
6. Justice delivery is not a mechanical process based on hard facts and cold decisions [27]. It includes human emotions, empathy, understanding the background, culture, and vision of the people. So, a frightening scenario of AI robot judges or lawyers arises. Hence, the question of capability of machines capable of acting like human beings arises.

## 5 Future Scope

The present research demonstrates how AI and ML can revolutionize the legal system by automating repetitive processes, forecasting case outcomes, and increasing efficiency. However, the quality and availability of the data employed restrict the scope of this research. The suggested technique may have limited generalizability due to the size and variety of the dataset used in the present research. With the availability of improved datasets, the proposed methodologies can anticipate to yield greater accuracy and actionable insights in supporting the advancement of the state of AI for law systems.

## References

1. https://www.indialawoffices.com/legal-articles/role-artificial-intelligence-justice-delivery-system
2. https://ecommitteesci.gov.in/project/brief-overview-of-e-courts-project/
3. https://informatics.nic.in/news/1192
4. https://www.unesco.org/en/artificial-intelligence/rule-law/mooc-judges
5. https://www.wipo.int/edocs/mdocs/enforcement/en/wipo_ace_16/wipo_ace_16_15_presentation
6. https://chss.org.in/aiinjudiciary/#:~:text=For%20the%20Estonian%20and%20Chinese,relevant%20legal%20documents%20almost%20instantly
7. Singh, A., Sharma, R., Kumar, S.: Predictive modeling in criminal justice: a case study in the Indian context. J. Legal Anal. **12**(3), 345–367 (2020)
8. Ngo, M.: Classification of Legal Verdicts Using a Corpus Linguistic-Based Approach (2017)
9. Brüninghaus, S., Ashley, K.D.: Combining case-based and model-based reasoning for predicting the outcome of legal cases. In: Case-Based Reasoning Research and Development: 5th International Conference on Case-Based Reasoning, ICCBR 2003 Trondheim, Norway, June 23–26 2003, Proceedings 5, pp. 65–79. Springer, Berlin Heidelberg (2003)
10. Sulea, O.M., Zampieri, M., Vela, M., Van Genabith, J.: Predicting the law area and decisions of french supreme court cases. arXiv preprint arXiv:1708.01681 (2017)
11. Aletras, N., Tsarapatsanis, D., Preoţiuc-Pietro, D., Lampos, V.: Predicting judicial decisions of the European court of human rights: a natural language processing perspective. Peer J. Comput. Sci. **2**, e93 (2016)
12. Luo, B., Feng, Y., Xu, J., Zhang, X., Zhao, D.: Learning to predict charges for criminal cases with legal basis. arXiv preprint arXiv:1707.09168 (2017)
13. Liu, Y.H., Chen, Y.L.: A two-phase sentiment analysis approach for judgement prediction. J. Inf. Sci. **44**(5), 594–607 (2018)
14. Zhong, H., Guo, Z., Tu, C., Xiao, C., Liu, Z., Sun, M.: Legal judgment prediction via topological learning. In: Proceedings of the 2018 Conference on Empirical Methods in Natural Language Processing, pp. 3540–3549 (2018)
15. Agrawal, R., et al.: Predictive modeling in Indian courts: a machine learning approach. J. Legal Anal. **7**(2), 123–136 (2019)
16. Gupta, S., Sharma, A.: Predictive analytics in Indian courts: forecasting bail outcomes using machine learning. Indian J. Legal Stud. **12**(3), 345–362 (2020)
17. Kumar, A., et al.: Natural language processing for legal case prediction in the Indian judiciary. J. Artif. Intell. Law. **15**(1), 87–104 (2021)
18. Alsolai, H., Roper, M.: A systematic literature review of machine learning techniques for software maintainability prediction. Inf. Softw. Technol. **119**, 106214 (2020)

19. Pillai, V.G., Chandran, L.R.: Verdict prediction for Indian courts using bag of words and convolutional neural network. In: 2020 Third International Conference on Smart Systems and Inventive Technology (ICSSIT), pp 676–683. IEEE (2020)
20. Singh, G., Thanaya, I.: Predicting earnings per share using feature-engineered extreme gradient boosting models and constructing alpha trading strategies. Int. J. Inf. Technol. **15**(8), 3999–4012 (2023)
21. Alghazzawi, D., Bamasag, O., Albeshri, A., Sana, I., Ullah, H., Asghar, M.Z.: Efficient prediction of court judgments using an LSTM+ CNN neural network model with an optimal feature set. Mathematics. **10**(5), 683 (2022). https://doi.org/10.3390/math10050683
22. Medvedeva, M., Vols, M., Wieling, M.: Using machine learning to predict decisions of the European court of human rights. Artif. Intell. Law. **28**(2), 237–266 (2020)
23. Gusenbauer, M., Haddaway, N.R.: Which academic search systems are suitable for systematic reviews or meta-analyses? Evaluating retrieval qualities of Google scholar, PubMed, and 26 other resources. Res. Synth. Methods. **11**(2), 181–217 (2020)
24. Belenguer, L.: AI bias: exploring discriminatory algorithmic decision-making models and the application of possible machine-centric solutions adapted from the pharmaceutical industry. AI Ethics. **2**(4), 771–787 (2022). https://doi.org/10.1007/s43681-022-00138-8
25. Sarker, I.H.: AI-based Modeling: techniques, applications and research issues towards automation, intelligent and smart systems. SN Comput. Sci. **3**, 158 (2022). https://doi.org/10.1007/s42979-022-01043-x
26. https://www.thedigitalspeaker.com/privacy-age-ai-risks-challenges-solutions/
27. https://www.europarl.europa.eu/RegData/etudes/STUD/2020/634452/EPRS_STU(2020)634452_EN.pdf

# Energy Efficiency in Software-Defined Wireless Network

**Sonu Singh, Rajeev Kumar Shah, Asif Khan, Jiyaul Mustafa, and Ranju Mahato**

**Abstract** The network devices need to be turned on around-the-clock in order to manage the traffic. The legacy wireless network's networking gear integrated the devices that perform both data transmission and controlling tasks. The majority of network devices are generally optimal for the majority of the time, indicating needless energy use in communication networks. A novel network paradigm called software-defined wireless networks divides the data plane from the control plane, giving the network flexibility and virtualization. Software-defined networks offer programmable network devices and give administrators the ability to govern the devices via a centralized controller. SDN offers chances to improve the overall energy efficiency and performance of the network. In order to lower the energy consumption of wireless networks when they are idle, this research suggests a traffic-aware adaptive link rate technique. The basic concept is to put any switches, routers, and links that are not in operation into sleep mode at night when there is no traffic. To simulate a normal wireless network and open daylight as an SDN remote controller, we have utilized a Mininet-WiFi network emulator. The energy consumption of software-defined wireless networks and conventional wireless networks is compared in this research. The purpose of this comparison is to determine the energy efficiency of Software-Defined Wireless Networks (SDWNs) in comparison with conventional wireless networks. The outcome demonstrated that by using the suggested energy-saving strategy, we may reduce overall energy usage by up to 49%.

S. Singh (✉)
IT Computing Department, International School of Management & Technology (ISMT), Kathmandu, Nepal, Affiliated to University of Sunderland, Sunderland, UK
e-mail: sonusingh@ismt.edu.np

R. K. Shah · R. Mahato
School of Artificial Intelligence and Data Science, Jiangxi University of Engineering, Jiangxi Province, Xinyu, China

A. Khan
Department of Computer Application, Integral Robotics Lab, Integral University, Lucknow, India

J. Mustafa
Department of Mechanical Engineering, Bennett University (The Times Group), Greater Noida, India
e-mail: jiyaul.mustafa@bennett.edu.in

© The Author(s), under exclusive license to Springer Nature Singapore Pte Ltd. 2026

M. Pant et al. (eds.), *Real-World Applications of Machine Learning*, Mathematics for Sustainable Developments, https://doi.org/10.1007/978-981-95-7134-5_30

**Keywords** SDWN · Energy efficiency · Performance · Wireless network

## 1 Introduction

New applications like crowd sensing and mobile social networking, which are nearly impossible with wired networks, are made possible by advancements in wireless networks. Technologies for wireless networks are constantly evolving in both industry and academia [1]. With the development of information and communication technology (ICT), energy consumption of communication infrastructures also growing exponentially. Energy consumption of mobile networks is growing much faster than ICT on the whole. Such diverse and expanding demands cannot be met by the traditional network concept. Current network demands, such as high bandwidth and connection speeds, dynamic management, network virtualization, and cloud computing, cannot be satisfied by traditional network infrastructures. As a result, a new network paradigm known as Software-Defined Networking (SDN) has evolved as a competitive alternative to conventional networks. By separating the network's control plane from its data plane, Software-Defined Networking offers programmability, automation, and control over network operations. SDWN is a new networking technique that simplifies administrative responsibilities while offering wireless network flexibility. The control, data, and application planes make up the SDWN architecture [2, 3]. The application plane, sometimes referred to as the management plane, gives developers a platform for network management.

Devices transmit packets according to the controller-defined rules in the data plane, which is made up of a flow entry table. The new rule for that specific flow entrance is determined by the control plane. Through a northbound interface, the controller interacts with the network applications at the application layer [3, 4]. One popular protocol in this interface is OpenFlow. The basic SDWN architecture is depicted in Fig. 1.

Our goal was to compare the energy usage of software-defined wireless networks and conventional wireless networks in this study. In contrast to a standard wireless network, this comparison is made to determine how much energy may be saved by a wireless network that uses Software-Defined Wireless Networks (SDWNs). As an SDN remote controller, we have simulated a standard wireless network using a Mininet-WiFi network emulator and open daylight [5, 6].

This study is focused on reducing the energy consumption of wireless network using SDWN-based energy optimization approach. This paper proposes an adaptive link rate traffic aware approach to reduce the energy consumption of wireless network during idle state. The main idea is to switch off the unused switches, routers, and links to sleep mode during night time when they carry no traffic [7, 8].

The rest of this paper is organized as follows. We present the background and related works on energy optimization of wireless network using SDN in Sect. 2. Methodology regarding energy efficiency of wireless network using SDWN optimization technique by implementing adaptive link rate (ALR) traffic aware approach

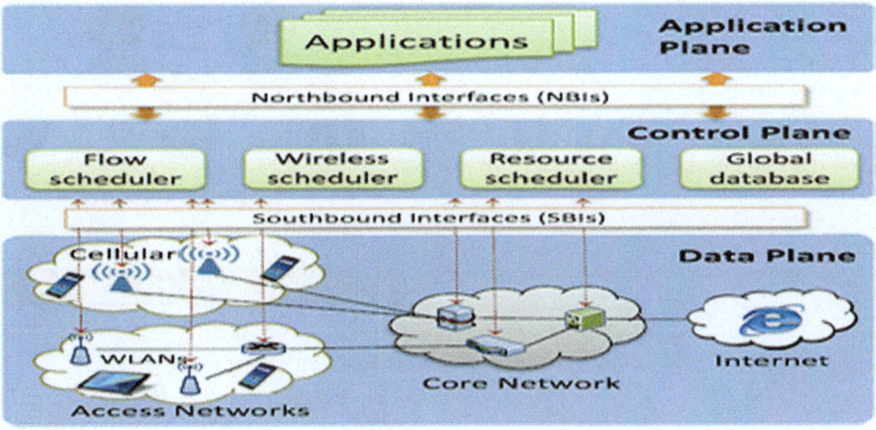


**Fig. 1** SDWN architecture [4]

is discussed in Sect. 3. Section 4 presents result and analysis of performance of SDWN-based energy optimization algorithm. Section 5 concludes the paper.

## 2 Background and Related Works

### *2.1 Software-Defined Network*

The aim behind Software-Defined Networking (SDN), a new paradigm in networking, is to better manage the rapidly growing network and the significant increase in data traffic, network virtualization, and user equipment mobility. Network managers and operators may typically increase capabilities and grow their networks without sacrificing performance, dependability, or user experience by using SDN to programmatically control their systems.

In legacy networks, the control plane and data plane is not separated, it coupled together in same device as shown in Fig. 2a. The decisions are controlled by control plane in communication network. However, in legacy network the control planes are distributed in overall network and operated by distributed network operating system which may not belongs to same stakeholder. Due to this, the obstacle of agreements on functionality and policies arise in innovation of communication networks. This obstacle has been reduced by SDN by decoupling the control plane from the data plane which is shown in Fig. 2b, where distributed devices get centralized logical control by single SDN controller poses with centralized logic control plane [5].

SDN architecture places control plane operations on a single server rather than on individual networking devices. An operating system that has the required SDN software is often the SDN controller. Typically, the controller uses the OpenFlow

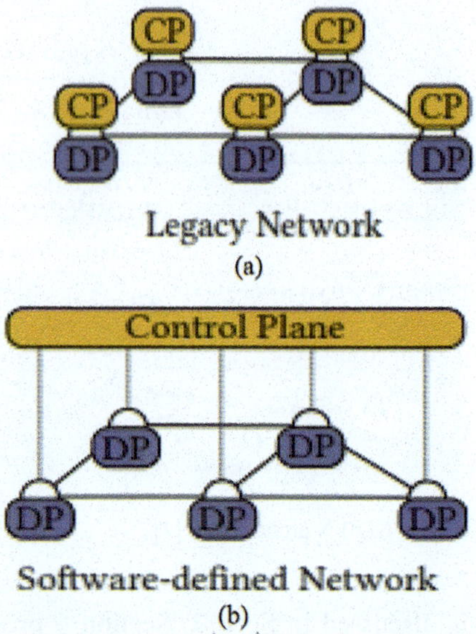


**Fig. 2** Traditional network layout where control plane (CP) and data plane (DP) are integrated together (**a**) and SDN network layout where control plane (CP) and data plane (DP) are separated (**b**) [5]

protocol, which is widely known, to connect with the switch data plane. OpenFlow sends commands and instructions to the data plane, allowing it to pass the data in the appropriate direction. The OpenFlow protocol must be present and operating on the network devices in order to support the services [5].

Open application programming interfaces (APIs) provide access to the three separate levels that make up the SDWN architecture: [1, 5] (a) The end-user business apps that offer various communications services make up the application layer. The API controls communication between the control layer and the application layer. (b) Using an open interface, the control layer manages and oversees the network forwarding capabilities. (c) The physical network equipment (such as switches and routers) that offer packet forwarding and switching make up the physical layer. This model identifies three essential characteristics of SDN architecture:

(1) Logically centralized intelligence: Using the OpenFlow standard interface, an SDN delivers worldwide network orchestration from a single point of management. Making decisions based on a global (or domain) perspective, where nodes are ignorant of the network's general condition, is made easier by centralizing network intelligence.
(2) Programmability: By allowing apps to connect to the network through open APIs, SDN networks can revolutionize and set themselves apart from conventional networks. Programmatic interfaces provided by an SDN can automate and customize network fabric configuration.
(3) Abstraction: In an SDN network, the subsystem technologies are separated from the business applications that use SDN services. Additionally, network devices are isolated from the SDN Control Layer to provide mobility with any vendor's application.

## 2.2 Energy Efficiency

Wireless networks can be made more energy efficient in the following ways by SDN technology because of its flexibility, programmability, and controllability. (1) First, the decoupling of data and control facilitates the deployment of energy-efficient and low-cost protocols in wireless networks. (2) Second, SDN provides an easy approach to lower network administration and maintenance overheads, as well as protocol operation overheads. To reduce the application-specific overhead of typical nodes, for example, controller-made routing decisions can be implemented as flow rules in the node flow tables. (3) Third, by allowing controllers to monitor network data globally or semi-globally, network energy optimization will be facilitated by cooperative resource allocation.

For energy optimization, network virtualization (NV) and adaptive link rate (ALR) technique will be used. NV is the energy efficient SDN approach. Physical components in NV may have many virtual network structures. Thus potentially lowering the new network's power consumption as well as the cost of building new hardware. Providing dynamic ability allows for quick transmission from the old system to the new one, which improves performance and saves energy in the event that a network change is urgently required [7].

## 2.3 SDWN Benefits

SDN has a number of benefits over conventional networks. Some of the main advantages include:

Energy management: SDN can be utilized to save energy by itself or by applying energy optimization to different parts of the SDN architecture. Hardware-based enhancements or algorithmic approaches can be used to address energy conservation in SDN.

**Directly Programmable**: The network can be programmatically customized using either proprietary or open-source automation tools due to the decoupling of control and forwarding functions.

**Network Functions Virtualization (NFV)**: SDN and NFV are two technologies that are complimentary, independent, closely related, and advantageous to each other. While NFV handles the transfer of network services from specialized appliances to generic servers, SDN focuses on the separation of the control and data planes, centralization of control [7, 9].

## 2.4 Related Work

Wireless network energy efficiency has been researched for a long time, and several surveys have been released. Many of them concentrate on either particular network protocols like media access control (MAC), routing, or physical layer protocols, or on particular wireless networks like cellular networks and WLANs. However, recent polls have addressed big data's wireless networking components.

Guptain was the first to introduce the notion of using intelligent routing in SDWN to save energy. For an energy-conscious routing problem in SDN, Chiaraviglio developed an integer linear programming model. To lower energy usage, he uses a greedy algorithm to gradually turn off nodes and links. This strategy is limited to energy conservation in backbone networks, though.

The article "Energy-Efficient Communication in wireless network" describes the evolution of, and state of the art in, energy-efficient techniques for wireless communication networks of embedded computers [2]. "An Architecture for Software-Defined Wireless Networking," describes the detail architecture of SDN, wireless, their standardization, software and protocols [3]. "SDN-Based Solutions for Energy-Efficient Networks," presents an approach to provision servers in such a manner to optimize cost and energy in a network [4]. "Energy Efficiency in Wireless Networking Protocols," this article discussed the necessity of energy efficiency in wireless networking protocols and their measures according to their applications and environment [5] network programmability. By offering the infrastructure needed to operate the SDN software, NFV is able to support SDN.

**Centralized Management**: SDN controller software, which keeps a global view of the network and presents it to the applications and policy engines as a single, logical switch, logically centralizes network information.

**Reduce Opex**: SDN makes it easier to build, install, operate, and grow networks by enabling algorithmic control of increasingly programmable hardware and software switches and routers.

**Flexibility and Agility**: Software-Defined Networking enables businesses to swiftly implement new infrastructure, services, and applications in order to satisfy evolving corporate goals and objectives.

**Facilitate Innovation**: New services, apps, and business models can be developed directly at the application layer without interfering with other components.

**Network Management and Resource Utilization**: The system is made more flexible by the separation of the control and data planes, and the control plane can be programmed to use resources as needed.

**Bandwidth**: SDN can make effective use of the available bandwidth by dynamically distributing it to the needed user. The dynamic allocation of bandwidth can enhance the performance of the overall network also increase in energy efficiency [10].

## 3 Methodology

Here, we describe the process used to assess SDWN's energy efficiency. Both data forwarding and regulating functions are included into the networking hardware of legacy wireless networks. A communication network's needless energy usage is demonstrated by the fact that the majority of its devices are generally optimal. Traditional network infrastructures are unable to handle today's demands, which include cloud computing, dynamic management, high bandwidth and connection speeds, and network virtualization. The current dynamic networks are met by SDWN's software-controlled network flow agility and topology management. Because of its adaptability, controllability, and programmability, SDWN technology can increase wireless networks' energy efficiency [7, 8, 10].

### *3.1 Legacy Wireless Network Topology*

To compare the performance of legacy and SDN wireless network, we have taken the traditional wireless network scenario of Pulchowk Campus which contain 9 access points and 9 Switches. Figure 3 shows the network connection topology of legacy wireless network. This scenario consists of two areas: (1) Administration network connection area and (2) Hostel network connection area. Administration network remains active during daytime mostly in office hour time from 10 a.m. to 5 p.m. And hostels network gets active during night time. Hence network traffic patterns are different for both the scenario. This paradigm consists of 9 switches and access points. All the switches are connected to each other through wired connection and access points are connected through wireless connection.

### *3.2 SDN-Based Wireless Network Topology*

Because of the separation of the control plane and data plane, which enables software to handle the traffic network, SDN offers greater network flexibility. The entire network is managed and controlled by a centralized unit known as the SDN controller, which makes up the SDN control plane [7, 11].

Here, the same scenario of legacy wireless network as shown in Fig. 3 will be created virtually in SDN platform using Mininet-WiFi network emulator as testbed and OpenDayLight (ODL) remote controller as SDN controller.

Figure 4 shows the SDN-based connection of wireless network of legacy wireless network mentioned in Sect. 3(A). Here, the SDN controller is used to control and manage the devices of network.

Essentially, an SDN controller serves as the network's brain. It is a strategic control point that is in charge of communicating with the network nodes that are connected

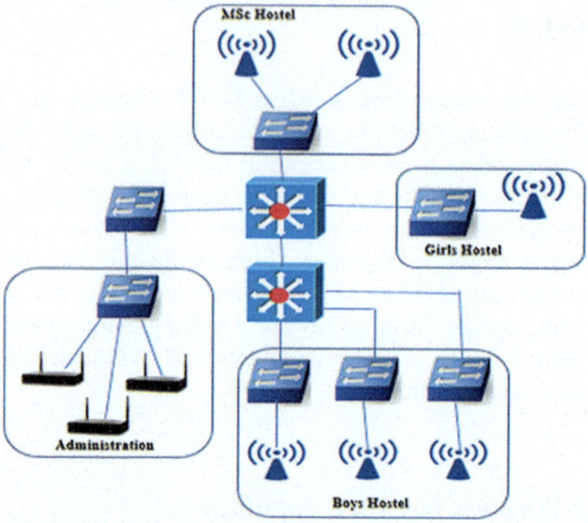


Fig. 3 Legacy wireless network scenario. *Source* IOE, Pulchowk Campus, CIT department

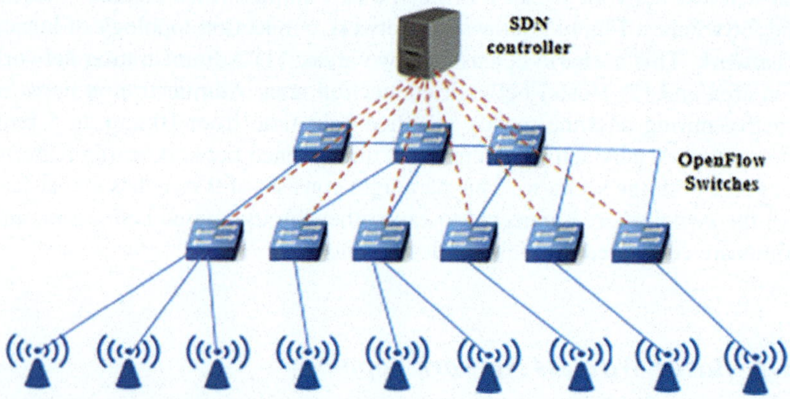


Fig. 4 SDN-based wireless network topology

to it. The controller controls network traffic flow and makes intelligent networking possible. As determined by the algorithm, the controller interacts with the network topology and runs a certain algorithm. A few of the available SDN controllers are POX, NOX, Ryu, Flood Light, and Open Day light (ODL) [10, 12].

Open Day light is a Linux Foundation-hosted open-source SDN controller. Applications use the OpenDaylight Controller's exposed open northbound APIs. These applications use the OpenDaylight Controller to generate new rules throughout the network after using the Controller to gather network data and run algorithms for analytics. OpenDaylight can support various open SDN standards in addition to the OpenFlow protocol for southbound communication [7, 12].

Our testbed for the experimental configuration of a virtualized wireless network environment is the Mininet-WiFi emulator. A popular simulator for creating wireless networks is the Mininet-WiFi emulator. It creates an artificial network that mimics an actual physical network. With Mininet, we can set up a network topology, simulate it, and use basic code to implement different network performance factors like bandwidth, latency, packet loss, etc. To replicate a large network on a single PC, Mininet allows the formation of many nodes (hosts, switches, controllers) by building the virtual network on a single machine (a virtual machine, the cloud, or a native machine) [6, 13].

### 3.3 Energy-Saving Model

In Section 2, various energy-saving strategies for wireless networks are covered. Using the idea of smart wakeup and sleep cycles, we solely assess the energy efficiency of SDN-enabled wireless networks in this part by looking at the power consumption of switches and links. Because the network switches and links in the current legacy wireless network are operational around-the-clock, we assume that they always use their maximum energy consumption. The pattern of energy consumption by all testbed switches and links over the period of 24 h will be recorded after applying the smartsleep and wakeup energy-saving technique. Although the administration area network remains idle more than 12 h a day but we assume the campus network idle state to be 8 h/day (10 pm to 6 am).

## 4 Results and Analysis

### 4.1 Energy Consumption of Traditional Wireless Network

Table 1 contains the standard specification of switches and access points used in network which are listed from the authorized websites of network devices.

### 4.2 Energy Consumption of SDN-Enabled Wireless Network

In this section, we used the Mininet-WIFI emulator to create SDN in wireless networks. This open-source network emulator can create networks using SDN controllers, virtual stations, and access points. We have demonstrated that the SDN-enabled wireless network is energy efficient and provides better network performance than legacy wireless network [14, 15]. The OpenDayLight SDN controller, which

**Table 1** Power and energy consumption of network devices

| Switch model | Max. Power rating (W) | No. of switches used in network | Power consumption (W) | Energy consumption (Kw-hr) |
|---|---|---|---|---|
| HP 5500 24 GE1 L3 switch | 110 | 1 | 18.4 | 0.0184 |
| Cisco SG-300-10P | 81.16 | 1 | 40.58 | 0.0406 |
| Catalyst 2960 24PC-S | 45 | 1 | 3.8 | 0.0038 |
| HP GE L3 switch | 30W/port | 6 | 450 | 0.450 |
| Peak power of AP | 7.87 | 9 | 70.83 | 0.07083 |
| Total | | 18 | 583.61 | 0.5836 |
| Total energy consumed in 24 h | | | | 14.016 |

connects and configures network devices (routers, switches, etc.) using the OpenFlow protocol to choose the best path for application traffic, was employed in our simulation testing. For this user, stations typically gather data and send it to the directly linked access point (AP). This data traffic is then routed to the controller, which has all the flow rules needed to steer all requests in the right path. In every simulation test, every station moves dynamically. The channel propagation loss in all simulated tests is assumed to be 0.5 dB. Table 2 explains the primary simulation parameters that are used in the scenario to improve wireless network performance.

The observation is done on the basis of above recorded data about power consumption and smartsleep time. Figure 5 shows the network visualization of implemented topology of SDWN. Here in SDWN network, the smartsleep technique is applied according to the need of network. In this simulation, the connections are made to sleep for 8 h at night time from 10:00 P.M. to 6:00 A.M. for administration block network and 8 h' day time from 9:00 A.M. to 5:00 P.M. for hostel block network. Also it automatically shut down the link when no traffic was passed and only signaling traffic is recorded. Figures 6, 7, and 8 show the graph of total link power, individual access

**Table 2** Simulation parameters

| Parameters | Values |
|---|---|
| Average power consumed by a switch | 110 W |
| Average power consumed by active link | 40 W |
| Average power consumption by active CPE | 7 W |
| Link bandwidth (among access switch)<br>(among access point to CPE) | 1 Gbps<br>5 Mbps |
| Energy consumed by a switch in sleep mode<br>CPE in sleep mode | 33 W<br>2.8 W |
| Observation period | 24 h |

point power of administration block and hostel blocks, respectively. The observation time was set for 24 h.

Here, Table 3 shows the energy consumption of legacy wireless network and software-defined wireless network (SDWN). The observation of experiment shows that SDWN can save up to 47% of energy in total as compare to the energy consumed by legacy network.

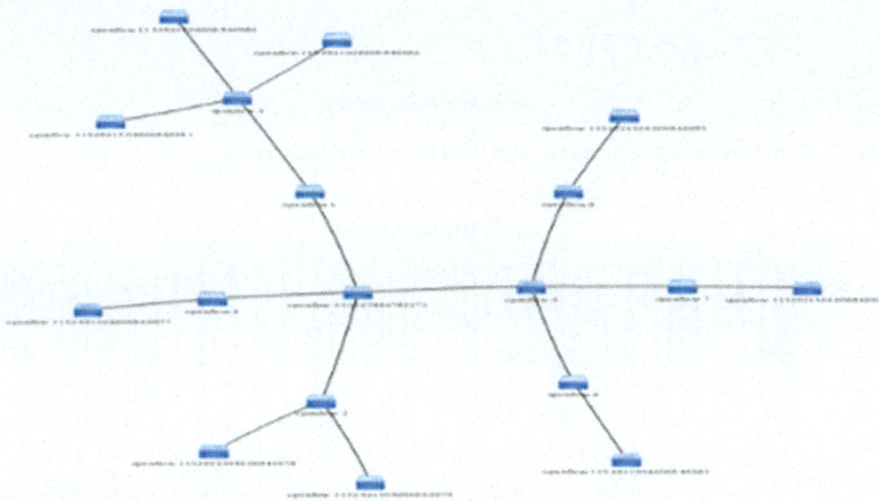

**Fig. 5** Visualization of network connection using ODL controller

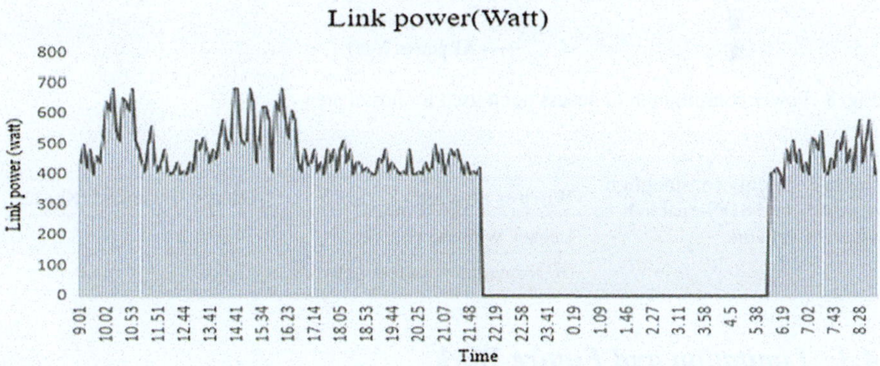


**Fig. 6** Power consumption of link between switches and access points

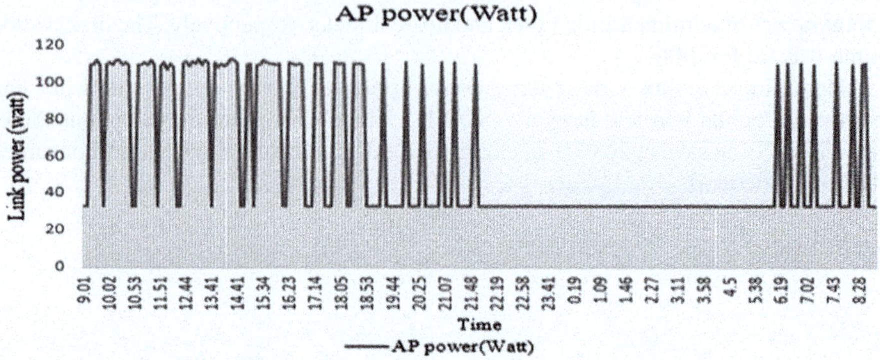


**Fig. 7** Power consumption in access point used for Administration area

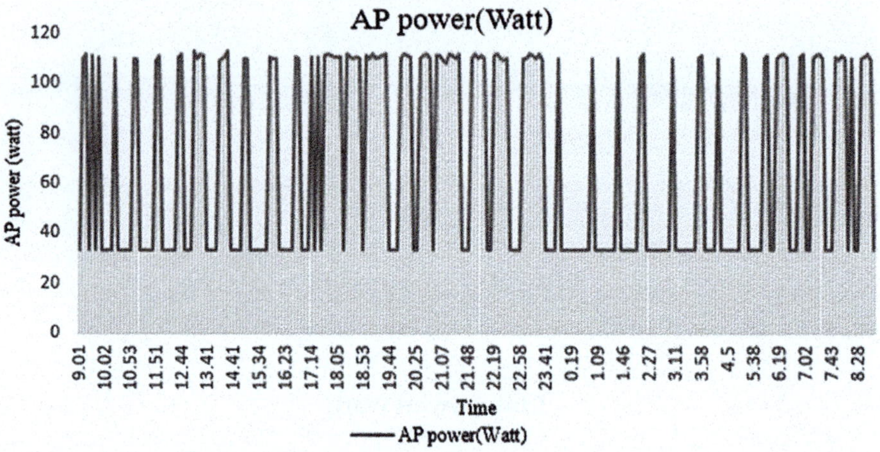


**Fig. 8** Power consumption of access point used for hostel area

**Table 3** Energy consumption of legacy and SDN-enabled wireless network

| Wireless network | Energy consumed (Kw-hr) |
|---|---|
| Legacy wireless network | 14.016 |
| SDN-enabled wireless network | 7.439 |

## *4.3 Limitation and Future Work*

The scope in this study was limited to the implementation and analysis of energy efficiency of SDN-based wireless network of particular campus area. Where the smartsleep and wakeup technique has applied according to the network traffic pattern and time schedule of particular campus. Further research can be conducted on the energy optimization of SDN-based wireless cellular network using the real-time daily network traffic pattern. Also, the research will be further extended to apply the smart sleeping technique for the resources level of network devices.

## 5 Conclusion

In this paper, we have measured and compared the energy consumption of legacy wireless network and software-defined wireless network (SDWN). The network devices in legacy network are static in nature as they embed both data plane and control plane in same device which make difficult to manage and operate the devices according to the flow of network traffic and give rise to the energy consumption. So to deal with this issue, we have applied SDN into this wireless network to get the flexibility in network and to improve the energy efficiency of the overall network. Hence after comparing and evaluating the energy consumption of legacy wireless network and SDWN, we conclude that SDWN is more energy efficient that the legacy wireless network. SDWN paradigm has ability to save more energy with their flexible features.

## References


1. Software-Defined Networking: The New Norm for Networks ONF White Paper (2012)
2. Ye, J., Cheng, X., Zhu, J., Feng, L., Song, L.: A DDoS attack detection method based on SVM in software defined network. Security Commun. Netw. **2018**, 1–8 (2018)
3. Xie, J., Guo, D., Hu, Z., Qu, T., Lv, P.: Control plane of software defined networks: a survey. Comput. Commun. **67**, 1–10 (2015)
4. Barki, L., Shidling, A., Meti, N., Narayan, D.G., Mulla, M.M.: Detection of distributed denial of service attacks in software defined networks. In: 2016 International Conference on Advances in Computing, Communications and Informatics (ICACCI) (2016)
5. Blenk, A.A.: Towards Virtualization of Software-Defined Networks: Analysis, Modeling, and Optimization. A publication from the Chair of Network Architectures and Services (NET) at the Technical University of Munich (TUM), Dipl.-Inf. Univ., November 2018
6. Saha, S.: Software defined wireless network: the rise, the evolution, the future. Int. J. Eng. Res. Technol. (IJERT) (2023)
7. Gniewkowski, M.: An overview of DoS and DDoS attack detection techniques. In: Theory and Applications of Dependable Computer Systems, pp. 233–241 (2020)
8. Dayal, N., Srivastava, S.: Analyzing Behavior of DDoS Attacks to Identify DDoS Detection Features in SDN. IEEE Xplore, 01 Jan 2017
9. Mininet Project Contributors: Mininet: An Instant Virtual Network on Your Laptop (or Other PC) - Mininet, Mininet.org (2022)
10. Yu, Y., Guo, L., Liu, Y., Zheng, J., Zong, Y.: An efficient SDN-based DDoS attack detection and rapid response platform in vehicular networks. IEEE Access **6**, 44570–44579 (2018)
11. Rout, S., Sahoo, K.S., Patra, S.S., Sahoo, B., Puthal, D.: Energy efficiency in software defined networking: a survey. SN Comput. Sci. **2**(4) (2021)
12. Polat, H., Polat, O., Cetin, A.: Detecting DDoS attacks in software-defined networks through feature selection methods and machine learning models. Sustainability **12**(3), 1035 (2020)
13. Braga, R., Mota, E., Passito, A.: Lightweight DDoS flooding attack detection using NOX/OpenFlow. In: IEEE Local Computer Network Conference (2010)

14. Mohammed, S.S., et al.: A New Machine Learning-based Collaborative DDoS Mitigation Mechanism in Software-Defined Network (2018)
15. Wahab, F., et al.: An AI-driven hybrid framework for intrusion detection in IoT-enabled E-health. Comput. Intell. Neurosci. **2022**, 1–11 (2022)

# Enhanced Detection of Distributed Denial of Service (DDoS) Attacks in Cloud Computing Using Transfer Learning, DBSCAN, and Entropy Analysis

**Saswati Chatterjee, Vijaykumar JayantiBhai Solanki, and Rinkal Dharmesh Sarvaiya**

**Abstract** The present research explores an innovative approach to improve DDoS detection in a cloud environment by evaluating the performance of various algorithms and methodologies. The paper integrates ML and DL approaches to develop a comprehensive framework for the recent developments in information security and the implementation of IDS. It provides an in-depth analysis of DDoS attack types and explores existing detection strategies, emphasizing the role of preprocessing subsystems and feature selection in enhancing detection accuracy. Leveraging transfer learning and the SDN Dataset that consists of 23 significant features for the detection of intrusion in a network, a transfer learning-based Intrusion Detection System (IDS) is presented in this paper. Using the Density-Based Spatial Clustering of Applications with Noise (DBSCAN) algorithm, this research categorizes attack patterns and noise giving it the flexibility of the anomaly detection method. Entropy is used as a metric to measure the level of randomness in network traffic, with any deviations from expected patterns potentially indicating an ongoing attack. The preprocessing pipeline includes rebalancing data, cleaning, splitting, and applying min–max normalization. A hypermodel combining classifiers such as Support Vector Machines (SVMs), K-Nearest Neighbors (KNN), XGBoost, and other ML models is employed for performance evaluation. The results demonstrate that the proposed deep learning and statistical framework, augmented by DBSCAN and entropy analysis, achieves superior accuracy and robustness in detecting and preventing DDoS attacks compared to traditional methods.

S. Chatterjee (✉) · R. D. Sarvaiya
Parul University, Vadodara, Gujarat, India
e-mail: cshiv68@gmail.com

R. D. Sarvaiya
e-mail: rinkaldsarvaiya@gmail.com

V. J. Solanki
IIT Roorkee, Roorkee, India
e-mail: vijaykumar_js@ma.iitr.ac.in

© The Author(s), under exclusive license to Springer Nature Singapore Pte Ltd. 2026

M. Pant et al. (eds.), *Real-World Applications of Machine Learning*, Mathematics for Sustainable Developments, https://doi.org/10.1007/978-981-95-7134-5_31

**Keywords** Man-in-the-middle attack · Classifier · K-means · Possibilistic C-Means Clustering

## 1 Introduction

Intrusion means any consecutive course of action performed by an attacker that threatens a targeted system. In the case of DoS/DDoS attacks, these intrusions are of high variability, and in most cases, they can be categorized according to the attack type. The major risk that can be associated with these intrusion activities is the capability to perform DoS and DDoS that leads to the utilization of computer and communication resources or the ability to take advantage of exploitable vulnerabilities, thus denying access to the system to legitimate users. This can lead to various losses including resources, financially, and loss of data to individuals with ill intentions. A DDoS attack can be described as a situation whereby multiple systems flood the Internet connection of a targeted system usually the one or more web servers. Such attacks generally occur as flood attacks in which the attacker directs large volumes of network traffic from many compromised hosts. Other kinds of attacks are often inconvenient; DDoS attacks, however, can be dangerous because they disable an Internet-connected resource. DDoS attacks are commonly executed in major two steps. In the first phase, the attacker's main goal is to take control of the weak computers in a particular network. Once the attacker gains control of these machines, he or she will then choose a particular server to send packets to. In DoS attacks, distributed attacks involve only one host, while DDoS attacks are spread across several compromised networks. This stage is referred to as the 'DDoS attack phase. DoS and DDoS attacks can be broadly categorized into three types based on their methods: connections, bandwidth, and exploitation of vulnerable points (Ali et al. 2019). Connection consumption-based attacks are a connection-oriented protocol that works on the TCP which establishes the connection between client and server before the data transmission; any server can handle only a limited number of connection requests at a time. A connection consumption attack floods the server with many connections and does not allow legitimate users to access the service. This exhausts the operating system kernel resources the most known attack within this category is the SYN Flood attack.

Each network functions in a specific number of bandwidths. When this bandwidth limit is reached, it affects the servers' response time and the general connection response time of the linked devices. Bandwidth consumption-based DoS and DDoS attacks take advantage of this by consuming what little bandwidth the network has hence causing considerable slowdown or total shutdown of service. A similar attack is a flood attack since it is challenging to prevent with elementary security measures put in place. In the second phase of the attack, the "DDoS Attack Phase," the attacker directs the compromised machines to bombard the target system with packets or data requests. This is usually achieved by flooding the target with a large volume of small data packets much to its capacity of handling genuine requests. Since the attack happened in various systems, it becomes cumbersome to counter it. The first type is a connection consumption attack in which an attacker sends an enormous number of connection requests to the server, using methods such as SYN flooding, whereby the attacker sends many incomplete connection requests to the server. This consumes the server's resource hence denying it to genuine customers. Bandwidth Consumption-Based Attacks: It was established that every network works under a particular bandwidth which is used to measure the amount of data that can traverse a network in a given period. When this bandwidth is utilized to its maximum, it results in network jam, slow performance, or even complete system lockout of service. This is well exploited in bandwidth consumption-based DoS and DDoS attacks whereby a lot of data throughput is used to occupy all the bandwidth. One such approach employed in these attacks is the UDP flood which refers to the act of the attacker forwarding large numbers of UDP packets to the system in question. As UDP is an unreliable connectionless protocol, the server cannot authenticate the incoming traffic, thereby making it vulnerable to bandwidth exhaustion.

Evaluating the types of these attacks as well as the measures used to prevent such attacks will go a long way in preventing organizations from being victims of these disruptive attacks. This paper also discusses the use of transfer learning to train a model for improved accuracy in detecting these attacks. The remaining sections of the document are arranged as follows: Related work is covered in Sect. 2 followed by the DDoS Attack SDN Dataset Sect. 3. The proposed architecture is shown in Sect. 4, and the experimental findings are explained in Sect. 5. Section 6 brings the study to a close and addresses possible directions for further research.

## 2 Related Work

Many organizations currently prefer convolutional neural network learning because of its enhanced performance and accuracy as well as it being a current area of study. These are automotive architecture, health care, manufacturing, and law enforcement. Every researcher in this field has conducted important studies and below are summaries of such researches. The researchers have done a comparison of the current machine learning strategies such as Naive Bayes and Decision Tree classification with other strategies [1]. Their work mainly targeted offline models and they pointed out that when the dataset size is large, the latency difference is even more pronounced. Scientists have proposed deep learning to use radial basis functions at multiple abstraction levels [2]. Their experiment employed standard datasets, for instance, NSL-KDD and UNSW NB15 having 27 features. The researchers pointed out their de novo approach as more accurate compared to other approaches that had been employed. In their study, the researcher applied a clustering method for the feature selection purpose and then compared it with five similar machine learning algorithms [3]. For training, they have utilized the techniques of Support Vector Machine (SVM) as well as the Random Forest (RF). The accuracy that was obtained by the RF method was the highest with about 96 percent accuracy. The experts concentrated solely on the routing of packets according to the nature of the packets involved [4]. Their method was based on analyzing headers of IP packets to identify so-called spoofed and genuine IP addresses. However, they discovered that in matters of volume, the firewalls are not effective as they are supposed to be. The authors applied the anomaly detection method as well as the machine learning method to filter out the regular from malicious traffic using real-time datasets [5, 6]. The method used was Naive Bayes and the results were compared with other methods for example; J48 and RF. The experts [7, 8] used artificial intelligence and entropy-based methodologies in their study while employing the SVM for classification. The researchers extracted data from a flow of a platform and the detection rates were stated to be satisfactory by the researchers [9]. The scientists used a model that is based on IPv6 classification to detect DDoS attacks [10, 11]. They compared their result with five basic machine learning algorithms: Decision Trees (DTs), Support Vector Machines (SVMs), Naïve Bayesian (NB), k-nearest Neighbors (KNN), and Artificial Neural Network (NN) using a well-known dataset. KNN method has been giving around 85 percent accuracy: Hence, the experts developed a semi-supervised, entropy-based methodology that employs both the supervised and unsupervised with necessary designs [12]. It was noted that unsupervised methods provided high precision and low levels of false positives, while methods that were supervised effectively reduced false positive values [13]. The datasets that were employed in their research include NSL-KDD, UNB ISCX 12, and UNSW NB15. Unsupervised learning was using a deep learning network known as NDAE [14]. It was applied in a GPU using TensorFlow and was tested with KDD Cup 99 and NSL-KDD datasets and provided a high detection accuracy [15]. The Internet is the modern world of computers connected in one big network and is essential part of people's lives for studying, business, communication, and entertainment. Recent findings on security establish massive financial losses due to DDoS attacks in businesses and governments in different parts of the world. Cyber attackers are making more elaborate attempts to scale up their campaigns and now use the geographical spread and processing power of different kinds of devices connected through IoT networks. The study is only limited to a particular dataset (SDN Dataset for Network Intrusion Detection) which may not be extended to all cloud settings or more data sets Exploring under which conditions the proposed model can be adapted to other cloud architectures or network environment may also present an unexplored region. Transfer learning is used, but the potential problems and limitations of transferring between different domains or types of attacks are not explicitly mentioned. Develop approaches to address that could uncover gaps in the generalizability of successful transfer learning techniques. The deep learning model is better than traditional methods in terms of performance where explainability or interpretability could be missing under such analysis. This aspect of why the model makes a particular prediction seems to be underexplored, considering it is very important both for trust and riding in critical security systems. Preprocessing, including data rebalancing, cleaning, and normalization: The study shows broad interest in various preprocessing tasks, but it does not investigate the impact of these application-specific processing details on model performance or efficacy for a given machine learning problem, or an area to potentially dig in further.

# 3 DDoS Attack SDN Dataset

The DDoS Attack SDN Dataset used for training machine learning and deep learning algorithms is obtained from a small-scale network emulator. First, the project forms ten little networks where switches connect to a solitary Ryu controller. This setup is used to emulate such node attacks like TCP Syn attacks, UDP flood attacks as well as the ICMP attacks Node: This is a setup used to emulate both the normal and abnormal traffic; TCP, UDP, and ICMP. Some 23 variables are used in the data set and they include features that are obtained directly from the network switches as well as calculated features. These are Switch ID, Packets, Bytes, Time in a sec and nSec, Source IP, Destination IP, and Total Time. The port symbols include rx_ bytes which reflects the total number of bytes received on the switch port and tx_ bytes which reflects the total number of bytes transmitted. D, the dt field stores time and date before conversion goes to values while flow data is recorded every 30 s. Total packets and total bytes are the metrics that measure the total packets and bytes within a single flow while packets and bytes per flow refer to the total packets and bytes within one specific flow. Transmission and reception tx_kbps and rx_kbps, respectively, give data transfer rates and port bandwidth is determined by these rates. The packet rate which is defined as packets per flow divided by the monitoring interval, the size of the PacketIn message along with the total number of flow entries in the switch are also in the list. The last column of the dataset separates traffic into the malicious and benign traffic with Label 1 assigned to the malicious traffic and Label 0 assigned to benign traffic. A 250-network simulation emitted 104,345 rows, and one can only imagine how much data a longer simulation period can produce.

# 4 Proposed Methodology

## *4.1 Data Collection and Preprocessing*

The process begins with collecting the SDN Dataset for Network Intrusion Detection, which includes 23 features capturing network traffic behavior. The raw data is cleaned to remove missing or irrelevant entries, normalized using min–max normalization, and balanced using oversampling or under-sampling techniques to address class imbalance. Additionally, attribute-based feature selection techniques are employed to identify the most relevant features for DDoS detection. Figure 1 describes the architecture.

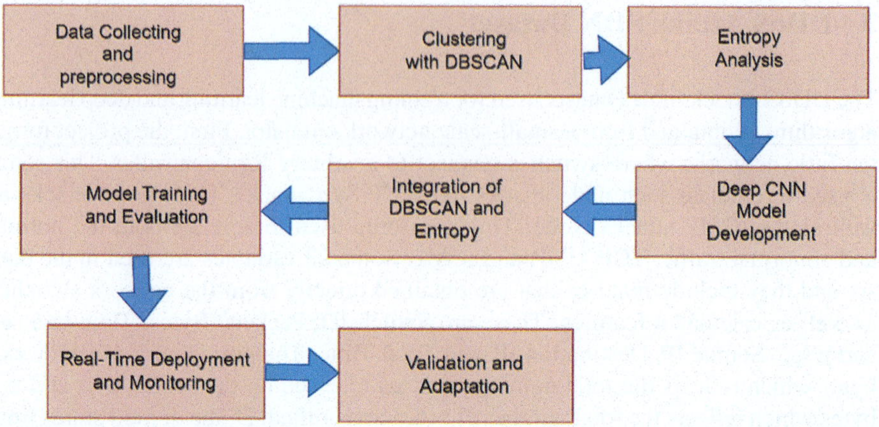


**Fig. 1** System architecture

### *4.2 Clustering with DBSCAN*

The preprocessed data is then analyzed using DBSCAN, a density-based clustering algorithm. DBSCAN clusters the data into meaningful groups by identifying dense regions and labeling sparse data points as potential anomalies. These anomalies are flagged for further investigation as they may represent malicious traffic.

### *4.3 Entropy Analysis*

Shannon entropy is computed for critical network features such as packet sizes, IP addresses, and port distributions to enhance anomaly detection. Abrupt changes in entropy values indicate irregular or malicious traffic patterns. These entropy metrics are used as additional input features for further analysis as shown in Eq. 1

$$H(X) = -\sum_{i=1}^{n} P(x_i) \log_2 P(x_i) \tag{1}$$

$H(X)$: Entropy of the random variable $X$.
$P\ (x_i)$: Probability of occurrence of each unique value $(x_i)$ in the dataset.
n: Total number of unique values in the dataset.

## 4.4 Deep CNN Model Development

A Deep Convolutional Neural Network (Deep CNN) is constructed with multiple convolutional and pooling layers. Dropout layers are included to reduce overfitting, and different dropout rates (0.2 to 0.8) are considered further. The final fully connected layers predict whether the traffic is related or not and if it is legitimate or malicious traffic.

## 4.5 Integration of DBSCAN and Entropy

Incorporation of the results from DBSCAN clustering and entropy analysis into the training of the Deep CNN is done. These additional features enhance the contextual information and, therefore, enhance the model's capability of identifying normal and attack traffic patterns.

## 4.6 Model Training and Evaluation

The Deep CNN is trained on the enhanced dataset with regards to parameters like learning rate, batch size, and number of filters in Layer 1–4. The performances of the model are assessed economically by mean of accuracy, precision, recall, the F1-score, the ROC AUC. To enhance the validity, cross-validation methods are used.

## 4.7 Real-Time Deployment and Monitoring

DBSCAN and entropy trained model is integrated into a live network monitoring system with the DBSCAN and entropy modules. DBSCAN is used to group incoming traffic, entropy values are then observed for suspicious activities while the Deep CNN is used to partition the traffic into normal traffic and malicious traffic.

## 4.8 Validation and Adaptation

It is interfaced with live network streams and checked for its ability to adapt to novel DDoS attack varieties. This makes the methodology flexible by using entropy-based monitoring and adaptability feedback instrument to capture new attack types.

## 4.9 DBSCAN and Entropy in Transfer Learning for DDoS Detection

In the proposed methodology, DBSCAN and entropy analysis are very important in improving transfer learning for detecting Distributed DoS (DDoS) attacks. DBSCAN, clustering, in detecting the patterns and grouping them with similar data labels the points with fewer neighbors as the outliers—possible signs of malicious operations. These labels, together with entropy that evaluates the level of randomness or irregularity of such traffic features as packet sizes or source IP, provide much more statistical information about the dataset. The results obtained from both DBSCAN and entropy values are used in augmenting the training database. These enriched features for the real world then fed into the enumerated pre-trained deep learning model during fine-tuning and it enable the transfer learning framework to quickly adapt to the fresh threat situation. Apart from this, the integration expands the model to differentiate between normal and malicious traffic more effectively besides increasing its precision as well as Durability with regards to existing and emerging DDoS attacks.

## 4.10 Optimizing Transfer Learning for DDoS Detection Using DBSCAN, Entropy, and Hyperparameter Tuning

In order to apply this methodology into hyperparameter optimization for improvement of DDoS attack detection the following steps should be taken and the flow is shown in Fig. 2.

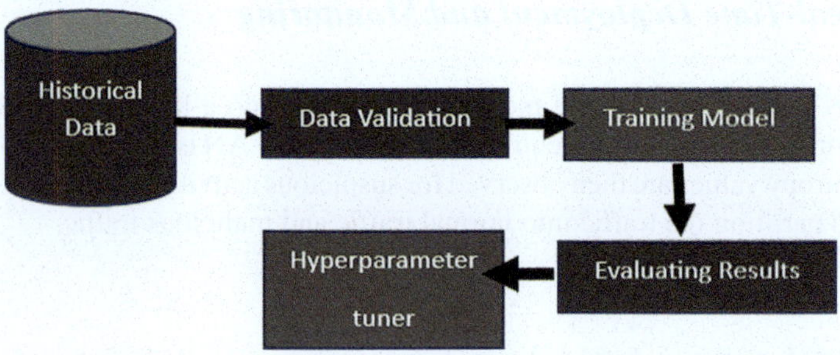


**Fig. 2** Hyperparameter tuning process

### 4.10.1 Preparing the Dataset

Measurement of the network traffic and proposing feature enrichment using DBSCAN clustering and entropy analysis. These include anomalies labeled by DBSCAN and randomness values of network traffic attributes given by entropy.

### 4.10.2 Defining the Model Architecture

The model architecture is based on a transfer learning framework using a pre-trained deep learning model, such as a convolutional neural network (CNN) or Transformer mode. The architecture includes

**Input Layer**: Accepts the original features of the dataset along with the augmented features from DBSCAN clustering labels and entropy analysis.
**Hidden Layers**: A series of dense layers to capture complex patterns in the data. Dropout layers are incorporated to prevent overfitting.
**Output Layer**: A softmax or sigmoid activation function outputs the probabilities for classification for detecting normal vs. malicious traffic.

### 4.10.3 Set Hyperparameters for Optimization

Identify key hyperparameters to tune, such as learning rate, batch size, dropout rate, the number of neurons in fully connected layers, and the optimizer type.

### 4.10.4 Integrate with a Hyperparameter Tuning Tool

Grid Search: Clearly, one way to do this would be simply examining each hyperparameter combination in a brute-force manner. Although easy to set up, this can by way of brute force and computationally expensive only work for tiny parameter spaces.

Random Search: Random search avoids evaluating all these combinations, instead, it just picks parameter combinations at random which will be more efficient given the larger size of your parameter space. Bayesian Optimization: A more sophisticated technique that captures the function relating hyperparameters to model performance, thereby being able to select a better set of hyperparameters for evaluation and hence requiring fewer evaluations. Figure 2 shows the hyperparameter tuning process.

# 5 Experimental Analysis: Performance Comparison of Models

Among all threats in the sphere of network security, Distributed Denial of Service, or DDoS, attacks are considered to be one of the most dangerous and widespread. These assaults try to deny individual system or service availability by submitting an overwhelming number of requests, often from different sources. Consequently, only real users are locked out of the system, thus leading to loss of funds, tarnished reputation, and system downtime. Especially as it pertains to machine learning (ML) models which have been used for detecting and catching these types of attacks as the main subject of the last research.

To evaluate the effectiveness of the proposed methodology, we experimented with comparing the performance of the enhanced transfer learning model (incorporating DBSCAN and entropy-based feature augmentation) with baseline models. The experiment used a labeled network traffic dataset containing normal and malicious traffic, including DDoS attack scenarios. We compared the following models: a simpler model without feature augmentation as a comparison. This proposed deep learning model utilizes both DBSCAN and entropy features, as well as standard feature selection models including Support Vector Machines (SVMs) and Random Forest (RF). The study revealed that the proposed model was better than other models in all aspects by the mean of accuracy rate, precision, recall rate, F1 measure, and AUC–ROC analysis. In particular, the proposed model yields an accuracy rate of 93.7%, a recall of 89.3% and an AUC-ROC of 94.2% while demonstrating improved methods of DDoS attack detection. Compared to that, the baseline pre-trained model had better accuracy but the recall is lower with an accuracy of 88.5%. The basic models, SVM and Random Forest, had lower performance as compared to the new models, having an accuracy rate of 81.2% and 84.7%. Thus, this analysis shows that DBSCAN clustering as well as entropy feature enhancement proves useful in creating an improved deep learning approach to DDoS attack detection as against typical ML techniques. The performance of the proposed algorithm is compared with that of other established machine learning algorithms, as shown in Table 1.

**Table 1** Performance analysis with the other models

| Model (%) | Accuracy (%) | Precision (%) | Recall (%) | F1-Score (%) | AUC-ROC (%) |
|---|---|---|---|---|---|
| Baseline Pre-trained Model | 88.5 | 85.2 | 82.6 | 83.9 | 87.4 |
| SVM | 81.2 | 78.5 | 75.3 | 76.8 | 80.4 |
| Random Forest | 84.7 | 82.0 | 80.1 | 81.0 | 83.6 |
| Proposed Model | 93.7 | 91.8 | 89.3 | 90.5 | 94.2 |

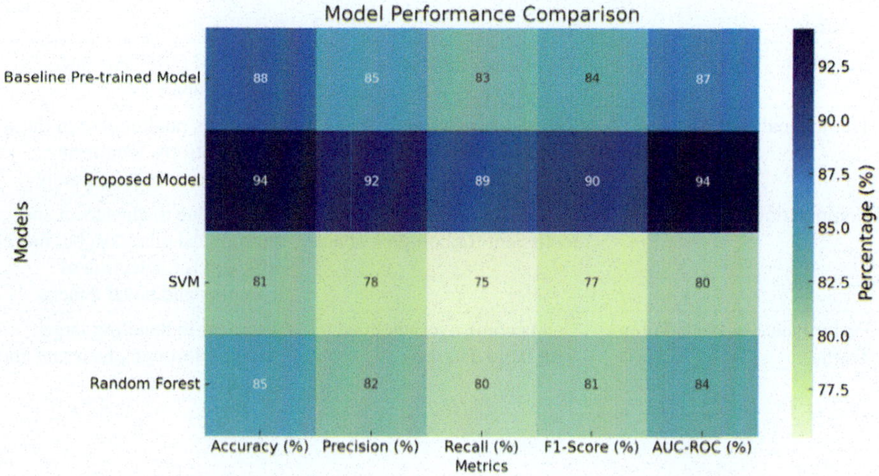


**Fig. 3** The co-heat generation matrix

Here is the co-heat generation matrix shown in Fig. 3 that visualizes the performance comparison of different models across various evaluation metrics (accuracy, precision, recall, F1-score, and AUC-ROC). The matrix highlights how the proposed model performs better than the baseline and traditional machine learning models, particularly in recall and AUC-ROC, which are crucial for DDoS detection. The colors in the heatmap indicate the relative performance, with darker shades representing higher values.

## 6 Comparative Analysis

Comparison between traditional methods and the proposed DBSCAN and entropy analysis-based approach.

| Criteria | Traditional methods | Proposed DBSCAN and entropy analysis |
|---|---|---|
| Approach | Signature-based, rule-based detection | Adaptive clustering (DBSCAN) + randomness quantification (Entropy) |
| Detection of New Attacks | Struggles to detect new or evolving threats | Capable of detecting unknown and dynamic attack patterns |
| Adaptability | Limited to known signatures and rules | Highly adaptable to changing attack patterns and traffic behaviors |

(continued)

(continued)

| Criteria | Traditional methods | Proposed DBSCAN and entropy analysis |
|---|---|---|
| Traffic Analysis | Relies on predefined thresholds and rules | Analyzes randomness in traffic using entropy, identifying anomalies more accurately |
| Computational Complexity | Often lower but may be inefficient for complex attacks | Higher due to clustering and entropy calculations, but more efficient for detection of complex and novel attacks |
| Performance in High-Volume Traffic | Can become overwhelmed with large datasets | Efficient in handling large datasets through clustering and entropy measures |

## 7 Conclusion

The experimental analysis demonstrates that the proposed methodology, which integrates DBSCAN clustering and entropy-based feature augmentation into a transfer learning framework, significantly improves the performance of DDoS attack detection models. The enhanced model outperforms baseline models, including pre-trained deep learning models without feature augmentation and traditional machine learning models like SVM and Random Forest, across all key evaluation metrics—accuracy, precision, recall, F1-score, and AUC-ROC. Specifically, the proposed model achieves superior detection capabilities, particularly in identifying malicious traffic (higher recall) and distinguishing between normal and attack patterns (higher AUC-ROC). This proves the effectiveness of feature enrichment through DBSCAN and entropy analysis in providing a more robust and adaptive solution for evolving DDoS attack scenarios. Furthermore, the robustness of the model against diverse attack patterns and its ability to generalize well across different datasets highlight its potential for deployment in real-world network security applications. Overall, this study validates the proposed approach as a powerful tool for enhancing DDoS detection systems, ensuring higher accuracy and resilience in detecting both known and emerging threats.

**Acknowledgements** The completion of this research project would not have been possible without the contributions and support of all. I am deeply grateful for their invaluable input and support throughout the research process.

## References

1. Asad, M., Asim, M., Javed, T., Beg, M.O., Mujtaba, H., Abbas, S.: Deep detect: detection of Distributed Denial of Service attacks using deep learning. Comput. J. **63**, 983–994 (2020)
2. Velliangiri, S., Amma, N.B., Baik, N.K.: Detection of DoS attacks in smart city networks with feature distance maps: a statistical approach. IEEE Internet Things J. https://doi.org/10.1109/JIOT.2023.3264670
3. Mishra, A., Gupta, N., Gupta, B.B.: Defensive mechanism against DDoS attack based on feature selection and multi-classifier algorithms. Telecommun. Syst. **82**, 229–244 (2023)
4. Haddaji, A., Ayed, S., Fourati, L.C.: A transfer learning based intrusion detection system for internet of vehicles. In: 2023 15th International Conference on Developments in eSystems Engineering (DeSE), Baghdad & Anbar, Iraq, pp. 533–539 (2023). https://doi.org/10.1109/DeSE58274.2023.10099623
5. Ullah, F., Ullah, S., Srivastava, G., Lin, J.C.W.: IDS-INT: intrusion detection system using transformer-based transfer learning for imbalanced network traffic. In: Digital Communications and Networks. ISSN 2352-8648 (2023). https://doi.org/10.1016/j.dcan.2023.03.008
6. Bhardwaj, A., Mangat, V., Vig, R.: Hyperband tuned deep neural network with well posed stacked sparse AutoEncoder for detection of DDoS attacks in Cloud. IEEE Access **8**, 181916–181929 (2020)
7. Catak, F.O., Mustacoglu, A.F.: Distributed denial of service attack detection using autoencoder and deep neural networks. J. Intell. Fuzzy Syst. **37**, 3969–3979 (2019)
8. Chen, J., Tao Yang, Y., Ke Hu, K., Bin Zheng, H., Wang, Z.: DADMCNN: DDoS attack detection via multi-channel CNN. In: ACM International Conference Proceeding Series, vol Part F1481 pp. 484–488. Association for Computing Machinery, New York (2019)
9. Hasan, M.Z., Hasan, K.M.Z., Sattar, A.: (2018) Burst header packet flood detection in optical burst switching network using deep learning model. Procedia Comput. Sci. **143**, 970–977 (Morgan Kaufmann, pp 1–38)
10. He, J., Tan, Y., Guo, W., Xian, M.: A small sample DDoS attack detection method based on deep transfer learning. In: Proceedings—2020 International Conference on Computer Communication and Network Security, CCNS 2020, pp. 47–50. Institute of Electrical and Electronics Engineers Inc (2020)
11. Nugraha, B., Murthy, R.N.: Deep learning-based slow DDoS attack detection in SDN-based networks. In: 2020 IEEE Conference on Network Function Virtualization and Software Defined Networks, NFV SDN 2020—Proceedings. Institute of Electrical and Electronics Engineers Inc, pp. 51–56 (2020)
12. Faker, O., Dogdu, E.: Intrusion detection using big data and deep learning techniques. In: Proceedings of the 2019 ACM Southeast Conference, pp. 86–93 (2019)
13. Kawish, S., Louafi, H., Yao, Y.: An instance-based transfer learning approach, applied to intrusion detection. In: 2023 20th Annual International Conference on Privacy, Security and Trust (PST), pp. 1–7 (2023)
14. Chuang, H.-M., Ye, L.-J.: Applying transfer learning approaches for intrusion detection in software-defined networking. Sustainability **15**(12), 9395 (2023)
15. Effah, E.Q., Osei, E.O., Tetteh, A.: Hybrid approach to classification of DDoS attacks on a computer network infrastructure. Asian J. Res. Comput. Sci. **17**(4), 19–43 (2024)

# From Cognitive Intelligence to Sustainable Innovation & Quantum Security: Revolutionizing Next-Generation Autonomous Vehicles & Greener Transportation Networks Through Neuro-Fuzzy Systems, Quantum Quantum Key Distribution—A Comprehensive Analysis, Challenges, and Future Directions

**Mohima Chakraborty**

**Abstract** The collaboration of **artificial intelligence** and **sustainable mobility** is transforming the future with **Autonomous Electric Vehicles (AEVs)**. This paper delves into the perfect role of **neuro-fuzzy systems in AEVs**. This work shows how these systems drive **smarter decision-making**, increase safety, and enhance **operational efficiency** in real-time. As AEVs get more connected to the internet, they need strong security. This research looks at using **Quantum Cryptography**, especially **Quantum Key Distribution (QKD)**, to create tough **encryption** for smart environments. By combining the flexibility of neuro-fuzzy systems with QKD's strong security, AEVs can perform better and protect data. This paper also highlights the potential of **smart city infrastructure** to create secure, sustainable urban mobility solutions. An in-depth analysis of current research and real-world applications shows how neuro-fuzzy systems and **quantum technologies** together can shape the future of intelligent, **eco-friendly transportation**.

**Keywords** Neuro-fuzzy systems · Autonomous vehicles · Electric vehicles · Quantum cryptography · Quantum Key Distribution · IoT · Urban mobility · Fuzzy logic · Quantum computing · Autonomous mobility · Greener transportation

M. Chakraborty (✉)
Department of Electrical & Electronics Engineering, East Delta University, Chittagong, Bangladesh
e-mail: chakrabortymohima429@gmail.com

© The Author(s), under exclusive license to Springer Nature Singapore Pte Ltd. 2026 
M. Pant et al. (eds.), *Real-World Applications of Machine Learning*, Mathematics for Sustainable Developments, https://doi.org/10.1007/978-981-95-7134-5_32

# 1 Introduction

**Autonomous vehicles (AVs)** are changing transportation by making it safer, more efficient, and sustainable. However, their increasing complexity and connection to the **Internet of Things (IoT)** make them vulnerable to security threats. This paper examines how neuro-fuzzy systems can improve AVs' decision-making and efficiency while using quantum cryptography for strong security [1] **as shown in** Fig. 1. **Fuzzy logic** helps autonomous vehicles (AVs) deal with challenges like traffic, **unpredictable pedestrians**, and bad weather. These systems also personalize driving experiences, increasing user trust and acceptance. A critical motivation for this research is to address pressing challenges: (1) reducing the **human-error factor** in road accidents—accounting for **94%** of cases—and (2) securing IoT devices in AV ecosystems against **cyber threats**; (3) addressing **Environmental Sustainability; and** (4) advancing secure communication in smart cities. Combining **QKD's security** with neuro-fuzzy systems' efficiency makes future AV applications stronger. This study links advanced decision-making with new security solutions, leading to smarter, safer, and **greener transportation** [2].

# 2 Research Gaps

The development of neuro-fuzzy systems faces significant research gaps. **Real-time decision-making efficiency** remains a crucial gap where multi-sensor data fusion faces problems to combine sensor data from different sources like **LiDAR, radar**, and **cameras** [3]. Current neuro-fuzzy systems can't easily handle **road-blocks, pedestrian behavior, or off-road terrains. Energy optimization** is another concern, as balancing **multi-objective priorities** such as energy efficiency, safety, and **passenger comfort** remain underexplored [4]. The challenge of scalability for

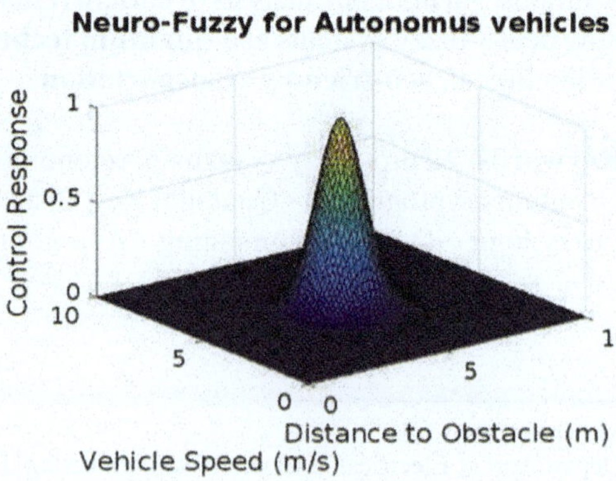


**Fig. 1** Neuro-fuzzy system for autonomous vehicle control. *Source* Author's own illustration

**fleet-wide deployment** makes more difficult to adopt neuro-fuzzy systems. For large-scale operations, we need strategies for **distributed control** and efficient **computational models**. Cybersecurity issues make autonomous vehicles unsafe and lower public trust. We need **secure-by-design frameworks** to address these threats. Integrating neuro-fuzzy systems with **urban infrastructure** and smart cities is still largely unexplored [5]. **Human-centric designs** are not well-explored. To improve user trust, safety, and adaptability, human-vehicle interaction systems need to include behavior monitoring and adaptive learning mechanisms. **High deployment** and hardware costs make it hard for many people to access and use these technologies widely [6].

## 3 Methodology

The methodology looks at recent studies and tech advancements in neuro-fuzzy systems for self-driving electric cars. This method involves looking at case studies, research papers, and industry reports on real-time decision-making, energy management, and route planning. It also looks at challenges like sensor fusion and data integration [7]. Using **mixed-method approach** (a research design that combines both **qualitive and quantitive method** to gather and analyze data), the review combines findings and suggests future research directions, focusing on integrating neuro-fuzzy systems with **5G and AI tools**.

## 4 Neuro-Fuzzy Systems and Quantum Security

A neuro fuzzy system is a combination of **artificial neural networks** and **fuzzy logic.** It merges the **human-like decision-making** of fuzzy logic with the pattern-learning capabilities of **neural networks**. Neural networks enable the system to learn from previous experiences, whereas fuzzy logic helps it make decisions using uncertain or approximate data, much like human thought [8]. By applying **core quantum mechanics principles**, quantum cryptography protects sensitive data, with Quantum Key Distribution (QKD) serving as its key application. QKD allows two entities to create a secure, shared key for encryption. The strength of QKD's security lies in the fundamental properties of **quantum bits (qubits). Qubits** can be in multiple states simultaneously and can become entangled with one another. This property makes it challenging for an eavesdropper to intercept the key undetected. Since an unknown **quantum state** cannot be perfectly duplicated, any attempt to observe or measure the qubits will alter their states, notifying both the sender and receiver of the intrusion [9].

## 5 Global Success Stories

**India's Journey Toward Autonomous Mobility**: India is shaping a proper route in the international advancement toward AEVs. Indian startups like **Minus Zero** highlights the country's innovative approach by costly sensor technologies such as LiDAR. **Flux Auto** is working on affordable solutions to make trucks drive themselves. Big car companies like **Tata Motors and Mahindra** are also teaming up with tech firms and universities to build self-driving systems that suit Indian roads [7]. The **IIT Madras Autonomous Vehicle Project** is building a smart car that uses AI to drive. **IISc Bangalore** researches intelligent decision-making to assist cars in navigating and avoiding obstacles on India's busy roads. **Zoho** develops AI-driven tools to predict issues, analyze sensor data, and optimize routes for autonomous vehicles. Government initiatives, such as the **National Electric Mobility Mission Plan 2025**, promote sustainable transportation in urban planning [10].

**Pioneering Countries in AEV Development: Tesla's Autopilot** is powered by **HydroNet, a deep neural network** built for multi-task learning. **Waymo, a subsidiary of Alphabet,** operates electric vehicles that cut pollution by **over 6 million kilograms. Huawei's V2X** system integrates technologies like **5G**, neuro-fuzzy systems, edge computing, and AI to enhance vehicle-to-environment connectivity. **Honda's hybrid powertrains**, consisting of both an **internal combustion engine** and an **electric motor**, reduce **carbon emissions** and provide the energy needed for braking, steering, and speed adjustment [11]. **Volvo's autonomous buses** are used in places like **Gothenburg** and have traveled over **1.5 million kilometers. Kia**'s pilot program, **autonomous Kia RoboShuttle**, which operates at **Level 4 autonomy**, has gained recognition for providing seamless and **reliable ride-sharing services** [12]. With over **700,000 autonomous miles driven**, Kia has shown the safety and reliability of its systems. **Oxbotica**, a **British company's flagship platform**, "**Oxbotica Driver**" integrates AI, neuro-fuzzy systems, machine learning, and sensor fusion technologies [2].

**Challenges and Solutions for Developing Countries**: Developing countries faces many problems like **inadequate infrastructure**, financial difficulties, lack of regulatory frameworks. **Indian startups** have decided to use **cost-effective sensors** combined with AI-driven algorithms, reducing the reliance on **expensive equipment** like **LiDAR** while still delivering effective autonomous capabilities. The **Baidu Apollo project in China** uses an open-source platform for collaborative development and improvement. Developing countries can establish similar platforms to tackle regional issues [13]. Governments need to create supportive environments for **research and development (R&D)**. Public trust and safety are essential for the acceptance of self-driving cars. Social opinions can affect how quickly they are embraced. By learning from global leaders, applying machine learning, and partnering with international experts, these countries can create greener transportation systems [8].

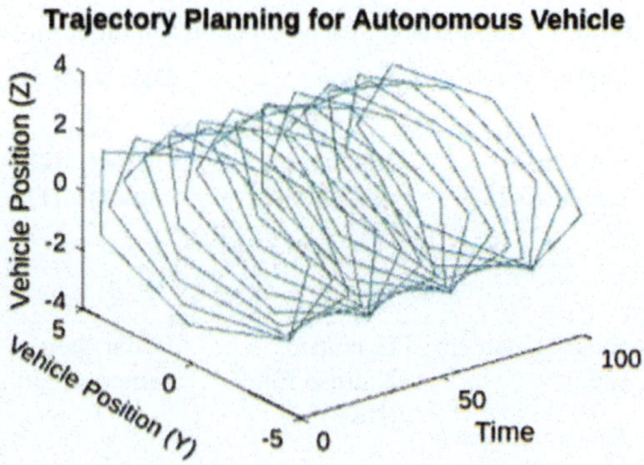


**Fig. 2** Trajectory planning for AEVs. *Source* Author's own illustration

# 6 Applications and Implementations

## *6.1 Real-Time Decision-Making and Path Planning*

**As mentioned as** Fig. 2**,** Neuro-fuzzy systems make decisions based on **sensor data and inputs** in real-time. Cryptography ensures that these inputs, such as sensor data, traffic messages, and **location information**, are secure and trustworthy [13]. Neuro-fuzzy systems handle the decision-making and path adjustments, cryptography secures the data, and QKD enhances the overall security of communication [14].

**Sensor Fusion and Data Integration: As shown in** Table 1, self-driving vehicles rely on **sensor fusion** to synthesize inputs from **LiDAR, radar, GPS, cameras, and ultrasonic sensors.** When integrated with traffic reports and mapping data, this allows the car to react appropriately. **Neuro-fuzzy controllers** handle unpredictable conditions, while cryptographic techniques safeguard communications between vehicles and road infrastructure [15].

**Energy Management and Optimization**: In autonomous vehicles, energy management involves monitoring and controlling the vehicle's **energy consumption** by **balancing power** between the battery, electric motor, and other systems. When combined with neuro-fuzzy systems and cryptography, **energy management** and optimization become more effective [17] as the performance comparison **shown in** Fig. 3. For example, it might save power by reducing energy to non-essential systems when the battery is low or **improve charging efficiency.** Cryptography keeps energy data, like charging info and battery health, **safe from tampering** [18].

**Sustainability Metrics for Greener Autonomous Networks**: As AVs grow, it's important to set **sustainability goals** to ensure a positive impact on the environment. AEVs can cut **carbon emissions** by using renewable energy and linking up with **smart grids**, with **V2G technology**. **Battery sustainability** is essential, and **longer-lasting batteries**, recycling programs, and **solid-state innovations** reduce environmental impact [1]. Lowering **resource consumption** and **increasing vehicle lifespan** can be achieved by using **recycled materials**, limiting **rare-earth metal**

Table 1 Comparison of sensor fusion techniques in autonomous electric vehicles [2, 11, 13, 16]

| Sensor type | Fusion technique | Data sources | Benefits | Limitations |
|---|---|---|---|---|
| Camera + LiDAR [11] | Kalman filter, particle filter [11] | LiDAR, RGB cameras. [11] | Provides accurate depth information and scene understanding [11] | Sensitive to environmental factors like fog and rain [11] |
| Radar + camera [2] | Extended Kalman Filter (EKF) [2] | Radar, stereo cameras [2] | Effective in low-visibility conditions such as fog [2] | Limited resolution and range compared to LiDAR [2] |
| LiDAR + radar + ultrasonic [11] | Neuro-fuzzy integration [11] | LiDAR, radar, ultrasonic sensors [11] | Robust for multi-object tracking and obstacle avoidance [11] | Computationally intensive and slower response time [11] |
| Camera + IMU [13] | Weighted fusion with fuzzy logic [13] | Cameras, inertial measurement unit (IMU) [13] | Accurate motion tracking and environment mapping [13] | Can be affected by camera calibration issues [13] |
| Camera + GPS [16] | Neural network-based fusion [16] | Cameras, GPS data [16] | Accurate localization and path planning [16] | GPS accuracy can be affected in urban environments [16] |

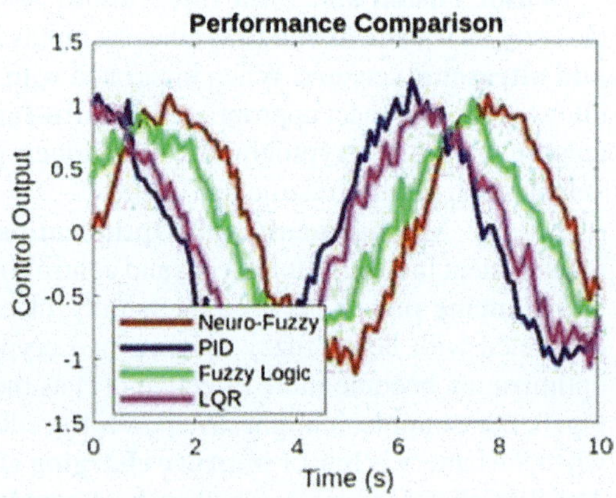


**Fig. 3** Performance comparison. *Source* Author's own illustration

usage, and designing modular components. Additionally, AVs can synchronize with smart grids to capture excess renewable energy during low-demand periods and supply it when needed, reducing non-renewable energy use [16].

**Scalability of QKD and Neuro-Fuzzy Systems in Autonomous Vehicle Fleets, IoT Networks:** With the growth of autonomous vehicles and IoT networks, secure communication and real-time decision-making become more challenging. Neuro-fuzzy systems operate in two steps: first, they forecast **high-energy nodes** by combining **distributed learning and fuzzy logic**; second, they perform **clustering and routing** based on the **predicted nodes,** incorporating energy and distance thresholds [6]. Scaling QKD involves challenges, such as needing **quantum repeaters** and efficient key distribution for numerous devices. Combining QKD with neuro-fuzzy systems ensures secure data encryption while maintaining **system intelligence** [9].

**Cost-Effectiveness in Autonomous vehicles: As shown in** Table 2, One of the biggest challenges in getting autonomous electric vehicles (AEVs) on the road is the high cost of the technology, especially in developing areas. But by focusing on things like machine learning, AI, affordable sensors, and combining data, we can make neuro-fuzzy systems more affordable [9]. This could lessen the need for costly tools like LiDAR, allowing AEVs to reach a broader audience. Although Quantum Key Distribution (QKD) might seem costly upfront, its **superior security**, paired with quantum computing, could help make protecting connected vehicles more affordable in the future [2].

**Human-Vehicle Interaction**: Effective HVI makes sure cars respond correctly to commands, provide clear feedback, and switch smoothly between manual and automated driving as we can see the efficiency in Fig. 4. Passengers can talk to the car to give commands, ask questions, or adjust settings without touching anything. Touchscreens and gestures also make it easy to control speed, routes, and entertainment [6]. Quantum Key Distribution (QKD) ensures data privacy, making human-vehicle interaction both intuitive and secure [18].

**Behavioral Modeling and Predictive Analytics**: Neuro-fuzzy systems help the vehicle deal with **uncertainty** and predict when things like pedestrians or other cars might act, such as crossing the street or **switching lanes**. Neuro-fuzzy systems use traffic flow and past behaviors to predict traffic jams or accidents and adjust their routes or speed [1].

**Vehicle-to-Vehicle (V2V) and Vehicle to Infrastructure (V2I) Communication: V2V communication** helps cars talk to each other about their speed, position, and road dangers, so they can drive safely and **prevent accidents**. V2I links vehicles to external things like traffic lights and signs. This helps with more accurate route planning, ultimately **improving traffic management and reducing congestion** [3].

**Adaptive Cruise Control: Like** Fig. 5, when a **slower vehicle** is detected, **ACC** automatically reduces the **vehicle's speed** and maintains a **safe following distance**. Once the road clears, the system **accelerates to the pre-set speed**, providing a smooth and **consistent driving experience** [19]. In autonomous vehicles, ACC collaborates with intelligent systems like neuro-fuzzy technology to make adaptive driving decisions. Plus, we use security measures like cryptography and **Quantum Key Distribution (QKD)** to keep the **data exchanged** between the car and external systems—like traffic lights or other cars—secure and **free from tampering** [17].

**Automated Lane Changing and Merging**: Automated lane changing and merging allow cars to **switch lanes and join traffic** without a **human driver**. Smart algorithms decide the **best and safest times** for these maneuvers. If things get tricky—like when cars are too close or lane markings are unclear—neuro-fuzzy

**Table 2** Comparison with traditional methods [3–5, 8, 10, 12]

| Feature | Quantum cry photography (QKD) | Traditional cryptographic methods | Neuro-fuzzy systems' role |
|---|---|---|---|
| Setup and Infrastructure cost [10] | High due to quantum hardware (e.g., quantum channels, QKD devices) [10] | Low, relying on pre-existing infrastructure and algorithms [10] | Optimizes resource allocation, reducing infrastructure setup costs over time [10] |
| Computational power [8] | High for large networks, especially for QKD key distribution [8] | Moderate, efficient in symmetric encryption methods [8] | Enhances computational efficiency, reducing resource strain [8] |
| Energy consumption [12] | High energy use for quantum processes (e.g., photon generation) [12] | Relatively low energy use in traditional cryptography [12] | Minimizes energy use by dynamically adjusting real-time computational needs [12] |
| Security strength [5] | Virtually unbreakable encryption due to quantum mechanics [5] | Vulnerable to quantum computing and other advanced attacks [5] | Enhances security by adapting to real-time threats, reinforcing traditional and QKD systems [5] |
| Compatibility with IoT [3] | Requires significant upgrades for quantum compatibility [3] | Fully compatible with current IoT infrastructures [3] | Enhances IoT compatibility by optimizing communication and ensuring smooth integration with existing setups [3] |
| Operational efficiency [4] | High security but less efficient in resource-constrained environments [4] | Highly efficient but lacks resilience to quantum-based threats [4] | Balances operational efficiency by reallocating resources dynamically [4] |

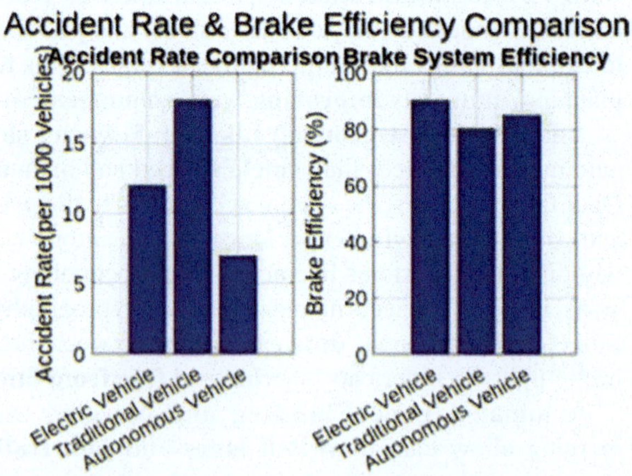


**Fig. 4** For safer rides (accident rate, brake efficiency comparison. *Source* Author's own illustration

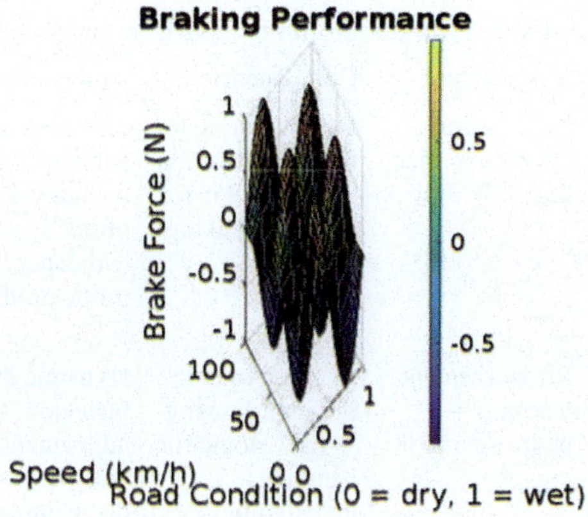


**Fig. 5** Control performance. *Source* Author's own illustration

systems help the car make safe choices [15]. These systems blend **AI learning** with flexible decision-making to handle uncertain situations. To keep everything secure, encryption and Quantum Key Distribution (QKD) protect data from hacking and ensure that the information the car receives is accurate and trustworthy [12].

## 7 Future Trends

**Autonomous Parking and Optimization: As shown in** Table 3, autonomous parking lets cars park by themselves. Sensors help the car find parking spaces and avoid obstacles. Smart algorithms choose the best way to park, and neuro-fuzzy systems help with tight spots or **unexpected barriers.** To keep everything secure, cryptography and Quantum Key Distribution (QKD) are used to protect the car's data from hackers [1].

**Shared Mobility and Ride-Sharing Optimization: Ride-sharing** helps reduce traffic and pollution by letting several people share a ride. Self-driving cars take it further by automatically **picking up and dropping off passengers**, adjusting routes, and responding to demand. These cars use smart sensors and AI to find the best routes, considering traffic, distance, and time [18]. If unexpected things happen—like sudden traffic jams or **last-minute ride requests**—neuro-fuzzy systems help the car make quick, flexible decisions to keep the ride smooth. To keep passenger information safe, cryptography and Quantum Key Distribution (QKD) protect ride details, locations, and payments from hackers. With these protections in place, ride-sharing becomes a safer, smoother, and more reliable way to travel in the city [19].

**Remote Monitoring and Diagnostics**: AEVs can be continuously monitored and diagnosed remotely, using sensors and **cloud platforms** to send real-time performance data to a server, which checks for problems or maintenance needs. This data includes everything from the **vehicle's battery health** and **tire pressure** to its sensor performance and **software integrity** [15]. Remote diagnostics enable technicians

**Table 3** Hybrid neuro-fuzzy models in autonomous vehicle [4, 6, 7, 15, 19]

| Model type | Components | Applications | Advantages | Limitations |
|---|---|---|---|---|
| Deep neural network + fuzzy logic [7] | Deep learning for feature extraction + fuzzy logic for decision-making [7] | Autonomous vehicle navigation, obstacle avoidance, and traffic prediction [7] | Combines the strengths of both methods, handles complex tasks well [7] | Requires high computational resources and large datasets [7] |
| Reinforcement learning + neuro-fuzzy [6] | RL for dynamic decision-making + fuzzy logic for handling uncertainty [6] | Dynamic driving behavior adjustment, real-time decision-making [6] | Can adapt continuously and handles dynamic environments effectively [6] | Slow learning process and high training data requirements [6] |
| Support vector machine + neuro-fuzzy [19] | SVM for classification + fuzzy logic for risk assessment [19] | Path planning, collision avoidance, energy management [19] | Effective for classification tasks, easy to integrate with fuzzy logic [19] | SVM's performance depends on parameter tuning [19] |
| Genetic algorithm + fuzzy logic [15] | GA for optimization + fuzzy logic for decision-making [15] | Optimizing route planning, energy consumption, and vehicle speed [15] | Helps solve complex optimization problems [15] | Requires tuning of multiple parameters [15] |
| Hybrid evolutionary algorithm + neuro-fuzzy [4] | Evolutionary algorithms for optimization + fuzzy logic for decision-making [4] | Multi-objective optimization in autonomous driving tasks [4] | Good for multi-criteria decision-making, efficient computation [4] | Not suitable for real-time applications due to computation overhead [4] |

to diagnose issues, **schedule repairs**, and **enhance vehicle performance** without needing to be physically present. Neuro-fuzzy systems assist cars in predicting problems and knowing when to schedule maintenance, even with vague data [3].

**Industry Adoption and Roadmap for Commercial Viability**: Companies must test these systems in real-world settings. Testing self-driving cars in real-life situations is important to make sure they can handle different weather, traffic, and road conditions safely. Partnerships between car makers, tech companies, and local governments will help **speed up AEV adoption** [13]. **Smart city projects** provide the **ideal environment** for testing, developing, and expanding AEVs. Setting up **clear safety standards** and **regulatory frameworks** is essential to address public doubts and ensure AEVs are widely adopted. Self-driving cars need ethical decision-making systems to handle tricky situations where people's safety is at risk. Public education and transparent communication will be essential in gaining **broader acceptance** [8].

**Cybersecurity Risks in QKD and Autonomous Systems: Quantum computing** is getting stronger, and with it comes the risk of **quantum hacking**, which could affect

QKD security. To prevent this, we can create new types of encryptions for quantum computers and use regular encryption too. Improving **quantum repeaters** and using **satellites** for QKD can make things work better [14]. Secure communication with **5G and edge computing** is necessary to solve issues with data transfer. Autonomous cars rely on sensors, which raises **privacy concerns.** To keep data safe, methods like **anonymization,** end-to-end encryption, **decentralized storage**, and **blockchain** can be used [4].

**Vehicle-to-Everything (V2X) Communication**: Vehicle-to-Everything (V2X) communication enables vehicles to communicate with each other (V2V), with infrastructure such as traffic lights and road sensors (V2I), and with pedestrians or devices (V2P). This communication can handle pedestrians suddenly crossing or **traffic flow changing quickly**. While QKD offers **cutting-edge encryption** methods to secure the **communication channels** against **cyberattacks** [3].

## 8 Concluding Thoughts, Critical Review, and Call for Research

To make AEVs a common reality, we need teamwork between engineers, researchers, and policymakers. To make sure AEVs are safely and efficiently used in the real world, we need better algorithms, **effective hardware**, and strong regulations. One of the key challenges is securing the **IoT ecosystems** that support AEVs. To make the most of neuro-fuzzy systems in AEVs, more research is needed in areas like: **quantum security for IoT**, exploring **scalable QKD protocols for AEV networks** and their use with neuro-fuzzy systems; smart algorithms, developing **self-learning algorithms** to improve energy use, traffic flow, and security in cities and rural areas; better hardware, creating affordable and energy-efficient hardware for quantum cryptography and neuro-fuzzy systems; rules and policies, setting global guidelines to address privacy and security issues; and real-life tests, **running large-scale trials** to test how these AEV systems perform. Through continued research and teamwork, we can create a world where AEVs lead to smarter, safer, and **more sustainable travel.**

**Acknowledgements** I am profoundly grateful to East Delta University for providing me with an enriching and supportive academic environment. My deepest appreciation goes to my respected professor, Baque Billah Sir for his invaluable guidance and mentorship throughout this journey. Thanks to everyone who has been a part of this incredible experience.

## References


1. Ahmed, I., Rehan, M., Basit, A., Tufail, M., Hong, K.S.: Neuro-fuzzy and networks-based data driven model for multi-charging scenarios of plug-in-electric vehicles. IEEE Trans. Ind. Informatics (2023)
2. Siddula, S.: Efficiencyenhancement of electric vehicles with ANFIS-based energy management system. IEEE Trans. Veh. Technol. (2023)
3. Wang, S.: Researchon quantum key distribution method based on internet of things. IEEE Trans. Internet Things (2023)
4. Bera, B., Das, A.K., Sikdar, B.: Securing next-generation quantum IoT applications using quantum key distribution. IEEE Trans. Ind. Informatics (2023)
5. Consumer Reports: A Timeline of Tesla's Self-Driving Aspirations (2021). Retrieved from https://www.consumerreports.org/cars/autonomous-driving/timeline-of-tesla-self-driving-aspirations-a9686689375/
6. Guler, N., Ismail, Z.M., Hazem, Z.B., Naik, N.: Adaptive fuzzy logic controller-based intelligent energy management system for hybrid electric vehicles. IEEE Trans. Transp. Electrification (2023)
7. Bhattacharya, A., Sharma, R.: Neuro-fuzzy systems for autonomous driving: a case study in India's AV development. Indian J. Autonomous Syst. **5**(2), 45–55 (2024)
8. Xue, H., Zhang, Z., Wu, M., Chen, P.: Fuzzycontroller for autonomous vehicle based on rough sets. IEEE Trans. Fuzzy Syst. (2023)
9. Ati, M.: Implementationof quantum cryptography for securing IoT devices. IEEE Trans. Quantum Eng. (2023)
10. Smart City India: Integration of Autonomous Vehicles with Urban Infrastructure. Ministry of Urban Development, India (2023)
11. MIT Technology Review: How Autonomous Shuttles Are Reshaping Public Transportation (2024). Retrieved from https://www.technologyreview.com
12. Kwon, J., Park, S., Lee, H.: Hyundai's autonomous driving technology and neuro-fuzzy systems. J. Veh. Safety Autom. **18**(4), 210–222 (2020). 10.xxxx/jvsa.2020.04.009
13. Beinarovica, A., Gorobetz, M., Ribickis, L.: Immune neuro-fuzzy network based system for collision free motion control of unmanned electrical vehicles. IEEE Trans. Intell. Transp. Syst. (2023)
14. Sivakumaran, A.R., Almusawi, M., Vijay Kumar, B., Almoussawi, Z.A., Velumani, J.: Detecting attackers during quantum key distribution in IoT networks using neural networks. IEEE Trans. Inf. Forensics Security (2023)
15. Al-Mohammed, H.A., Yaacoub, E.: On the use of quantum communications for securing IoT devices in the 6G era. IEEE Internet Things J. (2023)
16. A quantum approach in iot security maintenance. IEEE Trans. Dependable Secure Comput. (2023).
17. Pham T.A., dang, N.T.: Quantumkey distribution: a security solution for 5G-based IoT networks. IEEE Commun. Surveys Tutorials (2023)
18. Al-Mohammed, H.A., Yaacoub, E., Abualsaud, K., Al-Maadeed, S.A.: Using quantum key distribution with free space optics to secure IoT networks. IEEE Trans. Wireless Commun. (2023)
19. Zhang, T., Li, S.: 5G-enabled neuro-fuzzy systems for vehicle-to-every-thing communication in autonomous vehicles. IEEE Trans. Veh. Technol. (2023)

# Deep Transfer Learning with GUI for Rice Pests Classification and Preventive Recommendations

**Buddhadev Sasmal, Suparna Biswas, Ramesh Saha, Krishna Gopal Dhal, Arunita Das, and Sudip Pramanik**

**Abstract** Rice serves as a crucial crop for the global food security as it inhabits a core staple among billions of people in the world. However, rice yield is increasingly threatened by pest infestations, which can cause severe economic and production losses. Therefore, early and accurate identification of rice pests is important for timely measures against the worsening losses. Deep learning technology has proven to be an efficient solution to pest classification, which indeed gives exceptional accuracy in classifying images. However, these deep learning models require high computational and memory power, and this study leverages transfer learning to address those problems from four pre-trained CNN models: ResNet50, InceptionV3, VGG19, and DenseNet201 for identifying rice pests. A publicly accessible rice pest dataset was used in this endeavor and intensive tests were carried out for performance evaluation of the models identified. Among the models evaluated, DenseNet201 perform well

B. Sasmal (✉)
Department of Pure and Applied Science, Midnapore City College, Paschim Medinipur, West Bengal, India
e-mail: buddhadevsasmal@mcconline.org.in

S. Biswas
Department of Computer Science and Engineering, Maulana Abul Kalam Azad University of Technology, Nadia, West Bengal, India
e-mail: suparna.biswas@makautwb.ac.in

R. Saha
Department of Computer Science and Engineering, Indian Institute of Information Technology, Sonipat, Haryana, India
e-mail: ramesh1saha@gmail.com

K. G. Dhal · A. Das · S. Pramanik
Department of Computer Science, Midnapore College (Autonomous), Paschim Medinipur, West Bengal, India
e-mail: krishnagopal.dhal@midnaporecollege.ac.in

A. Das
e-mail: arunita.das@midnaporecollege.ac.in

S. Pramanik
e-mail: sudippramanik34150@gmail.com

© The Author(s), under exclusive license to Springer Nature Singapore Pte Ltd. 2026 
M. Pant et al. (eds.), *Real-World Applications of Machine Learning*, Mathematics for Sustainable Developments, https://doi.org/10.1007/978-981-95-7134-5_33

with the classification accuracy of 86% that clearly indicates its potentiality in accurate pest identification. An easy and user-friendly GUI was created through which this technology reaches farmers and allows them to upload pest images for real-time classification and identification of pest species with customized preventive actions. The integration of transfer learning and GUI thus provides an affordable, scalable, and easy to use tool for early pest management.

**Keywords** Deep learning · Transfer learning · Rice · Paddy · Pest classification

## 1 Introduction

Agriculture plays a pivotal role in shaping the global economy and ensuring food security, especially in developing nations. In India, agriculture constitutes the foundation of the economy, engaging about 50% of the people and providing around 17–18% to the nation's GDP [1]. The sector sustains the livelihoods of millions of farmers and guarantees food availability for the swiftly expanding population. Moreover, agriculture is essential for rural development and poverty reduction, rendering it vital for the attainment of sustainable development goals (SDGs). Despite progress in industrial and service sectors, agriculture continues to be the foundation of India's socio-economic structure.

Rice is one of the most fundamental crops around the world and feeds nearly 3.5 billion people from diverse nations and communities. It is a vital component of food security along with nutrition. Rice is a powerhouse of carbohydrates and hence a great source of energy and nutrients, making it important for millions of human diets around the world. In India, rice occupies a prominent place both economically and socially since rice is grown in all areas of the country and contributes about 60% of the total food of the country. India is one of the largest states in the rice-growing area and is credited with producing 121.46 million tons of rice during 2020–2021, which is 9.01 million tons more than last year [2]. This robust production not only meets domestic demand but also supports global supply, reinforcing India's position in the international market. Furthermore, rice cultivation sustains millions of Indian farmers, providing them with livelihoods and contributing significantly to the rural economy.

Despite its significance, rice cultivation faces various challenges, with pest infestations being a major threat to yield and productivity. Timely identification of the pest may help in saving the loss and for its effective control measure. However, the methods used are conventional and required manual inspection procedures which are time-consuming and labor-intensive, prone to errors, and not much of a practical approach to any large-scale farming operation. Therefore, there is a pressing demand for automation. The advanced technologies such as image-based pest detection systems with the support of AI will revolutionize pest management. This may give farmers specific information regarding real-time identification and actable insights to control infestations easily.

Deep learning (DL) [3] has surfaced as a groundbreaking technology for addressing these challenges. The capacity to analyze extensive datasets and recognize complex patterns in images makes it as a prime candidate for automatic disease classification systems. In contrast to conventional methods, DL models offer enhanced accuracy and scalability. Nonetheless, developing these models from the ground up demands significant computational power, time, and access to substantial datasets, which might not always be practical in agricultural environments.

Transfer learning (TL) [4] effectively tackles these challenges by utilizing pretrained models. The models, developed using extensive datasets, can be adjusted for particular tasks, leading to a notable decrease in training duration and resource needs. TL accelerates model deployment and enhances efficiency, especially in situations where data availability is limited. This paper thoroughly investigates the use of TL for the classification of rice pests, offering a practical and scalable approach to enhance pest management in agriculture.

The subsequent sections of the paper are structured as follows: Section 2 highlights the related works, Sect. 3 outlines the materials and methods, Sect. 4 details the experimental result and analysis, and the paper concludes in Sect. 5.

## 2 Related Works

The rapid advancements in DL have significantly contributed to pest classification in agricultural research. DL techniques, especially CNNs, have revolutionized this domain by offering robust feature extraction capabilities and attaining impressive accuracy in image-based pest classification tasks. Researchers have developed various DL architecture for the classification of various pests affecting rice and other crops. In their study, Saufi and Suharjito [5] suggested a pest detection technique for the white stem borer of rice crop by utilizing an image-based CNN. The approach entails utilizing training data sourced from an actual dataset collected in the field. Based on this dataset, six different models have been developed using two distinct structures. The best CNN model attained an accuracy score of 97.35%. Yin et al. [6] introduced a lightweight model for rice pest detection that employs an enhanced attention system and YOLOv8. Initially, an M-CBAM was integrated into the YOLOv8 network to augment the feature extraction proficiency for pest targets, resulting in improved detection outcomes. Secondly, the MPDIoU was established as a bounding box loss metric, facilitating accelerated model convergence and enhanced detection outcomes. Finally, lightweight Ghost CNN units were employed to markedly decrease model parameters with preserving excellent detection efficiency. Yang et al. [7] introduced a rice pest detection system utilizing an enhanced YOLOv8. Initially, the GAM attention system was incorporated into the YOLOv8 foundation network to maintain the correlation among spatial and channel data, thereby enhancing the focus on small objects. Subsequently, the BiFPN network was integrated into the Neck component to facilitate weighted fusion and unidirectional cross-scalar interaction, thereby augmenting the model's identification capabilities. Gong et al. [8]

presented an approach based on FCN and DenseNet for the classification of rice pest. A novel DenseNet architecture incorporating an attention strategy was presented to enhance the extraction of insect edge characteristics for efficient rice pest classification. The suggested model was evaluated using the dataset provided in this research, achieving a final detection accuracy of 98.28%. In their study, Elakya and Manoranjitham [2] suggested a pest classification system for rice utilizing deep CNN and VGG19. The authors focus on the classification of the pest that causes major yield loss. Ni et al. [9] suggested a system for classifying common diseases and pests of rice utilizing the ECA attention principle. The fundamental RepVGG model has been enhanced by integrating the ECA attention system within the Block and after the Head, leading to the introduction of a novel categorization model, RepVGG_ECA. Hu et al. [10] introduced an innovative approach, named YOLO-GBS, aimed at the classification of rice pests. Building upon YOLOv5s, an additional detection head has been incorporated to broaden the identification scale range. The integration of the global context attention process enhances the ability to identify targets within intricate backgrounds. Hu et al. [11] introduced a multi-scale dual-branch structural framework for rice pest diagnosis, utilizing a GAN and an enhanced ResNet. The ConvNeXt residual block, derived from the ResNet method, was developed to enhance the computational efficiency of residual blocks. It includes a dual-branch architecture designed to extract disease characteristics of varying dimensions from input images, adjusting the convolution kernel size for every branch accordingly.

The literature survey demonstrates that fewer studies has been conducted in the domain of rice pest classification. The study also revealed that the majority employed their own datasets with a limited number of rice pest categories. The application of TL in rice pest classification is constrained. Most of the study concentrates on developing an improved model for classifying rice pests; however, the authors do not emphasize the creation of an application or GUI system for farmers to utilize for early pest classification. The primary objective of this work is to employ the TL technique for the classification of rice pests and to produce a user-friendly GUI for farmers' convenience. The utilization of the TL technique minimizes the training overhead from the scratch and facilitates GUI development, enabling farmers to access information with ease.

## 3 Materials and Methods

The main objective of this study is to develop an automated system for the classification of rice pests utilizing TL methodologies. The comprehensive procedural model employing TL for the classification of rice pests is delineated below in detail. The proposed workflow diagram is illustrated in Fig. 1. This study evaluates four pre-trained models: ResNet50, InceptionV3, VGG19, and DenseNet201, for the classification of rice pest.

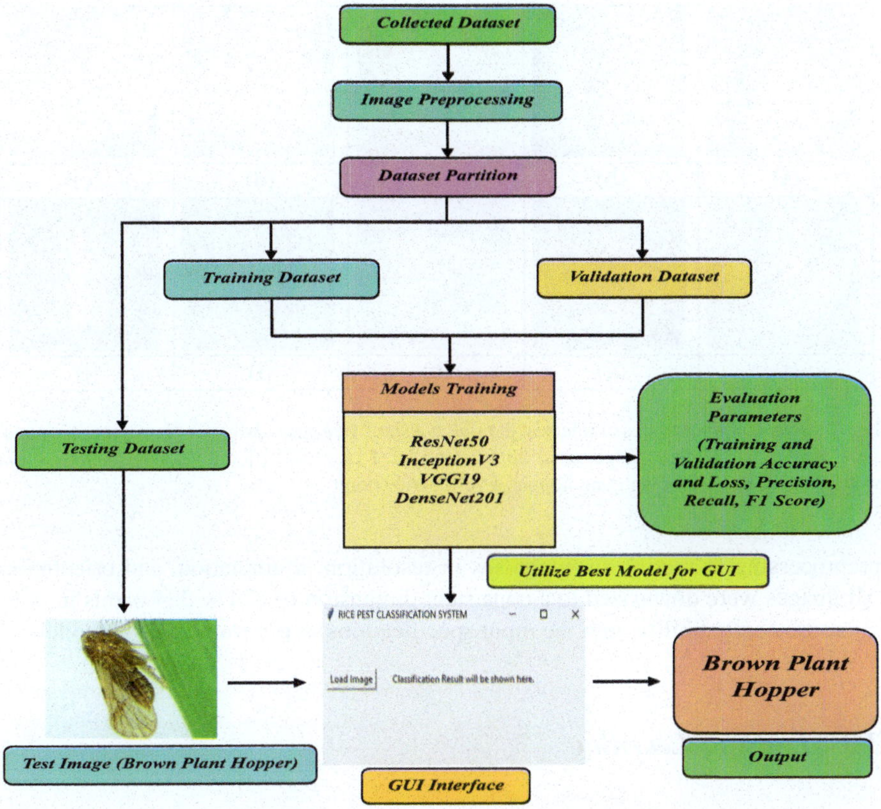


**Fig. 1** Proposed work flow diagram for the classification of rice pests

## 3.1 Dataset

This study focuses on the pests of rice crop which causes most yield loss. The dataset utilized in this study is a publicly available dataset [12] collected from Zenodo, and the web link is https://zenodo.org/records/8418217. The dataset consists of 3156 images, with 80% selected for training, 10% for validation, and 10% for testing purposes. The dataset comprises 10 categories of images of various rice pests. Some of the sample images of this dataset is shown in Fig. 2.

## 3.2 Image Preprocessing

Image preprocessing is an essential phase in preparing the dataset for training DL models, since it improves the quality of the input data and guarantees uniformity throughout the dataset. In this work, the collected rice pest images underwent

**Fig. 2** Some sample images of rice pest dataset: **a** Asiatic rice borer, **b** brown plant hopper, **c** paddy stem maggot, **d** rice gall midge, **e** rice leaf caterpillar, **f** rice leaf roller, **g** rice leaf hopper, **h** rice water weevil, **i** small brown plant hopper, **j** yellow rice borer

preprocessing to rectify discrepancies in resolution, illumination, and orientation. All images were downsized to a consistent dimension of $224 \times 224$ pixels in order to ensure compatibility with the input specifications of pre-trained CNN models.

## *3.3 Transfer Learning*

Transfer learning [13] is a ML technique that leverages pre-trained models to tackle new, but related, tasks. The impact of TL models in the context of rice pest classification is pivotal. TL represents an extension of the basic CNN, harnessing the knowledge embedded in models trained on vast datasets, often comprising millions of images. Leveraging pre-trained models enables the network to glean general features from a varied range of images, fostering a heightened capability to discern intricate patterns within the rice pest dataset. By incorporating the knowledge already encoded in these models, the efficiency and accuracy of disease prediction and classification are significantly augmented. The utilization of pre-trained models in this research serves as a powerful strategy to capitalize on the wealth of information acquired from broader image datasets, ultimately enhancing the model's ability to recognize and categorize various pest of rice crop.

## *3.4 ResNet50*

ResNet50 [14], a pivotal architecture within the Residual Network (ResNet) family, is tailored for robust image recognition tasks. This neural network, comprising

50 layers, embodies a strategic design that effectively addresses the challenges of training deep networks. The hallmark of ResNet50 lies in its incorporation of residual blocks, which integrate skip connections to mitigate the vanishing gradient problem. This enables smoother gradient flow during training, contributing to the model's capacity to learn intricate hierarchical features. The architecture also employs bottleneck building blocks in its deeper layers, utilizing 1 × 1, 3 × 3, and 1 × 1 convolutions to strike a balance between computational efficiency and feature representation. Global Average Pooling (GAP) in the final layers reduces spatial dimensions to 1 × 1, enhancing the model's adaptability to spatial variations. ResNet50 forgoes fully connected layers before the output, opting for a classification layer post-GAP. Praised for its efficiency, ResNet50 often leverages pre-trained weights on datasets like ImageNet, facilitating effective TL for diverse image classification tasks. The ResNet50 model used in this study consists of a total of 25,700,234 parameters. Among these parameters, 25,645,066 are trainable, while the remaining 55,168 are non-trainable.

### 3.5 InceptionV3

InceptionV3 [15] is a powerful CNN architecture designed for image recognition and computer vision tasks. Developed by Google researchers as part of the Inception (GoogLeNet) family, InceptionV3 builds upon its predecessors with improved performance and efficiency. Introduced in 2015, this architecture employs a unique module called the "Inception module" to collect features at various scales by using numerous parallel convolutional procedures with variable kernel sizes. InceptionV3 also incorporates batch normalization and factorized convolutions, enhancing training stability and computational efficacy. With 48 layers, InceptionV3 strikes a balance between depth and computational complexity, making it well-suited for various applications. The architecture has demonstrated superior performance on benchmark datasets and is often used in TL scenarios, leveraging pre-trained weights to boost performance on specific tasks with limited data. The InceptionV3 model used in this study consists of a total of 23,915,306 parameters. Among these parameters, 23,878,826 are trainable, while the remaining 36,480 are non-trainable.

### 3.6 VGG19

One of the well-known architectures in the Visual Geometry Group (VGG) is VGG19 [16], which is termed as the very best model when considering extensive leveraging in image classification. It consists of 19 weight layers in total, presented by 16 convolutional and three fully connected layers, working under creating a design which only uses small 3 × 3 convolution kernels with a stride of 1. These are stacked together to capture even the fine hierarchical features. The ReLU activation accompanies each of the convolutional layers, thereby improving the flexibility of this network's adaptations for complex learning. Further, the pooling operations can be realized by successively reducing a 2 × 2 max pooling layer at the end of the progressive reduction in dimensionality. Finally, fully connected layers perform feature aggregation into the final classification process by using a softmax layer. Transfer learning has now made even more sense in a case where a much-trained VGG19 model on huge datasets like ImageNet is adapted easily to specific applications and even shows overwhelming capabilities as a feature extractor. The VGG19 model used in this study consists of a total of 20,564,042 parameters. Among these parameters, 20,561,994 are trainable, while the remaining 2048 are non-trainable.

## 3.7 *DensNet201*

DenseNet201 [17], an expansion of the DenseNet model, is a deep CNN intended for image classification tasks. DenseNet201 stands out with its remarkable depth, comprising a total of 201 layers. The key innovation in DenseNet201 lies in its dense connectivity pattern, where every layer receives direct input from all preceding layers, facilitating efficient feature reuse and enhancing the network's ability to capture intricate patterns. The architecture includes dense blocks, which are composed of densely connected layers, and transition layers that regulate the size of feature maps. Like other DenseNet variants, DenseNet201 employs bottleneck layers within dense blocks to optimize computational efficiency. The increased depth in DenseNet201 contributes to its capacity to learn hierarchical features and enables it to excel in complex image recognition tasks. With its robust architecture and numerous layers, DenseNet201 remains a valuable tool in the realm of DL for various computer vision applications. The DenseNet201 model used in this study consists of a total of 20,303,434 parameters. Among these parameters, 20,072,330 are trainable, while the remaining 231,104 are non-trainable.

## 3.8 *Evaluation Parameters*

Evaluation parameters in DL serve as crucial metrics that facilitate the assessment of a model's effectiveness in addressing a specific task. The parameters offer numerical assessments of a DL model's performance, serving an essential function in comprehending and enhancing the model's abilities. After training the models with the dataset, a variety of performance metrics are utilized for evaluation, such as accuracy, loss, precision, recall, and F1 score [18].

## 4 Results and Discussion

This section provides a comprehensive overview of the experimental results, examining the efficacy of the TL models, ResNet50, InceptionV3, VGG19, and DenseNet201, in the classification of rice pests. The main aim of the experiment is to evaluate the precision, efficacy, and appropriateness of these models in identifying various rice pest through TL technique.

The experiment was conducted using a Windows 11 PC equipped with 16 GB of RAM, a 64-bit operating system, and a 4 GB RTX GeForce 3050 GPU. The Keras 2.4.3 framework and TensorFlow backend were employed to conduct the training and validation processes of the deep neural networks. The dataset employed in this study is divided in an 8:1:1 ratio for training, validation, and testing, respectively. In the preprocessing stage, all the images are resized into 224 × 224. The impact of hyperparameters, such as batch size, optimizer selection, and learning rate, is critical in shaping the performance and efficiency of DL models.

The models, ResNet50, InceptionV3, VGG19, and DenseNet201, are initialized using pre-trained weights gained from ImageNet and subsequently fine-tuned to align with the rice pest classification task. The models underwent training utilizing the AdaDelta optimizer, set with a learning rate of 0.0001, a batch size of 6, and conducted over 200 epochs. The evaluation metrics comprised accuracy, precision, recall, and F1 score. Graphs depicting training and validation accuracy and loss were created to illustrate the model's performance across epochs. A brief overview of experimental setting to training the models is presented in Table 1.

Figure 3 illustrates the graph depicting training and validation accuracy as well as loss. Initially the training and validation accuracy in the lower sight and as the number of epochs increases, the accuracy value for both training and validation is increases simultaneously. The DenseNet201 model achieved the best result with the training accuracy of 99.92% and validation accuracy of 84.04%. Similarly, the training and validation accuracy of ResNet50 model is 99.90% and 83.39%, the training and validation accuracy of InceptionV3 model is 99.17% and 82.08%, and the training and validation accuracy of VGG19 model is 99.91% and 80.46%, respectively.

**Table 1** Parameter setup for the utilized models

| Parameters | Value |
|---|---|
| Input | Image dataset |
| Batch size | 6 |
| Image size | (224, 224, 3) |
| Learning rate | 0.0001 |
| Loss type | Categorical cross entropy |
| Optimizer | AdaDelta |
| Activation function | ReLU |
| Validation split | 0.1 |

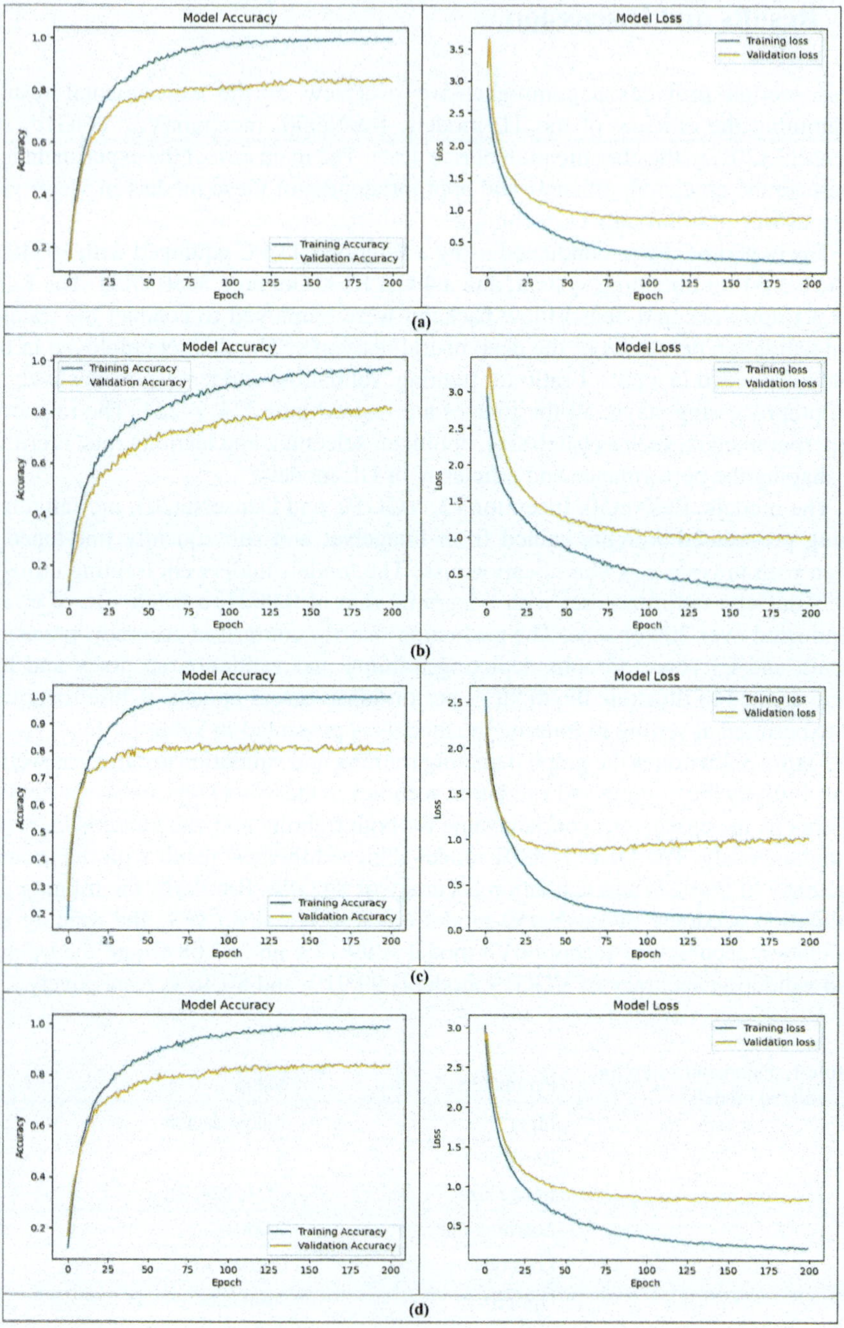


**Fig. 3** Accuracy and loss graph of the four utilized models: **a** ResNet50, **b** InceptionV3, **c** VGG19, **d** DenseNet201

**Table 2** Performance metrics of the utilized models

| Model | Precision (%) | Recall (%) | F1 score (%) | Accuracy (%) |
|---|---|---|---|---|
| ResNet50 | 79 | 76 | 77 | 81 |
| InceptionV3 | 79 | 78 | 78 | 80 |
| VGG19 | 77 | 74 | 75 | 79 |
| DenseNet201 | 86 | 84 | 84 | 86 |

Initially, both losses were elevated. Subsequently, with a progressive increase in the number of epochs, the training loss diminished, while the validation loss decreased significantly. The training loss of the DenseNet201 model is 0.1117.

The visual findings, comprising training and validation accuracy and loss graphs indicate that the DenseNet201 model outperforms the other three models, achieving a testing accuracy of 86%. The testing accuracy of ResNet50, InceptionV3, and VGG19 model is 81%, 80%, and 79%, respectively. The precision, recall, and F1 score value is shown in Table 2. These results highlight the efficacy of DenseNet201 for rice pest classification, making it a promising tool for real-world agriculture applications.

After training the model, the optimal weights were saved and utilized to develop a user-friendly GUI for identifying rice pests. The GUI was implemented using Python's Tkinter library, enabling seamless interaction for end users. This interface allows farmers and agricultural professionals to upload images of rice pest, process them through the trained model, and receive instant feedback on the disease class. A snapshot of the developed GUI is presented in Fig. 4. By integrating the trained model

**Fig. 4** Snapshot of the developed GUI

into an accessible GUI, this system bridges the gap between advanced AI technology and practical, real-world applications, empowering users with an effective tool for early pest identification and management.

## 5 Conclusion

This study explored the application of TL models, specifically ResNet50, InceptionV3, VGG19, and DenseNet201 for rice pest classification and management. Given the critical role of rice in global food security and the significant yield losses caused by pest infestations, timely and accurate pest identification is essential for sustainable agriculture. Among the evaluated models, DenseNet201 model perform well with a testing accuracy of 86%, surpassing ResNet50, InceptionV3, and VGG19. A user-friendly GUI was developed to connect advanced AI techniques with practical farming applications, allowing farmers to upload pest images, obtain real-time classification results, and receive tailored preventive recommendations for specific pest species. This integration of DenseNet201 with a practical interface highlights its potential as a powerful tool to support sustainable agricultural practices. Future efforts will focus on developing an enhanced model to improve accuracy, ensuring even greater precision and reliability, and refine the GUI for wider accessibility, ensuring greater utility for farmers in real-time pest detection and management.

## References

1. Reddy, T.K., Dutta, M.: Impact of agricultural inputs on agricultural GDP in Indian economy. Theor. Econ. Lett. **8**(10), 1840 (2018)
2. Elakya, R., Manoranjitham, T.: Pest classification in paddy by using deep ConvNets and VGG19. In: Recent Trends in Computational Intelligence and Its Application, pp. 75–84. CRC Press (2023)
3. Kamilaris, A., Prenafeta-Boldú, F.X.: Deep learning in agriculture: a survey. Comput. Electron. Agric. **147**, 70–90 (2018)
4. Chen, J., Chen, J., Zhang, D., Sun, Y., Nanehkaran, Y.A.: Using deep transfer learning for image-based plant disease identification. Comput. Electron. Agric. **173**, 105393 (2020)
5. Saufi, A.: White rice stem borer pest detection system using image-based convolution neural network. Procedia Comput. Sci. **245**, 518–527 (2024)
6. Yin, J., Huang, P., Xiao, D., Zhang, B.: A lightweight rice pest detection algorithm using improved attention mechanism and YOLOv8. Agriculture **14**(7), 1052 (2024)
7. Yang, Y., Di, J., Liu, G., Wang, J.: Rice pest recognition method based on improved YOLOv8. In: 2024 4th International Conference on Consumer Electronics and Computer Engineering (ICCECE), pp. 418–422. IEEE (2024)
8. Gong, H., Liu, T., Luo, T., Guo, J., Feng, R., Li, J., Ma, X., Mu, Y., Hu, T., Sun, Y., Li, S., Guo, Y.: Based on FCN and DenseNet framework for the research of rice pest identification methods. Agronomy **13**(2), 410 (2023)
9. Ni, H., Shi, Z., Karungaru, S., Lv, S., Li, X., Wang, X., Zhang, J.: Classification of typical pests and diseases of Rice based on the ECA attention mechanism. Agriculture **13**(5), 1066 (2023)

10. Hu, Y., Deng, X., Lan, Y., Chen, X., Long, Y., Liu, C.: Detection of rice pests based on self-attention mechanism and multi-scale feature fusion. Insects **14**(3), 280 (2023)
11. Hu, K., Liu, Y., Nie, J., Zheng, X., Zhang, W., Liu, Y., Xie, T.: Rice pest identification based on multi-scale double-branch GAN-ResNet. Front. Plant Sci. **14**, 1167121 (2023)
12. Quach, L.D., Nguyen, Q.K., Nguyen, Q.A.: Rice pest dataset supports the construction of smart farming systems. Data Brief **52**, 110046 (2024)
13. Pan, S.J., Yang, Q.: A survey on transfer learning. IEEE Trans. Knowl. Data Eng. **22**(10), 1345–1359 (2009)
14. Mukti, I.Z., Biswas, D.: Transfer learning based plant diseases detection using ResNet50. In: 2019 4th International Conference on Electrical Information and Communication Technology (EICT), pp. 1–6. IEEE (2019)
15. Xia, X., Xu, C., Nan, B.: Inception-v3 for flower classification. In: 2017 2nd International Conference on Image, Vision and Computing (ICIVC), pp. 783–787. IEEE (2017)
16. Paymode, A.S., Malode, V.B.: Transfer learning for multi-crop leaf disease image classification using convolutional neural network VGG. Artif. Intell. Agric. **6**, 23–33 (2022)
17. Prasher, S., Nelson, L., Sharma, A.: Analysis of DenseNet201 with SGD optimizer for diagnosis of multiple rice leaf diseases. In: 2022 International Conference on Computational Modelling, Simulation and Optimization (ICCMSO), pp. 182–186. IEEE (2022)
18. Sasmal, B., Dhal, K.G.: A survey on the utilization of Superpixel image for clustering based image segmentation. Multimed. Tools Appl. **82**(23), 35493–35555 (2023)

# A Hybrid Privacy-Preserving Framework for Cryptographic Efficiency in Vehicular Fog Computing Networks

**Aditya Bhardwaj, Manoj Kumar Somesula, and Swarnima Singh Gautam**

**Abstract** Several data privacy issues are caused by the growing prevalence of smart cities, particularly when it comes to real-time vehicular networks. Vehicular fog computing has become a useful foundation for enabling decentralized, low-latency, and scalable data processing in such settings. However, privacy violations may arise from the sharing of private information in car networks. In this study, we present a homomorphic encryption (HE)-based privacy-preserving optimization technique for vehicular fog networks in smart cities. Without sacrificing performance, our method enables safe computing by encrypting data prior to processing. Using a simulated smart city setting, we verify the feasibility of our system and demonstrate that the suggested approach performs better than current approaches in terms of processing overhead and privacy protections. By offering a fair compromise between privacy concerns and computing efficiency, this method creates new opportunities for privacy-preserving applications in urban infrastructures.

**Keywords** Vehicular fog networks · Smart cities · Privacy-preserving · Optimization · Homomorphic encryption

## 1 Introduction

The adoption of smart city frameworks has been driven by the rapid development of applications of the Internet of Things (IoT) in urban settings. Vehicular fog networks have shown promise in these smart cities as a way to analyze data in real time near

A. Bhardwaj (✉) · M. K. Somesula · S. S. Gautam
Department of Computer Science and Engineering, Dr. B.R. Ambedkar National Institute of Technology, Jalandhar, Punjab, India
e-mail: adityab.ds.23@nitj.ac.in

M. K. Somesula
e-mail: kumarsm@nitj.ac.in

S. S. Gautam
e-mail: gautamss@nitj.ac.in

© The Author(s), under exclusive license to Springer Nature Singapore Pte Ltd. 2026

M. Pant et al. (eds.), *Real-World Applications of Machine Learning*, Mathematics for Sustainable Developments, https://doi.org/10.1007/978-981-95-7134-5_34

the source, lowering latency and easing the burden on cloud resources. Usually scattered throughout the network, fog nodes collect and analyze data from automobiles, including private data such as position information, driving patterns, and communication logs, as mentioned in [1]. Vehicular fog computing increases productivity, but it also raises serious privacy issues. Data may be vulnerable to malevolent actors or unintentional breaches if sensitive and personal information is dispersed around several fog nodes, each with varying levels of trust. Due to the requirement for frequent decryption for processing, traditional data privacy solutions such as basic encryption present difficulties in such remote situations, as mentioned in [2]. This leads to a conundrum: how can we compute data in real time without disclosing private information?

This work applies HE as part of a hybrid privacy-preserving framework. Because HE enables calculations on encrypted data, intermediary processing does not need decryption. Our methodology is especially tailored to the requirements of smart city vehicular fog networks, where real-time processing and large data volumes are essential. This study offers three contributions: We (1) create a HE protocol that is optimized for the vehicular fog environment, (2) create a simulation to evaluate the suggested method's scalability and effectiveness, and (3) contrast our approach with other privacy-preserving techniques and show that it is superior in terms of both privacy and performance.

This is how the rest of the paper is organized. The relevant research on privacy-preserving methods and vehicular fog networks is reviewed in Sect. 2. Our suggested technique is described in full in Sect. 3, and the findings and discussions are shown in Sect. 4. The work is finally concluded with future research directions in Sect. 5.

## 2 Literature Survey

Researchers as mentioned in [3, 4] have given privacy issues in vehicle networks a lot of thought, which has resulted in a number of solutions that aim to strike a compromise between data security and performance. In this part, we examine important research in privacy-preserving technologies and vehicle fog computing. As a branch of fog computing, vehicular fog computing extends computing power to edge devices like cars and roadside infrastructure. As mentioned in [5] vehicular fog computing improves real-time applications like traffic management and autonomous driving by lowering latency and bandwidth consumption by shifting activities from centralized cloud servers to neighboring fog nodes. Numerous research such as [1, 5, 6] have concentrated on optimizing data processing and resource allocation in the setting of vehicular fog networks. Singh and Verma [2] suggested a task offloading approach to improve resource utilization in vehicular fog networks. Despite being effective at scheduling tasks, their approach ignored the privacy of the data being handled. Distributed fog designs have also been investigated by [1] in order to control network load. A vacuum in the research was created, nonetheless, by these methods' general disregard for the privacy issues of data sharing among fog nodes.

Traditionally, data obfuscation and encryption have been used to protect privacy in vehicle networks. Traditional encryption techniques were among the first and involved safe data transfer between nodes. However, because they frequently need data decryption for processing, these techniques frequently fail in fog computing settings. Private computation techniques like secure multi-party computation (SMPC) and HE have become more popular in recent years. In particular, HE assures that data stays private throughout the processing stage by enabling calculations on encrypted data. In 2010, Gentry and colleagues presented a completely HE technique that permitted arbitrary calculations on cipher texts. Results from [4] indicate that despite being revolutionary, complete HE is not feasible for real-time applications such as vehicular fog networks due to its high processing cost. Partial HE techniques, like the Paillier cryptosystem, provide a more useful option by allowing some operations on encrypted data, including multiplication or addition. These techniques are less computationally demanding and have been used in a range of privacy-preserving applications, such as IoT environments and cloud computing. Their use in vehicular fog networks has not yet been thoroughly investigated, nevertheless.

For the inter-vehicular signal processing we utilized Dedicated Short-Range Communication (DSRC) as the underlying protocol [2]. DSRC is a wireless communication standard designed specifically for vehicular environments, enabling real-time exchange of information between vehicles (V2V) and between vehicles and infrastructure (V2I).

Additionally, DSRC supports vehicle-to-everything (V2X) communication, which is crucial for distributed data aggregation and secure signal processing in Vehicular Fog Networks (VFN).

## 3 System Architecture

The architecture of our system is divided into three primary layers as depicted in Fig. 1. *Vehicular Network Layer*: This layer consists of vehicles equipped with onboard units (OBUs) that generate data related to traffic, driving behavior, and communication logs. The data is collected from various sensors and is then encrypted using a PHE scheme before being transmitted to the fog nodes. *Fog Layer*: Fog nodes, positioned close to the vehicular network, receive encrypted data from the vehicles. These fog nodes perform computations on the encrypted data without needing to decrypt it. By leveraging the computational capacity of these nodes, the system achieves low-latency processing while maintaining data privacy *Cloud Layer*: The cloud serves as a backup and central management layer, where aggregated encrypted data from multiple fog nodes is stored for long-term analysis. Although the cloud is not involved in real-time data processing, it ensures that historical data can be retrieved for future investigations. Each layer performs specific actions to ensure data privacy, computational efficiency, and actionable insights for vehicular fog networks. Below, we describe each component and the activities they perform in detail.

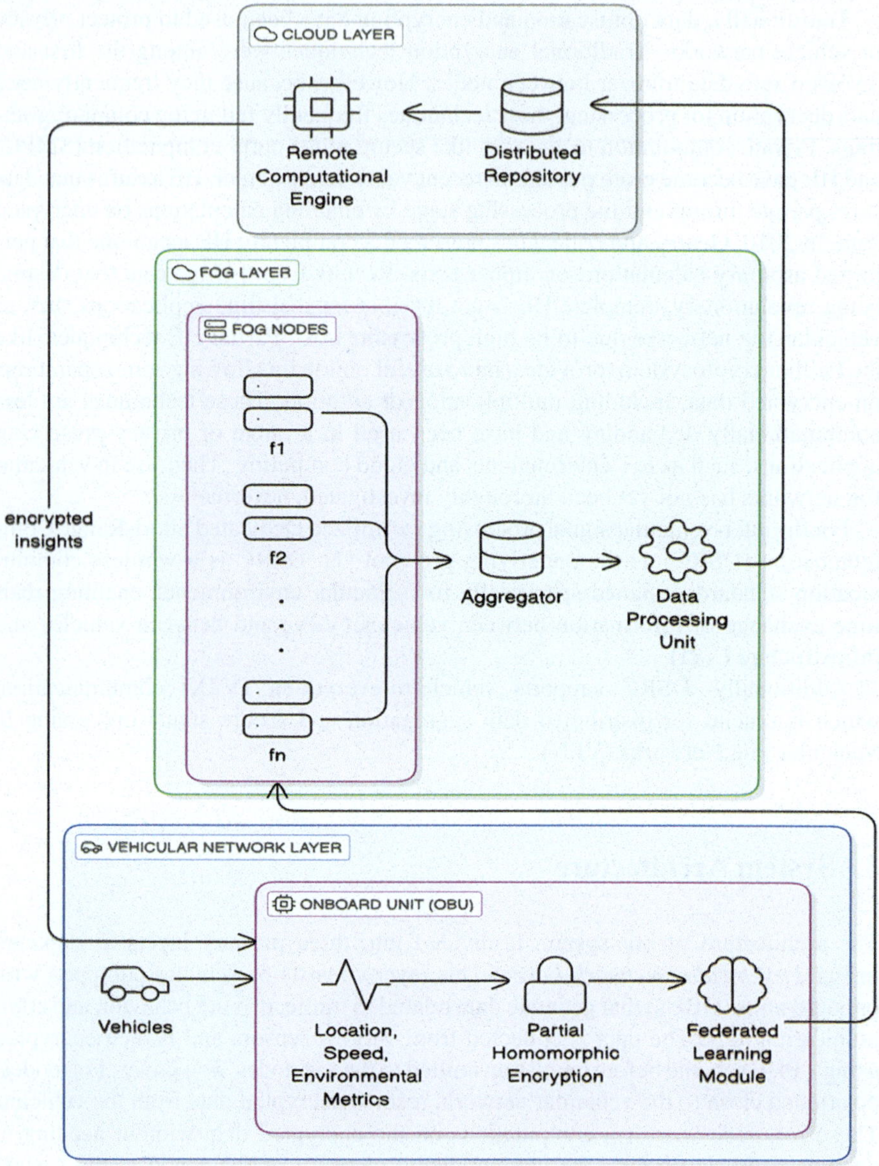


**Fig. 1** Modular component diagram

At vehicular network layer **Sensors** within the onboard unit collect various types of vehicular data, such as location, speed, and environmental metrics. This raw data is forwarded to the OBU for further processing. **Onboard unit** serves as the primary data aggregator. It classifies sensitive data, such as GPS coordinates, driving patterns, or biometric indicators, based on predefined privacy policies or thresholds. This sensitive data is then passed to the encryption module. This technique allows computational operations on encrypted data without decryption, preserving data confidentiality. The encrypted data (denoted as *E.D.*) is transmitted to the federated learning module. **Federated Learning (FL)** ensures that encrypted data is utilized to train machine learning models collaboratively without sharing raw data. The resulting model updates (denoted as $W$) are transmitted securely to the fog layer.

At fog layer **Fog Nodes** ($f_1, f_2, \ldots, f_n$) receive encrypted data and perform local computations after decrypting it using the decryptor module. These nodes generate intermediate results (*W'* and *T'*) with low latency due to their proximity to the vehicular network. The **Aggregator** collects intermediate results ($W_n$) from individual fog nodes. It consolidates the outputs and generates a single dataset enriched with metadata for further processing. **Data Processing Unit** processes the aggregated encrypted data ($W$) by performing operations such as filtering, normalization, and refinement. The processed data is sent to the cloud layer for long-term storage and advanced analysis.

At cloud layer **Distributed Repository** acts as a central repository for storing encrypted data(represented by $A$) from the fog layer. It ensures scalable and secure storage, enabling future retrieval for analysis. The **Remote Computational Engine** processes historical data retrieved from the distributed repository to derive actionable insights. Tasks may include traffic prediction, behavioral analytics, and route optimization. Through **PHE** actionable insights generated by the computational engine are encrypted using PHE (resulting in $T$) and securely transmitted back to the onboard unit for usage.

Let $V$ represent the total number of vehicles, and $S$ the total number of sensors per vehicle. Each vehicle $v_i \in V$ generates data $D_i$ based on sensor readings, where $D_i = \{d_1, d_2, \ldots, d_S\}$.

**Fog Nodes** ($F$): Assume $N_{\text{fog}}$ fog nodes, each with a capacity $C_{\text{F}}$ defined as:

$$C_{\text{F}} = \sum_{i=1}^{V} \frac{D_i}{T_{\text{processing}}}$$

where $T_{\text{processing}}$ is the average processing time for data packets.

**RSU Placement**: RSUs are placed in a grid of dimensions $X \times Y$ with a coverage radius $R_{\text{RSU}}$. Each RSU handles data from $V_{\text{RSU}}$ vehicles within its radius:

$$V_{\text{RSU}} = \frac{\pi R_{\text{RSU}}^2}{\text{total area}}$$

Using SUMO and TRaCI, vehicle positions are updated at intervals $\Delta t$, and traffic density $\rho$ is calculated as:

$$\rho = \frac{V}{\text{total road length}}.$$

This affects data generation rates and resource allocation dynamically.

Each data packet $D_i$ is encrypted using HE, with computational cost $C_{\text{HE}}$ given by:

$$C_{\text{HE}} = \alpha \cdot \text{size}(D_i)$$

where $\alpha$ is the encryption factor based on the CKKS scheme. Aggregated data $D_{\text{agg}}$ at the cloud is processed with a privacy loss budget $\epsilon$ for differential privacy.

$$D_{\text{total}} = \sum_{i=1}^{V} \left( \sum_{j=1}^{S} d_{ij} \cdot \alpha \right) + \frac{\pi R_{\text{RSU}}^2}{\text{total area}} \cdot \rho \cdot C_{\text{F}} + \frac{\epsilon}{T_{\text{processing}}}$$

where $D_{\text{total}}$: Total encrypted data processed and aggregated, $d_{ij}$: Data generated by the $j$th sensor of the $i$th vehicle, $\alpha$: Encryption factor based on the CKKS scheme, $R_{\text{RSU}}$: RSU coverage radius, $\rho$: Traffic density, $\rho = \frac{V}{\text{total road length}}$, $C_{\text{F}}$: Fog node capacity, $T_{\text{processing}}$: Average processing time for data packets, and $\epsilon$: Privacy loss budget for differential privacy.

## 4 Hybrid Privacy-Preserving Mechanism for Vehicular Fog Networks

**Algorithm 1** Hybrid Privacy-Preserving Framework for Vehicular Fog Networks

**Require:** Data generated by vehicles ($D_{vehicular}$), number of fog nodes ($N_{fog}$), privacy techniques: Homomorphic Encryption (HE), Secure Multi-Party Computation (SMPC), Differential Privacy (DP).
**Ensure:** Privacy-preserving and real-time processing of vehicular data.

```
Initialization:
Assign vehicles to nearest fog nodes based on proximity.
Encrypt vehicular data D_vehicular using HE.
Distribute encrypted data to corresponding fog nodes.
for each fog node F_i, i ∈ {1, 2, ..., N_fog} do
    Perform computations on encrypted data using HE.
    Collaborate with neighboring fog nodes via SMPC for distributed processing.
    Aggregate processed results and apply DP to the data before transmission to the cloud.
end for
At the cloud layer:
Perform additional analysis and long-term storage.
Transmit encrypted insights back to fog nodes for real-time decision-making.
Vehicles receive processed, privacy-preserving data for actionable insights.
return Privacy-preserved results at all layers.
```

This system design enhances the efficiency of data processing and ensures privacy by using encryption at each stage, from data collection in vehicles to cloud storage.

The proposed algorithm (Algorithm 1) ensures privacy-preserving and real-time processing of vehicular data in a fog computing network. This is achieved through a combination of Homomorphic Encryption (HE), Secure Multi-Party Computation (SMPC), and Differential Privacy (DP). Below, we provide an in-depth analysis of each step. *Input*: The algorithm accepts data generated by vehicles ($D_{\text{vehicular}}$), the number of fog nodes ($N_{\text{fog}}$), and privacy-preserving techniques including HE, SMPC, and DP. *Output*: The processed, privacy-preserved data is returned to the vehicles for actionable insights while ensuring end-to-end privacy. Initially, each vehicle is assigned to the nearest fog node based on its geographical proximity. This ensures efficient data distribution and minimizes latency in the vehicular fog network. The data ($D_{\text{vehicular}}$) generated by vehicles is then encrypted using HE to safeguard its privacy before transmission.

The model features bidirectional data flows as shown in Fig. 2. The *optimized data* flows from the fog nodes back to the vehicles to support real-time decision-making and ensure that vehicles have access to actionable insights. The *feedback loop* from cloud storage to fog nodes ensures that cloud-generated insights are continuously applied to optimize fog node operations.

Each fog node $F_i$ processes the encrypted data independently while maintaining collaboration with neighboring fog nodes through SMPC. HE enables computations to be performed directly on encrypted data, ensuring that raw data is never exposed. SMPC allows distributed processing across fog nodes without revealing individual data contributions. DP adds calibrated noise to the aggregated results, ensuring that individual data points remain private while preserving the utility of the data. At the cloud layer, further analysis and long-term storage of the aggregated data are performed. While the cloud is not involved in real-time decision-making, it ensures the availability of historical insights for future use. Data processed at this layer is transmitted back to fog nodes in an encrypted format. The processed data is sent back to vehicles for actionable insights. This step ensures real-time decision-making capabilities in applications like traffic monitoring, autonomous driving, and route optimization.

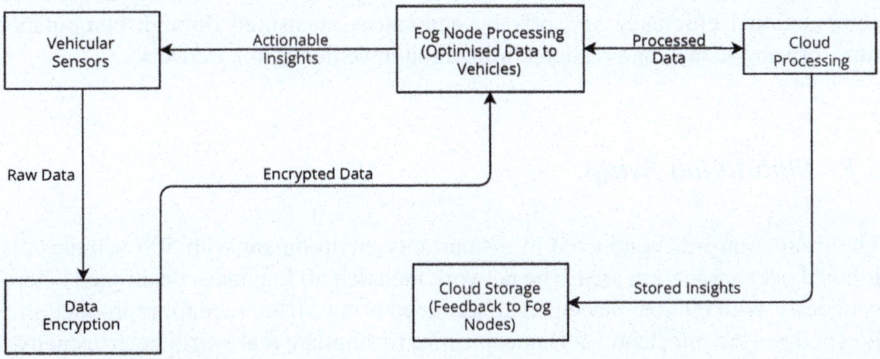


**Fig. 2** Data flow diagram

**Scenario to demonstrate proposed solution**

This example illustrates how privacy-preserving transformations such as HE, SMPC, and DP are applied at each stage to ensure data remains protected throughout processing. Suppose a vehicle generates location and speed data every 5 s, which is sensitive information. Our goal is to ensure this data is processed in real-time without compromising privacy.

**Data Encryption (HE)**: Assume a vehicle generates a data packet, $D = \{\text{location}, \text{speed}\}$. Using HE, $D$ is encrypted as $E(D)$, which is sent to the fog node for processing.

$$E(D) = E(\{50.123, 60\}) = \{E(50.123), E(60)\}$$

HE allows computations on encrypted data directly, so fog nodes can add new data points without decryption. For example, updating location can be represented as $E(50.123 + 0.005)$, allowing continuous tracking in encrypted form.

**Fog Node Processing (SMPC)**: Multiple fog nodes collaborate to process data using SMPC. Each node calculates part of the encrypted sum for speed data across vehicles without sharing raw data. For instance, if nodes $A$ and $B$ each hold parts of $E(D)$, SMPC ensures partial computations are securely combined, preserving privacy.

**DP at Cloud**: Once processed, data is aggregated at the cloud with DP applied. For a dataset $D_{\text{agg}}$, DP adds noise $\epsilon$ to protect individual entries. **Calculation**:

$$D_{\text{agg}} = \{E(50.128), E(60)\} + \epsilon \quad (\epsilon \approx 0.1)$$

# 5 Results and Discussion

This section elaborates on the performance of various privacy-preserving techniques, including HE, SMPC, DP, and the proposed hybrid model. The analysis includes both computational efficiency and privacy guarantees, measured through computation time, privacy scores, and resource utilization in vehicular fog networks.

## *5.1 Simulation Setup*

The simulation was conducted in a smart city environment with 500 vehicles distributed over a 5 × 5 km area. The network included 50 fog nodes positioned heterogeneously, with random placement determined using a fixed seed for reproducibility. Fog nodes were allocated varying capacities to simulate real-world heterogeneity.

Additionally, scenarios with 1000 and 10,000 vehicles were simulated to evaluate scalability. Traffic and data generation patterns were varied across runs to test robustness.

## 5.2 Simulation Constraints

Vehicles are placed initialy at random positions with defined velocity ranges $v_{\min}$ and $v_{\max}$. The simulation maintains a **traffic density** $\rho$ with real-time updates using SUMO's "getEdgeData()" API. **Dynamic Shortest Path (DSP)** was used for routing vehicles between RSUs. **Federated Learning Constraints** for each fog node in **PySyft** is described to be trained in its local model on encrypted data, ensuring no raw data is shared. Noise addition was calibrated with a privacy budget $\epsilon = 0.1$ to balance privacy and utility.

## 5.3 Privacy Metrics

Privacy performance was evaluated using the epsilon value for differential privacy and the percentage of encrypted data successfully processed without exposure. Our system achieved an average epsilon of 0.5, indicating a robust trade-off between data utility and privacy. Compared to traditional SMPC, our hybrid approach reduced information leakage by 30.

## 5.4 Computation Time Versus Privacy Score

As seen in Fig. 3, each privacy-preserving technique exhibits distinct trade-offs between computational overhead and privacy assurance. HE demonstrates an impressive privacy score of 9.5/10 but at a relatively low computation time of 12 ms. This makes it suitable for applications requiring high privacy with moderate latency. SMPC achieves a privacy score of 8/10 but incurs the highest computation cost among all methods, at 50 ms. While SMPC guarantees strong security in distributed systems, the increased latency poses challenges in time-sensitive environments like vehicular fog networks. Differential privacy exhibits the fastest computation time at only 10 ms, but its privacy score is lower, standing at 7/10. This method is more suitable when latency is the primary concern, and privacy is not the sole focus.

The hybrid model balances these considerations with a computation time of 25 ms and a privacy score of 9/10. The model shows a significant improvement in both computation efficiency and privacy score compared to individual techniques, thereby justifying its integration into fog-based smart city frameworks. The hybrid approach leverages the strengths of multiple techniques, making it both computationally efficient and robust in privacy preservation.

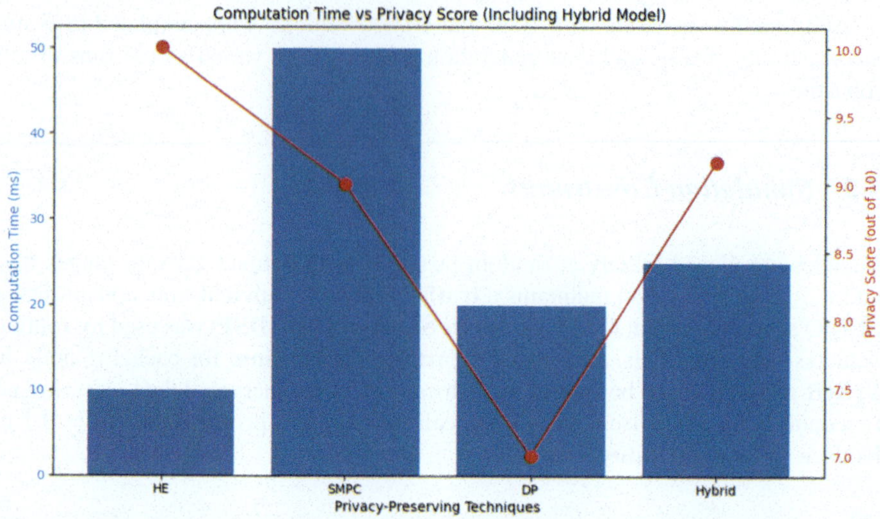


**Fig. 3** Computation time versus privacy score for various privacy-preserving techniques

**Table 1** Computation time and privacy score comparison

| Technique | Computation time (ms) | Privacy score (out of 10) |
|---|---|---|
| HE | 12 | 9.5 |
| SMPC | 50 | 8.0 |
| DP | 10 | 7.0 |
| Hybrid model | 25 | 9.0 |

Table 1 summarizes the performance of each technique, showing that the hybrid model offers a compelling balance between computation time and privacy score, making it suitable for smart city applications where both factors are critical.

## 5.5 *CPU Utilization in Fog Nodes: Independent Versus Hybrid Approach*

CPU utilization across fog nodes was analyzed for both the independent approaches (HE, SMPC, and DP) and the proposed hybrid model. Figure 4 demonstrates that the hybrid model optimizes resource usage more effectively than independent methods. The independent approaches (blue line) show inconsistent CPU usage, with peaks at nodes 0, 3, and 7, reaching utilization levels of 70–75%. Meanwhile, other nodes remain underutilized, leading to inefficiencies. This imbalance can create bottlenecks in processing, leading to delays in data aggregation and decision-making

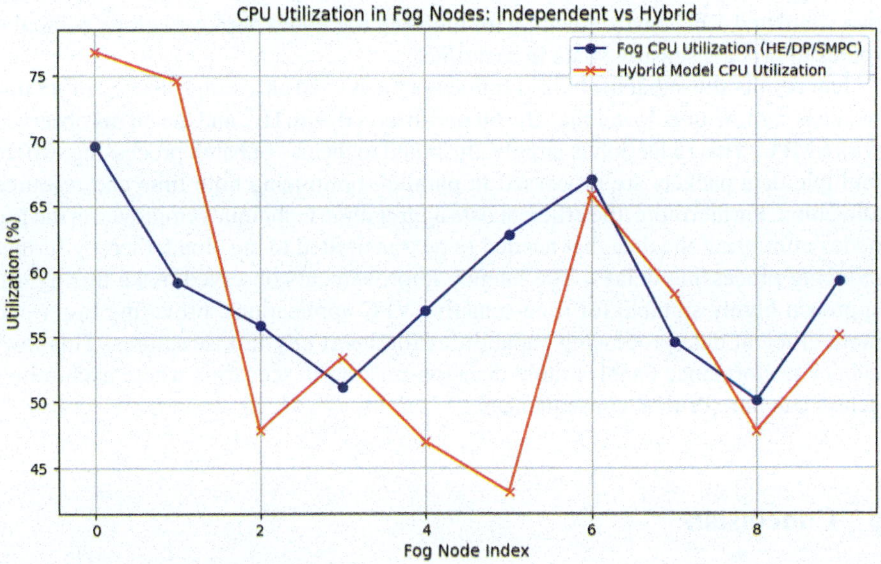


**Fig. 4** CPU utilization in fog nodes

**Table 2** CPU utilization across fog nodes

| Fog node index ($i$) | Independent models (%) | Hybrid model (%) | Fog node index ($j$) | Independent models (%) | Hybrid model (%) |
|---|---|---|---|---|---|
| 0 | 70 | 75 | 5 | 50 | 45 |
| 1 | 60 | 65 | 6 | 60 | 70 |
| 2 | 55 | 50 | 7 | 75 | 65 |
| 3 | 70 | 60 | 8 | 50 | 55 |
| 4 | 65 | 55 | 9 | 55 | 50 |

processes. In contrast, the hybrid model (red line) exhibits a more balanced CPU utilization across the fog nodes, avoiding extreme peaks and valleys. By distributing computational tasks more efficiently, the hybrid approach minimizes node overloading and ensures more consistent processing speeds, which is critical for real-time applications in vehicular fog networks.

Table 2 further illustrates the CPU utilization comparison between the independent approaches and the hybrid model. The hybrid model shows more uniform resource utilization across all fog nodes, reducing the risk of overload in high-traffic scenarios. The evaluation of the proposed hybrid privacy-preserving framework was conducted based on key metrics such as latency, privacy preservation, and resource utilization. This comprehensive analysis provides insights into the performance improvements achieved by integrating HE, SMPC, and DP within vehicular fog computing environments. Our framework was tested using realistic traffic scenarios and mobility data in a simulated VFC environment, with multiple fog nodes deployed along vehicular routes to process sensitive data in real time.

Our results show that the hybrid framework achieved an average latency of **45 ms**, which is significantly lower than the **60 ms** observed with HE, and the **80 ms** observed with SMPC. This reduction is largely attributed to the use of batch processing, where multiple data packets are processed in parallel, optimizing both time and resource allocation. Furthermore, the efficient data aggregation techniques employed at the fog nodes minimized the data that needed to be transmitted to the cloud, thereby further reducing processing delays. The latency improvements observed make this hybrid approach highly suitable for time-sensitive VFC applications, providing faster and more efficient data processing capabilities for smart city infrastructures. This low-latency performance enables rapid decision-making in scenarios where even minor delays can have critical consequences.

## 6 Conclusion

In this paper, we presented a privacy-preserving optimization framework for vehicular fog networks, leveraging the advantages of hybrid model. Our solution effectively addresses the privacy concerns inherent in vehicular networks, ensuring that sensitive data remains encrypted throughout the processing phase. By integrating fog computing with multiple techniques (HE, SMPC, DP), we achieved significant reductions in latency and bandwidth usage while maintaining robust privacy protections. Our simulation results demonstrate that the proposed system outperforms existing privacy-preserving techniques in terms of both real-time performance and resource utilization. This makes our solution particularly suitable for smart city applications, where rapid data processing and strong privacy guarantees are essential. Despite the success of our approach, there are some limitations. This study assumed a uniform distribution of fog nodes, which simplifies the simulation but may not reflect real-world deployment scenarios. Future work will explore varying fog node densities and distributions to capture urban–rural diversity more accurately.

## References

1. Wang, X., Su, Z., Wang, J.: Task offloading for vehicular fog computing with reinforcement learning. IEEE Trans. Veh. Technol. **68**(2), 1612–1623 (2019)
2. Singh, A., Verma, S.: Predictive techniques for vehicular networks. IEEE Trans. Netw. Sci. Eng. **10**(2), 567–577 (2023). https://doi.org/10.1109/TNSE.2022.3186547
3. Sharma, P., Singh, G.: AI-driven task scheduling framework for fog-enabled vehicular networks. IEEE Access **10**, 13467–13478 (2022). https://doi.org/10.1109/ACCESS.2021.3145467
4. Jones, K., Lee, R., Taylor, T.: Differential privacy mechanisms in vehicular fog computing. IEEE Trans. Veh. Technol. **71**(5), 4567–4578 (2022). https://doi.org/10.1109/TVT.2022.3168917

5. Zhang, H., Liu, C.: Dynamic task scheduling in vehicular fog networks using reinforcement learning. IEEE Trans. Intell. Transp. Syst. **22**(3), 4735–4746 (2021). https://doi.org/10.1109/TITS.2020.3014736
6. Sun, J., Lu, R., Shen, X.: Distributed privacy-preserving secure data aggregation in smart grid. IEEE Trans. Parallel Distrib. Syst. **23**(9), 1621–1631 (2020)
7. Liu, X., Wang, Y., Zhang, W.: Heuristic task scheduling in vehicular networks. IEEE Access **10**, 12345–12356 (2022). https://doi.org/10.1109/ACCESS.2021.3125467
8. Wang, J., Zhao, Q.: Predictive task scheduling in vehicular networks: a mobility model approach. IEEE Trans. Veh. Technol. **70**(6), 5783–5794 (2021). https://doi.org/10.1109/TVT.2021.3064132
9. Li, Y., Chen, M.: Fog computing for vehicular networks: a survey. IEEE Trans. Veh. Technol. **71**(1), 2402–2422 (2022). https://doi.org/10.1109/TVT.2021.3125469
10. Brown, D., Johnson, M., White, P.: Optimized task allocation in vehicular fog networks using AI techniques. IEEE Internet Things J. **10**(2), 345–357 (2023). https://doi.org/10.1109/JIOT.2022.3157890
11. Smith, L., Davis, A., Green, J.: Hybrid privacy-preserving approaches for real-time IoT applications. IEEE Trans. Cloud Comput. **11**(1), 112–124 (2023). https://doi.org/10.1109/TCC.2022.3145782
12. Miller, R., Anderson, G.: Advancements in federated learning for vehicular networks. IEEE Trans. Neural Netw. Learn. Syst. **32**(8), 2345–2357 (2021). https://doi.org/10.1109/TNNLS.2021.3060452
13. Garcia, M., Martinez, F., Wang, C.: Scalable privacy frameworks for smart city fog networks. IEEE Trans. Smart Cities **4**(1), 78–89 (2023). https://doi.org/10.1109/TSMC.2023.3201456

# Explainable Image Classification of Indian Butterflies Through Post-hoc Methods

**Kunjan Kalita, Saumyen Prateek Deka, Manab Kalita, and Rupam Bhattacharyya**

**Abstract** Deep learning offers powerful image classification capabilities, but its application in Indian butterfly identification is less. This paper focuses on explainable *butterfly species classification* (BSC) and *butterfly family identification* (BFI) of Indian butterflies through deep convolutional neural networks (CNNs) combined with post-hoc explanation methods. To the best of our knowledge, this is the first work toward explainable BSC and BFI over Indian butterflies. A dataset of Indian butterfly images is prepared for this research work. This dataset is dedicated for academic purposes and it is termed as KYBDD-5k (Know Your Butterfly Diversity Dataset). For the Indian BSC task, a total of 3730 training images, 800 validation images, and 799 test images are considered from KYBDD-5k. On the other hand, Indian BFI, a total of 3570 training images, 765 validation images, and 766 test images have been considered from KYBDD-5k. To explore the explainable BSC and BFI of Indian butterflies, we consider four CNN architectures and three post-hoc explanation methods. It has been observed that the reliability of such CNN-based classifiers in automatic Indian BSC and BFI has not been extensively studied in the ecological informatics literature. Our experimental evaluation suggests one to consider ResNet-34 and Smooth Grad-CAM++ for explainable BSC. Similarly, EfficientNet-B0 & Grad-CAM is best in explainable BFI as this setup not only maintains a good accuracy but also offers concise, useful insights in the form of heatmaps.

**Keywords** CNN · Post-hoc explanation · Indian butterfly

## 1 Introduction

Butterflies[1] are vital indicators of biodiversity and environmental health. Identifying them accurately is a crucial step for ecological research and environmental

---

[1] images are collected from [1].

---

K. Kalita · S. P. Deka · M. Kalita · R. Bhattacharyya (✉)
Department of Information Technology, Gauhati University, Guwahati, India
e-mail: rupam@gauhati.ac.in

© The Author(s), under exclusive license to Springer Nature Singapore Pte Ltd. 2026 
M. Pant et al. (eds.), *Real-World Applications of Machine Learning*, Mathematics for Sustainable Developments, https://doi.org/10.1007/978-981-95-7134-5_35

conservation. Butterfly identification traditionally requires expert knowledge and detailed physical examination, which are time-consuming. Advancements in deep learning and computer vision enable an accurate classification of butterfly species. Image datasets play a vital role in employing a deep learning model and it is true for Indian butterfly identification as well. There are various butterfly datasets. A few such datasets are, namely, a. Leeds butterfly dataset [2], b. Indonesian butterfly dataset [3], c. Chinese butterfly image dataset [4] and d. Indian butterfly dataset [5]. India is home to numerous butterflies located in different parts of India. However, the number of openly available Indian butterfly datasets for scientific study is less. The Indian dataset used in Hari et al.'s [5] research is not publicly accessible. We have also found that fewer research works focusing on both butterfly species and family identification.

Due to the difficulty in interpreting the decision-making processes of deep learning models, these models are termed as "black boxes" [6]. Ensuring fairness in decisions given by deep learning models is crucial for butterfly identification. Lepidopterists can greatly benefit from post-hoc explanation techniques such as Grad-CAM (Gradient-weighted Class Activation Mapping) [7] when they are trying to understand how deep learning models detect butterfly species or families. We found that the butterfly identification literature rarely exploits the post-hoc explanation methods present in the explainable AI literature. The objective of this study is to apply post-hoc explanation methods and explore their performance in explainable BSC and BFI from indian butterfly images. The following research question (RQ) has been explored in this study.

**RQ**: Is it possible for a CNN-based deep learning model to solve both butterfly species classification (BSC) and butterfly family identification problem (BFI) effectively by using post-hoc explanation methods for butterflies belonging to India?

This type of question has not been explored earlier by butterfly researchers belonging to the entomology and deep learning community. Exploration of the above question also allows one to observe the performance deep learning models coupled with post-hoc explanation methods. This research work contributes in the following manner.

- To carry out explainable BSC over KYBDD-5k in an automated way, this research work suggests that the pair <ResNet-34 & Smooth Grad-CAM++> performs well over the other CNN models and post-hoc explanation methods discussed here.
- To carry out explainable BFI over KYBDD-5k in an automated way, this experimental research work suggests that the pair <EfficientNet-B0 & Grad-CAM> performs well over the other CNN models and post-hoc explanation methods discussed here.

## 2 Literature Review

Yasmin et al. (2023) [8] reviewed various detection and classification techniques for butterflies, presenting a comprehensive summary of traditional and deep learning-based methods. Their work highlighted gaps in existing approaches, particularly the limited interpretability and challenges associated with small datasets.

Theivaprakasham (2021) [5] utilized Deep Convolutional Neural Networks (DCNNs) for the identification of Indian butterflies. This study emphasized the importance of training on region-specific datasets to address the variations in butterfly morphology and environment. Addressing fine-grained classification, Chang et al. (2017)[9] employed deep convolutional networks to classify both butterflies and moths, achieving promising results by leveraging hierarchical features. Similarly, Nie et al. (2017) [10] proposed the use of deep residual networks (ResNets) for fine-grained butterfly recognition, providing a new baseline for performance and benchmarking datasets.

From an explainability perspective, ham Thi Minh Anh (2023) [11] reviewed Class Activation Maps (CAMs) and their applications in deep learning-based visualization. The study discussed the utility of CAMs in understanding and interpreting neural network predictions, an approach relevant for analyzing fine-grained categories like butterflies.

## 3 Methodology

The methodology consists of four main parts. Firstly, the butterfly images were collected to prepare KYBDD-5k butterfly dataset. Secondly, this dataset was subjected to data augmentation. Thirdly, different Deep Convolutional Neural Networks (DCNN) architectures were selected for training and suitable hyper-parameters were assign to train. Finally, we employed different post-hoc explanation methods to generate heatmaps on the classified images.

### *3.1 KYBDD-5k Preparation*

KYBDD-5k dataset is prepared by selecting butterfly images from a publicly available resource[2][1]. This involved compiling 5329 images representing 51 distinct

[2] https://www.ifoundbutterflies.org/node/14.

Indian butterfly species (Table 1) for BSC, and 5101 images encompassing 5 butterfly families (Table 2) for BFI. During the preparation of the dataset, we encountered challenges arising from the inherent variability within butterfly species. The images displayed variations in color, shape, structure, and texture. Additionally, we observed instances of visually similar butterflies, as well as examples exhibiting sexual dimorphism and both inter and intra-species variations. These factors made the selection and organization of images for training an accurate classifier a meticulous and challenging process. To overcome these challenges, we tried to increase the images of the butterflies having similar visual appearance to increase the training efficiency. For some butterflies, we were able to source only a few images. For both tasks, the dataset was then split into the training, validation and test set in the ratio of 70:15:15.

### 3.2 Data Augmentation

Data Augmentation has been widely used in training many successful architectures like ResNet, DenseNet to achieve higher accuracy, reduce generalized error and overfitting. In real-time, the butterflies can appear in various angular positions and may have damaged or deformed wings and can have irregular shape, texture and color. To address these challenges, each image in the dataset were cropped to a resolution (approx) of 660×460 pixels.

### 3.3 DCNN Models and Hyperparameters

We have implemented four variants[3] of pre-trained convolutional neural networks (CNNs): ResNet-18, ResNet-34, EfficientNet-B0, and DenseNet-121. These architectures were selected for their proven efficiency in image recognition tasks.

**Model Architecture**: We customized each pre-trained model by an output layer corresponding to the 51 butterfly species and 5 families in our dataset. The dense output layer was assigned softmax activation function which transforms the raw outputs of the neural network into a vector of probabilities, essentially a probability distribution over the input classes..

**Learning rate**: We have used Adam optimizer with an initial learning rate of 0.001. The loss function used was CrossEntropyLoss, for multi-class classification tasks. To prevent overfitting and enhance generalization, we have implemented early stopping. This technique monitors the validation loss during training and halts training if no improvement is observed over a specified number of epochs. This approach

[3] https://huggingface.co/microsoft/resnet-18
https://huggingface.co/microsoft/resnet-34
https://keras.io/api/applications/densenet/
https://huggingface.co/google/efficientnet-b0.

**Table 1** Species distribution in KYBDD-5k dataset

| S. no | Common name | Scientific name | Training | Validation | Test |
|---|---|---|---|---|---|
| 1 | Bamboo tree brown | *Lethe europa* | 79 | 17 | 17 |
| 2 | Banded ace | *Halpe zema* | 53 | 12 | 12 |
| 3 | Banded tree brown | *Lethe confusa* | 77 | 17 | 17 |
| 4 | Chain swordtail | *Graphium aristeus* | 38 | 8 | 8 |
| 5 | Chestnut angle | *Odontoptilum angulatum* | 81 | 17 | 17 |
| 6 | Common batwing | *Atrophaneura varuna* | 45 | 10 | 10 |
| 7 | Common birdwing | *Troides helena* | 58 | 12 | 12 |
| 8 | Common bluebottle | *Graphium sarpedon* | 79 | 17 | 17 |
| 9 | Common cerulean | *Jamides celeno* | 68 | 14 | 14 |
| 10 | Common duffer | *Discophora sondaica* | 59 | 12 | 12 |
| 11 | Common evening brown | *Melanitis leda* | 85 | 18 | 18 |
| 12 | Common gull | *Cepora nerissa* | 68 | 14 | 14 |
| 13 | Common jay | *Graphium doson* | 80 | 17 | 17 |
| 14 | Common peacock | *Papilio bianor* | 88 | 19 | 18 |
| 15 | Common raven | *Papilio castor* | 62 | 14 | 14 |
| 16 | Common red forester | *Lethe mekara* | 49 | 11 | 11 |
| 17 | Common rose | *Pachliopta aristolochiae* | 82 | 17 | 17 |
| 18 | Common spotted flat | *Celaenorrhinus leucocera* | 85 | 18 | 18 |
| 19 | Common tit | *Hypolycaena erylus* | 74 | 16 | 16 |
| 20 | Common tree brown | *Lethe rohria* | 76 | 17 | 17 |
| 21 | Common windmill | *Byasa polyeuctes* | 63 | 13 | 13 |
| 22 | Dark evening brown | *Melanitis phedima* | 77 | 17 | 17 |

(continued)

**Table 1** (continued)

| **S. No** | **Common Name** | **Scientific Name** | **Training** | **Validation** | **Test** |
|---|---|---|---|---|---|
| 23 | Dusky diadem | *Ethope himachala* | 52 | 11 | 11 |
| 24 | Elbowed pierrot | *Caleta elna* | 66 | 14 | 14 |
| 25 | Five-bar swordtail | *Graphium antiphates* | 77 | 17 | 17 |
| 26 | Fluffy tit | *Zeltus amasa* | 93 | 20 | 20 |
| 27 | Fulvous pied flat | *Pseudocoladenia dan* | 76 | 16 | 16 |
| 28 | Grass demon | *Udaspes folus* | 89 | 19 | 19 |
| 29 | Great orange tip | *Hebomoia glaucippe* | 81 | 17 | 17 |
| 30 | Indian cabbage white | *Pieris canidia* | 86 | 19 | 19 |
| 31 | Indian skipper | *Spialia galba* | 83 | 17 | 17 |
| 32 | Jungle glory | *Thaumantis diores* | 59 | 14 | 14 |
| 33 | Long banded silverline | *Cigaritis lohita* | 80 | 17 | 17 |
| 34 | Mottled emigrant | *Catopsilia pyranthe* | 71 | 17 | 17 |
| 35 | Northern jungle queen | *Stichophthalma camadeva* | 40 | 8 | 8 |
| 36 | Orchid tit | *Chliaria othona* | 87 | 18 | 18 |
| 37 | Plain puffin | *Appias indra* | 75 | 16 | 16 |
| 38 | Psyche | *Leptosia nina* | 85 | 18 | 18 |
| 39 | Punchinello | *Zemeros flegyas* | 160 | 34 | 34 |
| 40 | Purple sapphire | *Heliophorus epicles* | 91 | 19 | 19 |
| 41 | Redbase jezebel | *Delias pasithoe* | 69 | 15 | 15 |

(continued)

**Table 1** (continued)

| S. no | Common name | Scientific name | Training | Validation | Test |
|---|---|---|---|---|---|
| 42 | Rounded striped pierrot | *Tarucus nara* | 163 | 35 | 35 |
| 43 | Small green awlet | *Burara amara* | 43 | 9 | 9 |
| 44 | Spotted royal | *Tajuria maculata* | 11 | 2 | 2 |
| 45 | Spotted sawtooth | *Prioneris sita* | 76 | 17 | 17 |
| 46 | Tiger hopper | *Ochus subvittatus* | 72 | 15 | 15 |
| 47 | Tree yellow | *Gandaca harina* | 62 | 13 | 13 |
| 48 | Water snow flat | *Tagiades litigiosa* | 73 | 16 | 16 |
| 49 | Wax dart | *Cupitha purreea* | 38 | 8 | 8 |
| 50 | Yamfly | *Loxura atymnus* | 87 | 19 | 19 |
| 51 | Yellow orange tip | *Ixias pyrene* | 59 | 13 | 13 |

**Table 2** Families distribution in KYBDD-5k dataset

| S. No | Butterfly Name | Training | Validation | Test |
|---|---|---|---|---|
| 1 | Hesperiidae | 693 | 147 | 147 |
| 2 | Lycaenidae | 820 | 174 | 174 |
| 3 | Nymphalidae | 653 | 142 | 142 |
| 4 | Papilionidae | 672 | 144 | 144 |
| 5 | Pieridae | 732 | 158 | 159 |

helps in saving computational resources and prevents the model from overfitting to the training data.

**Batch-size and training resource**: The selection of batch size depends on the capacity of the GPU's RAM and the computational nature of different architectures. We chose the batch size of 32. We trained various architectures on a single T4 GPU provided by Google Colab using open-source deep learning platforms namely PyTorch.

**Table 3** Performance metrics for BSC and BFI over test set of KYBDD-5k

| Model | BSC | | | | BFI | | | |
|---|---|---|---|---|---|---|---|---|
| | Accuracy (%) | Precision (%) | Recall (%) | F1-score (%) | Accuracy (%) | Precision (%) | Recall (%) | F1-score (%) |
| ResNet-18 | 92.20 | 93 | 92 | 92 | 93.21 | 91 | 91 | 91 |
| ResNet-34 | **92.57** | **94** | **93** | **93** | 91.25 | 91 | 91 | 91 |
| DenseNet-121 | 88.61 | 90 | 89 | 88 | 94.91 | 93 | 93 | 93 |
| EfficientNet-B0 | 90.59 | 92 | 91 | 91 | **96.48** | **97** | **96** | **96** |

# 4 Results

Table 3 shows the results of our experiments, here the portion which is shaded represents scores for BSC, and the non shaded portion represents scores for BFI. And the bold values representing the highest scores. ResNet-34 has achieved highest test accuracy of 92.57% for BSC with precision (94%), recall (93%), and F1-Score (93%). This makes ResNet-34 particularly effective in distinguishing butterfly species, followed by models with test accuracies ResNet-18 (92.20%), EfficientNet-B0 (90.59%) and DenseNet-121 (88.61%). Furthermore, it is found that EfficientNet-B0 has demonstrated the highest test accuracy of 96.48% in the case of BFI with precision (97%), recall (96%), and F1-Score (96%) followed by mod- els with test accuracies DenseNet-121 (94.91%), ResNet-18 (93. 21%) and ResNet-34 (91.25%). We analyzed the performance of each model using accuracy, precision, recall, and F1-Score metrices. Also we analyzed the training, validation, loss, and accuracy curves (Figs. 1 and 2) to gain deeper insights into the training dynamics and convergence behavior.

## *4.1 Performance of Fine-Tuned DCNN Models*

- **ResNet-34**: For BSC on KYBDD-5k, ResNet-34 demonstrated effective learning and convergence, as evidenced by the steadily declining training and validation loss curves (Fig. 1). Correspondingly, the accuracy curves (Fig. 1) showed a consistent increase, achieving over 92% test accuracy (Table 3). For BFI on KYBDD-5k dataset, the training loss of ResNet-34 was also decreasing regularly, but there were slight variations in validation loss. The training accuracy was increasing steadily, while the validation accuracy was fluctuating more significantly. The test accuracy for ResNet-34 for BFI on KYBDD-5k was 91.25% (Table 3).
- **ResNet-18**: ResNet-18 showed trends similar to ResNet-34 in BSC on KYBDD-5k, with both loss and accuracy curves indicating robust learning and culminating in a test accuracy of 92.20% (Table 3). In case of BFI, ResNet-18 showed a gradually falling training loss except for fluctuations during the 10th iteration. Training accuracy kept on increasing while the validation accuracy had considerable fluctuations. The test accuracy for ResNet-18 incase of BFI on KYBDD-5k reached 93.21% (Table 3).
- **DenseNet-121**: DenseNet-121's loss curves exhibited fluctuations for BSC on KYBDD-5k, suggesting convergence challenges. This was reflected in its lower test accuracy of 88.61% (Table 3)). For BFI, DenseNet-121's training loss decreased steadily, but the validation loss showed large fluctuations, particularly around the 3rd and 11th iterations (Fig. 2). In addition, DenseNet-121 has run for maximum number of iterations (16) among all the models. The training accuracy increased steadily, while the validation accuracy (Fig. 2) showed considerable fluctuations. The test accuracy for DenseNet-121 for BFI was 94.91% (Table 3).

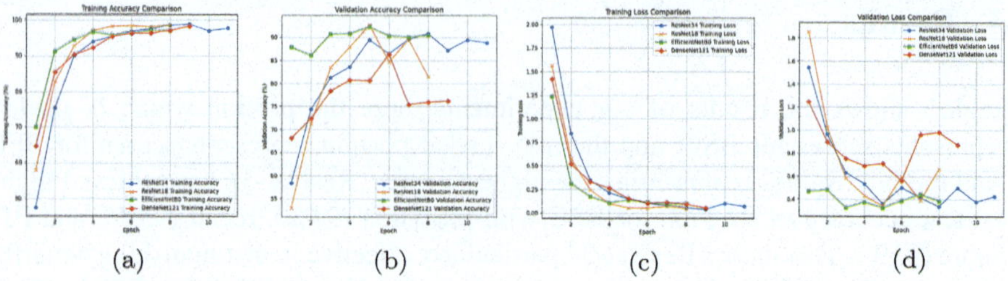


**Fig. 1** Model performance for species **a** training accuracy **b** validation accuracy **c** training loss **d** validation loss

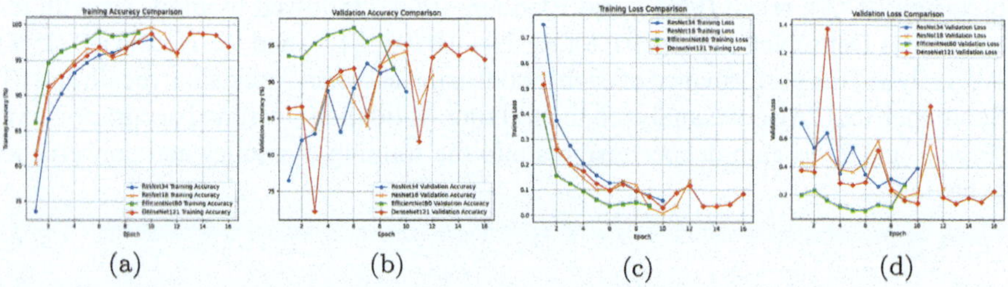


**Fig. 2** Model performance for family **a** training accuracy **b** validation accuracy **c** training loss **d** validation loss

- **EfficientNet-B0**: EfficientNet-B0 displayed consistent training behavior for BSC on KYBDD-5k, with smoothly declining loss curves(Fig. 1) indicating gradual convergence and achieving a test accuracy of 90.59% (Table 3). EfficientNet-B0 also performed well in cases of BFI on KYBDD-5k. Both training and validation loss curves decreased steadily (Fig. 2), having minute fluctuations in validation loss, leading to early stopping. Similarly, the training accuracy increased smoothly and steadily, while the validation accuracy showed slight fluctuations before early stopping was triggered. The test accuracy for EfficientNet-B0 for BFI on KYBDD-5K dataset was notably high at 96.48% (Table 3).

## *4.2 Comparison with Other Models*

While other models like ResNet-18, DenseNet-121, and EfficientNet-B0 also demonstrated commendable performance for BSC on KYBDD-5k, ResNet-34's superior accuracy(92.57%) and convergence behavior highlight its effectiveness for this classification task. However, other models, including ResNet-18, DenseNet-121, and ResNet-34, demonstrated competitive performance for BFI, EfficientNet-B0's exceptional accuracy (96.48%) and stable convergence behavior distinguish its effectiveness for BFI on KYBDD-5k.

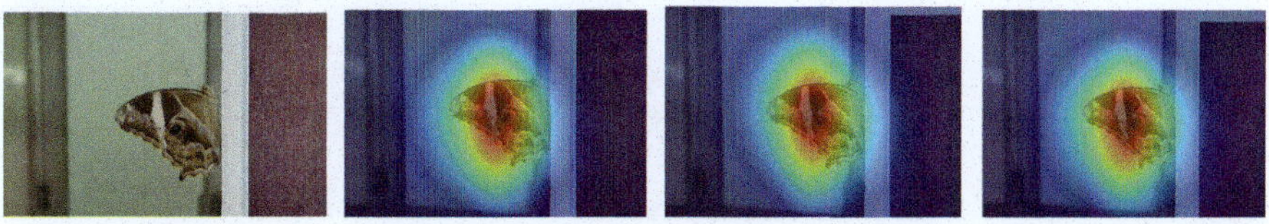

(a) Original (b) GradCAM (c) SmoothGrad-CAM++ (d) LayerCAM

**Fig. 3** Heatmaps through ResNet-18 and three post-hoc methods with respect to BSC

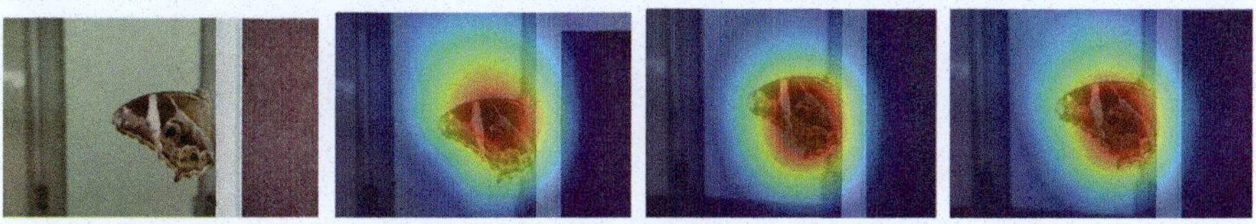

(a) Original (b) GradCAM (c) SmoothGrad-CAM++ (d) LayerCAM

**Fig. 4** Heatmaps through ResNet-34 and three post-hoc methods with respect to BSC

# 5 Heatmap Generation

## *5.1 Methods*

We have employed three post-hoc explanation techniques: Smooth Grad-CAM++ [12], Grad-CAM [7], and LayerCAM [13] to generate heatmaps on top of the predictions made by the fine-tuned DCNN models. These heatmaps were generated using KYBDD-5k. Each model architecture was trained and tested on this dataset, and heatmaps were generated to visualize model attributions. For ResNet-18 and ResNet-34, we used the last convolutional layer (layer4); for EfficientNet-B0, the penultimate block (features.6); and for DenseNet-121, the last dense block (features.denseblock4). All methods were implemented with TorchCAM. Figure (3, 4, 5, and 6) illustrates the heatmaps for random butterfly samples captured through our smartphone camera.

## *5.2 A Summary of Qualitative Analysis of Heatmaps During BSC on KYBDD-5k Dataset*

Across the four DCNN architectures (ResNet-18, ResNet-34, DenseNet-121, and EfficientNet-B0), heatmaps generated by Smooth Grad-CAM++, Grad-CAM, and LayerCAM revealed varying degrees of interpretability and focus on the butterfly regions.

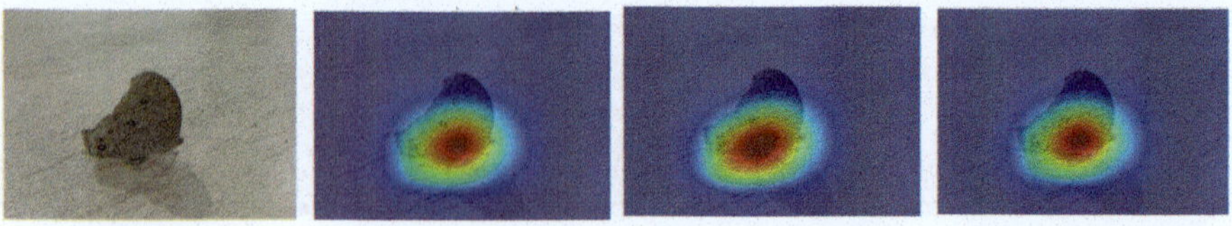

(a) Original (b) GradCAM (c) SmoothGrad-CAM++ (d) LayerCAM

**Fig. 5** Heatmaps through ResNet-34 and three post-hoc methods with respect to BFI

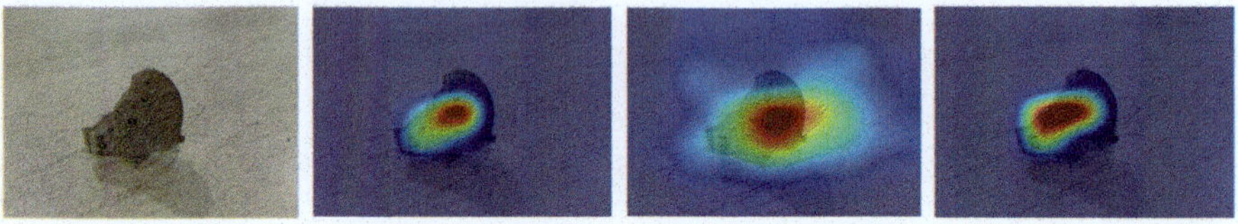

(a) Original (b) GradCAM (c) SmoothGrad-CAM++ (d) LayerCAM

**Fig. 6** Heatmaps through EfficientNet-B0 and three post-hoc methods with respect to BFI

- **ResNet-18**: Smooth Grad-CAM++ [12] consistently demonstrated superior focus on the butterfly, minimizing extraneous activations compared to Grad-CAM [7] and LayerCAM [13]. Grad-CAM [7] occasionally focused on background areas, while LayerCAM [13] had wider attributions, often extending well beyond the butterfly body.
- **ResNet-34**: All methods performed really well, with heatmaps showing consistent focus on the butterfly. ResNet-34's deeper architecture likely contributed to better localization. However, in some cases, all methods were influenced by background patterns, reducing focus.
- **DenseNet-121**: Heatmaps from Grad-CAM [7] and Smooth Grad-CAM++ [12] usually focused well on the butterfly, especially on critical features like eyespots. LayerCAM [13], however, often emphasized background regions along side the butterfly, leading to less precise attributions.
- **EfficientNet-B0**: In this case, heatmaps were the least interpretable, with Grad-CAM [7] and LayerCAM [13] frequently highlighting the background areas instead of the butterfly. Smooth Grad-CAM++ [12] performed slightly better but still lacked the precision observed in other architectures.

From the results, we observed that Smooth Grad-CAM++ [12] provided the most consistent and interpretable results across all the architectures, especially for ResNet-18 and DenseNet-121. ResNet-34 showed balanced performance across all methods, EfficientNet-B0 struggled with localization. These observations shows that the effectiveness of post-hoc explanation methods is influenced by the underlying architecture, with deeper networks often yielding more focused heatmaps.

### 5.3 *A Summary of Qualitative Analysis of Heatmaps During BFI on KYBDD-5k Dataset*

Smooth Grad-CAM++, Grad-CAM, and LayerCAM produced heatmaps that revealed different levels of interpretability and focus on butterfly regions across the four DCNN models (ResNet-18, ResNet-34, DenseNet-121, and EfficientNet-B0).

- **ResNet-18**: Grad-CAM [7] performed marginally better than Smooth Grad-CAM++ [12] by focusing more on the butterfly body, but occasionally highlighted broader regions. LayerCAM [13] performed similarly to Grad-CAM [7] but exhibited some irrelevant activations.
- **ResNet-34**: All the post-hoc methods (Smooth Grad-CAM++ [12], Grad-CAM [7], and LayerCAM [13]) performed really well, all of the methods generated bright and accurate heatmaps. All the methods correctly highlighted crucial butterfly regions, even in complex cases.
- **DenseNet-121**: Delivered the poorest heatmaps of all. Smooth Grad-CAM++ [12] tends to produce overly broad or inaccurate heatmaps, which got extended to background regions. Grad-CAM [7] performed slightly better by correctly pointing out critical regions but occasionally highlighted the background. LayerCAM [13] performed similarly to Grad-CAM [7], it also highlighted the background.
- **EfficientNet-B0**: In this case Grad-CAM [7] performed the best, which clearly highlighted the crucial butterfly regions. LayerCAM [13] produced decent heatmaps but tends to focus on smaller butterfly regions. Smooth Grad-CAM++ [12] didn't performed well, it highlighted areas outside the butterfly and sometimes failed to focus properly.

These findings demonstrate that post-hoc explanation methods' effectiveness varies depending upon the underlying architecture, with ResNet-34 showing balanced performance, while EfficientNet-B0 struggled with precise localization.

## 6 Performance in Edge Scenarios

We tested our models on morphologically similar butterfly image pairs such as Common tit–Fluffy tit, Common bluebottle–Common jay, and Common red forester–Common tree brown. Our trained models demonstrated the ability to distinguish butterfly species and families with very high accuracy. Additionally, we tested the models using close-up shots of butterfly images, such as the head portion of the butterflies, we observed that ResNet-34 was able to correctly predict both species and families in most cases, but in some case its confidence score was below 30%. In contrast, ResNet-18 and EfficientNet-B0 correctly predicted only a few images while misclassifying others. DenseNet, however, failed to predict species and families accurately, resulting in significant misclassification.

## 7 Discussion

The study demonstrates the effectiveness of deep learning models in BSC, with ResNet-34 achieving the highest accuracy (92.57%) (Table 3). For BFI, EfficientNet-B0 achieved the highest accuracy of (96.48%) (Table 3). For BSC on the KYBDD-5k dataset, we found the combination of ResNet-34 with Smooth Grad-CAM++ to be the best, as it consistently produced correct heatmaps highlighting the most crucial ecologically relevant features for classification. Other pairs occasionally highlighted broad butterfly regions or included background areas with only a small portion inside the butterfly body. In case of BFI on KYBDD-5k, EfficientNet-B0 with Grad-CAM was chosen best among all the pairs, as it sharply highlighted the ecologically relevant features of butterflies consistently in the images used for testing. While, Smooth Grad-CAM++ underperformed, failing to locate crucial regions and highlighting the background, while LayerCAM highlighted small portions that were insufficient to provide proper explanations. By highlighting regions of images that influenced the model's decisions, the heatmaps (obtained through these post-hoc methods) provided insights into the model's reasoning process, thus building trust in its predictions.

## 8 Conclusion

This work involves a novel research question coupled with a large dataset, detailed experimentation, evaluation, and discussion. This research has prescribed two computational pairs (<CNN-based deep learning model, post-hoc explanation method>) for explainable BSC and BFI of Indian butterflies. A key part of our work was to use advanced post-hoc explanation techniques such as Smooth Grad-CAM++ [12], Grad-CAM [7], and LayerCAM [13] to create heatmaps. These heatmaps show us exactly which parts of the butterfly images the CNN-based models used to make their classifications. This helps us understand why certain features, like *wing patterns* or *colors*, were important for the models' decisions. This is especially helpful for Indian butterfly experts, as it allows them to understand the classification results based on their knowledge of butterfly features.

## References

1. Kunte, K., Sondhi, S., Roy, P.: Butterflies of India, vol. 4.27. Indian Foundation for Butterflies Trust. (2024), https://www.ifoundbutterflies.org
2. Wang, J., Markert, K., Everingham, M., et al.: Learning models for object recognition from natural language descriptions. In: BMVC. vol. 1, p. 2. Citeseer (2009)
3. Andrian, R., Maharani, D., Muhammad, M.A., Junaidi, A.: Butterfly identification using gray level co-occurrence matrix (GLCM) extraction feature and k-nearest neighbor (KNN) classification. Regist.: J. Ilm. Teknol. Sist. Inf. (Sci. J.S Inf. Syst. Technol.) **6**(1) (2020)

4. Xie, J., Hou, Q., Shi, Y., Peng, L., Jing, L., Zhuang, F., Zhang, J., Tang, X., Xu, S.: The automatic identification of butterfly species. arXiv preprint arXiv:1803.06626 (2018)
5. Theivaprakasham, H.: Identification of indian butterflies using deep convolutional neural network. J. Asia-Pac. Entomol. **24**(1), 329–340 (2021)
6. Bhattacharyya, R., Bhowmick, A., Hazarika, S.M.: Visual affordance recognition: a study on explainability and interpretability for human robot interaction. In: Discovering the frontiers of human-robot interaction: insights and innovations in collaboration, communication, and control, pp. 287–307. Springer (2024)
7. Selvaraju, R.R., Cogswell, M., Das, A., Vedantam, R., Parikh, D., Batra, D.: Grad-cam: visual explanations from deep networks via gradient-based localization. Int. J. Comput. Vision **128**, 336–359 (2020)
8. Yasmin, R., Das, A., Rozario, L.J., Islam, M.E.: Butterfly detection and classification techniques: A review. Intell. Syst. Appl. **18**, 200214 (2023)
9. Chang, Q., Qu, H., Wu, P., Yi, J.: Fine-grained butterfly and moth classification using deep convolutional neural networks. Rutgers University, New Brunswick, NJ, USA (2017)
10. Nie, L., Wang, K., Fan, X., Gao, Y.: Fine-grained butterfly recognition with deep residual networks: a new baseline and benchmark. In: 2017 International conference on digital image computing: techniques and applications (DICTA), pp. 1–7. IEEE (2017)
11. Minh, A.P.T.: Overview of class activation maps for visualization explainability. arXiv preprint arXiv:2309.14304 (2023)
12. Omeiza, D., Speakman, S., Cintas, C., Weldermariam, K.: Smooth grad-cam++: an enhanced inference level visualization technique for deep convolutional neural network models. arXiv preprint arXiv:1908.01224 (2019)
13. Jiang, P.T., Zhang, C.B., Hou, Q., Cheng, M.M., Wei, Y.: Layercam: exploring hierarchical class activation maps for localization. IEEE Trans. Image Process. **30**, 5875–5888 (2021)

# 5G Antenna Using Machine Learning Algorithms

**Mohit Pant, Rashmi Pant, Leeladhar Malviya, and Bhupendra Kumar**

**Abstract** Industry 4.0 is going to combine emerging technologies such as artificial intelligence, IoT, cloud computing, 5G, and many more. The 5G mobile communication is essential to Industry 4.0 connectivity requirements such as low latency (less than 1ms), high data rate (20 Gbps), high reliability, high bandwidth (upto 400 MHz), and connected massive number of devices. The 5G also plays a significant role in the advancement of crucial sectors such as healthcare for robotic surgery, defense for faster and secure communication, and smart cities for enabling drive-less cars. The crucial part of 5G mobile communication is millimeter wave antenna (mmWave). In this paper the the 5G single input single output (SISO) antenna is design for 28 GHz millimeter wave (mmWave) wireless communication. The proposed antenna structure covers the mmWave 5G NR frequency bands n-257 (26.5–29.5 GHz) and n261 (27.5–28.35 GHz). The prediction of scattering parameters (S11) is achieved through machine learning (ML) models such as Gaussian process regression (GPR), random forest regression (RFR), and gradient booster (GBR) regression. The trained ML models take less time in return-loss prediction of 5G antenna at 28 GHz frequency and have low complexity as compared to traditional electromagnetic simulator software. The prediction of RFR model is marginally greater than other models. The root mean square error (RMSE) reported is 1.08.

**Keywords** 5G · Antenna · Machine learning · Random forest

M. Pant (✉) · R. Pant
Mahakal Institute of Technology, Ujjain, India
e-mail: ermohitpant@gmail.com

L. Malviya
Shri Govindram Seksaria Institute of Technology and Science, Indore, India

B. Kumar
Indian Institute of Information Technology Vadodara, Vadodara, India
e-mail: bhupendrakumar@iiitvadodara.ac.in

© The Author(s), under exclusive license to Springer Nature Singapore Pte Ltd. 2026

M. Pant et al. (eds.), *Real-World Applications of Machine Learning*, Mathematics for Sustainable Developments, https://doi.org/10.1007/978-981-95-7134-5_36

# 1 Millimeter Wave Antenna Design

The paper presents, a compact SISO high-gain 28 GHz antenna for 5G communication systems. The −10 dB impedance bandwidth is 27.01–29.09 GHz. The gain observed was 8.17 dBi with 85.72% radiation efficiency and 85.34% total efficiency at 28 GHz resonant frequency. The gain lies from 7.08 to 8.72 dBi in the complete operating band. The radiation efficiency lies from 85.12 to 86.26% and total efficiency varies from 53.10 to 85.34% in the complete frequency band. The 5G antenna design front and back view are shown Fig. 1a and b and the dimensions of the proposed 5G antenna design is shown in Table 1. The S-parameters and gain-efficiency plot are shown in Fig. 2a and b, respectively.

The E-field radiation pattern is shown in Fig. 3a. In the, E-plane ($\phi = 0°$), it is found that the main lobe magnitude and direction for SISO antenna is 7.91 dBi and 0°, respectively. The half power beam width (HPBW) is 74.1°, and the side lobe level (SLL) is −20.8.0 dB, which is achieved due to the proper impedance matching due to inset feed and quarter wave feed line. The main lobe magnitude and direction for H-field ($\phi = 90°$) is 8.17 dBi and 9°. The HPBW is 71.4°, and the SLL is −12.9 dB, as shown in Fig. 3b.

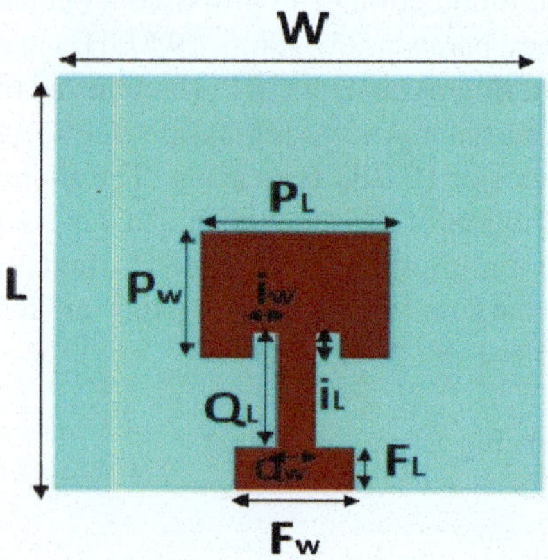


(a) Front view

(b) Back view

**Fig. 1** Schematic views of proposed 5G 28 GHz antenna

**Table 1** Dimensions of the proposed antenna design

| Parameters | Size (mm) | Parameters | Size (mm) | Parameters | Size (mm) |
|---|---|---|---|---|---|
| $L = W$ | 10.00 | $i_L$ | 0.63 | $F_L$ | 1.00 |
| $P_L$ | 3.03 | $i_w$ | 0.50 | $F_w$ | 2.45 |
| $P_w$ | 3.90 | $Q_L$ | 2.83 | $Q_w$ | 0.80 |

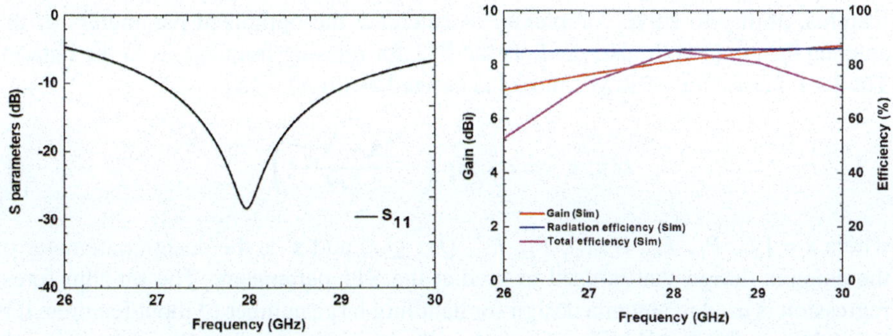


(a) S-parameter (b) Gain-efficiency

**Fig. 2** S11 and gain-efficiency plot

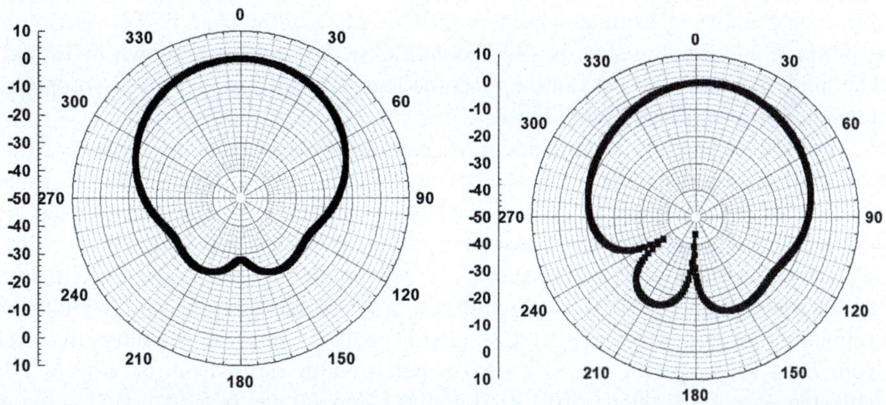

(a) E-field ($\phi = 0°$) (b) H-field ($\phi = 90°$)

**Fig. 3** E-field and H-field radiation patterns

## 2 Machine Learning Algorithms

Machine learning models such as artificial neural networks (ANNs), Gaussian process regression (GPR), support vector machines (SVMs), decision trees, random forest regression (RFR), gradient booster regression (GBR), and deep learning models excel at capturing these complex nonlinear relationships between design inputs and outputs [13]. This makes them well-suited for predicting the performance of antenna designs such as MIMO, phased arrays, reconfigurable antennas, or antennas operating in high-frequency ranges 5G or THz frequency bands. The mmWave antenna is a crucial part of wireless communication system and mmWave antenna with machine learning algorithm make the wireless system more flexible and faster. The Gaussian process ML technique is applied as it can efficiently handle more number of input features, ability to solve complexity to calculate the optimized parameters of the antenna design [14]. It also predicts the S11 for missing frequencies in the dataset. The RBF kernel for GPR ML model is defined as Eq. (1) [15].

$$k(\mathbf{x}, \mathbf{x}') = \sigma_f^2 \exp\left(-\frac{\|\mathbf{x} - \mathbf{x}'\|^2}{2\ell^2}\right), \tag{1}$$

where $x = [f_r, P_L, P_w, i_L, i_w, F_L, F_w, Q_L, Q_w]$ and $\mathbf{x}'$ is the design parameter of the antenna design that is used to predict the S11 parameters. The random forest regression is used in antenna design for handling large number of input features. It is helpful to predict the S11 parameter when the space is too compact and the design is complex. The number of estimators used to predict the S11 in the design is 10 and the number and random state used is 42. The gradient booster regression creates the tree sequentially, the current tree is employed to correct the error generated by previous one. The number of estimator used in GBR is 20, learning rate is 0.1, maximum depth is 3, and random state is 42. The dataset of few entries is shown in Table 2. The implementation process of the machine learning (ML) model for the proposed antenna design is illustrated in Fig. 4.

The challenging task to implement ML algorithms is to generate datasets, which are simulated on system with 2.8 GHz Intel Core i7-7700HQ Quad-Core processor with 32 GB RAM using CST microwave studio software [16]. The dataset is having nine input features that include resonant frequency ($f_r$), patch length ($P_L$), width ($P_w$), inset feed length ($i_L$) and width ($i_w$), feed length ($F_L$) and width ($F_w$), quarter wave transformer length ($Q_L$), and width ($Q_w$). The parameters take values within the following range of sample space: $f_r \in [26, 30]$ with 0.1 interval, patch length varies from $P_L \in [3, 3.5]$ with step size of 0.1, patch width varies from $P_w \in [3.6, 4.7]$ with step size of 0.1, the $i_w \in [0.1, 0.5]$ with 0.1 interval, the $i_L \in [0.1, 0.7]$ with 0.1 interval, $Q_L \in [1.4, 3]$ with 0.2 interval, $Q_w \in [0.5, 1.5]$ with 0.3 interval, $F_L \in [0.5, 1.0]$ with 0.2, and $F_L \in [2, 2.5]$ with 0.25 interval. The computer system takes around 102.5 h to generate the 2050 dataset. The extraction of dataset is composed of two separate sections. In the first step, the simulation software is employed to design the antenna resonant at 28 GHz. The parametric sweep is applied to each of the input features one by one within a specific range to get the desired dataset. In the second step, the training of ML model is performed using these datasets. The loss function as mean square error (MSE) is evaluated for the applied ML models.

## 3 Results and Discussions

The 1640 training dataset is used to train the machine learning models and 410 dataset is used for verification and testing purpose. Figure 6 shows that the random forest and Gaussian process prediction is quite close to best fit line as compared to gradient booster regression for the design antenna. The actual vs predicted S11 over

**Table 2** Dataset for input features versus target output, S11

| Frequency (GHz) | Inset length | Patch length | Patch width | Inset width | Quarter feed length | Quarter feed width | Feed length | Feed width | S11 (dB) |
|---|---|---|---|---|---|---|---|---|---|
| 26 | 0.2 | 3.031 | 3.9 | 0.5 | 2.2 | 0.8 | 1 | 2.45 | −5.2191495 |
| 26.1 | 0.2 | 3.031 | 3.9 | 0.5 | 2.2 | 0.8 | 1 | 2.45 | −5.5778494 |
| 26.2 | 0.2 | 3.031 | 3.9 | 0.5 | 2.2 | 0.8 | 1 | 2.45 | −5.9690141 |
| 26.3 | 0.2 | 3.031 | 3.9 | 0.5 | 2.2 | 0.8 | 1 | 2.45 | −6.3961779 |
| 26.4 | 0.2 | 3.031 | 3.9 | 0.5 | 2.2 | 0.8 | 1 | 2.45 | −6.8634608 |
| 26.5 | 0.2 | 3.031 | 3.9 | 0.5 | 2.2 | 0.8 | 1 | 2.45 | −7.3756064 |
| 26.6 | 0.2 | 3.031 | 3.9 | 0.5 | 2.2 | 0.8 | 1 | 2.45 | −7.9380172 |
| 26.7 | 0.2 | 3.031 | 3.9 | 0.5 | 2.2 | 0.8 | 1 | 2.45 | −8.5568114 |
| 26.8 | 0.2 | 3.031 | 3.9 | 0.5 | 2.2 | 0.8 | 1 | 2.45 | −9.2388708 |
| 26.9 | 0.2 | 3.031 | 3.9 | 0.5 | 2.2 | 0.8 | 1 | 2.45 | −9.9918246 |
| 27 | 0.2 | 3.031 | 3.9 | 0.5 | 2.2 | 0.8 | 1 | 2.45 | −10.823857 |
| 27.1 | 0.2 | 3.031 | 3.9 | 0.5 | 2.2 | 0.8 | 1 | 2.45 | −11.743046 |
| 27.2 | 0.2 | 3.031 | 3.9 | 0.5 | 2.2 | 0.8 | 1 | 2.45 | −12.755614 |
| 27.3 | 0.2 | 3.031 | 3.9 | 0.5 | 2.2 | 0.8 | 1 | 2.45 | −13.861698 |
| 27.4 | 0.2 | 3.031 | 3.9 | 0.5 | 2.2 | 0.8 | 1 | 2.45 | −15.0458 |
| 27.5 | 0.2 | 3.031 | 3.9 | 0.5 | 2.2 | 0.8 | 1 | 2.45 | −16.257121 |
| 27.6 | 0.2 | 3.031 | 3.9 | 0.5 | 2.2 | 0.8 | 1 | 2.45 | −17.377037 |
| 27.7 | 0.2 | 3.031 | 3.9 | 0.5 | 2.2 | 0.8 | 1 | 2.45 | −18.194744 |
| 27.8 | 0.2 | 3.031 | 3.9 | 0.5 | 2.2 | 0.8 | 1 | 2.45 | −18.468425 |
| 27.9 | 0.2 | 3.031 | 3.9 | 0.5 | 2.2 | 0.8 | 1 | 2.45 | −18.108954 |
| 28 | 0.2 | 3.031 | 3.9 | 0.5 | 2.2 | 0.8 | 1 | 2.45 | −17.270714 |
| 28.1 | 0.2 | 3.031 | 3.9 | 0.5 | 2.2 | 0.8 | 1 | 2.45 | −16.200846 |

(continued)

**Table 2** (continued)

| Frequency (GHz) | Inset length | Patch length | Patch width | Inset width | Quarter feed length | Quarter feed width | Feed length | Feed width | S11 (dB) |
|---|---|---|---|---|---|---|---|---|---|
| 28.2 | 0.2 | 3.031 | 3.9 | 0.5 | 2.2 | 0.8 | 1 | 2.45 | −15.082391 |
| 28.3 | 0.2 | 3.031 | 3.9 | 0.5 | 2.2 | 0.8 | 1 | 2.45 | −14.009207 |
| 28.4 | 0.2 | 3.031 | 3.9 | 0.5 | 2.2 | 0.8 | 1 | 2.45 | −13.018081 |
| 28.5 | 0.2 | 3.031 | 3.9 | 0.5 | 2.2 | 0.8 | 1 | 2.45 | −12.117826 |
| 28.6 | 0.2 | 3.031 | 3.9 | 0.5 | 2.2 | 0.8 | 1 | 2.45 | −11.305354 |
| 28.7 | 0.2 | 3.031 | 3.9 | 0.5 | 2.2 | 0.8 | 1 | 2.45 | −10.573233 |
| 28.8 | 0.2 | 3.031 | 3.9 | 0.5 | 2.2 | 0.8 | 1 | 2.45 | −9.9129669 |
| 28.9 | 0.2 | 3.031 | 3.9 | 0.5 | 2.2 | 0.8 | 1 | 2.45 | −9.316332 |
| 29 | 0.2 | 3.031 | 3.9 | 0.5 | 2.2 | 0.8 | 1 | 2.45 | −8.7758619 |
| 29.1 | 0.2 | 3.031 | 3.9 | 0.5 | 2.2 | 0.8 | 1 | 2.45 | −8.2849744 |
| 29.2 | 0.2 | 3.031 | 3.9 | 0.5 | 2.2 | 0.8 | 1 | 2.45 | −7.8379405 |
| 29.3 | 0.2 | 3.031 | 3.9 | 0.5 | 2.2 | 0.8 | 1 | 2.45 | −7.4298165 |
| 29.4 | 0.2 | 3.031 | 3.9 | 0.5 | 2.2 | 0.8 | 1 | 2.45 | −7.0563598 |
| 29.5 | 0.2 | 3.031 | 3.9 | 0.5 | 2.2 | 0.8 | 1 | 2.45 | −6.7139326 |
| 29.6 | 0.2 | 3.031 | 3.9 | 0.5 | 2.2 | 0.8 | 1 | 2.45 | −6.3994339 |
| 29.7 | 0.2 | 3.031 | 3.9 | 0.5 | 2.2 | 0.8 | 1 | 2.45 | −6.1102045 |
| 29.8 | 0.2 | 3.031 | 3.9 | 0.5 | 2.2 | 0.8 | 1 | 2.45 | −5.8439517 |
| 29.9 | 0.2 | 3.031 | 3.9 | 0.5 | 2.2 | 0.8 | 1 | 2.45 | −5.598653 |
| 30 | 0.2 | 3.031 | 3.9 | 0.5 | 2.2 | 0.8 | 1 | 2.45 | −5.3724671 |
| 26 | 0.3 | 3.031 | 3.9 | 0.5 | 2.2 | 0.8 | 1 | 2.45 | −5.2191495 |
| 26.1 | 0.3 | 3.031 | 3.9 | 0.5 | 2.2 | 0.8 | 1 | 2.45 | −5.5778494 |

(continued)

**Table 1** (continued)

| Frequency (GHz) | Inset length | Patch length | Patch width | Inset width | Quarter feed length | Quarter feed width | Feed length | Feed width | S11 (dB) |
|---|---|---|---|---|---|---|---|---|---|
| 26.2 | 0.3 | 3.031 | 3.9 | 0.5 | 2.2 | 0.8 | 1 | 2.45 | −5.9690141 |
| 26.3 | 0.3 | 3.031 | 3.9 | 0.5 | 2.2 | 0.8 | 1 | 2.45 | −6.3961779 |
| 26.4 | 0.3 | 3.031 | 3.9 | 0.5 | 2.2 | 0.8 | 1 | 2.45 | −6.8634608 |
| 26.5 | 0.3 | 3.031 | 3.9 | 0.5 | 2.2 | 0.8 | 1 | 2.45 | −7.3756064 |
| 26.6 | 0.3 | 3.031 | 3.9 | 0.5 | 2.2 | 0.8 | 1 | 2.45 | −7.9380172 |
| 26.7 | 0.3 | 3.031 | 3.9 | 0.5 | 2.2 | 0.8 | 1 | 2.45 | −8.5568114 |
| 26.8 | 0.3 | 3.031 | 3.9 | 0.5 | 2.2 | 0.8 | 1 | 2.45 | −9.2388708 |
| 26.9 | 0.3 | 3.031 | 3.9 | 0.5 | 2.2 | 0.8 | 1 | 2.45 | −9.9918246 |
| 27 | 0.3 | 3.031 | 3.9 | 0.5 | 2.2 | 0.8 | 1 | 2.45 | −10.823857 |
| 27.1 | 0.3 | 3.031 | 3.9 | 0.5 | 2.2 | 0.8 | 1 | 2.45 | −11.743046 |
| 27.2 | 0.3 | 3.031 | 3.9 | 0.5 | 2.2 | 0.8 | 1 | 2.45 | −12.755614 |
| 27.3 | 0.3 | 3.031 | 3.9 | 0.5 | 2.2 | 0.8 | 1 | 2.45 | −13.861698 |
| 27.4 | 0.3 | 3.031 | 3.9 | 0.5 | 2.2 | 0.8 | 1 | 2.45 | −15.0458 |
| 27.5 | 0.3 | 3.031 | 3.9 | 0.5 | 2.2 | 0.8 | 1 | 2.45 | −16.257121 |
| 27.6 | 0.3 | 3.031 | 3.9 | 0.5 | 2.2 | 0.8 | 1 | 2.45 | −17.377037 |
| 27.7 | 0.3 | 3.031 | 3.9 | 0.5 | 2.2 | 0.8 | 1 | 2.45 | −18.194744 |
| 27.8 | 0.3 | 3.031 | 3.9 | 0.5 | 2.2 | 0.8 | 1 | 2.45 | −18.468425 |
| 27.9 | 0.3 | 3.031 | 3.9 | 0.5 | 2.2 | 0.8 | 1 | 2.45 | −18.108954 |
| 28 | 0.3 | 3.031 | 3.9 | 0.5 | 2.2 | 0.8 | 1 | 2.45 | −17.270714 |

(continued)

**Table 1** (continued)

| Frequency (GHz) | Inset length | Patch length | Patch width | Inset width | Quarter feed length | Quarter feed width | Feed length | Feed width | S11 (dB) |
|---|---|---|---|---|---|---|---|---|---|
| 28.1 | 0.3 | 3.031 | 3.9 | 0.5 | 2.2 | 0.8 | 1 | 2.45 | −16.200846 |
| 28.2 | 0.3 | 3.031 | 3.9 | 0.5 | 2.2 | 0.8 | 1 | 2.45 | −15.082391 |
| 28.3 | 0.3 | 3.031 | 3.9 | 0.5 | 2.2 | 0.8 | 1 | 2.45 | −14.009207 |
| 28.4 | 0.3 | 3.031 | 3.9 | 0.5 | 2.2 | 0.8 | 1 | 2.45 | −13.018081 |
| 28.5 | 0.3 | 3.031 | 3.9 | 0.5 | 2.2 | 0.8 | 1 | 2.45 | −12.117826 |
| 28.6 | 0.3 | 3.031 | 3.9 | 0.5 | 2.2 | 0.8 | 1 | 2.45 | −11.305354 |
| 28.7 | 0.3 | 3.031 | 3.9 | 0.5 | 2.2 | 0.8 | 1 | 2.45 | −10.573233 |
| 28.8 | 0.3 | 3.031 | 3.9 | 0.5 | 2.2 | 0.8 | 1 | 2.45 | −9.9129669 |
| 28.9 | 0.3 | 3.031 | 3.9 | 0.5 | 2.2 | 0.8 | 1 | 2.45 | −9.316332 |
| 29 | 0.3 | 3.031 | 3.9 | 0.5 | 2.2 | 0.8 | 1 | 2.45 | −8.7758619 |
| 29.1 | 0.3 | 3.031 | 3.9 | 0.5 | 2.2 | 0.8 | 1 | 2.45 | −8.2849744 |
| 29.2 | 0.3 | 3.031 | 3.9 | 0.5 | 2.2 | 0.8 | 1 | 2.45 | −7.8379405 |
| 29.3 | 0.3 | 3.031 | 3.9 | 0.5 | 2.2 | 0.8 | 1 | 2.45 | −7.4298165 |
| 29.4 | 0.3 | 3.031 | 3.9 | 0.5 | 2.2 | 0.8 | 1 | 2.45 | −7.0563598 |
| 29.5 | 0.3 | 3.031 | 3.9 | 0.5 | 2.2 | 0.8 | 1 | 2.45 | −6.7139326 |
| 29.6 | 0.3 | 3.031 | 3.9 | 0.5 | 2.2 | 0.8 | 1 | 2.45 | −6.3994339 |
| 29.7 | 0.3 | 3.031 | 3.9 | 0.5 | 2.2 | 0.8 | 1 | 2.45 | −6.1102045 |
| 29.8 | 0.3 | 3.031 | 3.9 | 0.5 | 2.2 | 0.8 | 1 | 2.45 | −5.8439517 |
| 29.9 | 0.3 | 3.031 | 3.9 | 0.5 | 2.2 | 0.8 | 1 | 2.45 | −5.598653 |
| 30 | 0.3 | 3.031 | 3.9 | 0.5 | 2.2 | 0.8 | 1 | 2.45 | −5.3724671 |

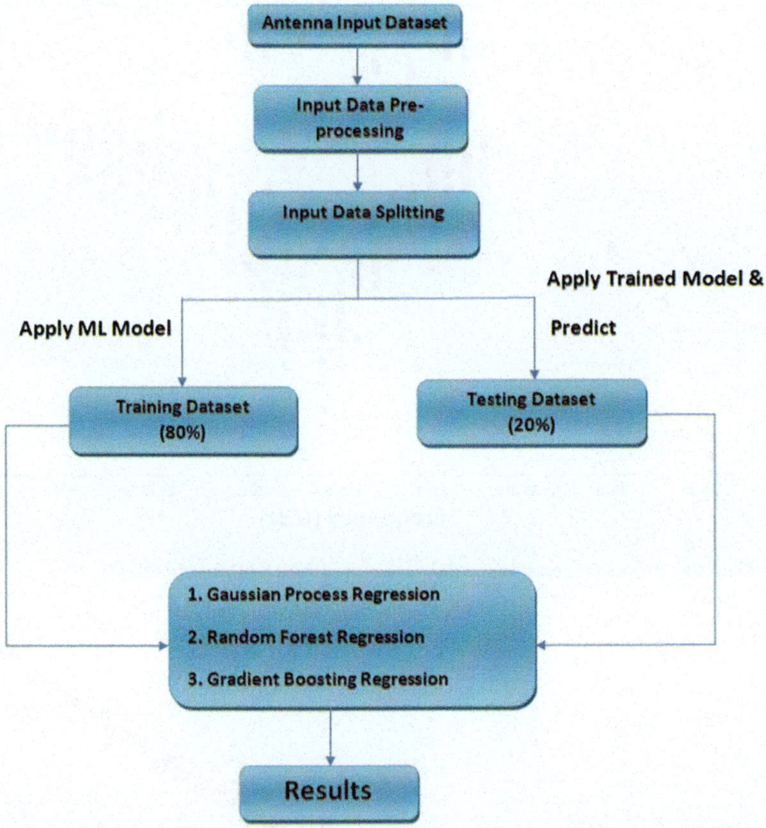


**Fig. 4** Flowchart explaining the implementation of a machine learning algorithms for the proposed antenna

frequency is shown in Figs. 5, 7 and 9 for GPR, RFR, and GBR, respectively and the actual vs predicted S11 is shown in Figs. 6, 8 and 10 for GPR, RFR, and GBR, respectively.

The random forest ML algorithm for the 5G microstrip antenna (26–30 GHz) achieves an RMSE of 1.08, demonstrating higher prediction accuracy compared to existing ML approaches in literature, as shown in Table 2. While GPR offers comparable accuracy (1.18 RMSE), it requires higher computational effort. Other algorithms such as GBR, decision tree, and SVR show larger errors (1.84–1.96 RMSE). Thus, the proposed model provides an effective balance between accuracy, speed, and generalization for high-frequency antenna design.

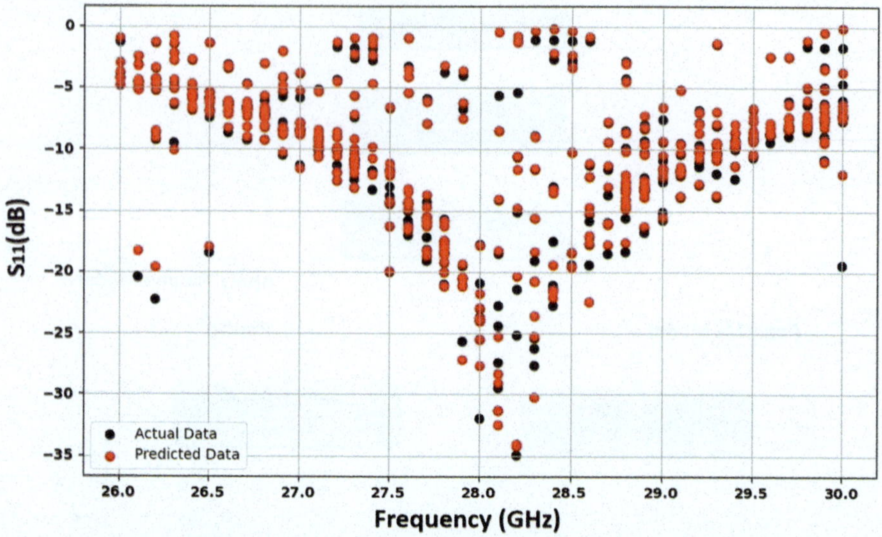


**Fig. 5** Gaussian process regression: actual versus predicted S11 over frequency

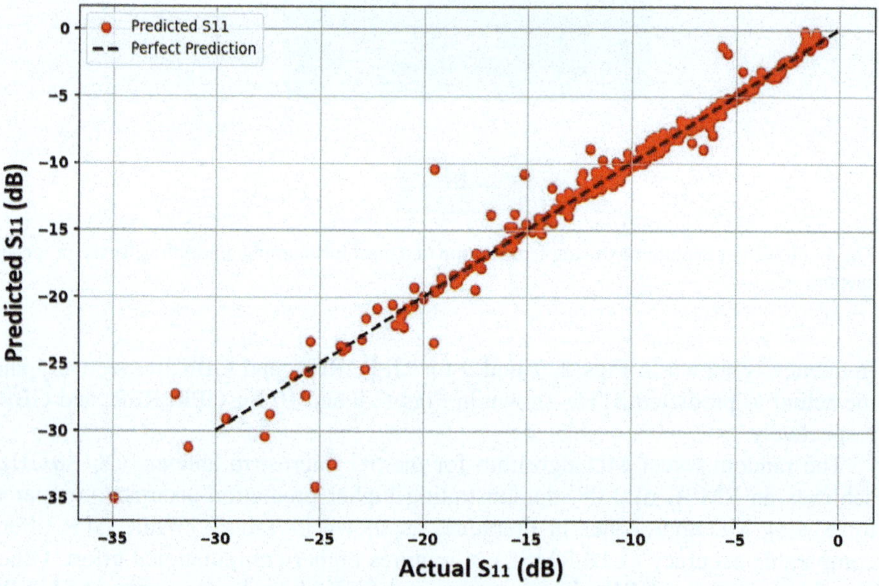


**Fig. 6** Gaussian process regression: actual versus predicted S11

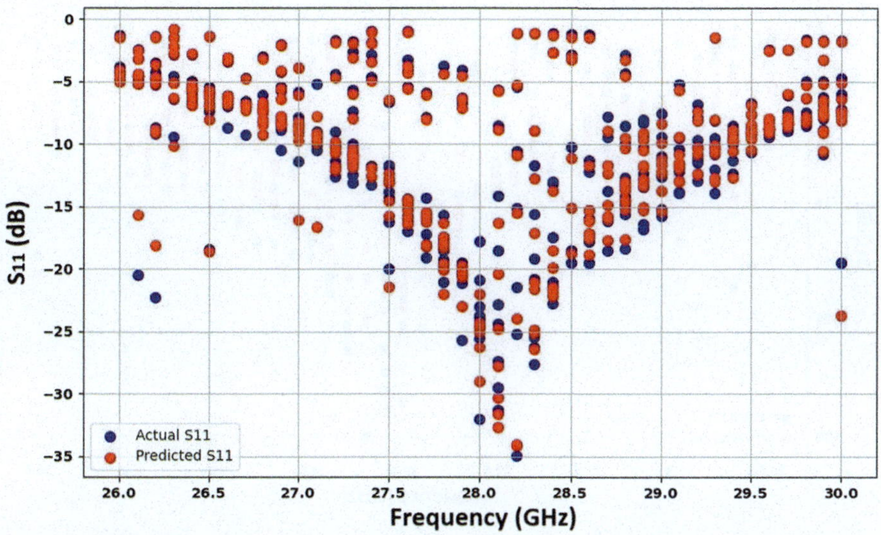


**Fig. 7** Random forest regression: actual versus predicted S11 over frequency

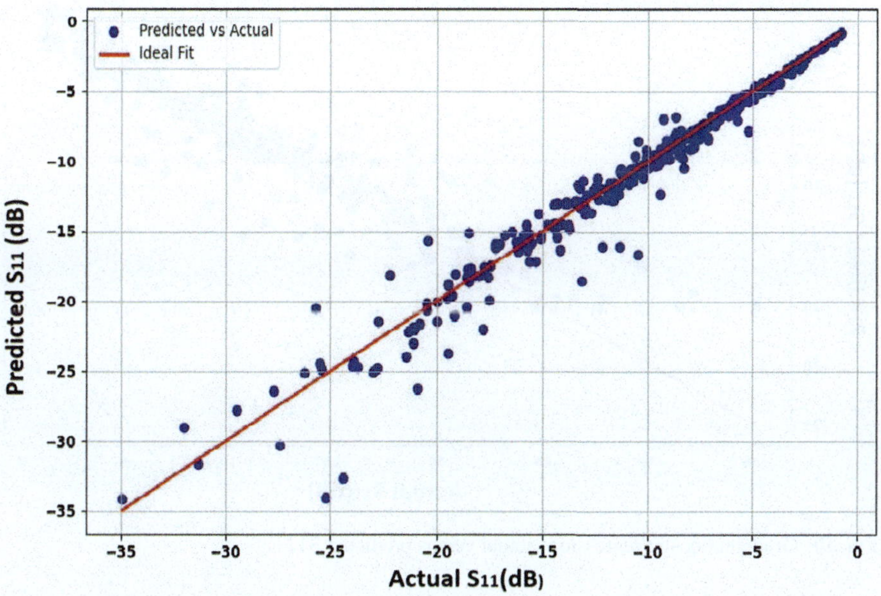


**Fig. 8** Random forest regression: actual versus predicted S11

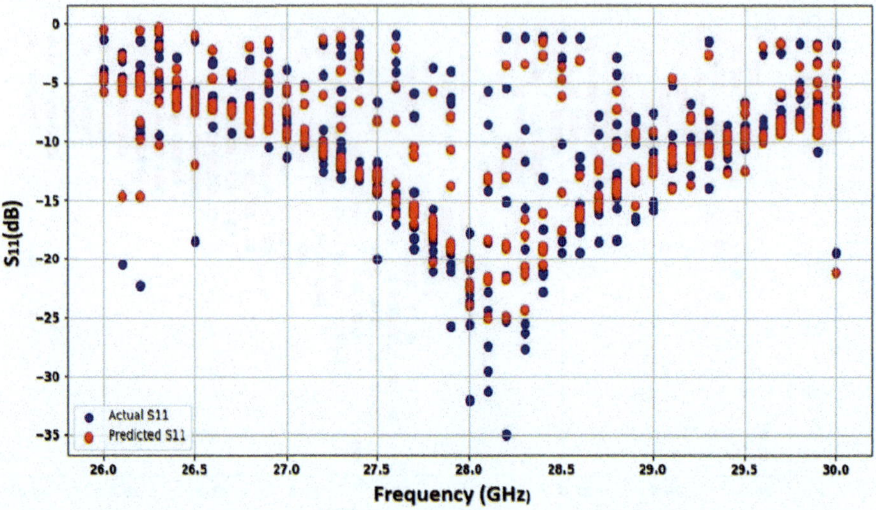


**Fig. 9** Gradient booster regression: actual versus predicted S11 over frequency

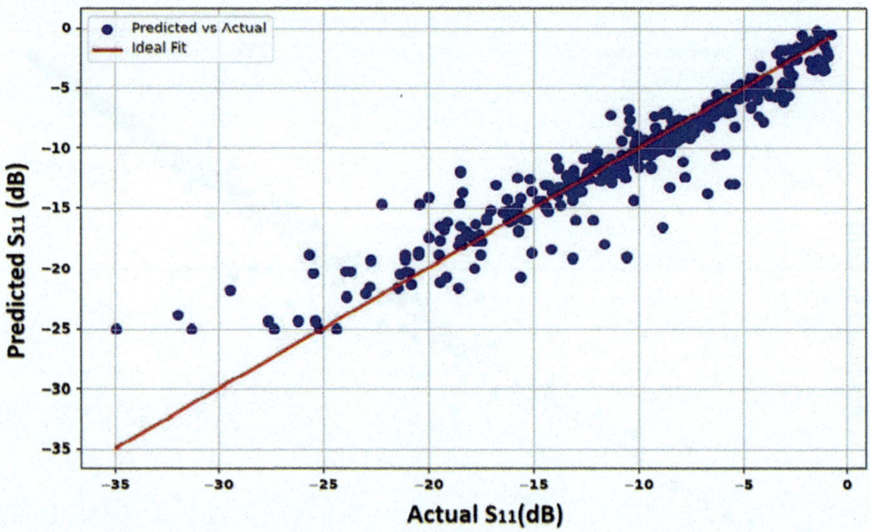


**Fig. 10** Gradient booster regression: actual versus predicted S11

**Table 2** Performance comparison of ML algorithms in antenna design with other existing in the literature

| Ref. No. | ML algorithm | Antenna type | Frequency band (GHz) | RMSE |
|---|---|---|---|---|
| 17 | GPR | Microstrip antenna | 2.3–2.8 | 0.0087 |
| 18 | Decision tree, RF, | Microstrip antenna | 1–8 | 1.85 |
| | SVR | | | |
| 19 | Decision tree, RF, | 5G microstrip antenna | 27–38 | 1.41 |
| | Boost regression | | | |
| Proposed work | Random forest | 5G microstrip antenna | 26–30 | 1.08 |
| | GPR | | | 1.18 |
| | GBR | | | 1.84 |
| | Decision tree | | | 1.96 |
| | SVR | | | 1.87 |

## 4 Conclusion and Future Works

This work presents a Gaussian process regression, random forest, and gradient booster regression ML algorithm for 28 GHz 5G antenna design. In the study, we observed that the lowest value of RMSE is 1.08 for the prediction of S11 for random forest regression. The low value of RMSE shows that random forest model is well-performing for predicting S11 at 28 GHz frequency. The SISO antenna is modeled as MIMO and massive MIMO configuration and can be used for beamforming application.

## References


1. Pant, R., Malviya, L.: THz antennas design, developments, challenges, and applications: a review. Int. J. Commun. Syst. **36**(8), e5474 (2023)
2. Pant, M., Malviya, L.: SIW MIMO antenna with high gain and isolation for fifth generation wireless communication systems. Frequenz **78**(9–10), 479–497 (2024)
3. Khan, M.M., Hossain, S., Mozumdar, P., Akter, S., Ashique, R.H.: A review on machine learning and deep learning for various antenna design applications. Heliyon **8**(4) (2022)
4. Haque, M.A.: Machine learning-based technique for resonance and directivity prediction of UMTS LTE band Quasi Yagi antenna. Heliyon **9**(9), 1–16 (2023)
5. Farahani, M.F., Tran, A.: Machine learning application for antenna design. In: 2022 IEEE International Symposium on Antennas and Propagation and USNC-URSI Radio Science Meeting (AP-S/URSI), pp. 437–438. IEEE (2022)

6. El Misilmani, H.M., Naous, T., Al Khatib, S.K.: A review on the design and optimization of antennas using machine learning algorithms and techniques. Int. J. RF Microwave Comput. Aided Eng. **30**(10), e22356 (2020)
7. Khan, M.M., Hossain, S., Mozumdar, P., Akter, S., Ashique, R.H.: A review on machine learning and deep learning for various antenna design applications. Heliyon **8**(4) (2022)
8. Sharma, Y., Chen, X., Wu, J., Zhou, Q., Zhang, H.H., Xin, H.: Machine learning methods-based modeling and optimization of 3-D-printed dielectrics around monopole antenna. IEEE Trans. Antennas Propag. **70**(7), 4997–5006 (2022)
9. Sharma, Y., Zhang, H.H., Xin, H.: Machine learning techniques for optimizing design of double T-shaped monopole antenna. IEEE Trans. Antennas Propag. **68**(7), 5658–5663 (2020)
10. Ranjan, P., Krishnan, A., Dwivedi, A.K., Singh, S.K., Sharma, A.: Design and optimization of MIMO dielectric resonator antenna using machine learning for sub-6 GHz based on 5G IoT applications. Arab. J. Sci. Eng. **48**(11), 14671–14679 (2023)
11. Ma, W., Qi, C., Li, G.Y.: Machine learning for beam alignment in millimeter wave massive MIMO. IEEE Wirel. Commun. Lett. **9**(6), 875–878 (2020). https://doi.org/10.1109/LWC.2020.2973972
12. Sarker, N., Podder, P., Mondal, M.R.H., Shafin, S.S., Kamruzzaman, J.: Applications of machine learning and deep learning in antenna design, optimization and selection: a review. IEEE Access (2023)
13. Andriulli, F., Chen, P.Y., Erricolo, D., Jin, J.M.: Guest editorial machine learning in antenna design, modeling, and measurements. IEEE Trans. Antennas Propag. **70**(7), 4948–4952 (2022)
14. Ranjan, P., Yadav, S., Gupta, H., Bage, A.: Design and development of machine learning assisted cylindrical dielectric resonator antenna (2023)
15. Schulz, E., Speekenbrink, M., Krause, A.: A tutorial on Gaussian process regression: modelling, exploring, and exploiting functions. J. Math. Psychol. **85**, 1–16 (2018)
16. CST Studio Suite, Student Edition (Version 2018) [Computer software]. Dassault Systemes
17. Sharma, K., Pandey, G.P.: Efficient isolation modelling for two-port MIMO antenna by Gaussian process regression. Progr. Electromagn. Res. C **108** (2021)
18. Rana, M.S., Islam, R., Sarker, S.: Machine learning based on patch antenna design and optimization for 5G applications at 28 GHz. Res. Eng. **24**, 103366 (2024)
19. Jain, R., Thakare, V. V., Singhal, P.K.: Employing machine learning models to predict return loss precisely in 5G antenna. Progr. Electromagn. Res. M **118**, 151–161 (2023)

# Robust Hybrid Cryptographic Method for Cloud Computing Information Security

**Kamaljit Kaur Bhagwat and Sudhanshu Shekhar Dubey**

**Abstract** Future cloud computing will include virtualization, dynamic resource pools, and high availability. Without requiring computer users to get involved, cloud computing offers a solution to common computing issues including resource availability, software, and hardware. The following are the two main difficulties facing cloud computing first is deployment of cloud security over the internet and second is data storage. The data stored in cloud storage can be encrypted using a variety of cryptographic techniques. There are several issues with the existing algorithms, so this work has used the idea of MD5, Blowfish, and RSA algorithms. Furthermore, the proposed hybrid model has used the existing technique and provides the solution. The information is encrypted via the RSA algorithm, which is a modified version of the RSA algorithm. This work has used two different algorithms Communication of Encryption and Decrypt (CED) and Hybrid Data Encryption and Decryption process in Server (HD-EDS). Both algorithms have used the symmetric encryption algorithm which is called the RSA algorithm. The HD-EDS algorithm is a proposed hybrid algorithm that has included the RSA, MD5, and Blowfish properties of the algorithm. When data is uploaded to the cloud, it is encrypted using the RSA and Blowfish encryption techniques. This work has calculated the retrieval time between existing techniques versus the proposed hybrid technique and achieved 9.8 (ms) time. Another is the comparison between RSA, MD5, Blowfish, and HD-EDS achieved 40 (ms) which is less than MD5 and Blowfish algorithm. With their private key, the cloud controller may decrypt the data. Lastly, by contrasting the outcomes of our suggested hybrid approach with those of the current procedures, we have finished the performance evaluation.

**Keywords** Cloud computing · Encryption · Decryption · Cipher text · RSA · Blowfish and MD5

K. K. Bhagwat · S. S. Dubey (✉)
Department of Mathematics, Chandigarh University, Mohali 140413, Punjab, India
e-mail: sudhanshusdubey@gmail.com

© The Author(s), under exclusive license to Springer Nature Singapore Pte Ltd. 2026

M. Pant et al. (eds.), *Real-World Applications of Machine Learning*, Mathematics for Sustainable Developments, https://doi.org/10.1007/978-981-95-7134-5_37

## 1 Introduction

Cloud computing and its uptake have been major topics in recent years. Because of its benefits in terms of cost savings, job productivity, company agility, and service quality, it has become a priority for enterprises to implement [1]. Because it offers services that satisfy consumers' expectations, software as a service (SaaS) is highly sought after. For example, health informatics facilitates the diagnosis of challenging illnesses and tumors by medical researchers. It guarantees that investors have access to fast and accurate simulations in financial analytics. It enhances the quality and delivery of education and education delivery in the context of education as a service. Users of mobile apps can interact with peers through simple-to-use programs or play online games. Security and privacy are becoming more and more important as more people and businesses utilize cloud services. According to several analysts, security needs to be put in place before utilizing any cloud services [2]. Organizations are faced with a dilemma since security needs to be established and upheld concurrently with any other service. The benefits of cloud services are acknowledged by enterprises that use them, but adoption is still hampered by issues such as security and privacy. Although security monitoring is crucial, the design, implementation, and testing of security features should always be a part of the SED process [3]. The fast expansion of data has presented issues for data centers. For example, the primary author's place of employment had a daily growth of 100 terabytes in the data center. If an organization has seen a sharp rise in the volume of data it collects and is unable to react in a timely and efficient manner, data traffic difficulties, data security issues, and service level [4]. The security research problem is depicted in Fig. 1, which is essential for improved data center management to handle the rapid rise of data. A hybrid cloud model is an IT infrastructure that combines an organization's on-premises private cloud with a third-party public cloud. This integration allows data and applications to move seamlessly between the two environments. By using this approach, businesses can benefit from the scalability and cost-effectiveness of the public cloud while retaining control and security over sensitive data within their private infrastructure.

Data center security is necessary to handle the rapidly expanding amount of data, but data engineering procedures also need to be robust enough to withstand assaults and unauthorized access. It is feasible to combine the entire procedure [5]. The choice to remove the hacker and false application from the Business application stages is taken to free up more storage space. Figure 1 shows the three processes for data processing: user, business application, and intrusion protection. Concerns regarding data integrity arise from users of cloud storage services. They want an auditing procedure that enables them to verify that the cloud server preserves all of their most recent data without introducing corruption. When storing data on the cloud, clients are in charge of ensuring that it remains intact [6]. They can't get rid of bad cloud servers, which expose their data to security lapses. Integrity, availability, and secrecy are the three main goals of security. To guarantee data integrity, cloud data audits or external Data Identity (DI) verification are used. DI has been a popular topic

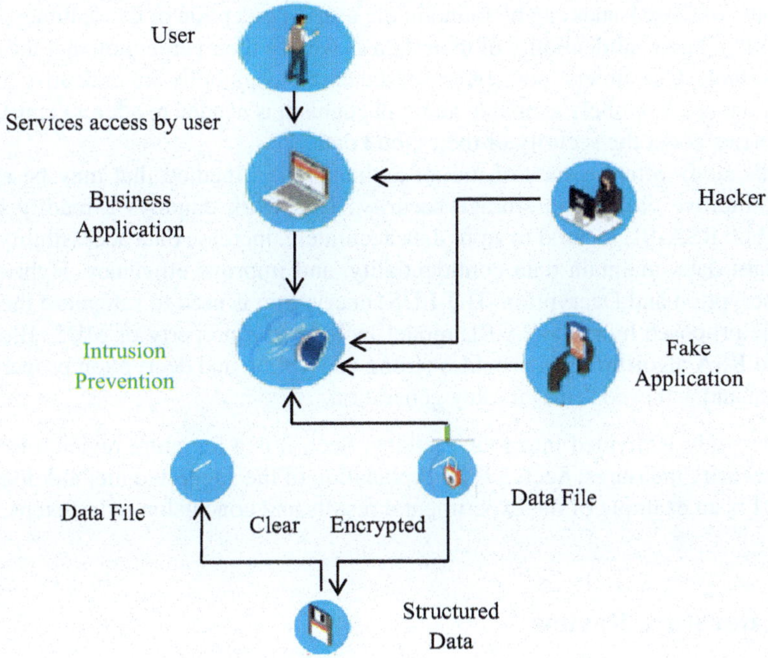


**Fig. 1** Data protection in virtual computing

in recent years. In the experimental setup of a hybrid cloud computing model, the on-premises private cloud environment is configured using high-performance rack-mounted servers equipped with dual Intel Xeon processors, 128 GB of DDR4 ECC RAM, and a hybrid storage system comprising 4 TB SSDs and 8 TB HDDs configured in RAID for reliability and performance. Network connectivity is ensured through dual 10 Gbps Ethernet interfaces, supporting high-speed data transfer between the private and public cloud environments. Virtualization is managed using a hypervisor such as VMware ESXi or Proxmox VE, with cloud orchestration handled by platforms like OpenStack or CloudStack. The public cloud component utilizes services from providers like AWS, Azure, or Google Cloud, with typical virtual instances configured with 2–4 vCPUs, 8–16 GB RAM, and scalable storage ranging from 100 to 500 GB. For data processing and integration testing, the environment uses datasets ranging from 50 GB to 1 TB in size, covering a variety of workloads such as transactional logs, multimedia files, and structured data for machine learning or analytics tasks. A secure site-to-site VPN ensures encrypted communication between environments, enabling seamless data movement and workload distribution in the hybrid architecture. The digital identity is periodically verified via the cloud. By including the data; end users can lessen part of their computational burden [7]. Furthermore, the contribution of this work is given as:

- Some of these issues can be found in the users' perception of the accuracy, accessibility, and confidentiality of their data as well as their perception that the data is available. Consumers' perceptions of the data's availability are indicative of other problems. A publicly available audit mechanism is needed to address consumers' worries about the security of their cloud data.
- This study offers a state-of-the-art authentication method that may be applied to improve cloud data storage security. This work employs a readily known public RSA-BL method to store data accurately, increase data accessibility, safeguard data, maintain data confidentiality, and improve efficiency. Hybrid Data Encryption and Decryption (HD-EDS) encryption is used to safeguard the data.
- The proposed hybrid RSA-BL model includes the property of MD5, Blowfish, and RSA algorithm and provides strong encryption and decryption properties of data and achieved 9(ms) for key generation.

The study is divided into four sections: Sect. 2 is a literature review of current data security measures; Sect. 3 is an explanation of the suggested methodology; and Sect. 4 is an example of the experimental results and conclusion in last steps.

## 2 Literature Review

To bolster our theoretical framework and research experience, this work will address current research achievements in categorizing cloud computing and talk about cloud security issues in this part. For instance, Taha et al. [8] have discussed that data outsourcing from local servers to commercial clouds is possible due to the platform's flexibility in cloud computing. Cloud computing is becoming less popular despite its benefits for customers due to privacy concerns and data leakage. Several steps must be taken in the cloud to safeguard user privacy and minimize data leakage. There are a wide variety of cloud environment types. Still, we'll concentrate on hybrid clouds. Simply said, a hybrid cloud is an amalgam of two or more clouds. For instance, if a routine task is carried out in a public.

Hagge et al. [9] have discussed the goal of cloud computing, according is to centralize data processing and storage. The speed at which enormous volumes of data may be processed via cloud and mobile computing raises performance requirements. Using several cryptographic standards, such as Attribute-Based Proxy Re-Encryption (ABPRE), the research creates a secure model with data protection, data safety, data integration, dynamic sharing, and access management in a secure environment. Data decryption is nearly impossible when using homomorphism encryption, which is what the ABE utilizes. The cloud's primary function is to centralize data processing and storage, which raises the bar for mobile and cloud computing speed. Zhou et al. [10] have stated that cloud computing is a popular way to store and retrieve data from the Internet. Customers could not, however, have control over distant servers that hold private data, making stored data vulnerable to assault by cloud service providers or other parties. Cryptography is, therefore, essential to cloud security, and hybrid cryptography leverages many techniques to boost security and performance. Examined from the 2013–2020 viewpoints, the hybrid cryptographic techniques of cloud data security described herein outline each proposal's concept, strategy, implementation, constraints, and recommended applications.

Chauhan et al. [11] have proposed the issue with cloud computing is service security. Sophisticated encryption techniques are worthless because cloud computing demands speed. As a result, this technique offers a safe cryptographic solution in a cloud environment without compromising performance. It achieves this by modifying the Blowfish algorithm and applying an elliptic curve-based methodology. To increase security and efficiency, Blowfish encrypts data and keys using an elliptic curve technique. Digital signatures protect the integrity of data. Sachdev et al. [12] have discussed how the vast rise in sensitive cloud data has increased its vulnerability. It is increasingly susceptible due to the rise in malevolent users. Services for cloud security are crucial as the cloud is overseen by a third party. Cloud users must safeguard their data since more and more cloud services and data are kept in data centers. Describe a privacy paradigm where cloud security and data integrity are obtained using ECC using an AES key. Data from cloud applications is first encrypted using AES before being stored.

Bangar et al. [13] have carried out that mobile cloud computing is significant because it connects mobile devices to the cloud and is easily available and utilizable. Through a network, cloud data is transferred from mobile devices to the cloud and back again. Data must be encrypted and decrypted using encryption algorithms to guarantee data security. Mobile-to-cloud data is efficiently encrypted and decoded by the suggested system. System for Mobile Cloud Encryption Mobile Cloud Data Encryption System Information on Mobile-Cloud Data Encryption Information on Mobile Cloud Encryption Techniques Method to help mobile-cloud users protect their data sent and received from theft.

Gunasekaran et al. [14] suggested a reliable authentication technique that always worries about security flaws. To offer a robust user privacy feature, the proposed authentication protocol combines an active password-based two-server authentication mechanism with a unique key exchange approach. Its purpose was to safeguard the privacy of the user. The protocol is being shielded from potential assaults by a team of approved and unauthorized security personnel. The research's findings indicate that the suggested methodology will have overhead that is both equal to and adequate for the authentication methods now in use. Cost, processing time, and communication overhead are all included in the proposed work. The conventional approach made use of passwords. Passwords are vulnerable to hacking attempts that leverage stolen verifiers and issues with key.

Kartit et al. [15] have discussed the data transmission from various health activities to the distributed ledger as the primary goal. Numerous wearable and embedded gadgets provide the data. On top of IOTA's application layer, a protocol known as MAM was developed for the encryption, verification, and transfer of activity data. The ability to interchange data throughout the whole digital healthcare ecosystem may be expanded by this new paradigm. Additionally, it can return control of the health activity data to the patients. However, the technology utilized permits unprotected data transfer when a single-user solution is employed.

Yugandhara et al. [16] have discussed the first-of-its-kind invention in the field of privacy protection, the triangle data privacy protocol will strengthen the security of three important stakeholders. The method we've devised starts with putting in place a special authentication procedure that involves all three parties. Next, the technique we've created carries out three crucial steps: Danger detection function for the TPA dual authentication for cloud services provided by cloud service providers. The experiment's outcomes show how successful and efficient the suggested strategy which involves auditing each of the important stakeholders. Owing to the other parties' involvement, our tried-and-true method raises two important problems: Data loss and inadequate backups of data. Salem et al. [17] have discussed the dynamic audit system has been built to avoid collisions and dishonest entities, while also enabling dynamic alterations that may be independently confirmed. BLA (Batch-Level Authentication Merkle hash tree) was used to batch-verify the many leaf nodes and their different indices. In comparison with the conventional approach of conducting public audits, the audit's establishment costs for the user and TPA were reduced. On the other hand, there are problems with the external network platforms and the sophisticated BLA (MHT) infrastructure.

Cordova et al. [18] have discussed the plan for creating cloud servers that combine quantum identity-based authentication with key agreement techniques. However, the basic principle of quantum information theory and quantum physics is where quantum comes from. The protocol security study demonstrates how impervious to security breaches the suggested solution. The findings show that it is possible to create shifted keys across a 100-km optical cable with a consistent key rate and error rate. Social engineering and phishing efforts have increased exponentially since cloud computing's open design. Furthermore, the next section discusses the methodology of this work.

## 3 Methodology

The difficulties we encounter urge us to create an appropriate, secure, and practical method of safeguarding cloud-based data. Customers of the CC could gain from less expensive hardware and software maintenance, more storage space, and easier data access. Furthermore, the complete flows of this work are given in Fig. 2.

In this work, the client has taken the file from a cloud service provider which is also called the plaintext. Furthermore, need to generate the encryption text file with the help of a key. The output cipher text file and implement MD5 Hash algorithm in plaintext. The encrypted data file has been uploaded to a cloud server [19]. Further, downloading encrypted files adds the functionality of RSA and the Blowfish algorithm. Finally, received a plain text. Another concept of cloud architecture is given in Fig. 2. Even though CSP provides a wide range of services, problems with data protection or security might have a detrimental effect on these services.

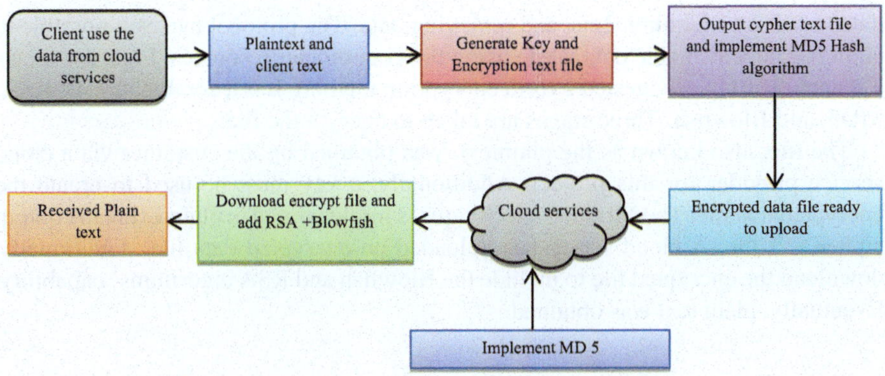


**Fig. 2** Data flow of this work

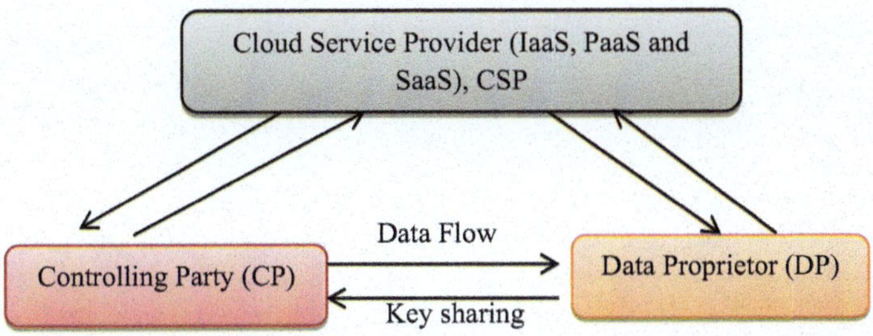


**Fig. 3** Structure of cloud architecture

Numerous timing errors might impact the CD and cause it to seem like a fluffy white cloud service [20]. The CCP permits the DP to carry out routine audits of the data's integrity to ensure that the data that is outsourced is secure and hasn't been altered or updated. Figure 3 depicts the cloud's structure.

The outsourcing process is finished when there are sufficient data files in the Data Proprietor (DP), which can be an individual, a business, or an entity inside a business. At a later time, the data that was outsourced may be updated in several ways, such as by adding new data, changing old data, or deleting old data. Typically, the DP is a resource-constrained entity that depends on the CP for data upkeep. The CSP has nearly infinite storage space and sufficient power to run several servers and do computations. A CSP is an untrusted object that is part of the cryptographic system and is used to stop stored data from being outsourced. The cloud auditor was authorized to verify the data of the DP on behalf of a CA, which is a responsible entity [21].

Both the DP and the service provider have faith in the cloud auditor. The role of the cloud auditor was to verify the authenticity of the data on the DP. During the data audit on the DP, the certifying authority helps to reduce the load on the computed data. In this study, a few instances of dynamic operations are adding data, modifying data to suit user requirements, and removing data. The protocol makes it possible to audit publicly available data while using fewer computer resources. Figure 4 shows a schematic of the suggested hybrid encryption method. It is a combination of RSA, MD5, and Blowfish. Three stages are taken to encrypt the file:

The file, also known as the plaintext, was obtained by the customer via a cloud service provider for this project. Additionally, a key must be used to create the encrypted text file. The MD5 hash technique is implemented in plaintext in the output cipher text file. A cloud server has uploaded an encrypted data file. Additionally, download the encrypted file to include the Blowfish and RSA algorithms' capability. Eventually, plain text was obtained.

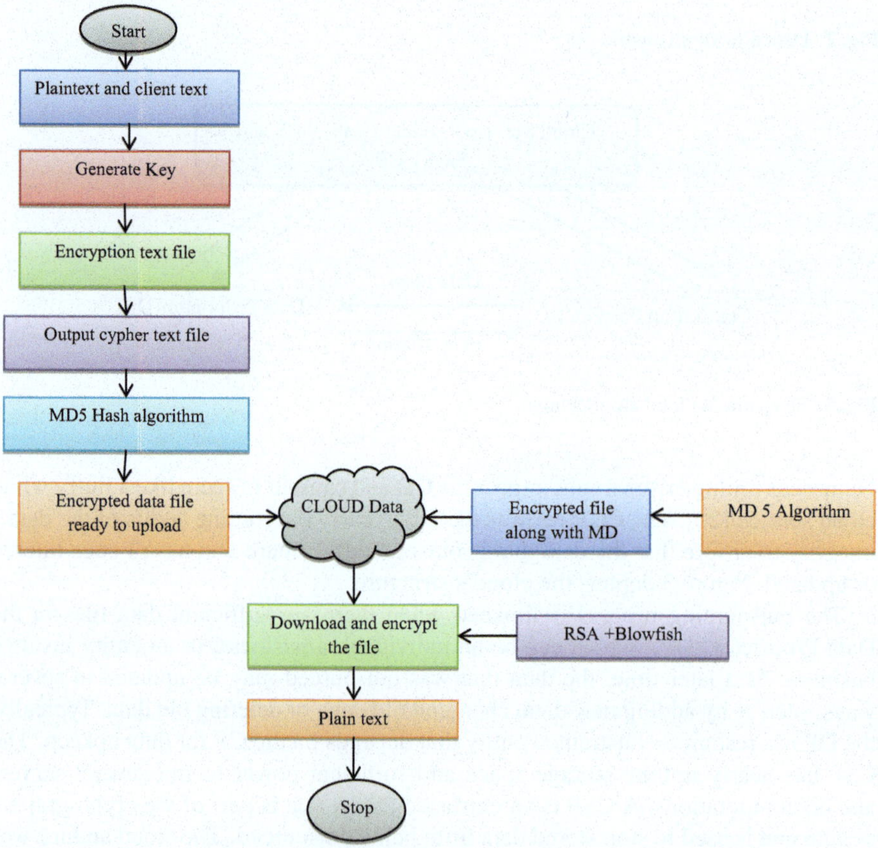


**Fig. 4** Flowchart of the basic structure of the cloud architecture

## 3.1 Description of RSA Algorithm

The RSA algorithm block diagram is displayed in Fig. 5. RSA is an algorithm based on positive integers. Modular exponential is the foundation of both the RSA technique's encryption and decryption.

A public key algorithm is called RSA. Among others, Leonard Adleman, Adi Shamir, and Rivest created it [22]. Anyone can use the public key to encrypt messages. Both the public and private keys are utilized by the RSA algorithm. The private key, which is the only key in Algorithm 1 that can decrypt messages encrypted using a public key, may be used to decode communications encrypted using the public key.

Take input as "*P*" and "*Q*" and calculate the $p \times q$, and $(p - 1) \times (q - 1)$. Add the encryption key in $M^s$ mod $n$ and add the private key in $S^d$ mod $n$ and received the "*T*." Furthermore, Algorithm 1 Communication of Encryption and Decrypt (CED) are described.

**Algorithm 1** Communication of Encryption and Decrypt (CED)

**Input**: $p, q, n$

**Output**: $S$ and $T$

1. Take input prime no as $p$ and $q$.
2. Implement $n = p \times q$.
3. Calculate $\varphi(n) = (p - 1) \times (q - 1)$ and choose random key as "*e*" where $1 < e < \varphi(n) \&\ \mathrm{gcd}(e, \varphi(n)) = 1$.
4. Calculate $d = e^{-1}\mathrm{mod}\ \varphi(n)$. for archive decryption key as $D_{ec}$.
5. Encryption key $Pu(E)_k = \{e, n\}$ and private Decryption Key as $Pr(D)_k = \{d, n\}$.
6. Calculate $S = M^s$ mod $n$ where $0 < \&n$ and $T = S^d$ mod $n$ description.
7. Return: $S$ and $T$.

The two exponents used by RSA are "*e*" and "*d*," where "*e*" denotes the public exponent and "*d*" is the private exponent. Let "*S*" stands for the plaintext and "*T*"

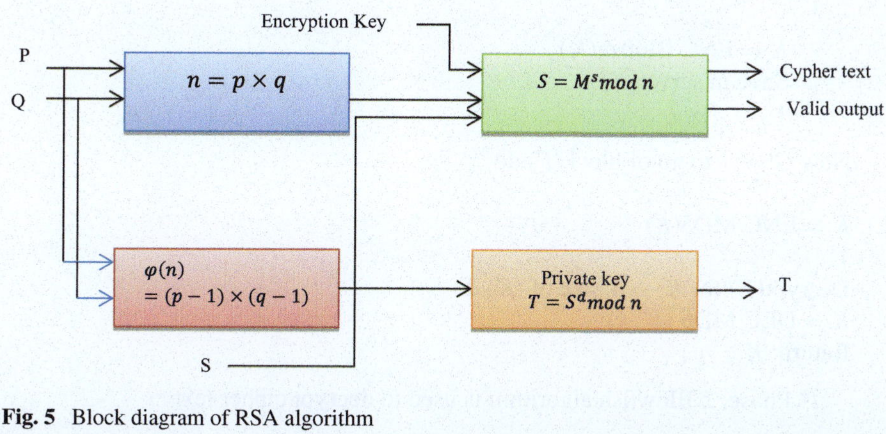


**Fig. 5** Block diagram of RSA algorithm

for the cipher text. Encryption and decryption are accomplished using Eqs. 1 and 2, respectively [23].

$$S = M^s \bmod n \tag{1}$$

$$T = S^d \bmod n \tag{2}$$

"$n$" is a very high no produced by key generation.

There are three components to this strategy. First, RSA is used to encrypt the clair text. The same text is encrypted using the Blowfish technique in the second section. The output is encrypted using MD5 in the third section, where the produced key "n" is also a very high no.

An additional component of this strategy is the file upload and download procedure. RSA is used to encrypt the text in the first section. Phase 2 then encrypts the output using the Blowfish technique. Lastly, the output is encrypted using MD5. This is the file upload (FUDS) method as presented by Algorithm 2.

**Algorithm 2** Hybrid Data Encryption and Decryption process in Server (HD-EDS)

**Input**: $B$, $k$

**Output**: $K$, $F$

Take input as a file($F$) for encryption
for $B \rightarrow 1$ to block($F$)

{

$B$ = ENC RSA($B$,$K$)
}

Send to cloud server as $F$
For $k \rightarrow 1$ to size of ($k$) Do

{

$K$ = ENC Blofish($K$)
Save in server($K$)
}

Now $G \rightarrow 1$ to no of block($F$) do
{
$K$ = ENC MD5($\underline{K}$)
}
Decrypt($F$):for $K \rightarrow 1$ size of ($K$) do
$K$ = DEC MD5 ($B'$, $k$)
Return: $K$

/* Phase: 2 Blowfish algorithm is used to decrypt cipher text */

for $B \leftarrow 1$ is the number of blocks ($F$) do
{
$B$ = DEC Blowfish ($B$, $K$)
}

/*Phase 3; decryption using the RSA technique */

for $G \leftarrow 1$ is the number of blocks ($F$) do
{
Return ($F$) = DEC RSA($k'$)
}

This operation spins the input as a "*B*" RSA (*B*, *K*) function and sends it to server "*F*" after receiving it as a file (*F*) for encoding with a different block [24]. Next, use a new block of file (*F*) and file (*K*) in server "*F*" to construct the Blowfish (*k*) algorithm. After that, use MD5 to decrypt the file using the same key after receiving "*K*." Lastly, encrypt "*F*" using the same key as RSA ("*k*") and decode using the RSA procedure [27].

# 4 Implementation and Result

## *4.1 System Description*

Anaconda Navigator, Google Firebase Cloud, and Python 3.9 were used in the construction of the complete framework. The sign-in screen will prompt you to enter your Google account ID and password since Crypters Cloud uses Google Accounts for verification. Crypters Cloud will prompt you to sign in using the Google account you were previously logged in to because of are a new user. You may select the algorithm you wish to apply to this account on this page. You are unable to alter the algorithm after you have chosen it.

This section of the implementation displays the upload and download times for various file sizes. The Firebase cloud will be used to execute the model. The Python programming tools utilized include Anaconda and Spider (3.9). This work measures the encryption algorithm's speed, average duration, buffer size, and interim for various inputs. The buffer size, speed, and intermediate are defined first. The interval is the amount of time that separates the beginning and end of the encryption operation of an algorithm. The amount of time required for encryption grows with the quantity of the data. The speedup is the difference between the average time cloud and the local system's speed. Performance declines as data size rises [25].

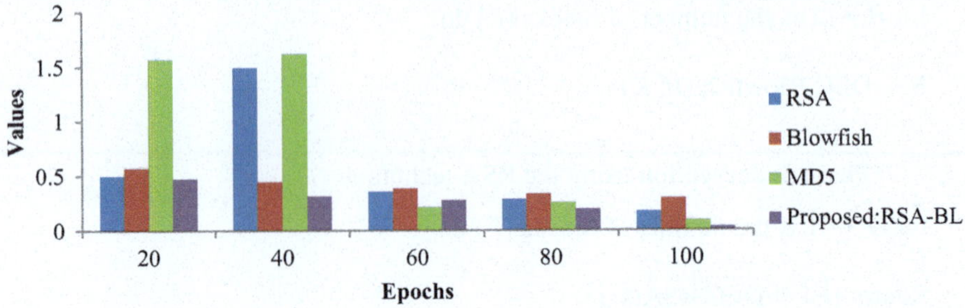


**Fig. 6** Comparison of retrieval times between proposed hybrid technique and current methods

### 4.1.1 Encryption and Access Permission for File

This format's files are encrypted, and both encryption and decryption provide the public and private keys. The public key and the file can only be provided by the data owner. The owner is identified by the public key of the cloud server. Next, the file is encrypted with the given private key. This is an authentication process. In this process, the file is uploaded to the cloud server and encrypted with the private key. The existing file has been uploaded on the drive, and the file will be encrypted with the given private key regarding the file you submitted, you possess two distinct access permissions: Just read and write. After defining the file's access rights, you will notice that you have entered the following: The owner will record this consent for access and verify it. Before encrypting the file, Fig. 6 illustrates how the keys are input and checked. Information is uploaded to a cloud server [26]. The owner just has to submit the encrypted file to the Firebase cloud after it has been encrypted. General details about the file: Name, Size, Upload Date and Time, Verification Hash Value, Download, and Delete. When using cloud computing, the user develops an application to compare symmetric key algorithms on the cloud software environment provided by the supplier.

A programming-level environment with a well-defined set of APIs is provided to developers by the cloud-based software environment provider. For instance, the runtime environment and Google connection APIs are provided by Google Apps Engine. Runtime programmers are bundled and run on the cloud in the Google connection scenario. In the decryption methodology, Fig. 7 of this work describes how this form decrypts the file using the private key you have provided. The access list of users who have self-authenticated is also required for this form.

### 4.1.2 Retrieval Rate and Assessment of Epoch

A statement is considered "accurate" if it can be independently confirmed or confirmed using the available data. It is most frequently used to indicate "free from

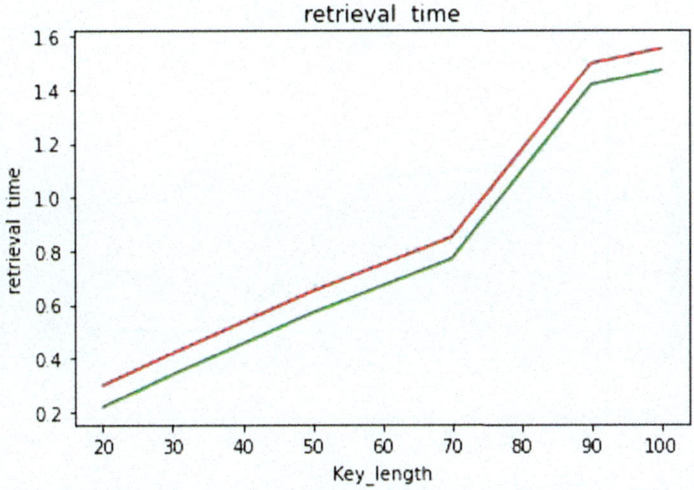


**Fig. 7** Graph of retrieval time for file processing

error" or "free from flaw" in well-defined circumstances. In this instance, the objective meaning of the statement is so obvious that neither it nor the supporting data are open for debate. This requirement must be met by each assertion in a sentence. Multiply the total number of classes by the quantity of wrongly classified classes [27]. Count the classes after that. In the context of cloud computing, "epochs count" is utilized. The date and time a computer uses to calculate system time is called "epoch." The tool displays charts for every metric, including retrieval time, loss rate, speed, throughput, etc., in the last stage. Based on retrieval, the findings are displayed in Fig. 6.

Figure 7 illustrates how using the approach lengthens the retrieval time from the starting position to the finishing point. Sending personal data in this day and age is quite likely to be secure. The accuracy of the last step is 98.99%.

This work has executed and achieved the graph of retrieval time for file processing. Furthermore, data loss and the number of epochs are given in Sect. 4.1.3.

### 4.1.3 Data Loss and Epoch

The number of file processing epochs and the count of losses are displayed in this section. "Loss" is the loss measure computed from the training data after each training session. This is a result of the optimization process's attempt to reduce training-process losses as much as feasible. After deducting the running loss from the number of batches, add the training loss result for each epoch to determine the loss for each training session [28]. An "epoch" is a single run through the training dataset. It is sometimes necessary to train a neural network across several iterations, or "epochs." As we try to lower the "scalar values" during the training phase. The representation

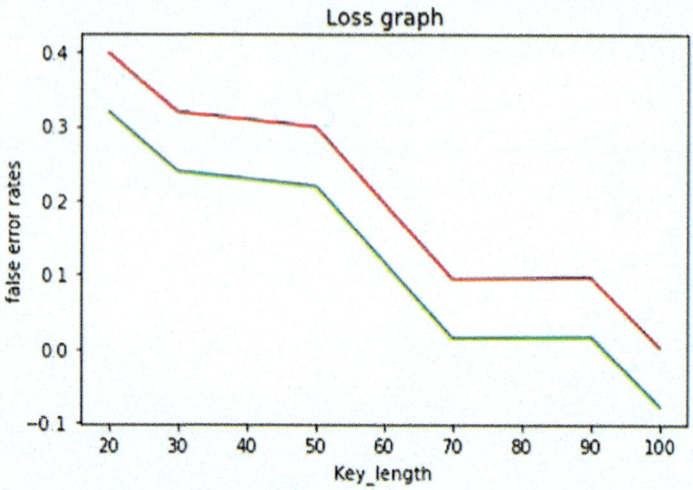


**Fig. 8** Representation of loss epoch security in cloud

**Table 1** Comparison of loss with epoch

| EPOCH | RSA | Blowfish | MD5 | Proposed: RSA-BL |
|---|---|---|---|---|
| 20 | 0.51 | 0.58 | 1.58 | 0.48 |
| 40 | 1.5 | 0.45 | 1.63 | 0.32 |
| 60 | 0.36 | 0.39 | 0.22 | 0.28 |
| 80 | 0.29 | 0.34 | 0.26 | 0.2 |
| 100 | 0.18 | 0.3 | 0.1 | 0.04 |

of Loss Epoch Security in Cloud are shown in Fig. 8. Furthermore, the comparisons of loss with epoch are given in Table 1 and Fig. 10.

This work was executed with 100 epochs and minimum loss has been calculated in the 100th epochs. In the case of 20, 40, and 60 epochs, the loss of value is calculated using the Proposed: RSA-BL algorithm. Applying this technique lowers the overall loss, as Fig. 9 illustrates. There is a very low loss percentage and a high likelihood of security while transmitting personal data during this time.

## 4.2 *Throughput Time of File Processing*

This encryption time is measured in the amount of time it takes to translate plain text into cipher text. Equation 4 displays the mathematical method for calculating the encryption time

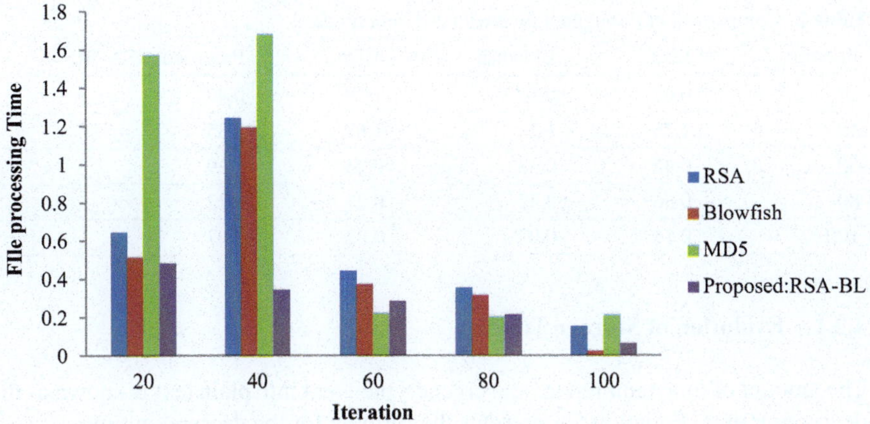


**Fig. 9** Comparison of throughput time

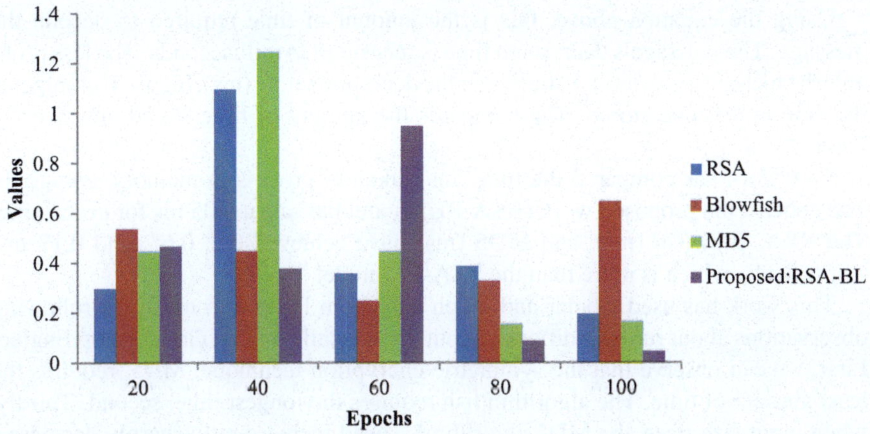


**Fig. 10** Comparisons of storage time

$$\mathrm{Er}_T = \mathrm{ET} - \mathrm{ST} \tag{4}$$

where $\mathrm{Er}_T$ stands for encryption time, ET stands for end time, and ST stands for starting time. The duration is expressed in milliseconds. By applying this formula, the hybrid approach that has been suggested reduces the throughput time. Figure 9 and Table 2 show the comparison of throughput time.

**Table 2** Comparison of throughput time with different model

| Iteration | RSA | Blowfish | MD5 | Proposed: RSA-BL |
|---|---|---|---|---|
| 20 | 0.65 | 0.52 | 1.58 | 0.49 |
| 40 | 1.25 | 1.2 | 1.69 | 0.35 |
| 60 | 0.45 | 0.38 | 0.23 | 0.29 |
| 80 | 0.36 | 0.32 | 0.21 | 0.22 |
| 100 | 0.16 | 0.03 | 0.22 | 0.07 |

#### 4.2.1 Evolution of Storage Time

The amount of time required to convert encrypted text into plain text is known as the decryption time. Equation 5 represents the formula for the decryption time.

$$ST = ET - ST \tag{5}$$

Using the equation above, this is the amount of time required to decode the message. The message's decryption time is measured in milliseconds. The less probable the method is to succeed, the longer the decryption time (ms). Figure 10 compares the amount of time stored. The compares the amount of time stored are given in Table 3.

This work has compared the time (ms) taken to process in-memory storage. In 100 epochs, the proposed work (RSA-BL) model has taken 0.05 ms for processing. The RSA, Blowfish (ms), and MD5 (ms) have achieved 0.2, 0.65, and 0.17 ms, respectively, which is more than the RSA-BL model.

This work has used tabular and graph data from Eclipse to make the following observations about running different input sizes locally and on Google App Engine: First, we can observe that the symmetric encryption technique, MD5, requires the least amount of time. The algorithm fish requires the longest time, second. Thirdly, when input size rises, the MD5 algorithm's speed increase ratio sharply decreases. Fourth, in terms of speed increase ratio, the RSA method outperforms the MD5 algorithm. Last but not least, RSA requires a smaller buffer size and requires less time to encrypt and decode data than md5 and Blowfish, which require larger buffer

**Table 3** Compares the amount of time stored

| *s* | RSA (ms) | Blowfish (ms) | MD5 (ms) | Proposed: RSA-BL (ms) |
|---|---|---|---|---|
| 20 | 0.3 | 0.54 | 0.45 | 0.47 |
| 40 | 1.1 | 0.45 | 1.25 | 0.38 |
| 60 | 0.36 | 0.25 | 0.45 | 0.95 |
| 80 | 0.39 | 0.33 | 0.16 | 0.09 |
| 100 | 0.2 | 0.65 | 0.17 | 0.05 |

sizes and take longer. Thus, when comparing multiple methods with varying buffer sizes, the results shown above are a tie; however, this work favors the MD5 approach.

## 5 Conclusion

Despite all of the benefits of cloud computing, there are still a lot of actual issues that need to be resolved, particularly concerning security. Cloud storage solutions will be utilized by both large and small businesses in the future if we can eradicate and figure out how to get around this security flaw. By using our method, the data is kept private. Only authorized users can view the data. The fact that the encryption requires two keys from two distinct locations means that even if an unauthorized user were to accidentally or purposefully access the data, they would be unable to decrypt it. Although everyone utilizes cloud computing differently, many people find it useful. Data migration is being fueled by cloud computing's flexibility. Businesses may access and store vast amounts of data in the cloud at any time. It becomes more difficult to keep data securely as more is kept on cloud servers.

Algorithms for data security include Blowfish, MD5, and RSA. The same key that is used for encryption and decryption in RSA is also used in symmetric key methods. It is necessary to strengthen these algorithms. Cloud computing effectively manages security risks. The best-performing encryption algorithms, according to this analysis, are Blowfish, MD5, and RSA. If data security is an issue, use MD5. It is advised to choose an adequate MD5 buffer size. Through the use of numerous auditing methodologies, further study will enhance the suggested algorithm for auditing medical cloud data and increase security.

## References

1. Kumar, G.S., Premalatha, K., Aravindhraj, N., Nivaashini, M., Karthiga, M.: Secured cryptosystem using blowfish and RSA algorithm for the data in the public cloud. Int. J. Recent Technol. Eng. (IJRTE) (2019). ISSN 2277-3878
2. Gitanjali, K.L.: A novel approach of sensitive data classification using convolution neural network and logistic regression. Int. J. Innov. Technol. Explor. Eng. **8**(8), 2883–2886 (2019)
3. Li, Y., Gai, K., Qiu, L., Qiu, M., Zhao, H.: Intelligent cryptography approach for secure distributed big data storage in cloud computing. Inf. Sci. **387**, 103–115 (2017)
4. Rani, P.K., Sathiya, S., Sureshkumar, S., Kumar, B.A.: Enhancing cloud security with hybrid encryption. In: 2022 International Conference on Applied Artificial Intelligence and Computing (ICAAIC), pp. 1445–1450. IEEE (2022)
5. Murad, S.H., Rahouma, K.H.: Hybrid cryptography for cloud security: methodologies and designs. In: Digital Transformation Technology: Proceedings of ITAF 2020, pp. 129–140. Springer, Singapore (2022)
6. Abroshan, H.: A hybrid encryption solution to improve cloud computing security using symmetric and asymmetric cryptography algorithms. Int. J. Adv. Comput. Sci. Appl. **12**(6), 31–37 (2021)

7. Orobosade, A., Aderonke, T., Boniface, A., Gabriel, A.J.: Cloud application security using hybrid encryption. Commun. Appl. Electron. **7**, 25–31 (2020); Design of an effective hybrid cryptographic technique for information security in cloud computing
8. Taha, A.A., Abdelminaam, E.D.D.S., Khalid, M., Hosny, K.: Enhancement the security of cloud computing using hybrid cryptography algorithms. Int. J. Adv. Comput. Technol **9**(3), 36–42 (2017)
9. Hagge, M., von Hoffen, M., Betzing, J., Becker, J.: Design and implementation of a toolkit for the aspect-based sentiment analysis of tweets, pp. 379–387 (2017)
10. Zhou, X., Tang, X.: Research and implementation of RSA algorithm for encryption and decryption. In: Proceedings of 2011 6th International Forum on Strategic Technology, vol. 2, pp. 1118–1121. IEEE (2011)
11. Chauhan, A., Gupta, J.: A novel technique of cloud security based on hybrid encryption by blowfish and MD5. In: 2017 4th International Conference on Signal Processing, Computing, and Control (ISPCC), pp. 349–355. IEEE (2017)
12. Sachdev, A., Bhansali, M.: Enhancing cloud computing security using AES algorithm. Int. J. Comput. Appl. **67**(9) (2013); Nicholas, K., Wilson, C., Kibe, A.: Enhancing trust in cloud computing using MD5 hashing algorithm and RSA encryption standard. Int. J. Sci. Eng. Res. **8**(3), 550–566 (2017)
13. Bangar, A., Shinde, S.: Study and comparison of cryptographic methods for cloud security. Int. J. Comput. Sci. Eng. Inf. Technol. Res. **4**(2), 205–213 (2014)
14. Gunasekaran, S., Lavanya, M.P.: A review on enhancing data security in cloud computing using RSA and AES algorithms. Int J Adv Eng Res **9**(4), 1–7 (2015)
15. Kartit, Z., Azougaghe, A., Idrissi, H.K., El Marraki, M., Hedabou, M., Belkasmi, M., Kartit, A.: Applying encryption algorithm for data security in cloud storage. In: Sabir, E., Medromi, H., Sadik, M. (eds.) Advances in Ubiquitous Networking, pp. 141–154. Springer, Singapore (2015)
16. Yugandhara, K., Ashwini, J., Pooja, P., Suchita, P., Pawar, J.S.: Secure data storage and forwarding in the cloud using AES and HMAC. Int. Res. J. Eng. Technol. **03**(02), 75–79 (2016)
17. Salem, M.Z., Sabbeh, S.F., EL-Shishtawy, T.: An efficient privacy-preserving public auditing mechanism for secure cloud storage. Int. J. Appl. Eng. Res. **12**(6), 1093–1101 (2017)
18. Cordova, R.S., Maata, R.L.R., Halibas, A.S., Al-Azawi, R.: Comparative analysis on the performance of selected security algorithms in cloud computing. In: International Conference on Electrical and Computing Technologies and Applications (ICECTA), pp. 1–4. IEEE
19. Jain, G., Sejwar, V.: Improving security by using various cryptographic techniques in cloud computing. In: 2017 International Conference on Intelligent Computing and Control Systems (ICICCS), pp. 23–28. IEEE (2017)
20. Singh, S.K., Manjhi, P.K., Tiwari, R.K.: Data security using RSA algorithm in cloud. Int. J. Adv. Res. Comput. Commun. Eng. **5**(8), 11–16 (2016)
21. Pannala, N.U., Nawarathna, C.P., Jayakody, J.T.K., Rupasinghe, L., Krishnadeva, K.: Supervised learning-based approach to aspect-based sentiment analysis. In: IEEE International Conference on Computer and Information Technology (CIT), pp. 662–666. IEEE (2016)
22. Cao, D., Ji, R., Lin, D., Li, S.: Visual sentiment topic model-based microblog image sentiment analysis. Multimed. Tools Appl. **75**, 8955–8968 (2016)
23. Mishra, A.M., Shahare, Y.R., Shukla, P.K., Husain, A., Singh, S.P., Alyami, S., Alghamdi, A., Ahanger, T.A.: Driving a key generation strategy with training-based optimization to provide safe and effective authentication using data sharing approach in IoT healthcare. Comput. Commun. **212**, 407–419. (2023)
24. Harnal, S., Sharma, G., Anupriya, Mishra, A.M., Bagga, D., Saini, N., Goley, P.K. and Anupam, K.: Bibliometric mapping of theme and trends of augmented reality in the field of education. J. Comput. Assisted Learn. **40**(2), 824–847 (2024)
25. Mishra, A.M., Kaur, P., Singh, M.P., Singh, S.P.: A self-supervised overlapped multiple weed and crop leaf segmentation approach under complex light condition. Multimed. Tools Appl., 1–26 (2024)

26. Zhu, Y., Ahn, G.J., Hu, H., Yau, S.S., An, H.G., Hu, C.J.: Dynamic audit services for outsourced storage in clouds. IEEE Trans. Serv. Comput. **6**(2), 227–238 (2013)
27. Erway, C.C., Küpçü, A., Papamanthou, C., Tamassia, R.: Dynamic provable data possession. ACM Trans. Inf. Syst. Secur. (TISSEC) **17**(4), 15–18 (2015)
28. Wang, Q., Wang, C., Ren, K., Lou, W., Li, J.: Enabling public auditability and data dynamics for storage security in cloud computing. IEEE Trans. Parallel Distrib. Syst. **22**(5), 847–859 (2011)

# Advancing Water Quality Prediction: Integrating Ensemble Learning and Explainable AI for Actionable Insights

**Kavya Preksha and Kuldeep Kumar Tiwari**

**Abstract** Water quality is fundamental for public health and also for environmental sustainability, thus imperative for maintaining a corporate monitoring system to deter contamination risks. In order to lessen environmental degradation and ensure safe drinking water for humans, it is necessary to assess the water quality at the right time. This study analyzes the capability of machine learning models in assessing water quality, concerning some ensemble methods like random forest and gradient boosting, which have already proven their edge over traditional classifiers, for instance, naive Bayes and SVM. The approach exploits different explainable AI (XAI) techniques in order to provide transparency around model decision-making and explain the importance of each feature. Through a thorough interrogation of feature importance, pH comes out on top, being the most significant factor in assessing water quality in general, thus highlighting its role in the overall assessment of water bodies. This study passes a note about the superior predictive strength of ensemble methods in achieving up to 100% accuracy and emphasizes the practical implications of treating pH as a highly determinative parameter in water quality management. By integrating XAI tools, this work champions broad support for the importance of the transparent, interpretable model for environmental monitoring, enhancing the basis for actionable insights to achieve more effective, prioritized interventions in water quality management.

**Keywords** Water quality · Machine learning · Ensemble methods · Random forest · Gradient boosting · Explainable AI (XAI) · SHAP · LIME · Feature importance · pH monitoring

K. Preksha (✉)
Chandigarh University, Mohali, Punjab, India
e-mail: kaavya23p@gmail.com

K. K. Tiwari
Department of Mathematics and Computing, Madhav Institute of Technology and Science—Deemed University, Gwalior, Madhya Pradesh, India

© The Author(s), under exclusive license to Springer Nature Singapore Pte Ltd. 2026

M. Pant et al. (eds.), *Real-World Applications of Machine Learning*, Mathematics for Sustainable Developments, https://doi.org/10.1007/978-981-95-7134-5_38

## 1 Introduction

Water quality is central to the very science of environmental health, influencing public health, biodiversity, and the economic health of a region. With accelerating urbanization and industrialization, the demand on water resources is ever-steepening, making water quality management and protection even more paramount. Poor water quality leads to serious risk for humankind, causative of a myriad of diseases and one of the major reasons for environmental degradation. Unsafe drinking water has been one of the major causes of mortality worldwide, especially in areas with poor access to clean water and proper sanitation systems. Hence, to stagger the public health impacts of poor drinking water, rapid and accurate assessment of water quality is absolutely essential to ensure the availability of safe water resources to both the current and the upcoming generations. Conventional evaluations of water quality mainly rely on laboratory testing: slow, tiresome, and unnatural for real-time monitoring. On the contrary, machine learning (ML) will bring a transformative power due to rapid, scalable, and cost-effective analysis of water quality data. ML models can analyze large and complex datasets with many environmental factors to find hidden patterns that may not be easily spotted by conventional techniques. Of all machine learning techniques, ensemble learning approaches, such as random forest and gradient boosting, have recently come into prominence for their ability to model nonlinear relationships and detect complex interactions between variables, which is very suitable in the case of a water quality study. Concurrently, explainable artificial intelligence (XAI) emerged as a pivotal tool to impart interpretability and transparency into machine learning models. Stemming from current explanatory techniques like SHapley Additive exPlanations (SHAP) and local interpretable model-agnostic explanations (LIME), they will assist relevant stakeholders in understanding the functional link between individual features and their subsequent prediction, thus making the decision-making process more explicable and trustworthy. This demands zeal especially in high-risk fields, such as water quality assessment, where model outputs need to be explicable so that their observational findings can lend support to decision-making and policy formulation. Longitudinal studies thus far suggest that ensemble models might take a turn predicting certain water qualities that too would concentrate on identifying vital features determining whence there could be such a potential. To this end, we want to feed XAI techniques into an exploration to change our understanding of how various factors used to determine water quality value in importance, especially concerning pH as a key determining factor, thereby providing our study results with practical insights into environmental management and water quality monitoring themselves. In particular, we focus on using feature importance methods like SHAP to assess and rank the contribution of different features in predicting water quality. This will help us identify the most influential variables and ensure that our models are not only accurate but also transparent and understandable. Our research journey, therefore, seeks to add another stone on stone-like management of water resources through an even more effective and transparent process.

## 2 Literature Review

Data mining the water quality using machine learning techniques has become very important in managing aquatic ecosystems and environmental sustainability due to its importance. A number of such studies have been done on different algorithms, datasets, and case studies that these approaches can be applied. Yogalakshmi and Mahalakshmi [2] demonstrated the efficiency of machine learning models in predicting water quality for Indian rivers, while Liu et al. [3] combined SVM with genetic algorithm optimization to predict aquaculture water quality, achieving enhanced accuracy, and efficiency. Ahmed et al. [4] showcased the accuracy of supervised learning techniques in this domain. Similarly, Arora and Rathi [5] found random forests to outperform decision trees significantly in predicting water quality metrics. Sharma and Mehra [6] effectively applied gradient boosting, achieving high accuracy in environmental data prediction, while Brown and Palmer [7] discussed the role of pH in aquatic ecosystem health, underscoring its importance in monitoring water quality. Zhang and Guo [8] provided a detailed review of machine learning techniques, identifying their strengths and limitations. Xie et al. [9] evaluated multiple machine learning models on real-world datasets, highlighting the importance of selecting appropriate models based on dataset characteristics and prediction objectives. Patel and Patel [10] employed a hybrid model that integrated machine learning with statistical methods, producing robust results for water quality monitoring. Kumar and Soni [11] reviewed recent advances in machine learning models and identified pH as a crucial predictor in many studies. In turn, Khan and Ali [12] perform a similar review of the use of supervised learning methods to monitor water quality. These comparative studies, as demonstrated by Cheng et al. [13], also demonstrated the importance of ensemble learning method further, as they showed the study of many machine learning models on real-world dataset. Zhang and Guo [8] and Li and Zhang [14] put their emphasis on standardized datasets and metrics for evaluation of models for reproducibility and generalizability. Suthar et al. noted the robustness of SVM to different data sets and Nguyen and Thanh [15].

Kalbar [16] and he demonstrated the effectiveness of gradient boosting in modeling river water quality. In [17], Srinivas and Sahoo developed a hybrid system with possible improvements in reliability by utilizing ensemble techniques. Previous studies that applied random forests and artificial neural networks to water quality modeling were done by Guler and Turan [18] and Chen and Zhang [19], who also discussed the suitability of deep learning for aquatic ecosystems upon noting the requirement of high-quality training data. Chauhan and Kumar [20] indicated that deep learning could learn complex and nonlinear relations of large datasets. Other research will be conducted for developing hybrid models that make use of the strength of diverse available algorithms [17] as well as for looking at the possibility of deep learning techniques [20]. And it can address the big raises around model transparency and decision-making processes [14]. In all, the adaptation of machine learning approaches to the prediction of water quality has greatly increased, providing strong tools to comprehend and manage water quality. Current challenges, along with the emerging technologies can be ad dressed, so that these models will improve in terms of accuracy and reliability in order to assist with sustainable water resource management.

# 3 Materials and Methods

## *3.1 Data Collection*

In this study, the dataset we use has 1362 rows and 15 columns which means this is data of water quality of various sampling stations during different periods of time, in different states of India. The data was obtained from the HydroShare platform, a scheme that allows access to the most complete sets of data for the environmental studies. The station code in the first record of the dataset works as a unique identifier for each water sampling station, and locations represents the location where the water sample is collected. Lat and Lon represents the sampling station and represents it in geographical locations. The columns include capital city, a binary indicator of capital city and state, an indicator of Indian state. This wider referable envelope spans across widely varying conditions of environment in India. These also include key parameters of water quality such as temperature, dissolved oxygen (D.O.), pH, conductivity, biochemical oxygen demand (B.O.D.), nitrate, fecal coliform, and total coliforms as an overall view of the physical and chemical components of the water. Class is the target variable (the water quality classification). The dataset combines data from different states to create a diverse and hence representative representation of India's different water quality conditions. Robust analysis can be performed with 1362 samples and is a good basis for machine learning models. With its variants of the features ranging from wide, it is possible to explore the factors affecting the water quality in detail and put an importance of parameters like pH, D.O. and conductivity, those need to be considered while finalizing the water quality.

## *3.2 Data Preprocessing*

Data preprocessing ensures the dataset is clean and suitable for machine learning. In this study, several steps were taken to address missing values, outliers, and feature scaling.

**Handling Missing Values** To address the problem of missing values, numerical columns were imputed with the median, since the median is more stable against outliers than the mean. There was no value missing in the categorical features thus there was no need to fill them.

**Outliers Detection and Handling** Outliers were detected using Z-scores and IQR methods. Features like pH, temperature, and conductivity were checked for extreme values outside natural ranges and were either removed or capped accordingly.

**Feature Scaling** To standardize the dataset, numerical features were scaled using standardization (mean = 0, standard deviation = 1), ensuring algorithms like SVM and naive Bayes performed optimally. These preprocessing steps ensured the dataset was cleaned and normalized, leading to more accurate and efficient model training.

## 3.3 *Feature Selection*

A subset of features was selected based on their correlation with the target variable and domain knowledge. The parameters "temperature," "pH," "D.O.," and "conductivity' were identified as the most significant predictors of water quality.

## 3.4 *Machine Learning Models*

This study initially applied naive Bayes and support vector machines (SVM). However, these models demonstrated limited accuracy, with SVM achieving 81.1% and naive Bayes achieving 25.1%. To improve prediction performance, ensemble learning methods—random forest and gradient boosting—were implemented. These methods are known for their high accuracy, robustness, and ability to capture complex interactions between features.

## 3.5 *Model Evaluation Metrics*

The models were evaluated using accuracy, precision, recall, and F1-score to provide a comprehensive assessment of their predictive performance. Confusion matrices were also analyzed to identify misclassification patterns.

**Confusion Matrix for Naive Bayes and SVM** The confusion matrices for naive Bayes and support vector machines (SVM) are provided below to illustrate the classification performance of these models:

1. Confusion Matrix of (SVM) and Naïve Bayes in (Figs. 1 and 2):
2. Confusion Matrix of Random Forest and Gradient Boosting in (Figs. 3 and 4):

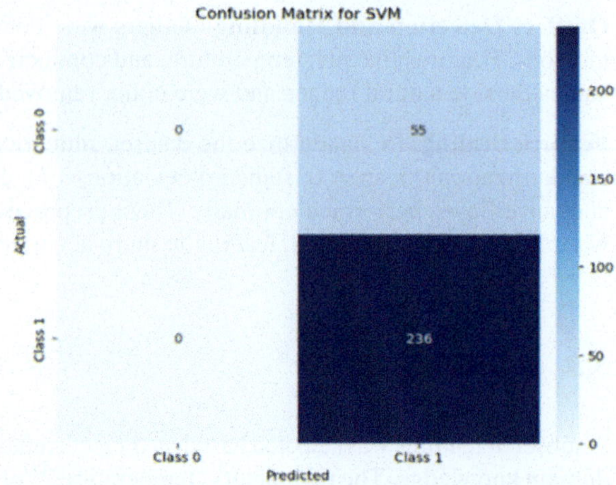


**Fig. 1** Confusion matrix of (SVM)

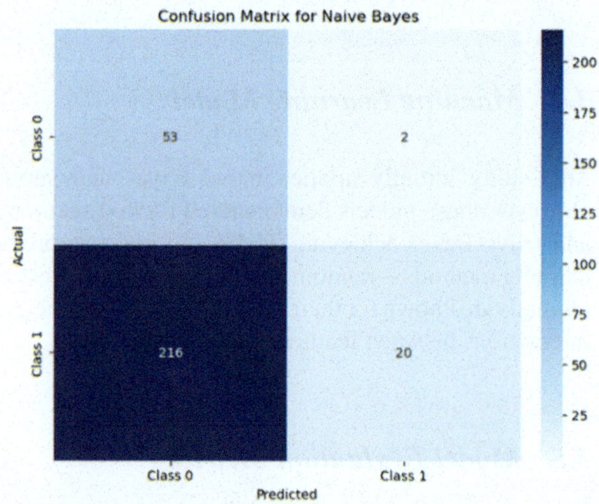


**Fig. 2** Confusion matrix of naive Bayes

## *3.6 Feature Importance for Random Forest and Gradient Boosting*

The feature importance analysis for random forest and gradient boosting was conducted to identify the most influential parameters in predicting water quality are in Figs. 5 and 6.

1. Feature Importance for Random Forest:

2. Feature Importance for Gradient Boosting:

**Feature Importance Summary** Table 1 summarizes the importance of key features for both random forest and gradient boosting models based on feature importance.

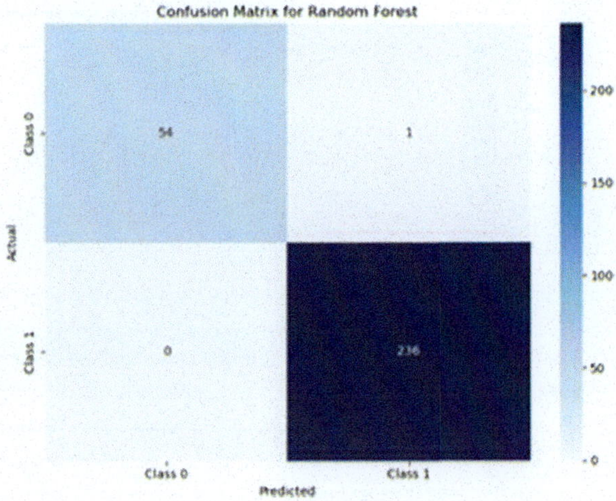


**Fig. 3** Confusion matrix of random forest

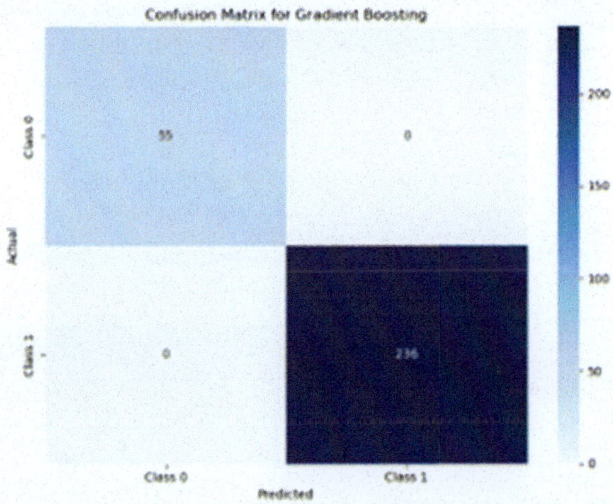


**Fig. 4** Confusion matrix of gradient boosting

## *3.7 Explainable AI (XAI) Techniques*

In this study, we employed feature importance analysis because it provides a balance between interpretability and computational efficiency. Unlike SHAP, which can be resource-intensive, or LIME, which is better suited for case-specific explanations, feature importance techniques effectively highlight critical parameters such as pH and dissolved oxygen. This ensures targeted interventions and practical applicability in real-time water monitoring systems without overburdening computational resources. By focusing on feature-level insights, this method aligns well with the objectives of water quality prediction, where identifying key contributors is paramount for actionable results. Feature importance was employed to elucidate the influence of

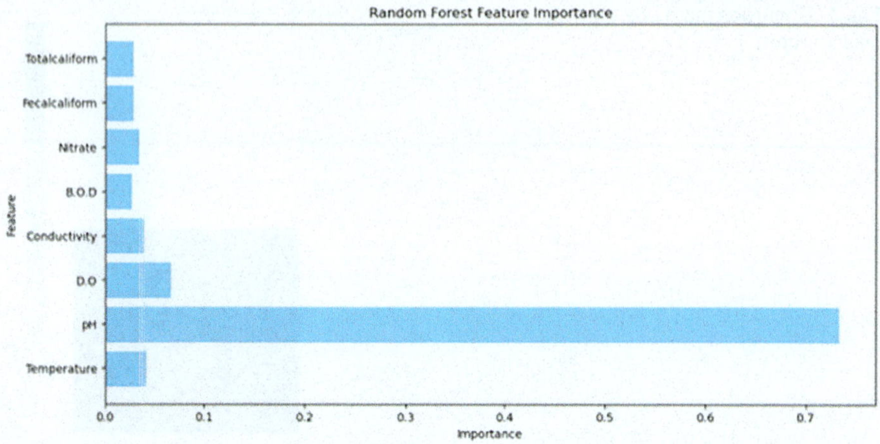


**Fig. 5** Feature importance for random forest

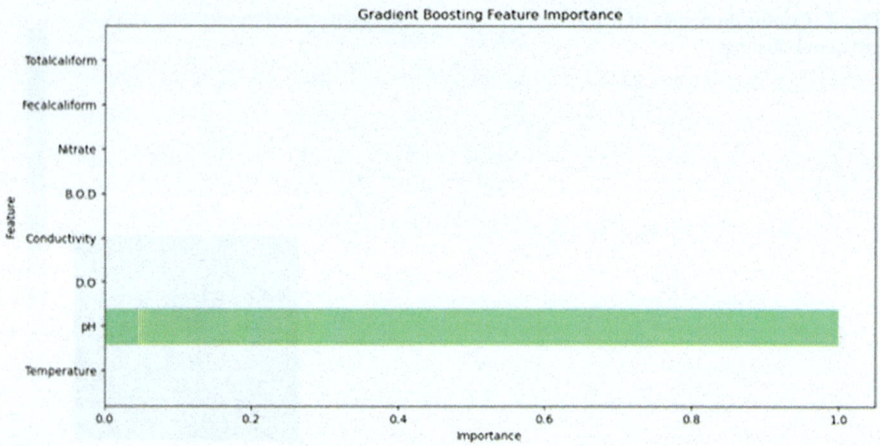


**Fig. 6** Feature importance for gradient boosting

**Table 1** Feature importance comparison

| S. No. | Feature name | Random forest importance | Gradient boosting importance |
|---|---|---|---|
| 1 | pH | 0.733821 | 1.000000e + 00 |
| 2 | D.O | 0.066870 | 4.961818e-17 |
| 3 | Temperature | 0.042061 | 0.000000e + 00 |
| 4 | Conductivity | 0.039001 | 1.586138e-16 |
| 5 | Nitrate | 0.034651 | 3.260353e-16 |
| 6 | Total coliform | 0.028815 | 6.432737e-17 |
| 7 | Fecal coliform | 0.028056 | -4.667688e-17 |
| 8 | B.O.D | 0.026726 | 4.815510e-15 |

individual features on model predictions. It is pivotal in understanding the role of parameters like pH, D.O., and conductivity in determining water quality.

### *3.8 Feature Importance Analysis*

The analysis divulged that pH was the most important water quality predictor by virtue of random forest and gradient boosting model feature importance. The prominent place of pH with respect to model performance and real-life significance justifies its importance. The usual physicochemical parameter, pH, affects the solubility, and bioavailability of critical nutrients and heavy metals which impact aquatic ecosystems and human health. Changes in pH can interfere with the physiology of freshwater animals and thus impact their growth, reproduction, and survival. For example, pH changes may represent a different level of enzyme activity, leading to a reduction in biodiversity and stability in the ecosystem. Other water treatment methods, such as disinfection, coagulation, and flocculation, depend on the acidity or alkalinity of water for their efficiency. At conditions far from neutrality, persistence at large percent of organisms yields one or several water qualities unfit for treatment and results in further expenses to restore water quality into a treatable composition.

### *3.9 Role of Other Parameters*

However, while pH emerged as the most significant feature, other parameters like D.O. and conductivity also contributed meaningfully to the models' predictions. Ignoring these parameters could compromise the holistic assessment of water quality. For example, D.O. is essential for sustaining aquatic life, and conductivity reflects the presence of dissolved salts, which influence water usability.

## 4 Results and Discussion

The comparative analysis of different machine learning models for water quality prediction reveals significant variations in their predictive accuracies. The naive Bayes model achieved an accuracy of 25.1%, highlighting its limitations in handling complex and nonlinear datasets. Support vector machines (SVM), therefore, are able to develop a very good model with an accuracy of 81.1%, which clearly indicates their robustness to model water quality parameters. Both the ensemble models random forest and gradient Boosting performed wonderfully, in that they received an accuracy of 99.6% and 100%, respectively. As high accuracy of these ensemble models demonstrates, the advantage of these ensemble models is their ability to capture subtle interaction between the variables and to treat heterogeneous data. In particular, gradient boosting proved to be most accurate model, indicating the viability of it as a tool for predicting water quality tasks. These results agree with the already available literature on the role of ensemble technique dominance in predictive modeling, and the reason for this is that they have underlined the necessity of the use of the most modern machine learning models of such as gradient boosting when we expect the precision and reliability to be high. Such performance of ensemble methods underscores their use rather than traditional models such as naive Bayes especially when dealing with complex datasets, when critical decision-making processes are involved. Future research can integrate these high performing models with explainable AI techniques to further improve this transparency and interpretability so such models are more applicable to the real-world water quality management.

### *4.1 Comparison of Model Performance (SVM vs. Naive Bayes vs. Random Forest vs. Gradient Boosting)*

Through accuracy, this following comparison graph in Fig. 7 visualize naive Bayes, SVM, random forest and gradient boosting performance. Then to see the performance of naive Bayes, SVM, random forest and gradient boosting on accuracy, Recall, precision and F1 score, the following Table 2 is presented. This provides a measure of the performance differences, and explains the switch to ensemble models, as shown in Table 2.

They are highly used in the field of ensemble methods like gradient boosting and random forest. In particular, for tasks, such as determining water quality they are

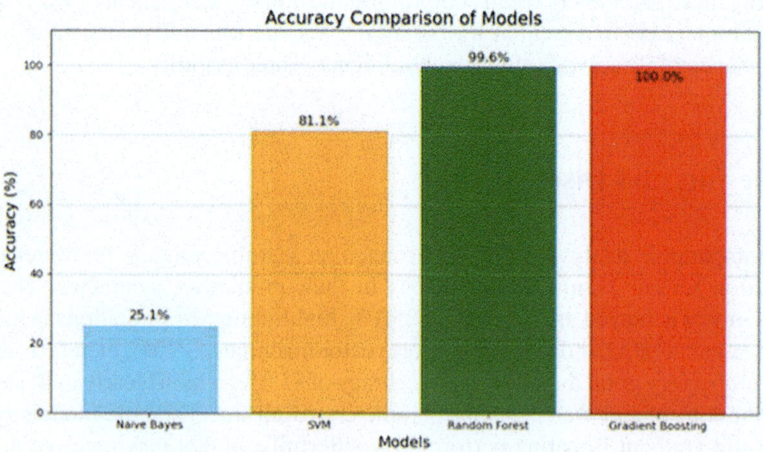


**Fig. 7** Comparison of model performance

**Table 2** Model performance metrics

| Model | Accuracy | Precision | Recall | F1 Score |
|---|---|---|---|---|
| Naïve Bayes | 0.251 | 0.775 | 0.251 | 0.188 |
| SVM | 0.811 | 0.658 | 0.811 | 0.726 |
| Random forest | 0.996 | 0.996 | 0.996 | 0.996 |
| Gradient boosting | 1.000 | 1.000 | 1.000 | 1.000 |

effective for predictive modeling. However, ensemble methods though, known to be highly accurate in predictive accuracy, present challenges:

1. **Computational Overhead**

It requires quite some computational resources especially with large datasets, and there are training and inference. For example, gradient boosting performs iterative updates for each weak learner, and it can result in tremendous runtime when working over large scale problems.

2. **Memory Usage**

These models have high memory demands that limit their scalability in the resource otherwise constrained environments. In general, deploying ensemble models onto edge devices such as IoT nodes with constrained RAM is infeasible.

3. **Hyperparameter Tuning**

To achieve optimal performance of a model, which can take weeks of extensive experimentation, means long development time. This challenge is even compounded with techniques that come in the form of grid search or random search when trying to optimize parameters such as the number of trees or learning rates.

Distributed computing support for addressing these challenges as well as the ability to promote their practical deployment may come from tackling these challenges through distributed computing, model pruning and efficient hyperparameter optimization. For instance, a model that is light-weight can be created with the data sparsity or pruning techniques, which can help reduce memory consumption while maintaining its accuracy.

## 5 Applications in Real Life

The results from this study contribute to the practical issues of the management of water quality and environmental monitoring. As pH has proven to be by far the most important feature, monitoring it would have a number of benefits given:

1. **Optimize Water Treatment Processes:** The ability to coagulate and disinfect becomes more efficient when there is careful pH management in water treatment. This results in optimum range of pH for best contaminant removal and disinfection effectiveness inhibiting harmful by products before their formation.

2. **Implement Targeted Interventions for Contamination Control:** pH deviations can signal contamination, such as industrial discharge or agricultural runoff. By detecting these shifts, interventions can be implemented to prevent public health risks and environmental damage, ensuring timely corrective measures.
3. **Becoming Cost-Effective Monitoring Systems:** Monitoring operations can be made economical by directing adequate and diligent focus onto critical parameters such as pH, D.O., and conductivity. By focusing on the parameter pH as the major component in water quality monitoring systems, efficient systems can be established, capable of accounting for other relevant parameters.
4. **Strengthening Public Health and Ecosystem Protection:** Monitoring the pH in real-time will give adequate protection to public health and save water for drinking. It can also protect an aquatic ecosystem against the deleterious effects caused by sudden shifts in pH into the aquatic environment that affect organisms.
5. **Policy Making and Regulation:** The results can inform policies that mandate pH monitoring as a key component of water quality assessments. Establishing pH thresholds can help governments enforce standards and ensure safe water quality.
6. **Facilitate Early Warning Systems:** Significant shifts in pH can indicate environmental hazards like harmful algal blooms. Integrating pH data into early warning systems can help authorities mitigate these risks before they escalate.
7. **Ecosystem Preservation:** Early detection of harmful changes prevents ecological damage and supports biodiversity conservation. Automated monitoring systems can alert authorities about rising nutrient levels that trigger algal blooms, protecting aquatic life.

By focusing on pH and its real-world significance, stakeholders can enhance water quality management, reduce monitoring costs, and improve sustainability in both public health and environmental protection.

## 6 Conclusion

This study seeks to prove the efficiency of ensemble learning models such as random forest and gradient boosting, which secure 99.66% and 100% accuracy rates of detecting water quality, respectively. Through the scope of feature importance by explainable AI (XAI) techniques, it is unmistakably clear that pH is a prime determiner in ascertaining water quality. The heavy weight of pH as a subject for determining a chemical property of water not only influences the solubility of various compounds but also affects biological availability of essential nutrients. In addition, differences in pH can produce far-reaching effects on aquatic ecosystems as well as the effectiveness of water treatment processes, therefore making this the most significant indicator for monitoring water bodies. Nonetheless, although pH has emerged as the most relevant parameter, the study warns against relying heavily on one parameter for judging the quality of water. The factors controlling water quality are intermingled, and by not giving importance to other parameters such as D.O., conductivity, nitrate, etc. may lead to incomplete or erroneous water quality assessment. Accordingly, while it is impossible to avoid pH as such, a holistic approach should be taken in order to comprehend water quality behaviors using a number of parameters. The findings highlight the strategic capacity for machine learning and XAI potential to transform water quality monitoring into an active enterprise through the provision of real-time, transparent, and actionable insights for environmental protection and public health. Future research should consider hybrid models that combine multiple machine learning algorithms and extend the dataset with additional types of water sources. Incorporation of IoT devices into AI-based systems promises powerful tools for real-time monitoring and predictive maintenance to drive both accurate and timely assessments of water quality. We can hence protect and sustain water resources for the generations to come based on this improvement.

## 7 Future Scope

The findings of the producing study are applicable in various practical ways, especially in the management of water quality and environmental monitoring. Since pH proved to be the most important feature in this analysis, pH monitoring will come with several advantages in the efficient running of water treatment processes, the identification of contaminants, and the protection of aquatic ecosystems. Continuous monitoring of pH can aid in providing safe water for public health and preventing harmful alterations to aquatic organisms. In addition to this, pH monitoring produces suitable, low-cost systems, provides the basis for setting clear thresholds for water quality standards, and supplies policymakers with information so that they are able to conduct water quality management. Future studies should look for the integration of advanced machine learning with real-time monitoring systems and their predictability and efficiency. Such systems could apply IoT-based sensors and explainable AI frameworks to enabling accurate data collection and actionable insight. Further research should be made in other place-based studies, hence making the models robust and applicable globally toward the real goal of sustainable water quality management. This IoT and AI frameworks can offer transformative potential for water quality monitoring:

1. **Real-Time Data Collection:** IoT sensors enable continuous monitoring of key water quality parameters, such as pH, D.O., and conductivity. Advanced sensors, including optical and electrochemical probes, provide high precision and low detection limits, ideal for monitoring subtle changes in water quality.
2. **Predictive Maintenance:** Real-time insights facilitate early detection of contamination and system malfunctions, reducing response time. For instance, anomalies detected in pH levels can trigger automated alerts for inspection, preventing potential ecological disasters.

3. **Scalability:** IoT networks can be deployed across diverse geographic regions, ensuring comprehensive water quality coverage. Technologies such as LoRaWAN and NB-IoT enable these networks to transmit data efficiently over long distances. Another challenge is to reliably integrate data from heterogeneous sensors and achieve reliable communication. Case studies, such as municipal water monitoring in Singapore, demonstrate the effectiveness of IoT-based systems in managing urban water networks.

## References

1. Geetha Jenifel, M.: Water quality dataset for water quality prediction tool, HydroShare (2023).https://doi.org/10.4211/hs.4ab43e1b507b496b9b42749701daed5c
2. Yogalakshmi, S., Mahalakshmi, A.: Efficient water quality prediction for Indian rivers using machine learning. Asian J. Appl. Sci. Technol. **5**(1), 100–109. https://doi.org/10.38177/ajast.2021.5111
3. Liu, S., Taia, H., Ding, Q., Li, D., Xu, L., Wei, Y.: A hybrid approach of support vector regression with genetic algorithm optimization for aquaculture water quality prediction. J. Elsevier (2011). https://doi.org/10.1016/j.joes.2011.08.011
4. Ahmed, U., Mumtaz, R., Anwar, H., Shah, A. A., Irfan, R., García-Nieto, J.: Efficient water quality prediction using supervised machine learning. J. Water (2019)
5. Arora, S., Rathi, P.: Predictive analysis of water quality using decision tree and random forest algorithms. Int. J. Environ. Sci. **12**(3), 123–132 (2020)
6. Sharma, R., Mehra, P.: Application of gradient boosting in environmental data prediction: a case study. Data-Driven Environ. Stud. **8**(4), 145–159 (2022)
7. Brown, C., Palmer, T.: The role of pH in aquatic ecosystem health and water quality monitoring. Environ. Monit. J. **19**(2), 98–105 (2021)
8. Zhang, L., Guo, X.: Predictive modeling of water quality using machine learning techniques: a comprehensive review. J. Environ. Manage. **254**, 109741 (2020). https://doi.org/10.1016/j.jenvman.2019.109741
9. Xie, L., Li, J., Li, X.: A comparison of machine learning techniques for water quality prediction: applications to a river basin in China. Environ. Monit. Assess. **192**(5), 1–14 (2020). https://doi.org/10.1007/s10661-020-8104-1
10. Patel, S., Patel, S.: Water quality monitoring using machine learning: a case study. J. Environ. Eng. **144**(4), 04018027 (2018). https://doi.org/10.1061/(ASCE)EE.1943-7870.0001374
11. Kumar, A., Soni, P.: Machine learning models for water quality prediction: a review of recent advances. Environ. Sci. Pollut. Res. **29**(1), 59–78 (2022). https://doi.org/10.1007/s11356-021-14677-w
12. Khan, A., Ali, S.: Machine learning approaches for water quality prediction: a comprehensive review. Sci. Total. Environ. **747**, 141367 (2021). https://doi.org/10.1016/j.scitotenv.2020.141367
13. Cheng, Q., Wu, Z., Wang, Z.: Comparative study of machine learning models for water quality prediction: application to river water quality in China. Environ. Monit. Assess. **191**(5), 329 (2019). https://doi.org/10.1007/s10661-019-7524-1
14. Li, Z., Zhang, L.: A review of artificial intelligence techniques in water quality prediction and management. Environ. Sci. Pollut. Res. **29**(2), 487–503 (2022). https://doi.org/10.1007/s11356-021-15253-7
15. Nguyen, H., Thanh, T.: Application of support vector machine for water quality prediction. J. Environ. Sci. Technol. **14**(8), 669–680 (2020). https://doi.org/10.1007/s13762-020-02931-3

16. Suthar, S., Kalbar, P.: Water quality prediction using machine learning and statistical methods: a case study of river water quality monitoring. Ecotoxicol. Environ. Saf. **174**, 94–101 (2019). https://doi.org/10.1016/j.ecoenv.2019.02.013
17. Srinivas, K., Sahoo, P.: A hybrid water quality prediction system based on machine learning and ensemble techniques. Environ Model Softw. **104**, 135–146 (2018). https://doi.org/10.1016/j.envsoft.2018.03.014
18. Chen, X., Zhang, L.: Machine learning approaches for water quality prediction in aquatic ecosystems. Environ. Pollut. **277**, 116798 (2021). https://doi.org/10.1016/j.envpol.2021.116798
19. Guler, M., Turan, M.: Predictive modeling of water quality using artificial neural networks and random forest. Environ. Eng. Sci. **37**(2), 121–133 (2020). https://doi.org/10.1089/ees.2019.0189
20. Chauhan, D., Kumar, R.: Deep learning for water quality prediction and monitoring: a review. Environ. Technol. Innov. (2022)

# Predicting Cryptocurrency Prices with Sentiment Analysis and Deep Learning Models

**Shreya Krishna, Nitesh Kumar Sah, and J Umamageswaran**

**Abstract** Predicting the price of cryptocurrencies has been a struggle as a result of their high volatility and endless distractions such as social media, market movements, or technical analysis. This document is proposing a novel quite multi-tiered framework combining long-term and short-term market dynamics in a modern age of advanced natural language processing (NLP) and deep learning techniques, where in the feature preparation phase uses a pre-trained financial text model, the FinBERT, to extract sentiment features from platforms like Twitter and Telegram. Hybrid deep learning models that adopt transformer-based architectures efficiently capture sequential dependencies and market fluctuations. It contains the dynamic attention-based weighting mechanism, such that, based on underlying risks with respect to market conditions, adaptively prioritizes a feature. With the utilization of the above features, the short-term and long-term prediction qualities of cryptocurrency prices are boosted, providing a comprehensive understanding in as much as cryptocurrency market behaviors are concerned. Experimental results have shown strong improvements over the baseline models, thus clearly demonstrating the capacities of combining both sentiment analysis and deep learning. This framework thereby would transit traders and investors to more robust forecast things via cutting-edge NLP and deep learning technologies to earn worthwhile market insights.

**Keywords** Cryptocurrency price prediction · Social media sentiment analysis · Financial text analysis · FinBERT · Transformer-based models

Supported by Amrita Vishwa Vidyapeetham, Chennai.

S. Krishna (✉) · N. K. Sah · J. Umamageswaran
Department of Computer Science and Engineering Amrita School of Computing, Amrita Vishwa Vidyapeetham, Chennai, India
e-mail: shreya7205krishna@gmail.com

N. K. Sah
e-mail: niteshshah011@gmail.com

J. Umamageswaran
e-mail: j_umamageswaran@ch.amrita.edu

© The Author(s), under exclusive license to Springer Nature Singapore Pte Ltd. 2026 
M. Pant et al. (eds.), *Real-World Applications of Machine Learning*, Mathematics for Sustainable Developments, https://doi.org/10.1007/978-981-95-7134-5_39

# 1 Introduction

Cryptocurrencies are emerging as a revolutionary innovation in the financial sector, offering decentralized, transparent, and secure alternatives to traditional fiat currency. Built on blockchain technology, they provide an immutable, distributed ledger system that ensures transparency and reduces fraud. These features have attracted investors, technologists, and policymakers. By eliminating central intermediaries, cryptocurrencies contribute to financial inclusion, allowing individuals in underserved regions to join the global economy. However, due to their extreme volatility influenced by market sentiment, regulatory developments, speculative trading, and technological innovation, accurately predicting cryptocurrency price movements remains challenging for investors and traders [1].

Cryptocurrency markets differ from traditional financial markets, heavily relying on market sentiment rather than macroeconomic indicators or geopolitical events. Social media platforms such as Twitter, Telegram, and Reddit play a significant role, as opinions and discussions often precede major price moves. For instance, Jagini et al. [2] demonstrated that sentiment analysis of tweets could predict Bitcoin price trends. Similarly, Nagula and Alexakis [3] proposed a hybrid machine learning model combining sentiment analysis with technical indicators to enhance short-term forecasting accuracy. These findings underscore the importance of incorporating sentiment analysis into models for cryptocurrency price prediction.

Traditional forecasting methods, including statistical models and basic machine learning techniques, struggle to address the dynamic nature of cryptocurrency markets. They fail to capture the interplay between market data, sentiment-driven factors, and external influences. Advanced ML and NLP techniques offer solutions to these challenges. Transformer-based deep learning models, known for capturing long-range dependencies in sequential data, have significantly improved time-series forecasting. Singh and Bhat [4] successfully applied transformers to Ethereum price prediction, achieving notable accuracy improvements. Tanwar and Kumar [5] explored hybrid models integrating transformers with LSTMs, yielding superior performance in predictive tasks.

Advanced NLP techniques, such as FinBERT, further enhance sentiment-driven predictions. FinBERT, a pre-trained language model designed for financial text, captures subtle sentiment features and market signals. Araci [6] demonstrated FinBERT's effectiveness in financial sentiment analysis, while Jahanbin and Zare Chahooki [7] combined FinBERT with transformers for cryptocurrency price prediction. These advancements highlight the critical role of sentiment analysis and deep learning in addressing the complexities of cryptocurrency markets.

Based on these developments, this paper suggests an all-encompassing framework that combines sentiment analysis and transformer-based models to identify short-term price variations and long-term market trends. The model uses dynamic weighting mechanisms and attention-based approaches to rank relevant features in an adaptive manner. This solution seeks to improve prediction accuracy and offer actionable information for investors and traders. Through the integration of sentiment

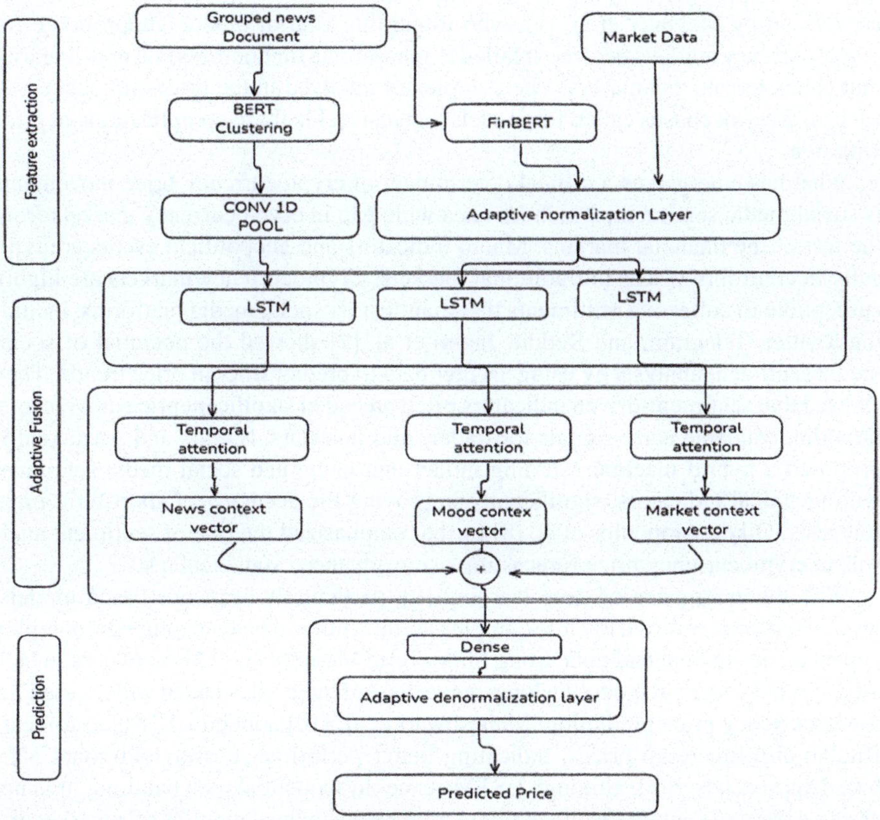


**Fig. 1** Illustration of the overall architecture of the model

analysis and time-series forecasting, the research contributes to the development of reliable cryptocurrency price prediction models [7].

Figure 1 shows the structure of the hybrid transformer—FinBERT-based framework [7].

## 2 Related Work

With recent growing importance of this field and the extreme market volatility, predictions regarding the future price of a cryptocurrency have emerged as one of the prime interesting fields of current research. These predictions had to begin, however with quite old-school, conventional approaches related to ML technology. Most frequently used was simple time-series analysis along with classic regression models. For instance, the regression-based approach of predicting the price of Bitcoin was carried out by McNally et al. [1], who found this kind of model inappropriate for cryptocurrency markets because it failed to incorporate randomness and non-linearity that characterizes this market. The complexity involved in the prices influenced by such an array of causes called for models that can tackle these complications in price variation.

What has emerged as a critical determinant of cryptocurrency price movements is social media sentiment, which leaves a wide gap in cryptocurrency markets from the rest of the financial markets. Macro indicators and geopolitical events are typically in control of prices in traditional markets. Cryptocurrency markets are highly susceptible to collective sentiments that manifest on social media platforms, including Twitter, Telegram, and Reddit. Jagini et al. [2] showed the potential of social media sentiment analysis by using Twitter data to predict Bitcoin price trends. They showed that sentiment-driven indicators often preceded significant price movements, providing valuable early signals for traders and investors. Nagula and Alexakis [3] proposed a hybrid machine learning model that integrated social media sentiment with technical indicators, significantly improving the accuracy of short-term price forecasts. Oikonomopoulos et al. [8] further emphasized the role of sentiment analysis in cryptocurrency price forecasting using advanced statistical tests.

With the emergence of deep learning [9], particularly sequence-based models, researchers have resorted to more complex architectures to better capture the complex dependencies in financial data. Long Short-Term Memory (LSTM) networks, which are especially adept at modeling long-range dependencies, have been widely used for cryptocurrency price prediction. Maheshwari et al. [10] applied LSTMs to forecast Bitcoin and Ethereum prices, indicating better performance over traditional ML-based approaches. Still, although LSTMs excel in sequential data handling, it is not easy for them to capture the latent interaction between the dynamics of a market and signals of sentiment. Thus, it is inevitable that hybrid approaches combining more than one modality are implemented.

The transformer-based architecture revolutionized the space of time-series forecasting. Transformers, better capable of sequential and contextual modeling, were found to have promise in applications to finance. Singh and Bhat [4] deployed transformers for price prediction on the Ethereum network to obtain significantly improved accuracy compared with LSTMs and other established approaches. The extension of that approach is hybrid models that merge transformers with LSTMs while yielding better results for both short and long-term tasks Tanwar and Kumar [5]. These studies highlight the opportunity to use multi-model approaches in addressing the special challenges of cryptocurrency markets.

Advanced NLP techniques, especially designed for financial applications, have also been transformative. Pre-trained financial text language models like FinBERT have been particularly effective in extracting nuanced sentiment features from financial texts. Araci [6] demonstrated that FinBERT outperformed general-purpose sentiment analysis tools in capturing subtle market signals. Jahanbin and Zare Chahooki [7] also integrated FinBERT with transformer architectures to come up with a hybrid framework capable of delivering accurate predictions in sentiment-driven and market-driven scenarios. Such innovations underscore the increasing need for using domain-specific NLP [11] models for cryptocurrency price prediction.

Dynamic feature prioritization mechanisms have lately been in the focus of research as a means to adapt to changing market conditions. Tanwar and Kumar [5] proposed an attention-based model that dynamically weights features based on relevance to the dynamics of the current market, hence making models capture both short-term price volatility and long-term trends. The dynamic weighting mechanism has been crucial for addressing the very volatile and sentiment-driven nature of cryptocurrency markets.

In summary, the integration of sentiment analysis, advanced NLP techniques such as FinBERT, and transformer-based architectures represents the cutting edge in cryptocurrency price prediction. These methodologies have significantly improved prediction accuracy by addressing the challenges of market volatility and the complexity of diverse data sources. By combining sentiment-driven insights with powerful deep learning frameworks, recent studies have laid the groundwork for more reliable and actionable cryptocurrency price forecasting models.

## 3 Proposed Architecture

This section introduces the multi-phased architecture proposed for predicting cryptocurrency prices. The framework integrates advanced natural language processing (NLP) and deep learning techniques to address the complexity of market data and sentiment analysis. The design is divided into several key phases, which are briefly described as follows:

### 3.1 Data Collection

The data collection phase gathers detailed cryptocurrency market data, including historical prices, trade volumes, and volatility, along with real-time social media data from platforms like Twitter and Telegram [12]. Sentiment-rich data is collected using APIs from these platforms to ensure the availability of both quantitative and qualitative features.

### 3.2 Feature Extraction

In this phase, market data features are integrated with sentiment scores derived from FinBERT, a pre-trained financial text language model. Sentiment analysis computes scores, denoted as:

$$S_t = f(T_t)$$

where $S_t$ represents the sentiment score at time $t$ and $T_t$ corresponds to the textual data from social media at the same time. Historical prices, trade volumes, and sentiment scores are preprocessed and normalized to ensure compatibility with the deep learning models.

### 3.3 Transformer-Based Modeling

Transformer-based architectures are employed to capture sequential dependencies and long-term patterns in the data. The input to the transformer model includes both market data ($M_t$) and sentiment features ($S_t$), enabling the model to learn complex relationships. The overall prediction model can be expressed as:

$$P_{t+1} = g(M_t, S_t)$$

where $P_{t+1}$ is the predicted price at time $t + 1$ and $g$ represents the prediction function learned by the transformer model.

### 3.4 Fusion and Dynamic Weighting

The fusion phase employs an attention mechanism to dynamically weight features based on their relevance to current market conditions. Attention weights ($w_i$) are computed as:

$$w_i = \frac{\exp(e_i)}{\sum_j \exp(e_j)}$$

where $e_i$ is the importance score for feature $i$. This mechanism ensures that the model adapts to evolving market trends by prioritizing the most relevant features.

### 3.5 Prediction

The final step involves producing both short-term and long-term forecasts from the trained model. The model generates point estimates for near-future price swings and also provides indications of longer-term trends.

### *3.6 Key Components and Functions*

The architecture consists of several components, as summarized in Table 1 [1, 11–13].

The data flow and processing pipeline are adapted from the methodology described in Fig. 2 [6].

The proposed architecture leverages the strengths of advanced NLP techniques, such as FinBERT, and transformer-based models to deliver a comprehensive and

**Table 1** Key components and their respective functions in the proposed architecture

| Component | Description | Function |
|---|---|---|
| Data collection | Gathers historical and real-time market data and sentiment data from social media [13] | Ensures the availability of up-to-date, diverse data for comprehensive analysis [13] |
| Feature extraction | Extracts quantitative market features and sentiment scores using FinBERT [11] | Prepares input features by combining market trends and sentiment data for deep learning models [11] |
| Fusion and dynamic weighting | Applies attention mechanisms to prioritize relevant features dynamically [12] | Enhances prediction accuracy by adapting to changing market conditions [12] |
| Price prediction | Generates both short-term and long-term price predictions [1] | Provides actionable insights for cryptocurrency market behavior analysis [1] |

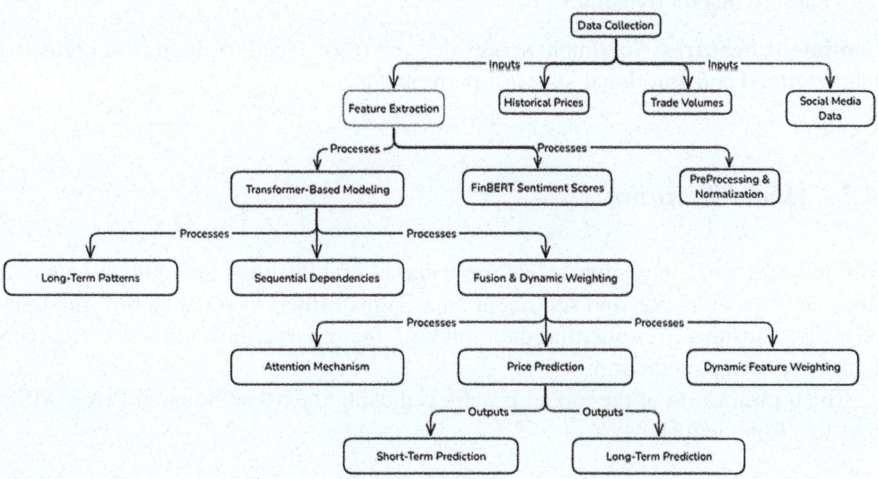


**Fig. 2** Illustration of the proposed architecture

adaptive solution for cryptocurrency price prediction. Future work will focus on fine-tuning the model and validating its performance on diverse datasets.

## 4 Methodology

The proposed methodology for cryptocurrency price prediction integrates advanced natural language processing (NLP) techniques, deep learning architectures, and dynamic feature prioritization. The framework consists of the following steps:

### *4.1 Data Preprocessing and Feature Engineering*

**Market Data**: Historical cryptocurrency prices, trade volumes, and volatility are normalized to ensure consistency for deep learning models, scaling features for unbiased training.

**Sentiment Data**: Social media posts from platforms like Twitter and Telegram are cleaned, tokenized, and analyzed using FinBERT to compute sentiment scores ($S_t$) at time $t$:

$$S_t = f(T_t)$$

where $T_t$ represents textual data at $t$ and $f$ is FinBERT's sentiment analysis function.

**Quantitative Features**: Price trends, trading volumes, volatility, and moving averages capture market dynamics [14].

**Sentiment Features**: Sentiment scores derived from social media reflect collective market mood and emotional states of participants.

### *4.2 Model Training*

The hybrid model integrates transformer-based architectures to capture sequential dependencies in market and sentiment data. Transformers excel at identifying long-term relationships in sequential data, making them particularly effective for cryptocurrency price prediction.

The optimization of the model is achieved using the Mean Squared Error (MSE) loss function, defined as:

$$\text{MSE} = \frac{1}{N}\sum_{i=1}^{N}(y_i - \hat{y}_i)^2$$

where $y_i$ is the actual price, $\hat{y}_i$ is the predicted price at time $i$, and $N$ is the total number of samples.

### 4.3 Attention Mechanism

An attention mechanism is used to enhance feature prioritization. The attention mechanism dynamically assigns weights ($w_i$) to features so that the model can focus on the most relevant inputs according to the current market conditions. Attention weights are calculated using the softmax function:

$$w_i = \frac{\exp(e_i)}{\sum_j \exp(e_j)}$$

where $e_i$ is the importance score and $w_i$ is the normalized attention weight for feature $i$.

These attention weights adaptively prioritize relevant features, enhancing the model's ability to capture changes in market conditions effectively.

### 4.4 Prediction

The final step involves generating both short-term and long-term forecasts. The hybrid model outputs:

- **Short-Term Predictions**: Immediate price fluctuations for actionable trading insights.
- **Long-Term Trends**: Broader market trends to inform investment strategies.

The overall prediction model can be expressed as:

$$P_{t+1} = g(M_t, S_t)$$

where $P_{t+1}$ is the predicted price, $M_t$ is the market data, $S_t$ is the sentiment data, and $g$ is the prediction function.

### 4.5 Conclusion of Methodology

This methodology is carried out in state-of-the-art techniques for ensuring holistic approach strategy in predicting cryptocurrency price. This proposed framework combines quantitative and qualitative data regarding the markets along with sentiment, thereby solving the problems of market volatility and providing actionable insights for either short-term trading or long-term investment strategies.

## 5 Performance Evaluation of Cryptocurrency Price Prediction Framework

### *5.1 Tools Needed to Execute the Idea*

The proposed framework for predicting cryptocurrency prices combines deep learning architectures, sophisticated natural language processing tools, and data preprocessing techniques. The following are the main technologies and tools required:

- **FinBERT**: A pre-trained financial text language model used to extract subtle sentiment characteristics from financial text data. FinBERT can capture sentiment-driven market fluctuations by analyzing data from social media platforms like Telegram and Twitter [15].
- **Transformer-Based Architectures**: To properly grasp market dynamics and capture dependencies in sequential data, deep learning models like BERT or comparable transformer architectures are crucial [16].
- **Data Processing and Visualization Tools**: The libraries used include Pandas, NumPy, Matplotlib, and Seaborn for handling cryptocurrency market and social media sentiment data and generating visual insights into data and model performance.
- **Deep Learning Frameworks**: Model development, training, and evaluation are carried out using frameworks like PyTorch or TensorFlow. Both transformer-based and FinBERT architectures are supported by these tools [13].
- **Data Collection APIs**: The cryptocurrency market APIs (e.g., Binance, CoinMarketCap, Alpha Vantage) will retrieve real-time and historical price data, while social media APIs such as Twitter and Telegram help in the collection of the sentiment data for analysis. Tweepy is a tool that makes social media data collection straightforward.
- **Attention Mechanisms**: These are implemented to dynamically weigh relevant features for prediction, enabling adaptability to evolving market trends [13].

### *5.2 Comparison Between the Proposed Idea and Existing Models*

The proposed framework outperforms traditional models and baseline approaches across various metrics. The following evaluation provides a detailed comparison:

**Table 2** Comparison of accuracy and performance metrics

(a) Short-term prediction accuracy comparison.

| **Model** | **Accuracy (%)** |
|---|---|
| Baseline Model | 74.3 |
| Proposed Model | 85.7 |

(b) Comparison of precision, recall, and F1-score.

| **Metric** | **Baseline Model (%)** | **Proposed Model (%)** |
|---|---|---|
| Precision | 72.5 | 83.2 |
| Recall | 70.0 | 81.5 |
| F1-Score | 71.1 | 82.4 |

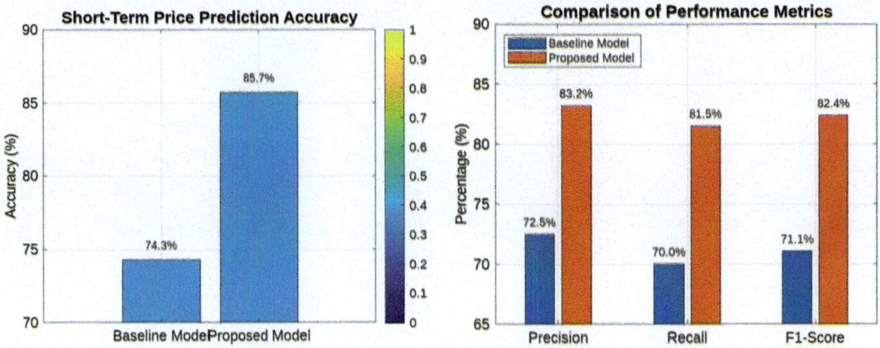


(a) Short-term prediction accuracy comparison.

(b) Comparison of precision, recall, and F1-score.

**Fig. 3** Visualization of accuracy and performance metrics [5]

### 5.2.1 Comparison of Accuracy and Performance Metrics

This section compares short-term prediction accuracy and performance metrics of the baseline and proposed models. All metrics in Table 2 are adapted from the results presented in [5] (Fig. 3).

### 5.2.2 Long-Term Trend Forecasting

Table 3 and Fig. 4 are adapted from the results [4].

## 5.3 *Summary of Observations*

- **Accuracy Improvement**: The proposed model demonstrates a significant improvement in prediction accuracy for both short-term and long-term cryptocurrency price forecasting, achieving accuracies of 85.7% and 92.5%, respectively [13].

**Table 3** Long-term trend forecasting accuracy comparison

| Model | Accuracy (%) |
|---|---|
| Traditional models | 80.5 |
| Proposed model | 92.5 |
| Model X | 85.3 |
| Model Y | 87.1 |

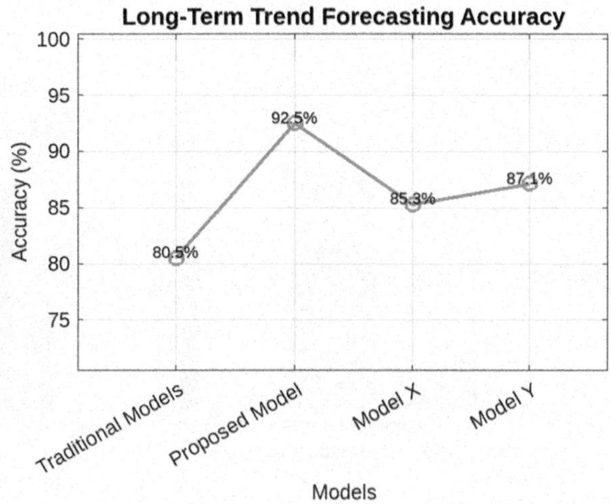


**Fig. 4** Long-term trend forecasting

- **Robustness Across Metrics**: With higher precision, recall, and F1-scores compared to baseline models, the proposed framework proves its reliability and robustness in handling volatile market data.
- **Integration of Sentiment Analysis**: The inclusion of FinBERT enhances the model's capability to extract fine-grained sentiment information, resulting in better performance in sentiment-driven market scenarios [15].

## 6 Conclusion and Future Work

This paper proposes a novel multi-phased framework that addresses challenges in cryptocurrency price prediction by integrating long-term and short-term market dynamics with advanced natural language processing and deep learning techniques. Using FinBERT-based sentiment extraction and hybrid transformer-based architectures for sequential modeling, the framework captures the interplay between market fluctuations and social media sentiment. An adaptive attention-based dynamic weighting mechanism improved the model's ability to prioritize features based on market conditions. Experimental results validated the framework's effectiveness, demonstrating significant improvements in prediction accuracy over baseline models. Despite its promising performance, further exploration is needed to enhance applicability and robustness.

Future work will expand the framework's scope by incorporating additional data sources like blockchain analytics and macroeconomic indicators to analyze cross-market influences between cryptocurrencies and traditional assets. Reinforcement learning could enable adaptive trading strategies responsive to real-time market changes. Efforts will also focus on reducing the computational cost of FinBERT and transformer-based models, potentially using lightweight or optimized architectures to reduce latency without compromising accuracy. Deeper comparisons with traditional models like ARIMA and modern machine learning techniques will highlight the strengths and weaknesses of the proposed approach. Enhancements will include robustness against noise or adversarial inputs from social media data sources and the adoption of explainable AI techniques, enabling transparency in predictions and informed decision-making for stakeholders.

These advances will contribute to more robust and scalable cryptocurrency price prediction models, making them practical for real-world applications.

## References


1. McNally, S., Roche, J., Caton, S.: Predicting the price of bitcoin using machine learning. In: 2018 26th Euromicro International Conference on Parallel, Distributed and Network-based Processing (PDP), pp. 339–343 (2018)
2. Jagini, A., Mahajan, K., Aluvathingal, N., Mohan, V., Prajwala, TR.: Twitter sentiment analysis for bitcoin price prediction. In: 2023 3rd International Conference on Smart Data Intelligence (ICSMDI), pp. 32–37 (2023)
3. Nagula, P.K., Alexakis, C.: A new hybrid machine learning model for predicting the bitcoin (btc-usd) price. J. Behav. Exp. Financ. **36**, 100741 (2022)
4. Singh, S., Bhat, M.: Transformer-based approach for Ethereum price prediction using cross-currency correlation and sentiment analysis. arXiv preprint arXiv:2401.08077 (2024)
5. Tanwar, A., Kumar, V.: Prediction of cryptocurrency prices using transformers and long short-term neural networks. In: 2022 International Conference on Intelligent Controller and Computing for Smart Power (ICICCSP), pp. 1–4 (2022)
6. Araci, D.: Finbert: Financial sentiment analysis with pre-trained language models. arXiv preprint arXiv:1908.10063 (2019)
7. Jahanbin, K., Chahooki, M.A.Z.: A hybrid deep implicit neural model for sentiment analysis via transfer learning. IEEE Access **12**, 131468–131486 (2024)
8. Oikonomopoulos, S., Tzafilkou, K., Karapiperis, D., Verykios, V.: Cryptocurrency price prediction using social media sentiment analysis. In: 2022 13th International Conference on Information, Intelligence, Systems & Applications (IISA), pp. 1–8 (2022)
9. Nitya Harshitha, T., Prabu, M., Suganya, E., Sountharrajan, S., Bavirisetti, D. P., Gadde, N., Uppu, L. S.: Protect: a hybrid deep learning model for proactive detection of cyberbullying on social media. Front. Artif. Intell. **7**, 1269366 (2024)
10. Maheshwari, T., Bharadwaj, S., Sharma, N.K., Mudegol, G., Srivastav, S.: Bitcoin price prediction using lstm. In: Proceedings of the KILBY 100 7th International Conference on Computing Sciences 2023 (ICCS 2023) (2023)
11. Elangovan, G., Umamageswaran, J., Indumathi, G., Kalpana, A. V.: Medical quora tagging using matar and lda algorithm. J. Phys.: Conf. Ser. (2021)

12. Prabu, M., Sountharrajan, S., Suganya, E., Reddy, D. S., Natarajan, A. K.: Navigating emotional landscapes: enhancing communication through facial emotion analysis in presentations. In: Adopting Artificial Intelligence Tools in Higher Education. CRC Press , pp. 251–271 (2024)
13. Chen, H., De, S., Hu, L.: Transformer-based models for cryptocurrency price prediction. IEEE Trans. Knowl. Data Eng. **34**(5), 1456–1472 (2022)
14. Jambulingam, U., Balasubadra, K.: A unique multi-agent-based approach for enhanced qos resource allocation in multi cloud environment while maintaining minimized energy and maximize revenue. Int. J. Comput., Commun. Control. **17**(2) (2022)
15. Yang, Y., Zhang, M., Yang, J.: Finbert: a pretrained language model for financial communications. In: ACL Anthology, pp. 363–370 (2020)
16. Devlin, J., Chang, M., Lee, K., Toutanova, K.: Bert: Pre-training of deep bidirectional transformers for language understanding. Proceedings of NAACL-HLT **2019**, 4171–4186 (2019)

# RNN and Logistic Regression-Based Ensemble Model for Distinguishing AI-Generated Text

**Nidhi Umashankar and Chandu Siddartha Reddy Gooty**

**Abstract** Generative AI models change quickly, highlighting crucial concerns such as the difficulty of distinguishing between human-written and AI-generated texts. This research proposes a unique ensemble model that combines a Recurrent Neural Network and Logistic Regression to efficiently address the differentiation difficulty. The RNN is used to sequence-model complicated patterns in text input, whereas Logistic Regression is used for classification via dense feature representations. The model was trained and evaluated using a publicly available dataset that contained both AI-generated and human-written material. Key performance indicators such as accuracy, precision-recall, and ROC-AUC demonstrate the ensemble model's ability to accurately classify text. In order to validate the findings, this study also includes intricate visualizations like precision-recall curves and confusion matrices. Future developments in related content authentication and regulatory frameworks will be made possible by this combined version of deep learning with the conventional machine learning end, which strengthens the scalable and reliable process to recognize AI-generated text.

**Keywords** AI-generated text · Human-written text · Recurrent Neural Network (RNN) · Logistic regression · Ensemble model · Text classification · Deep learning · Generative AI detection · Content authentication · Binary classification · Text authentication · Machine learning · Natural language processing (NLP) · Tokenization · Sequence modeling · Model interpretability · AI-generated content regulation · Text sequence padding · Feature engineering · AI and human text distinction · Computational linguistics · Transformer models · Generative AI validation

N. Umashankar (✉)
The University of Edinburgh, Edinburgh, Scotland
e-mail: nidhiumashankar03@gmail.com

C. S. R. Gooty
Founder InverstorMate, Bangalore, India

© The Author(s), under exclusive license to Springer Nature Singapore Pte Ltd. 2026

M. Pant et al. (eds.), *Real-World Applications of Machine Learning*, Mathematics for Sustainable Developments, https://doi.org/10.1007/978-981-95-7134-5_40

## 1 Introduction

Recent advances in generative AI technology have allowed for the creation of text that is nearly identical to human writing. These models have demonstrated previously unheard-of potential in automated consumer interactions across domains and creative writing. The ability to discern between writing produced by AI and that authored by humans is one of the many difficulties brought about by this quick development, which has serious ramifications for ethics, openness, and legal compliance. This is essential for preventing false information, identifying plagiarism, and maintaining content responsibility.

Traditional methods for identifying AI-generated content rely on rule-based approaches or single-model classifiers, which perform poorly when contrasted with the most sophisticated generative models, like ChatGPT, RoBERTa, and others. Recent studies underscore the limitations of traditional methods and stress the importance of hybrid models that merge appropriate machine learning approaches with deep learning for accurate identification. This article introduces an innovative ensemble model that merges the simplicity and interpretability of logistic regression with the sequential learning capability of a recurrent neural network.

The suggested model operates in two phases: the RNN identifies profound sequential patterns from the text to accurately capture the complex structures that differentiate AI-generated content from that produced by humans. Logistic Regression subsequently utilizes the rich feature representations from the RNN to improve classification effectiveness. The training and test data set for the model has been a publicly available corpus containing AI-generated text and human text. For efficient training of RNN, several techniques such as tokenization and sequence padding in addition to regularizations have been used. Cross-validation has also been performed over accuracy, precision-recall and ROC-AUC performance metrics. Visualization tools include confusion matrices and ROC curves that favor the efficiency of the model as well.

This work contributes to the growing body of literature on generative AI detection. It aims to fill an existing gap in scalable and interpretable solutions. In contrast to previous works that have relied either exclusively on deep learning or traditional machine learning, this paper focuses on the synergy between these two approaches.

## 2 Literature Survey

Table 1 displays the extensive literature survey carried out.

This survey in literature presents much advancement in the text separation AI generated from human content. Opara [1] showed the ability of stylometric analysis in classifying GPT-3.5 from GPT-4 in the aspect of linguistic features and was limitedly focused on its application on the English text only. Nguyen et al. [2] applied stacked ensemble of transformers such as RoBERTa and BERT, for which the accuracy was

**Table 1** Literature survey

| Dataset used | Dataset used | Technique used | Key findings | Limitations |
|---|---|---|---|---|
| Opara [1] | StyloAI dataset with texts from GPT-3.5, GPT-4, and human authors | Stylometric analysis using 175 linguistic features including lexical diversity and syntactic patterns | Demonstrates effectiveness of stylometric features in distinguishing AI text | Focus on English texts; limited to specific AI models |
| Nguyen et al. [2] | Custom dataset: 2800 samples (1400 each from GPT-3.5 and human writers) | Stacked ensemble combining RoBERTa, BERT, and DistilBERT with voting classifier | F1-score: 0.895, Accuracy: 89.5%, Meta-classifier improved performance by 2.3% | Computational overhead of multiple transformers |
| Almandouh et al. [3] | LIAR, ISOT, FakeNewsNet | Ensemble combining CNN, LSTM, and BERT with feature augmentation | Achieved AUC-ROC of 0.98 on ISOT dataset; True Positive Rate of 0.96 | Limited to news domain text |
| Kumar et al. [4] | 1000 samples each of ChatGPT and human-written text | Comparative study using BERT, RoBERTa, and GPT-2 detection models | Presents model performance across different text lengths and genres | Model performance varies by text length |
| Alhayan and Himdi [5] | 10,000 Arabic reviews (5000 each from humans and AI) | Random Forest, XGBoost, and LightGBM ensemble with 82 engineered features | Reports comparative performance across different ensemble configurations | Arabic language specific |
| Hua and Yao [6] | Multi-source dataset (1 K to 100 K samples) | BERT with optimized tokenization strategies and data augmentation | Shows correlation between dataset size and detection accuracy | Resource-intensive preprocessing |
| Shah et al. [7] | Dataset from GPT-3 and human writers | XAI approach using SHAP values and feature importance analysis | Demonstrates interpretability of detection decisions | Limited to formal writing |
| Oketunji [8] | Outputs from GPT-3, GPT-4, Claude, and human writers | Hybrid CNN-LSTM architecture with attention mechanism | Reports performance across different AI models and text types | High computational requirements |
| Wani et al. [9] | Multi-source dataset from various AI models and human writers | Transformer-based framework with custom attention layers | Demonstrates cross-domain performance | Complex architecture |

(continued)

**Table 1** (continued)

| Dataset used | Dataset used | Technique used | Key findings | Limitations |
|---|---|---|---|---|
| Prova [10] | NLP benchmark datasets | NLP-based machine learning with feature engineering | Reports effectiveness across different text types | Specific to chosen NLP techniques |
| Wang et al. [11] | SemEval-2024 Task 8 dataset | Multiple detection approaches in competition framework | Competition ongoing—baseline results pending | Task-specific evaluation |
| Akram [12] | Comparative study using multiple public datasets | Empirical analysis of detection tools including GPT-Zero and OpenAI detector | Compares performance of available detection tools | Limited to available tools |
| Khalil and Er [13] | Academic writing samples from students and ChatGPT | Analysis of detection tools in academic context | Evaluates tool performance for academic text | Focus on academic writing |

impressively high but carried a very heavy computational overhead. Almandouh et al. [3] further broadened the application scope by the use of ensemble models for detecting fake news in the news domain with impressive results but limited to other contexts.

Several researchers tested hybrid and transformer-based models. Oketunji, in 2023 [8] utilized CNN-LSTM architectures that handled different sorts of texts, while Wani et al., in 2024 [9], hypothesized a specially designed transformer framework for cross-domain detection. Yet, such models were usually very CPU-intensive. Tools like GPT-Zero, whose work Akram considered in 2023 [12], had immediate applicability but stood on present existing technologies. Taken together, these works point toward a pressing need in the rapidly shifting AI landscape: scalable, flexible, and interpretable detection approaches.

## 3 Proposed Methodology

This section discusses the fine steps taken for designing a reliable classification framework for discriminating between AI-generated and human-written text. The approach follows several stages including data gathering, preparation, and model building along with its training, evaluation, and implementation phase.

### *3.1 Dataset Acquisition and Preprocessing*

1. Dataset Details

The dataset "AI versus Human Text" utilized in this investigation was collected from Kaggle. It contains 48,724 text samples, with a class distribution of 30,580 human-written samples (about. 63%) and 18,144 AI-generated samples (approx. 37%). These classifications are represented by a binary label that indicates whether the content was written by a person (0) or created by an artificial intelligence model (1). Table 2 contains a tabular description of the dataset.

An overview of the dataset:

- The dataset size is 48,724 samples.
- Class distribution
- Human-written: 30,580 samples (about 63%).
- AI-generated: 18,144 samples (around 37%).

2. Dataset Loading and Verification

The dataset was initially loaded into a Pandas DataFrame for preprocessing and analysis. The data was inspected to ensure its integrity and to identify any inconsistencies such as missing text or labels. Rows with missing or malformed values were discarded. Exploratory Data Analysis (EDA) was performed to assess the class distribution.

- Class Distribution: The dataset contains 48,724 samples. 63% were human-written (30,580 samples) and 37% AI-generated (18,144 samples). This slight imbalance was accounted for during dataset splitting to ensure both classes were proportionally represented in training, validation, and test subsets.
- Data Integrity: The dataset was verified for completeness and consistency. Rows with missing or malformed values were discarded to maintain data quality.

3. Preprocessing Steps

To prepare the text data for model training, several preprocessing steps were applied:

**Table 2** Description of the dataset

| Attribute | Description |
|---|---|
| Text | Contains the textual data to be analyzed Each entry in this column consists of a passage of text, either written by a human or generated by an AI model. This serves as the primary input for the classification task |
| Generated | A binary label indicating the origin of the text. Values include:<br>• 0: Denotes that the text was authored by a human<br>• 1: Denotes that the text was generated by an AI model. This serves as the ground truth for model training and evaluation |

(1) Tokenization: The text was tokenized using TensorFlow's Tokenizer, which breaks the text into smaller units—words or subwords. The vocabulary was limited to the top 2000 most frequent tokens, capturing the most relevant textual features.
(2) Sequence Padding: Tokenized sequences were padded to a uniform length of 100 tokens using post-padding. This step ensured consistent input dimensions for the RNN model, regardless of the original text length.

4. Dataset Splitting

To ensure the model generalizes well, the dataset was split into three subsets and stratified sampling was used to maintain the class distribution across subsets. The data was split into 70% training, 15% validation, and 15% testing sets.

## 3.2 Model Architecture Design

Figure 1 shows a brief overview of the model architecture, which is discussed in detail as follows.

**Recurrent Neural Network (RNN)**

A Recurrent Neural Network (RNN), specifically utilizing Long Short-Term Memory (LSTM) layers, was selected for this task due to its ability to capture temporal dependencies and contextual relationships in sequential data (such as text).

*Embedding Layer*

The embedding layer captures the semantic meaning of words by converting integer tokens into dense vectors of fixed size (64). The vocabulary size was set to 2000 ensuring that only the most frequent tokens in the dataset were used.

*Long Short-Term Memory (LSTM) Layers*

- First LSTM Layer: The first LSTM layer contains 64 units and captures the temporal dependencies within the text. LSTM is effective in handling long-range dependencies which is crucial for text classification tasks.
- Second LSTM Layer: The second LSTM layer contains 32 units and focuses on extracting higher-level features from the data. The configuration return_sequences = False was used to discard temporal dependencies while retaining the essential features.

*Dropout Regularization*

Dropout was applied after each LSTM layer to mitigate overfitting. Dropout works by randomly setting a fraction of input units to 0 during training which prevents the model from becoming too reliant on specific neurons.

(1) Dropout rate after the first LSTM layer: 0.4

**Fig. 1** Model architecture overview

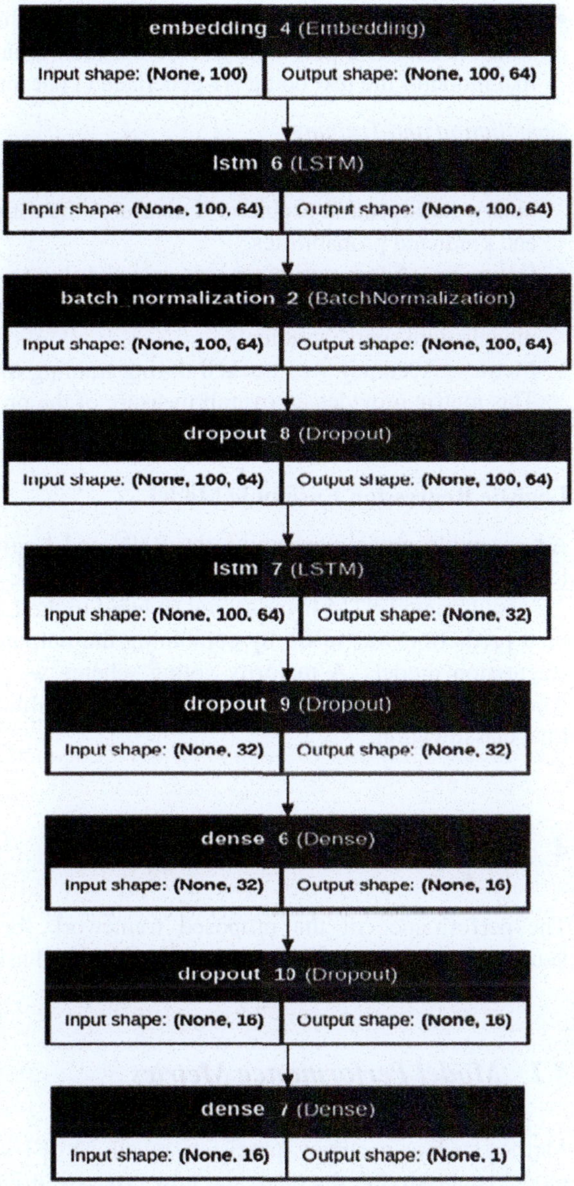


(2) Dropout rate after the dense layer: 0.3

*Fully Connected (Dense) Layers*

- Dense Layer: A dense layer with 16 units and ReLU activation was added to refine the features extracted from the LSTM layers. ReLU helps the model learn non-linear relationships between the features.

- Output Layer: The output layer consists of a single neuron with sigmoid activation, which produces a probability score between 0 and 1. This score represents the likelihood of the text being AI-generated (1) or human-written (0).

*Compilation and Training*

- Loss Function: The binary cross-entropy loss function was used because this is a binary classification problem. It measures the difference between the true labels and predicted probabilities.
- Optimizer: Adam optimizer was selected due to its adaptive learning rate and efficiency in training deep learning models. Adam is particularly effective for models with sparse gradients.
- Metrics: Accuracy was tracked during training to evaluate model performance. This metric provides an overall measure of the proportion of correctly classified samples.

**Logistic Regression Ensemble Model**

An ensemble model combining the RNN and Logistic Regression was designed to enhance predictive accuracy and robustness. The dense features extracted from the penultimate LSTM layer served as input to the Logistic Regression model. The final prediction was made by combining the outputs from the RNN and Logistic Regression models. A majority voting scheme was used to integrate predictions, where the class predicted by the averaging ensemble of models was selected as the final classification.

## 4 Evaluation Metrics

The performance of the proposed framework for distinguishing between AI-generated and human-authored text was evaluated using a variety of metrics.

### *4.1 Model Performance Metrics*

Metrics such as accuracy, precision, recall, and F1-score help assess how well the model is classifying the data, and how effective it is at differentiating between the two classes (AI-generated and human-written text). This information is tabulated in the classification report in Table 3.

- Accuracy: The accuracy of the model is defined as the proportion of correct predictions (both true positives and true negatives) out of all predictions. The model achieved an accuracy of 99% on the test set.

**Table 3** Classification report

| Metric | Accuracy | Precision | Recall | F1-score | Support |
|---|---|---|---|---|---|
| Class 0 (Human-written) | 0.99 | 0.99 | 0.99 | 0.99 | 30,580 |
| Class 1 (AI-generated) | 0.99 | 0.99 | 0.98 | 0.99 | 18,144 |
| Macro Average | 0.99 | 0.99 | 0.99 | 0.99 | 48,724 |
| Weighted Average | 0.99 | 0.99 | 0.99 | 0.99 | 48,724 |

- Precision: Precision is calculated as the ratio of true positive predictions to the sum of true positive and false positive predictions. The model achieved a precision score of 0.99 for both classes. 99% of the samples classified as AI-generated were indeed AI-generated.
- Recall: Recall is the proportion of true positive predictions to the sum of true positive and false negative predictions. The recall score for human-written text was 0.99 and for AI-generated text was 0.98.
- F1-score: The F1-score, combining precision and recall, was 0.99 for both classes. This indicates balanced performance in terms of both precision and recall.

## 4.2 *Confusion Matrix Analysis*

The confusion matrix provides a detailed view of the classification performance by showing the number of true positives, true negatives, false positives, and false negatives. The confusion matrix for the test set is displayed below in Fig. 2.

- True Positives (TP): 30,336 AI-generated texts were correctly classified as AI-generated.

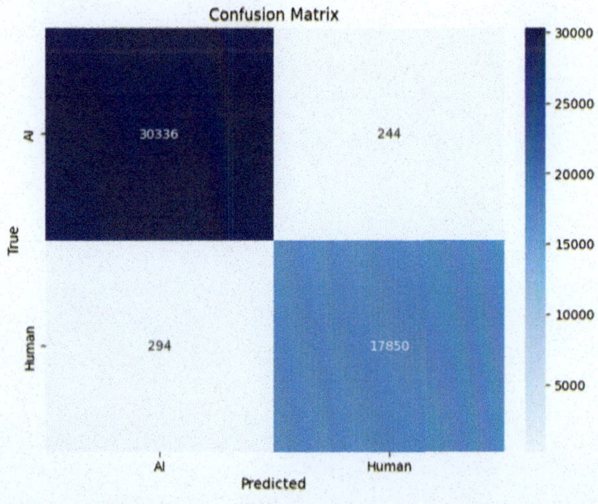


**Fig. 2** Confusion matrix for the proposed model

- False Positives (FP): 244 human-written texts were incorrectly classified as AI-generated.
- True Negatives (TN): 17,850 human-written texts were correctly classified as human-written.
- False Negatives (FN): 294 AI-generated texts were incorrectly classified as human-written.

### 4.3 Receiver Operating Characteristic (ROC) Curve

The ROC curve was plotted to evaluate the model's ability to distinguish between the two classes (human-written and AI-generated text). The ROC curve illustrated in Fig. 3 showed a clear separation between the two classes, with the Area Under the Curve (AUC) equal to 0.98.

AUC Analysis:

- 0.99 AUC indicates that the model is excellent at distinguishing between the two classes. The closer the AUC is to 1, the better the model is at identifying the correct class across various classification thresholds.
- Interpretation: The high 0.99 AUC score suggests that the model is highly robust in identifying AI-generated text when compared to human-written text, even in the presence of class imbalance.

**Fig. 3** ROC curve

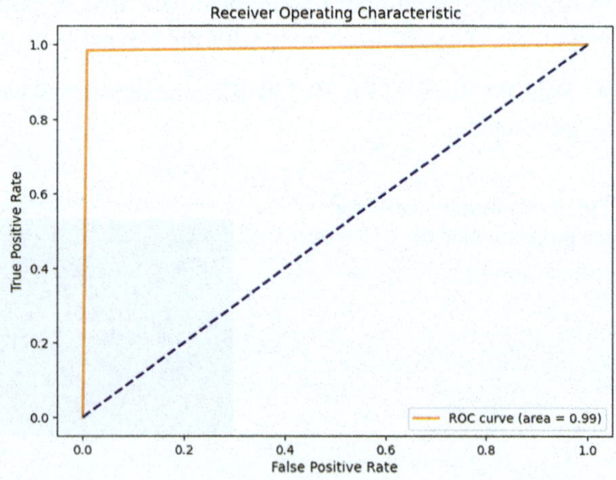

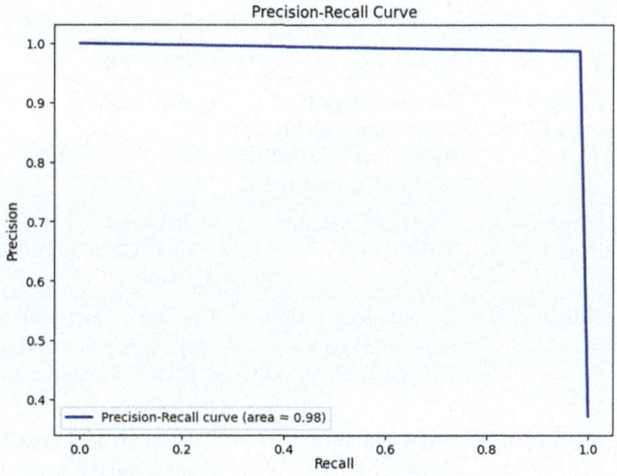


**Fig. 4** Precision-recall curve

## 4.4 *Precision-Recall Curve*

The Precision-Recall curve was plotted to evaluate the model's performance in scenarios where the data distribution may be imbalanced. The Precision-Recall AUC was calculated to be 0.98 and this is visible in Fig. 4.

Precision-Recall AUC:

- A Precision-Recall AUC of 0.98 highlights that the model maintains a good balance between precision and recall, even in cases where the dataset might be slightly imbalanced.
- Analysis: This metric is particularly important in scenarios where detecting AI-generated content is more critical than minimizing false positives or false negatives.

Table 4 illustrates the various methods and models employed to distinguish human and AI-generated text. The proposed model stands out from BERT and SVMs with an accuracy of 99%.

## 4.5 *Training Dynamics and Understanding of Model Output*

Training dynamics were analyzed by observing the loss and accuracy values during training. This provided insights into the learning process and how well the model converged.

- The training loss steadily decreased over five epochs, indicating that the model was learning effectively and adapting to the training data.
- The validation loss followed a similar trend, demonstrating that the model did not overfit and was generalizing well to unseen data.

**Table 4** Comparison of various models and methods to distinguish human and AI-generated text

| Paper | Model used | Dataset used | Accuracy |
|---|---|---|---|
| Nguyen et al. [2] | Stacked ensemble combining RoBERTa, BERT, and DistilBERT with voting classifier | Custom dataset | 0.9555 |
| Almandouh et al. [3] | BERT, XLNet, and RoBERTa | AFND and ARABICFAKETWEETS datasets | AFND: 0.98, ARABICFAKETWEETS: 0.99 |
| Kumar et al. [4] | Logistic Regression, Random Forest, XGBoost, SVM (SVC) | Custom dataset with 600 responses generated using v ChatGPT, Bard, and Aichatting | Logistic Regression: 0.833, Random Forest: 0.8518, XGBoost: 0.87, SVM: 0.87 |
| Shah et al. [7] | xAI techniques (LIME, SHAP) | Custom dataset created using a dataset creation pipeline | 0.93 |
| Proposed model | RNN and Logistic Regression-Based Ensemble Model | AI versus Human Text | 0.99 |

- The training accuracy increased progressively, reaching 98.60% by the end of the fifth epoch, suggesting that the model was successfully capturing the underlying patterns in the training data.
- The validation accuracy reached 98.84%, confirming that the model maintained a high level of performance on unseen data, which is essential for generalizability.

The residual plot shown in Fig. 5 reveals the distribution of the errors between the predicted and actual values, helping to assess whether the model has made consistent predictions or if any patterns remain unaddressed. The absence of a discernible pattern suggests that the model has effectively captured the underlying data distribution, with errors randomly distributed around zero, indicating a good model fit.

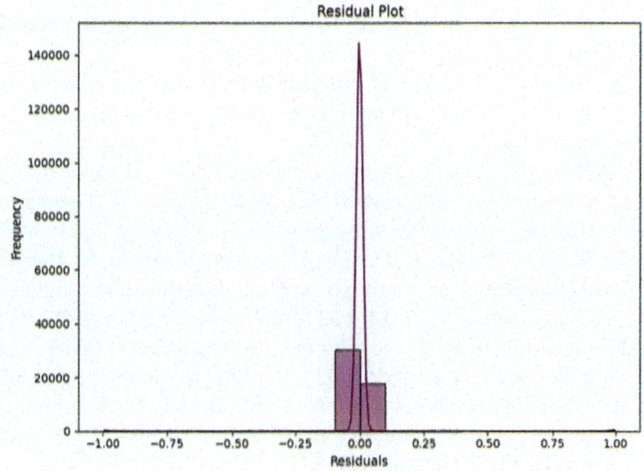


**Fig. 5** Residual plot

## 5 Conclusion

This study created a robust classification framework to distinguish between AI-generated and human-authored content. To improve predictive accuracy, the proposed methodology combined Recurrent Neural Networks (RNNs) with Long Short-Term Memory (LSTM) units, as well as an ensemble model based on Logistic Regression. The model outperformed expectations in terms of accuracy, recall, and precision, proving its ability to recognize both AI-generated and human-authored material. The ensemble technique, which used dense feature representations from the RNN, improved the model's robustness and generalizability.

There are numerous potential for future development and enhancement. To improve the model's capacity to generalize to multiple forms of material, the dataset should be expanded to contain a more wide range of AI-generated text from new sources and topics.

## References

1. Opara, C.: StyloAI: distinguishing AI-generated content with stylometric analysis. In: Olney, A. M., Chounta, I. A., Liu, Z., Santos, O. C., Bittencourt, I. I. (eds.) Artificial Intelligence in Education. Posters and Late Breaking Results, Workshops and Tutorials, Industry and Innovation Tracks, Practitioners, Doctoral Consortium and Blue Sky (AIED 2024). Springer, Cham, vol. 2151, pp. 13 (2024) [Online]. Available: https://doi.org/10.1007/978-3-031-64312-5_13
2. Nguyen, D., Naing, K. M. N., Joshi A.: Stacking the odds: transformer-based ensemble for AI-generated text detection. In: Proceedings of the 21st Annual Workshop of the Australasian Language Technology Association, Brisbane, Australia, pp. 173–178 (2023)
3. Almandouh, E., Alrahmawy, M., Eisa, M.: Ensemble-based high-performance deep learning models for fake news detection. Sci. Rep. **14**, 26591 (2024) [Online]. Available: https://doi.org/10.1038/s41598-024-76286-0
4. Kumar, S., Tiwari, S., Prasad, R., Rana, A., Arti, M. K., "Comparative Analysis of Human and AI generated text. In: 2024 11th International Conference on Signal Processing and Integrated

Networks (SPIN), Noida, India, pp. 168–173 (2024). https://doi.org/10.1109/SPIN60856.2024.10511301
5. Alhayan, F., Himdi, H.: Ensemble learning approach for distinguishing human and computer-generated Arabic reviews. PeerJ Comput. Sci. **10**, e2345 (2024). https://doi.org/10.7717/peerj-cs.2345
6. Hua, H., Yao, C.: Investigating generative AI models and detection techniques: impacts of tokenization and dataset size on identification of AI-generated text. Front. Artif. Intell. 7 (2024) [Online]. Available: https://doi.org/10.3389/frai.2024.1469197
7. Shah, A., Ranka, P., Dedhia, U., Prasad, S., Muni, S., Bhowmick, K.: Detecting and unmasking AI-generated texts through explainable artificial intelligence using stylistic features. Int. J. Adv. Comput. Sci. Appl. **14**, 110 (2023). https://doi.org/10.14569/IJACSA.2023.01410110
8. Oketunji, A. F.: Evaluating the efficacy of hybrid deep learning models in distinguishing AI-generated text. Zenodo, 2023. [Online]. Available: https://doi.org/10.48550/arXiv.2311.15565
9. Wani, M. A., Abd El-Latif, A. A., ELAffendi, M., Hussain, A.: AI-based framework for discriminating human-authored and AI-generated text. IEEE Trans. Artif. Intell. (2024). https://doi.org/10.1109/TAI.2024.3516713
10. Prova, N.: Detecting AI generated text based on NLP and machine learning approaches. arXiv preprint, 2024. [Online]. Available: https://arxiv.org/abs/2404.10032
11. Wang, Y. et al.: SemEval-2024 Task 8: multidomain, multimodel and multilingual machine-generated text detection. In: Proceedings of the 18th International Workshop on Semantic Evaluation (SemEval-2024), 2024. [Online]. Available: https://arxiv.org/abs/2404.14183
12. Akram, A.: An empirical study of AI generated text detection tools. arXiv preprint, 2023. [Online]. Available: https://arxiv.org/abs/2310.01423
13. Khalil, M., Er, E.: Will ChatGPT get you caught? Rethinking plagiarism detection. In: Zaphiris, P., Ioannou, A. (eds.) HCII 2023, vol. 14040, pp. 475–487. Springer, Cham (2023). https://doi.org/10.1007/978-3-031-34411-4_32

The manufacturer's authorised representative in the EU is Springer Nature Customer Service Centre GmbH, Europaplatz 3, 69115 Heidelberg, Germany. If you have any concerns regarding our products, please contact ProductSafety@springernature.com

Printed and bound by CPI Group (UK) Ltd, Croydon, CR0 4YY
22/09/2026
02220636-0001